# Human Adaptation and Accommodation

Enlarged and Revised Edition of *Human Adaptation*

A. Roberto Frisancho

Ann Arbor
The University of Michigan Press

This book is dedicated to my wife Hedy Frisancho
and my sons Roberto Javier and Juan Carlos Frisancho.


Published in the United States of America by
The University of Michigan Press
Manufactured in the United States of America

1996 1995 1994 1993 4 3 2 1

*A CIP catalogue record for this book is available from the British Library.*

Library of Congress Cataloging-in-Publication Data

Frisancho, A. Roberto, 1939–
Human adaptation and accommodation / A. Roberto Frisancho.
p. cm.
"Enlarged and revised edition of Human adaptation".
Includes bibliographical references and index.
ISBN 0-472-09511-0 (alk. paper)
1. Adaptation (Physiology) 2. Man—Influence of environment.
I. Title.
QP82.F74 1993
573—dc20 92-36234
CIP

# Preface

This book follows the conceptual framework of *Human Adaptation* previously published by the University of Michigan Press. What initially started as a fine tuning of the material to bring it in line with expanding scientific knowledge has resulted in significant changes that warrant to be presented as a new book rather than simply a revised edition. Its primary goal is to integrate basic concepts and relevant scientific information to provide the foundation for understanding the processes whereby humans are able to function normally under stressful environmental conditions. The book is divided into five main areas: Part 1: Principles and Definitions for the Study of Human Adaptation (chap. 1); Part 2: Thermoregulation (chaps. 2 to 5); Part 3: Skin Color and Adaptation to Solar Radiation (chaps. 6 and 7); Part 4: Adaptation to High Altitude (chaps. 8 to 12); Part 5: Accommodation to Energy Variability (chaps. 13 to 18). In each section or chapter the initial responses observed in laboratory studies with humans and experimental animals are discussed. Subsequently, the short- and long-term adaptive mechanisms that enable organisms to acclimatize themselves to natural, stressful environmental conditions are discussed. In addition, the book includes an appendix section giving conversions to metric and SI units and a glossary of terms and definitions.

This book, like the previous one, follows the conceptual framework of *developmental adaptation,* which maintains that intra- and interpopulation differences in biological phenotypic traits are related to the effects of environmental stress and the adaptive responses that an organism makes during growth and development and, to a lesser extent, during adulthood. In some cases the evidence is only suggestive, and I hope that the postulated hypotheses will stimulate researchers to prove or disprove them. If further research is so motivated, my efforts will have been worthwhile.

I very much hope that the book will continue to be valued by my friends, colleagues, and students worldwide. I wish to express a deep appreciation to all my colleagues and students who provided valuable criticism and assisted in various phases of the book preparation.

# Contents

PART 1

# Principles and Definitions

# CHAPTER 1 The Study of Human Adaptation

## THE SUBDISCIPLINE OF HUMAN ADAPTATION

Ever since the first appearance of the genus *Homo* to the emergence of the species *sapiens* frequent adaptations must have occurred to allow survival in the changing environment. In search of the mechanisms of how humans adapted to the changing environments, biological anthropologists, physiologists, epidemiologists, nutritionists, and human geneticists studied humans living under diverse and extreme environments. This type of research centers upon studying indigenous populations and comparing them to Western samples. Along with the study of indigenous populations, experimental studies of how humans respond to controlled laboratory conditions are performed. A common denominator of these investigations was that evolutionary selection processes had produced the human species and that the processes had produced a set of genetic characteristics that adapted our evolving species to the environment. Although genetic research shows that many of the morphologically distinct groups differ in the frequencies of identifiable genes, population genetics theory provided little explanation for the genetic and morphological differences between the groups other than showing that at least some of the gene frequency differences might have arisen through chance factors. Instead, current investigations demonstrated that the phenotype measured morphologically, physiologically, or biochemically was the product of genetic plasticity operating during development. Within this framework it is not assumed that all the biological traits are a result of genetic plasticity nor is it assumed that the causal relationships are all one-way. That is, some of the biological adjustments or adaptations people made to our natural and social environments have also modified how we adjusted to subsequent environments. The adjustments we have made to improve our adaptations to a given environment have produced a new environment to which we, in turn, adapt in an ongoing process of new stress and new adaptation. These endeavors stimulated the development of the subdiscipline of human adaptation with which this book is concerned.

### The Concept of Functional Adaptation

The term *adaptation* is used in the broad generic sense of functional adaptation, and it is applied to all levels of biological organization from individuals to populations. A basic premise of this approach is that adaptation is a process whereby the organism has attained a beneficial adjustment to the environment (1–34). This adjustment can be either temporary or permanent, acquired either through short-term or lifetime processes, and may involve physiological, structural, behavioral, or cultural changes aimed at improving the organism's functional performance in the face of environmental stresses. If environmental stresses are conducive to differential mortality and fertility, then adaptive changes may be established in the population through changes in genetic composition, and thus attain a level of genetic adaptation. In this context functional adaptation, along with cultural and genetic adaptation, is viewed as part of a continuum in an adaptive process that enables individuals and populations to maintain both internal and external environmental homeostasis. Therefore the concept of adaptation is applicable to all levels of biological organization, from unicellular organisms to the largest mammals, and from individuals to populations. This broad use of the concept of adaptation is justified not only in theory but also because it is currently applied to all areas of human endeavor, so that no discipline can claim priority or exclusivity in the use of the term (20). In this chapter the various forms of adaptation will be defined, and their relationship to each other, as well as their applicability to the study of human adaptation to environmental stress, will be discussed.

Functional adaptation involves changes in organ system function, histology, morphology, biochemical composition, anatomical relationships, and body composition, either independently or integrated in the organism as a whole. These changes can occur through acclimatization, acclimation, or habituation.

**Acclimatization.** Acclimatization refers to changes occurring within the lifetime of an organism that reduce the strain caused by stressful changes in the natural climate or by complex environmental stresses (21, 22, 34). If the adaptive traits are acquired during the growth period of the organism, the process is referred to as either *developmental adaptation* or *developmental acclimatization* (23, 32).

**Acclimation.** Acclimation refers to the adaptive biological changes that occur in response to a single experimentally induced stress (21, 22) rather than to multiple stresses as occurs in acclimatization. As with acclimatization, changes occurring during the process of growth may also be referred to as developmental acclimation (32).

**Habituation.** Habituation implies a gradual reduction of responses to, or perception of, repeated stimulation (21, 22). By extension, habituation refers to the diminution of normal neural responses, for example, the decrease of sensations such as pain. Such changes can be generalized for the whole organism (general habituation) or can be specific for a given part of the organism (specific habituation). Habituation necessarily depends on learning and conditioning, which enable the organism to transfer an existing response to a new stimulus. The extent to which these physiological responses are important in maintaining homeostasis depends on the severity of environmental stress. For example, with severe cold stress or low oxygen availability, failure to respond physiologically may endanger the well-being and survival of the organism.

**Acclimatization vs. Acclimation.** Studies on acclimatization are done with reference to both major environmental stresses and several secondary, related stresses. For example, any difference in the physiological and structural characteristics of subjects prior to and after residence in a tropical environment are interpreted as a result of acclimatization to heat stress. In addition, because tropical climates are also associated with nutritional and disease stresses, individual or population differences in function and structure may also be related to these factors. On the other hand, in studies of acclimation any possible differences are easily attributed to the major stress to which the experimental subject has been exposed in the laboratory. For understanding the basic physiological processes of adaptation, studies on acclimation are certainly better than those of acclimatization. However, since all organisms are never exposed to a single stress but instead to multiple stresses, a more realistic approach is that of studying acclimatization responses. Thus, both studies on acclimation and acclimatization are essential for understanding the processes whereby the organism adapts to a given environmental condition. This rationale becomes even more important when the aim is to understand the mechanisms whereby humans adapt to a given climatic area, since humans in a given area are not only exposed to diverse stresses but have also modified the nature and intensity of these stresses, as well as created new stresses for themselves and for generations to come.

## Cultural and Technological Adaptation

Cultural adaptation refers to the nonbiological responses of the individual or population to modify or ameliorate an environmental stress. As such, cultural adaptation is an important mechanism that facilitates human biological adaptation. It may be said that cultural adaptation both during contemporary times and in evolutionary perspective represents humanity's most important tool. It is through cultural adaptation that humans have been able to survive and colonize far into the zones of extreme environmental conditions. Human beings have adapted to cold environments by inventing fire and clothing, building houses, and harnessing new sources of energy. The construction of houses, use of clothing in diverse climates, certain behavioral patterns, and work habits represent biological and cultural adaptations to climatic stress. The development of medicine from its primitive manifestations to its high levels in the present era and the increase of energy production associated with agricultural and industrial revolutions are representative of human cultural adaptation to the physical environment.

Culture and technology have facilitated biological adaptation, but they have also created and continue to create new stressful conditions that require new adaptive responses. A modification of one environmental condition may result in the change of another. Such a change may eventually result in the creation of a new stressful condition. A classic example of such an interaction of culture and biology is the development of malaria. In West Africa malaria became hyperendemic when the *Anopheles* mosquito, the major vector of malaria, was propagated because of the development of agriculture in the tropical rain forest of West Africa (7). It was in response to this new stress that the adaptive qualities of the abnormal hemoglobins, such as sickle-cell and thalassemia, became important to the survival of humans in tropical climates. In the same manner, advances in the medical sciences have successfully reduced infant and adult mortality to the extent that the world population is growing at an explosive rate, and unless world food resources are increased, the twenty-first century will witness a world famine. Western technology, although upgrading living standards, has also created a polluted environment that may become unfit for good health and life. If this process continues unchecked, environmental pollution will eventually become another selective force to which humans must adapt through biological or cultural processes or face extinction. Therefore, the ability to adapt to the unforeseeable threats of the future remains an indispensable condition of survival and biological success (29). Adaptation to the world of today may be incompatible with survival in the world of tomorrow unless humans learn to adjust their cultural and biological capacities.

## Accommodation and Adaptation

The term *accommodation* as defined by Young and Marchini (35) is used to describe responses to environmental stresses that are not wholly successful because even though they favor survival of the individual they also result in significant losses in some important functions. For example, when exposed to low intake of leucine for 3 weeks subjects can achieve body leucine balance at the expense of reducing protein synthesis and protein turnover (35). Since a low protein synthesis and protein turnover diminish the individual's capacity to successfully withstand major stresses such as infectious diseases (36), under conditions of low dietary protein intake achieving body leucine balance represents only a temporary accommodation, which in the long run is not adaptive.

## Genetic Adaptation

Genetic adaptation refers to specific heritable characteristics that favor tolerance and survival of an individual or a population in a particular total environment. A given biological trait is considered genetic when it is unique to the individual or population and when it can be shown that it is acquired through biological inheritance. A genetic adaptation becomes established through the action of natural selection, the central theme of Darwinian evolution. Natural selection refers to the mechanisms whereby the genotypes of those individuals showing the greatest adaptation or ''fitness'' (leaving the most descendents through reduced mortality and increased fertility) will be perpetuated, and those less adapted to the environment will contribute fewer genes to the population gene pool. Natural selection favors the features of an organism that bring it into a more efficient relationship with its environment. Those gene combinations fostering the best-adapted phenotypes will be ''selected for,'' and inferior genotypes will be eliminated. The selective forces for humans, as for other mammals, include the sum total of factors in the natural environment. All the natural conditions, such as hot and cold climates and oxygen-poor environments, are potential selective forces. Food is a selective force by its own abundance, eliminating those susceptible to obesity and cardiac failures, or by its very scarcity, favoring smaller size and slower growth. In the same manner disease is a powerful selective agent, favoring in each generation those with better immunity. The natural world is full of forces that make some individuals, and by inference some populations, better adapted than others because no two individuals or populations have the same capacity of adaptation. The maladapted population will tend to have lower fertility and/or higher mortality than that of the adapted population.

The capacity for adaptation (adaptability) to environmental stress varies between populations and even between individuals. The fitness of an individual or population is determined by its total adaptation to the environment—genetic, physiological, and behavioral (or cultural). Fitness in genetic terms includes more than just the ability to survive and reproduce in a given environment; it must include the capacity for future survival in future environments. The long-range fitness of a population depends on its genetic stability and variability. The greater the adaptation, the longer the individual or population will survive and the greater the advantage in leaving progeny resembling the parents. In a fixed environment, all characteristics could be under rigid genetic control with maximum adaptation to the environment. On the other hand, in a changing environment a certain amount of variability is necessary to ensure that the population will survive environmental change. This requirement for variability can be fulfilled either genetically or phenotypically or both. In most populations a compromise exists between the production of a variety of genotypes and individual flexibility. Extinct populations are those which were unable to meet the challenges of new conditions. Thus, contemporary fitness requires both genetic uniformity and genetic variability.

Therefore, contemporary adaptation of human beings is both the result of our past and the present adaptability. It is this capacity to adapt that enables us to be in a dynamic equilibrium in our biological niche. It is the nature of the living organism to be part of an ecosystem whereby it modifies the environment and, in turn, is also affected by such modification. The maintenance of this dynamic equilibrium represents homeostasis, which, in essence, reflects the ability to survive in varying environments (20, 29). The ecosystem is the fundamental biological entity—the living individual satisfying its needs in a dynamic relation to its habitat. In Darwinian terms, the ecosystem is the setting for the struggle for existence, efficiency and survival are the measures of fitness, and natural selection is the process underlying all products (29).

## PURPOSE OF ADAPTATION: HOMEOSTASIS

The basic living unit of the body is the cell, and each organ is an aggregate of many different cells held together by intercellular supporting structures. Each type of cell is specially adapted to perform one particular function. For instance, the red blood cells transport oxygen from the lungs to the tissues. These cells along with the other numerous cells of the body have certain basic characteristics that are alike. For example, in all cells oxygen combines with carbohydrate, fat, or protein to release the energy required for cell function. Furthermore, the general mechanisms for changing nutrients into energy are basically the same in all cells, and all the cells also deliver the end-products of their chemical reactions into the surrounding fluids, and all cells have the ability to reproduce.

### The Internal Environment of the Cell and the Maintenance of Homeostasis

More than half of the human body consists of fluid, and most of this fluid is inside the cells (called *intracellular fluid*) and about one third is in the spaces outside the cells (called *extracellular fluid*). This extracellular fluid is transported throughout the body by the blood circulation and then mixed between the blood and the tissue fluids by diffusion through the capillary walls. In the extracellular fluid are the ions and nutrients needed by the cells for maintenance of cellular life. Therefore, all cells live in essentially the same extracellular fluid. For this reason, the extracellular fluid is called the *internal environment* of the body (15, 37). The extracellular fluid contains large amounts of sodium, chloride, and bicarbonate ions, plus nutrients for the cells, such as oxygen, glucose, fatty acids, and amino acids. It also contains carbon dioxide that is being transported from the cells to the lungs to be excreted and other cellular products that are being transported to the kidneys for excretion. On the other hand, the intracellular fluid contains large amounts of potassium, magnesium, and phosphate ions instead of the sodium and chloride ions found in the extracellular fluid. Special mechanisms for transporting ions through the cell membranes maintain these differences. Thus, so long as the proper concentrations of oxygen, glucose, the different ions, amino acids, and fatty substances are available in this internal environment, cells are capable of living, growing, and providing their special functions. Therefore, any conditions such as stress that disturb the equilibrium between the intra- and extracellular fluid affects the normal function of the cell and hence of the organism as a whole. The maintenance of this dynamic equilibrium constitutes the major objective of the various adaptive responses made by organisms. The necessity for the maintenance of homeostasis results from the fact that cellular functions are limited to a relatively small range of variations. For example, the chemical composition of the blood, lymph, and other body fluids varies within relatively narrow limits.

An environmental stress is defined as any condition that disturbs the normal functioning of the organism. Such interference eventually causes a disturbance of internal homeostasis. *Homeostasis* means the ability of the organism to maintain a stable internal environment despite diverse, disruptive, external environmental influences (29). On a functional level, all adaptive responses of the organism or the individual are made to restore internal homeostasis. These controls operate in a hierarchy at all levels of biological organization, from a single biochemical pathway, to the mitochondria of a cell, to cells organized into tissues, tissues into organs and systems of organs, to entire organisms. For example, the lungs provide oxygen to the extracellular fluid to continually replenish the oxygen that is being used by the cells, the kidneys maintain constant ion concentrations, and the gastrointestinal system provides nutrients.

Humans living in hot or cold climates must undergo some functional adjustments to maintain thermal balance; these may comprise the rate of metabolism, avenues of heat loss, heat conservation, respiration, blood circulation, fluid and electrolyte transport, and exchange. In the same manner, persons exposed to high altitudes must adjust through physiological, chemical, and morphological mechanisms, such as increase in ventilation, increase in the oxygen-carrying capacity of the blood resulting from an increased concentration of red blood cells, and increased ability of tissues to utilize oxygen at low pressures. Failure to activate the functional adaptive processes may result in failure to restore homeostasis, which in turn results in maladaptation of the organism and eventual incapacity of the individual.

Therefore homeostasis is a part and function of survival. The continued existence of a biological system implies that the system possesses mechanisms that enable it to maintain its identity, despite the endless pressures of environmental stresses (29). These complementary concepts of homeostasis and adaptation are valid at all levels of biological organization. They apply to social groups as well as to unicellular or multicellular organisms (29).

Homeostasis is a function of a dynamic interaction of feedback mechanisms whereby a given stimulus elicits a response aimed at restoring the original equilibrium. Several mathematical models of homeostasis have been proposed. In general, they show (as schematized in fig. 1.1) that when a primary stress disturbs the homeostasis that exists between the organism and the environment, to function normally the organism must resort either to biological or cultural-technological responses. Through the biological responses, the organism overcomes the environmental stress, and its physiological activities occur either at the same level as before the stress or take place at another level. For example, when faced with heat stress, the organism may simply reduce its metabolic activity so all heat-producing processes are slowed down, or may increase the activity of the heat-loss mechanisms. In either case the organism may maintain homeostasis, but the physiological processes will occur at a different set point. The attainment of full homeostasis or full functional adaptation, depending on the nature of the stress, may require short-term responses such as those acquired during acclimation or acclimatization, or may require exposure during the period of growth and development as in developmental acclimatization. In theory the respective contributions of genetic and environmental factors vary with the developmental stage of the organism—the earlier the stage, the greater the influence of the environment and the greater the plasticity of the organism (23, 29, 32). However, as will be shown in this book, the principle does not apply to all biological parameters; it depends on the nature of the stress, the developmental stage of the organism, the type of organism, and the particular functional process that is affected. For example, an adult individual exposed to high-altitude hypoxia through prolonged residence may attain a level of adaptation that permits normal functioning in all daily activities and, as such, we may consider him adapted. However, when exposed to stress that requires increased energy, such as strenuous exercise, this individual may prove to be not fully adapted. On the other hand, through cultural and technological adaptation humans may actually modify and thus decrease the nature of the environmental stresses so that a new microenvironment is created to which the organism does not need to make any physiological responses. For example, cultural and technological responses permit humans to live under extreme conditions of cold stress with the result that some of the physiological processes are not altered. However, on rare occasions humans have been able to completely avoid an environmental stress. Witness the fact that the Eskimos, despite their advanced technological adaptation to cold in their everyday hunting activities, are exposed to periods of cold stress and in response have developed biological processes that enable them to function and be adapted to their environment.

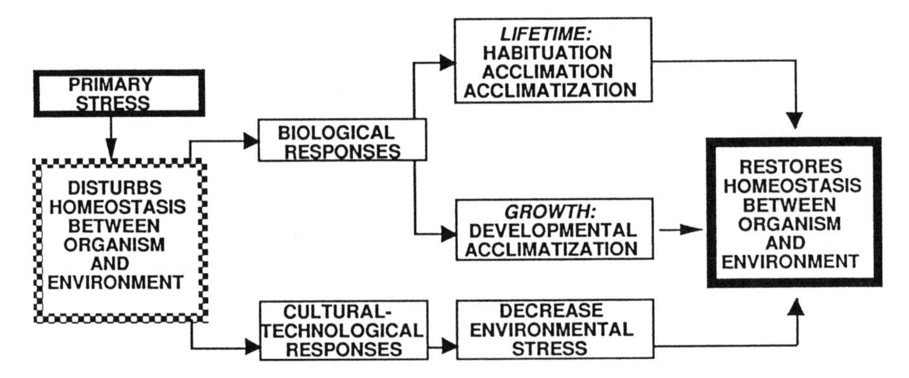


**Fig. 1.1.** Schematization of adaptation process and mechanisms that enable individual or population to maintain homeostasis in the face of primary disturbing stress.

Not all responses made by the organism can be considered adaptive. Although a given response might not be adaptive per se, through its effect on another structure or function it may prove beneficial to the organism's function. Conversely, a given adaptive response may aid the organism in one function but actually have negative effects on other functions or structures. Thus, within all areas of human endeavor a given trait is considered adaptive when its beneficial effects outweigh the negative ones. In theory this is a valid assumption, but in practice, because of the relative nature of adaptation, it is quite difficult to determine the true adaptive value of a given response. Every response must be considered in the context of the environmental conditions in which the response was measured and within the perspective of the length of time of the study and the subject population.

## The Process of Maintaining Homeostasis

It is said that adaptation is a form of long-range planning, whereby the organism gathers and stores information about the environmental conditions, and responds accordingly to future conditions. Because of this information storage capacity an organism is capable of progressive improvement in function. Although essentially all structures of the body are organized to help maintain the automaticity and continuity of life, for didactic purposes the process of homeostasis is considered with reference to a specific system.

### Transport of Extracellular Fluid

The maintenance of homeostasis is facilitated by the fact that the extracellular fluid is transported through all parts of the body. This transport occurs in two different stages. The first stage includes movement of blood around and through the circulatory system, and the second, movement of fluid between the blood capillaries and the cells. As illustrated in figure 1.2 the heart consists of two separate pumps, one of which propels blood through the lungs and the other through the systemic circulatory system. All the blood in circulation traverses the entire circuit on an average of once each minute at rest and as many as six times each minute during exercise.

As blood passes through the capillaries, because the capillaries are porous continual fluid exchange occurs between the plasma portion of the blood and the interstitial fluid that fills the intercellular spaces. In this manner, large amounts of fluid and its dissolved constituents can diffuse back and forth between the blood and the tissue spaces (see fig. 1.3). This process of diffusion is caused because the fluid and dissolved molecules are continually moving and bouncing in all directions within the fluid itself and also through the pores and through the tissue spaces. This process is facilitated by the fact that all cells and capillaries are close together. Therefore, almost any substance diffuses from the capillary to the cell within a few seconds. Thus, the extracellular fluid everywhere in the body, both that of the plasma and that in interstitial spaces, is continually being mixed, thereby maintaining almost complete homogeneity.

### Exchange of Nutrients in the Extracellular Fluid

**Respiratory System.** Each time the blood passes through the body it also flows through the lungs. The blood picks up oxygen in the alveoli to be used by the cells. The diffusion of oxygen is facilitated by the fact that the membrane between the alveoli and the lumen of the pulmonary capillaries is only 0.4 to 2.0 microns in thickness. Oxygen diffuses through this membrane into the blood in exactly the same manner that water and ions diffuse through tissue capillaries. At the same time that blood picks up oxygen in the lungs, carbon dioxide is released from the blood into the alveoli, and the respiratory movement of air into and out of the alveoli carries the carbon dioxide to the atmosphere. Carbon dioxide is the most abundant of all the end products of metabolism. Thus the respiratory system, operating in association with the nervous system, regulates the concentration of carbon dioxide in the extracellular fluid.

**The Gastrointestinal Tract.** A large portion of the blood pumped by the heart also passes through the walls of the gastrointestinal organs, where the carbohydrates, fatty acids, amino acids, and others, are absorbed into the extracellular fluid.

**The Liver.** Once the nutrients enter the liver their chemical composition is changed into more usable forms. The utilization and storage of these nutrients is assisted by other tissues of the body—the fat cells, the gastrointestinal mucosa, the kidneys, and the endocrine glands. The liver and the pancreas regulate the concentration of glucose in the extracellular fluid.

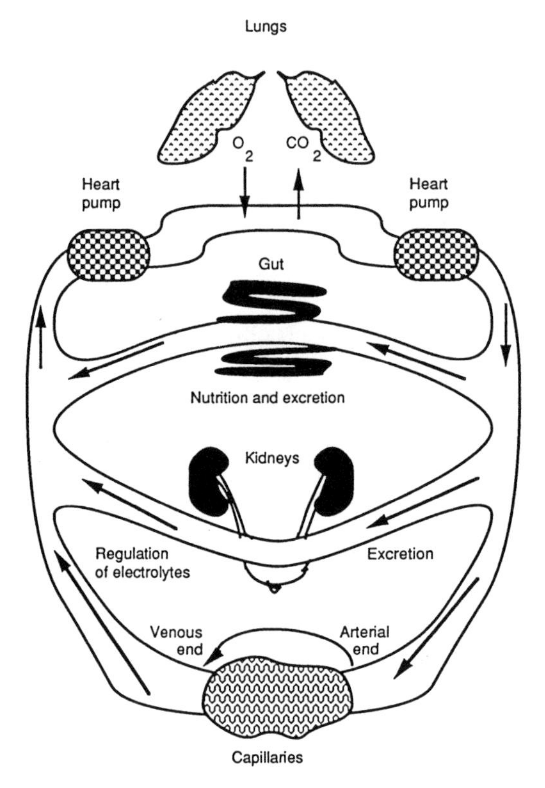


**Fig. 1.2.** General organization of the circulatory system. (Adapted from A. C. Guyton. 1986. Textbook of medical physiology, 7th ed. Philadelphia: W. B. Saunders Company.)

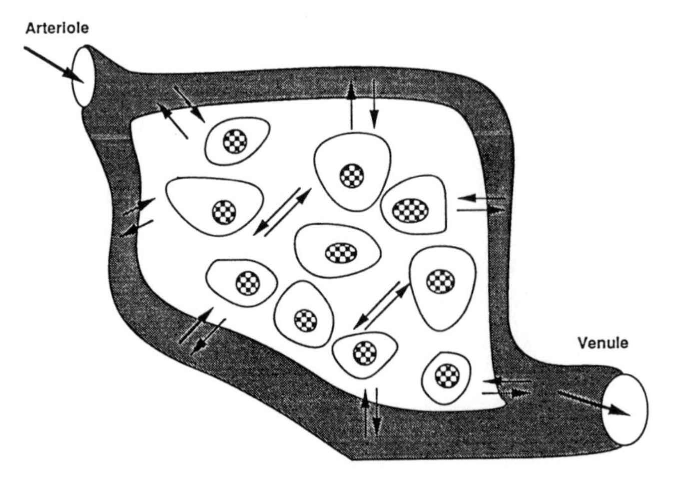


**Fig. 1.3.** Diffusion of fluids through the capillary walls and through the interstitial spaces. (Adapted from A. C. Guyton. 1986. Textbook of medical physiology, 7th ed. Philadelphia: W. B. Saunders Company.)

**The Kidneys.** After the nutrients are digested and utilized their by-products and waste are removed by the kidney. To achieve this goal the kidneys filter large quantities of plasma through the glomeruli into the tubules and then reabsorb into the blood those substances needed by the body such as glucose, amino acids, appropriate amounts of water, and many of the ions. The substances not needed by the body, especially the metabolic end products such as urea, are not reabsorbed but, instead, pass on through the renal tubules into the urine. In this manner, the kidneys regulate the concentrations of hydrogen, sodium, potassium, phosphate, and other ions in the extracellular fluid.

## Regulation of Body Functions: Role of the Nervous System and Hormonal System

Almost all functions of the organism and hence the maintenance of homeostasis depend upon the control and action of the nervous and hormonal system.

### The Nervous System

The nervous system is composed of three major parts: the sensory portion, the central nervous system (or integrative portion), and the motor portion.

**Sensory.** Sensory receptors detect the state of the body or the state of the surroundings. For instance, receptors present everywhere in the skin apprise one every time an object touches one at any point. The eyes are sensory organs that give one a visual image of the surrounding area. The ears also are sensory organs.

**Central Nervous System.** The central nervous system is composed of the brain and spinal cord. The brain can store information, generate thoughts, create ambition, and determine reactions the body performs in response to sensations.

**Motor.** The motor portion's function is to carry the signals that are transmitted by the nervous system. A large segment of the nervous system is called the autonomic system. It operates at a subconscious level and controls many functions of the internal organs, including the action of the heart, the movements of the gastrointestinal tract, and the secretion by different glands.

### The Hormonal System

The hormonal system regulates the metabolic functions and also complements the nervous system. The hormonal system consists of eight major endocrine glands that secrete the hormones. The hormones are transported in the extracellular fluid to all parts of the body to help regulate cellular function. Each hormone has a specific function. For example, thyroid hormone increases the rates of most chemical reactions in all cells, while insulin controls glucose metabolism, adrenocortical hormones control ion and protein metabolism, and parathormone controls bone metabolism.

## ADAPTATION RESEARCH

### Empirical and Experimental

The study of human adaptation involves a unique combination of field and laboratory methods, whereby the knowledge of ecologists, physiologists, geneticists, and cultural and physical anthropologists is pooled in an attempt to understand the human-environment relationship. To accomplish this objective, two different but related approaches are employed. The first is the geographical method, which may be called the indirect method. The indirect, or geographical, method attempts to establish the relationship between certain morphological or physiological characteristics and an environmental parameter. For example, anthropologists on a worldwide basis have established the relationship of various morphological features to climatic variables, such as temperature and humidity (11, 12). Because of the complex nature of a given climatic variable, the geographical method does not reveal the cause for the existence of this relationship. The explanation of the observed relationship requires the second, direct or experimental, approach. The experimental method collects and analyzes precise measurable changes in humans under reproducible and controlled experimental conditions, both in the laboratory and in the field. These conditions are the result of measurable environmental factors, which are generally of a chemical, physical, or biological nature and usually enable us to predict the qualitative and quantitative responses to each particular environment. The experimental approach requires a thorough understanding of the physiological and anatomical properties of the organism and the environmental parameters. It is a method designed to test a special theory, hypothesis, or assumption derived from the geographical, or indirect, method. In other words, research in adaptation requires the application of the geographical and experimental methods. The indirect method provides the pathway for experimental research. These methods permit the student of human adaptation to understand the mechanisms whereby a given population survives adverse environmental conditions such as heat, cold, altitude, disease, and malnutrition. The study of human adaptation is not oriented at determining biological or cultural differences among populations;

the goal is to identify the sources or causes that resulted in such adaptation and differences.

### Individuals vs. Populations

Whatever the method employed, geographical or experimental research in human adaptation is concerned with populations, not with individuals, although the research itself is based on individuals. There are two related reasons for this.

The first is a practical consideration. Studying all members of a given population, unless its size is small enough, is too difficult to be attempted by any research team. Therefore, according to the objectives of the investigation, the research centers on a sample that is considered representative of the entire population. Based on these studies, the researchers present a picture of the population as a whole, with respect to the problem being investigated.

The second reason is a theoretical one. In the study of adaptation, we usually focus on populations rather than on individuals because it is the population that survives and perpetuates itself. In the investigation of biological evolution, the relevant population is the breeding population because it is a vehicle for the gene pool, which is the means for change and hence evolution. The study of an individual phenomenon is only a means to understand the process. The adaptation of any individual or individuals merely reflects the adaptation that has been achieved by the population of which he/she is a member.

## OVERVIEW

The term *adaptation* encompasses the physiological, cultural, and genetic adaptations that permit individuals and populations to adjust to the environment in which they live. These adjustments are complex, and the concept of adaptation cannot be reduced to a simple rigid definition without oversimplification. The functional approach in using the adaptation concept permits its application to all levels of biological organization from unicellular to multicellular organisms, from early embryonic to adult stages, and from individuals to populations. In this context, human biological responses to environmental stress can be considered as part of a continuous process whereby past adaptations are modified and developed to permit the organism to function and maintain equilibrium within the environment to which it is daily exposed.

The mechanisms for attaining full functional adaptation include acclimation, acclimatization, habituation, and accommodation. The role played by each of these processes depends on the nature of the stress or stresses, the organ system involved, and the developmental stage of the organism. It is emphasized that the goal of the organism's responses to a given stress is to maintain homeostasis within an acceptable, normal range with respect to itself, other organisms, and the environment (as schematized in fig. 1.1). Such adaptations are usually reversible, but the reversibility depends on the developmental stage of the organism at which the adaptive response occurs and the nature of the environmental stress. This characteristic allows organisms to adapt to a wide range of environmental conditions. Furthermore, an adaptation is always a compromise between positive and negative effects. Every adaptation involves a cost. The process of adaptation is always positive—without it the organism would be worse off—but it has to pay a price for the benefit. The benefit derived from a given response depends on the circumstances and the conditions under which it occurs. As recently pointed out (38) every adaptation involves a choice. For example, if a man has 5 hours to walk 10 km and he walks slowly, he saves energy expenditure. This may be adaptive if the energy resources are limited, but, the man will have no time left to do anything else. On the other hand, if he walks fast he saves time at the cost of using more energy. Thus, the adaptive significance of a given type of response depends on the circumstances.

The study of individuals exposed to stressful conditions in natural and laboratory environments is one of the most important approaches for understanding the mechanisms whereby human populations adapt to a given environment. Knowledge of human adaptation is basic in our endeavors to understand past and present human variation in morphology and physiological performance.

## References


1. Bateson, G. 1963. The role of somatic change in evolution. Evolution 17:529–39.
2. Bock, W. J., and G. V. Wahlert. 1965. Adaptation and the form-function complex. Evolution 19:269–99.
3. Dobzhansky, T. 1968. Adaptedness and fitness. In R. C. Lewontin, ed., Population ecology and evolution. Proceedings of an International Symposium. Syracuse, NY: Syracuse University Press.
4. Lewontin, R. C. 1957. The adaptations of populations to varying environments. In Cold Spring Harbor Symposia on Quantitative Biology, vol. 22. Boston, MA: Cold Spring Harbor Laboratory of Quantitative Biology.
5. Mayr, E. 1956. Geographic character gradients and climatic adaptation. Evolution 10:105–8.
6. Slobodkin, L. B. 1968. Toward a predictive theory of evolution. In R. C. Lewontin, ed., Population biology and evolution. Proceedings of an International Symposium. Syracuse, NY: Syracuse University Press.

7. Livingstone, F. B. 1958. Anthropological implications of sickle-cell gene distribution in West Africa. Am. Anthropol. 60:533–62.
8. Wallace, B., and A. Sob. 1964. Adaptation. Englewood Cliffs, NJ: Prentice-Hall.
9. Wright, S. 1949. Adaptation and selection. In E. L. Jepson, E. Mayr, and G. G. Simpson, eds., Genetics, paleontology and evolution. Princeton, NJ: Princeton University Press.
10. Mayr, E. 1966. Animal species and evolution. Cambridge, MA: Belknap Press of Harvard University Press.
11. Schreider, E. 1964. Ecological rules, body heat regulation, and human evolution. Evolution 18:1–9.
12. Roberts, D. F. 1953. Body weight, race and climate. Am. J. Phys. Anthropol. 11:533–58.
13. Adolf, E. F. 1972. Physiological adaptations: hypertrophies and super functions. Am. Sci. 60:608–17.
14. Barcroft, J. 1932. La fixité du milieu intérieur est la condition de la vie libre. Biol. Rev. 7:24–87.
15. Bernard, C. 1878. Leçons sur les phénomènes de la vie. Paris: Baillière.
16. Baker, P. T. 1966. Human biological variation as an adaptive response to the environment. Eugen. Quart. 13:81–91.
17. Brauer, R. W. 1965. Irreversible effects. In O. G. Edholm and R. Bachrach, eds., The physiology of human survival. London: Academic Press.
18. Cannon, W. B. 1932. The wisdom of the body. New York: W. W. Norton and Co.
19. Cracraft, J. 1966. The concept of adaptation in the study of human populations. Eugen. Quart. 14:299.
20. Dubos, R. 1965. Man adapting. New Haven, CT: Yale University Press.
21. Eagan, C. J. 1963. Introduction and terminology. Fed. Proc. 22:930–32.
22. Folk, G. E., Jr. 1974. Textbook of environmental physiology. Philadelphia: Lea and Febiger.
23. Frisancho, A. R. 1975. Functional adaptation to high altitude hypoxia. Science 187:313–19.
24. Harrison, G. A. 1966. Human adaptability with reference to IBP proposals for high altitude research. In P. T. Baker and J. S. Weiner, eds., The biology of human adaptability. Oxford: Clarendon Press.
25. Hart, S. J. 1957. Climatic and temperature induced changes in the energetics of homeotherms. Rev. Can. Biol. 16:133–41.
26. Lasker, G. W. 1969. Human biological adaptability. Science 166:1480–86.
27. Mazess, R. B. 1973. Biological adaptation: aptitudes and acclimatization. In E. S. Watts, F. E. Johnston, and G. W. Lasker, eds., Biosocial interrelations in population adaptation. The Hague, Paris: Mouton.
28. McCutcheon, F. H. 1964. Organ systems in adaptation: the respiratory system. In D. B. Dill, E. F. Adolph, and C. G. Wilber, eds., Handbook of physiology, vol. 4. Adaptation to the environment. Washington, DC: American Physiological Society.
29. Proser, C. L. 1964. Perspectives of adaptation: theoretical aspects. In D. B. Dill, E. F. Adolph, and C. G. Wilber, eds., Handbook of physiology, vol. 4. Adaptation to the environment. Washington, DC: American Physiological Society.
30. Selye, H. 1950. The physiology and pathology of exposure to stress. Montréal: Acta.
31. Thomas, R. B. 1975. The ecology of work. In A. Damon, ed., Physiological anthropology. New York: Oxford University Press.

32. Timiras, P. S. 1972. Developmental physiology and aging. New York: Macmillan.
33. Rappaport, R. A. 1976. Maladaptation in social systems. In J. Friedman and M. Rowlands, eds., Evolution in social systems. London: Gerald Duckworth and Co.
34. Bligh, J., and K. G. Johnson. 1973. Glossary of terms for thermal physiology. J. Appl. Physiol. 35:941–61.
35. Young, V. R., and J. S. Marchini. 1990. Mechanisms and nutritional significance of metabolic responses to altered intakes of protein and amino acids, with reference to nutritional adaptation in humans. Am. J. Clin. Nutr. 51:270–89.
36. Frenk, S. 1986. Metabolic adaptation in protein-energy malnutrition. J. Am. Coll. Nutr. 5:371–81.
37. Guyton, A. C. 1986. Textbook of medical physiology, 7th ed. Philadelphia: W. B. Saunders Co.
38. Waterlow, J. C. 1990. Nutritional adaptation in man: general introduction and concepts. Am. J. Clin. Nutr. 51:259–63.

PART 2

# Thermoregulation

# CHAPTER 2 Principles of Temperature Regulation

---

## Overview

## BALANCE BETWEEN HEAT PRODUCTION AND HEAT LOSS

Humans, like most mammals, are *homeothermic*, which means they can maintain a relatively constant internal body temperature independent of environmental temperature. As shown in figure 2.1 the internal body temperature is in dynamic equilibrium with the factors that add or produce heat and the factors that facilitate heat loss. This balance is achieved through the integration of mechanisms that produce heat and regulate heat transfer from the internal organs to the shell or periphery, and mechanisms that facilitate the gain or loss of heat to the environment.

### Heat Production

#### Metabolic Rate and its Measurement

The metabolic rate (MR) represents the energy required to maintain all body functions at rest and under active conditions. The basal metabolic rate (BMR) is the minimum level of energy required to sustain the body's vital functions in the waking state. It is usually measured in the postabsorptive and fasting state whereby food is not eaten for at least 12 hours before the measurement. If the measurements are done under resting conditions but 3 or 4 hours after a meal it is referred to as resting metabolic rate (RMR), representing the minimum level of energy required to sustain the body's vital functions under resting conditions. The RMR is slightly higher than the BMR. However, for most purposes the terms of basal and resting metabolic rate are used interchangeably. The metabolic rate is measured in calories. A calorie is used to measure the quantity of heat required to raise the temperature of 1 g (1 ml) of water 1°C, from 14.5° to 15.5°C. Thus, a Calorie is usually referred to as kcal. The accepted international standard is the joule. To convert kcal to kilojoules (kj) multiply the kcal by 4.19. The metabolic rate as measured in kcalories or kilojoules can be obtained by direct and indirect calorimetry.

#### Direct Calorimetry

The metabolic rate can be obtained directly by measuring the total quantity of heat liberated from the body in a given time. For this purpose the subject is placed in a specially constructed and well-insulated air chamber with cool water pipes. The increase in temperature of the water pipes is equivalent to the heat liberated by the subject's body (see fig. 2.2). Obviously, direct calorimetry is not easily measured because of logistic factors.

Despite variability in the amount of energy available in food, the energy generated by a person does not vary a great deal according to the kind of

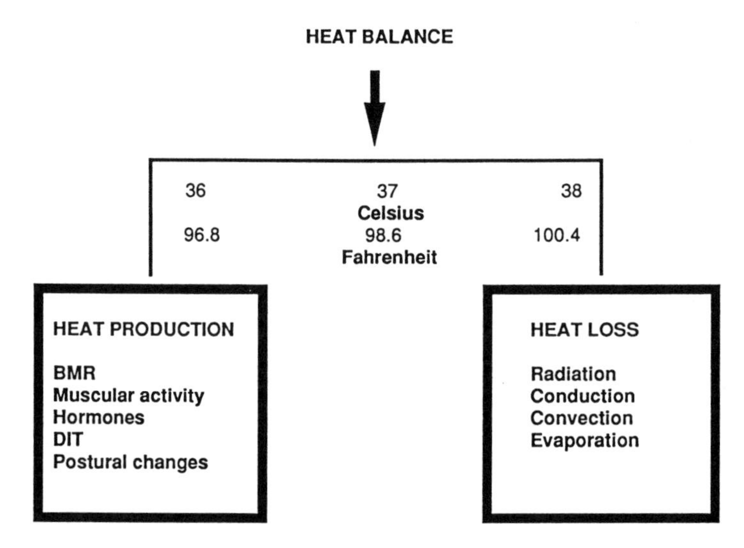


**Fig. 2.1.** Heat balance attained through the mechanisms of heat production and heat loss.

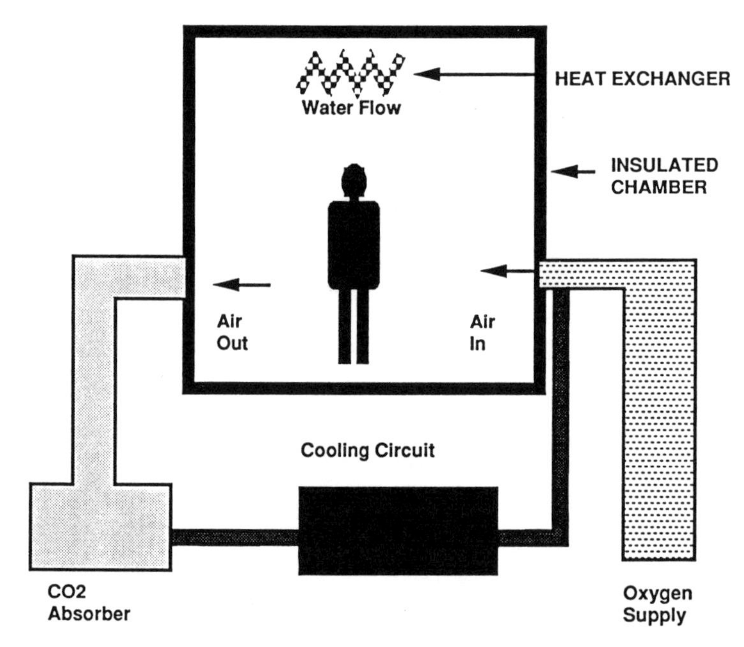


**Fig. 2.2.** Air-tight human calorimeter. The heat produced and radiated by the person is removed by a stream of cold water located in the heat exchanger. The change in temperature of the water indicates the amount of heat production.

food eaten. Thus, a subject at rest in a steady state, oxidizing an average diet, generates about 4.83 calories for every liter of oxygen he consumes. The metabolic rate (MR) in calories equals the volume of oxygen ($VO_2$) in liters consumed each minute multiplied by 4.83.

$$MR = 4.83 \times VO_2$$

### Indirect Calorimetry

Indirect calorimetry is based on the law of conservation of energy, equivalency of heat production, and the energy equivalent of oxygen.

**Law of Conservation of Energy.** According to the law of conservation ''when the chemical energy content of a system changes, the sum of all forms of energy given off or absorbed by the system must be equal to the magnitude of the change,'' which is expressed as:

$$dE = dQ + dW + dR,$$

where: dE = decrease in chemical energy, dQ = heat given off, dW = mechanical work performed, and dR = other forms of energy given off.

**Equivalency of Heat Production.** According to this principle the energy produced by oxidation of foodstuffs in the body is equivalent to that produced by combustion in a bomb calorimeter, which measures the heat generated when a foodstuff is burned in an excess of $O_2$. This assumption is valid even though in a bomb calorimeter oxidation occurs under high pressure and high temperature, while the oxidation in the body involves a series of enzyme-catalyzed reactions at a temperature of 37°C.

**Energy Equivalent of Oxygen.** Since virtually all the energy expended in the body is derived from reactions of oxygen with different foods, the metabolic rate can also be calculated from the rate of oxygen utilization. This is called the energy equivalent of oxygen. Using this energy equivalent one can determine the rate of heat liberation in the body from the quantity of oxygen utilized in a given period of time. For example, assuming a basal oxygen consumption of 250 ml/min, in 24 hours the total oxygen intake will equal 360 l ($250 \times 60$ min $\times 24$ hr = 360,000 ml), which equals 1734 kcal/24 hr ($4.83 \times 360 = 1734$ kcal). This is the amount of energy the body needs to maintain its function in absolute resting basal conditions, hence the name basal metabolic rate.

At present there are two methods for measuring oxygen consumption: closed-circuit and open-circuit spirometry.

#### *Closed-Circuit Spirometry*

In this method the subject breathes and rebreathes (and hence called closed circuit) from a prefilled container or spirometer of oxygen. Carbon dioxide in the exhaled air is absorbed by a canister of soda lime placed in the breathing circuit, and the amount of oxygen consumed is measured by a revolving drum attached to the spirometer. This method is usually used in laboratory settings and is not applicable for measurements during exercise.

#### *Open-Circuit Spirometry*

In this method the subject breathes ambient air. This method is based on the fact that the air has a constant composition of 20.99% oxygen, 0.03% carbon dioxide, and 79.04% nitrogen. Because oxygen is utilized during energy-yielding reactions and carbon dioxide is produced, the exhaled air contains less oxygen and more carbon dioxide than the inhaled air. Therefore, by measuring the amount of oxygen present in expired air one can calculate the amount of oxygen consumed. At present there are two techniques of open-circuit spirometry: (1) the portable spirometer, and (2) the "Douglas Bag."

**The Portable Spirometer.** The portable spirometer is a small instrument that weights about 4 kg and is usually worn on the back (see fig. 2.3). It has a two-way valve that allows one to breathe ambient air while the expired air passes through the gas meter that measures the volume and also collects a small gas sample. This sample is later analyzed for oxygen and carbon dioxide concentration, using either electronic oxymeters or the Scholander method. Although the method is very portable it does lose accuracy at high levels of activity.

**The Douglas Bag.** In this technique air is breathed through one side of the two-way valve while the exhaled air passes to the other side into a plastic or "Douglas" bag. The small sample of exhaled air that is collected in the bag is analyzed for its oxygen and carbon dioxide content.

Whatever the technique, the energy expenditure in kcal is computed from oxygen consumption using the corresponding transformation (see table 2.1).

## Reference Values of Basal Metabolic Rate: Resting Energy Expenditure

Since most measurements of oxygen consumption are done under resting rather than basal conditions the measurements of metabolic rate usually refer to Resting Metabolic Rate (RMR) or Resting Energy Expenditure (REE). The most widely used references for resting metabolic rate are those given in the equations of Harris and Benedict (1). The equations as given below permit the prediction of resting energy expenditure for subjects ranging in age from 6 to 70 years:

Males, REE (calories/day $= 66 + (13.7 \times$ weight, kg) $+ (5 \times$ height, cm) $- (6 \times$ age).

Females, REE (calories/day $= 665 + (9.6 \times$ weight, kg) $+ (1.7 \times$ height, cm) $- (4.7 \times$ age).

Solving the corresponding equation, the resting metabolic rate for a 25-year-old woman weighing 55 kg (120 lb) and 165 cm tall (5′4″) would equal 1323.5 calories per day. Since the surface area ($m^2$), derived by interpolating the height and weight in the nomogram given in figure 2.4 equals 1.62 $m^2$, the resting metabolic rate would equal 34.0 calories/$m^2$/hour ($1323.5/1.62/24 = 34.0$).

**Fig. 2.3.** Portable spirometer for measuring oxygen consumption and hence energy expenditure. (From W. MacArdle, F. I. Katch, and V. L. Katch. 1991. Nutrition and exercise physiology. Philadelphia: Lea and Febiger.)

## New Reference Values of Resting Energy Expenditure

Recent investigators have determined that the Harris-Benedict equations overestimated measured basal energy expenditure by 7–24% (2–4). These differences are usually attributed to changes in body size and composition, levels of physical activity, and diet that characterize modern populations. For these reasons, Mifflin et al. (5) have developed new predictive equations of resting energy expenditure rate (or resting metabolic rate) based on a sample of 498 healthy subjects ranging in age from 19 to 78 years, which are given below.

Males, REE (calories/day = (10 × weight, kg) + (6.25 × height, cm) − (5 × age) + 5.

Females, REE (calories/day = (10 × weight, kg) + (6.25 × height, cm) − (5 × age) − 161.

Solving the corresponding equation, the resting energy expenditure for a 25-year-old male weighing 75 kg (165 lbs) and 176 cm (5′9″) tall, with the surface area of 1.91 $m^2$, would equal 1730 calories per day or 37.74 calories/$m^2$/hour (1730/1.91/24 = 37.74), while for a female with the same characteristics would equal 1564 calories/day or 33.9 calories/$m^2$/hour.

Irrespective of the method of measuring BMR, calorie expenditure is usually expressed per square meter ($m^2$) of surface area. This is because heat exchange is proportional to surface area.

## Usual Factors Affecting Measurements of Metabolic Rate

In measuring the BMR, four factors are usually specified: (1) measurement is done in the morning; (2) the subject is at rest; (3) the subject is fasting; and (4) the ambient temperature is neutral. The four standard conditions of the BMR measurement indicate that several factors can affect the metabolic rate (MR), among which the most important are:

1. **Diurnal fluctuation.** Even if the other standard conditions are maintained, the metabolic rate is subject to diurnal fluctuation, with an increase in the morning and a decline during the night.

2. **Exercise.** During strenuous exercise the metabolic rate may increase to as much as 5 times the normal rate, due to increased activity of the musculature.

3. **Food consumption.** The consumption of nutrients and the subsequent digestion raises the metabolic rate, especially after protein has been eaten.

**TABLE 2.1. Calculation of resting metabolic rate from the oxygen consumption at standard conditions**

| Gender | Male |
|---|---|
| Age (yrs) | 20.00 |
| Height (cm) | 175.04 |
| Weight (kg) | 75.03 |
| Surface area ($m^2$) | 1.90 |
| Air ventilated in one hour (liters) | 420.01 |
| % oxygen present in inspired air (%) | 20.99 |
| Liters oxygen intake in one hour (liters) | 88.20 |
| % oxygen present in exhaled air (%) | 17.44 |
| Liters oxygen present in exhaled air (liters) | 73.20 (420.0 × 0.1740) |
| Liters oxygen consumed in one hour (liters) | 15.00 |
| Energy equivalent of oxygen (calories liberated/liter of oxygen consumed) | 4.825 |
| Calories liberated per hour (calories/hr) | 72.4 (15 × 4.825) |
| Resting Metabolic Rate (calories/day) | 1737.6 (72.4 × 24) |
| Resting Metabolic Rate (calories/$m^2$/hr) | 38.1 (72.4/1.90) |

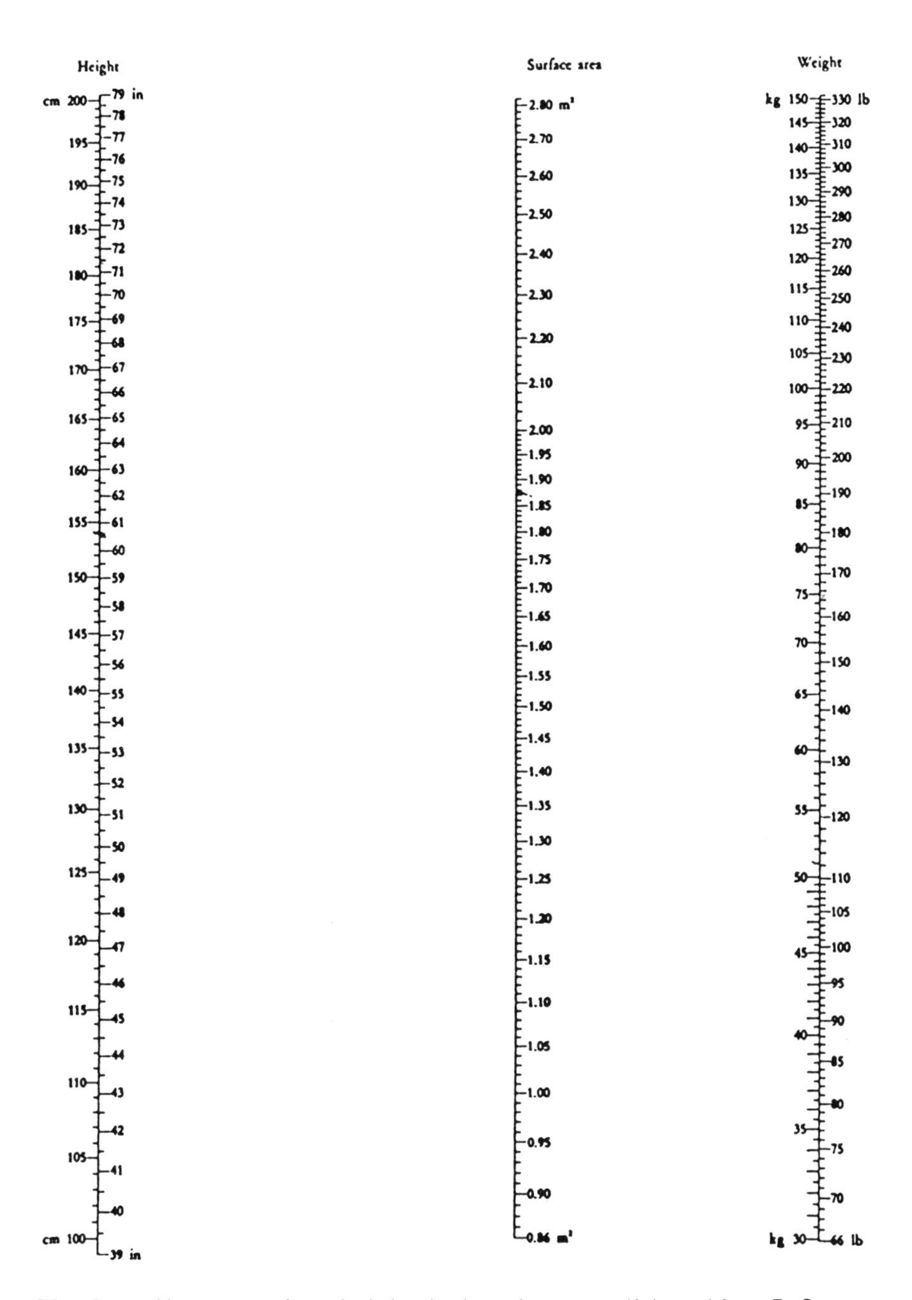


**Fig. 2.4.** Nomograms for calculating body surface area. (Adapted from B. S. DuBois and E. F. DuBois. 1916. Arch. Intern. Med. 17:863.)

This is called the specific dynamic action (SDA) of foodstuffs. The increase in MR after eating can last for 12 hours, or for as long as 18 hours after the consumption of large amounts of protein.

4. **Ambient temperature.** If the ambient temperature deviates from the-neutral range (25° to 27° C), the MR increases. It has been estimated that

each 1°C rise in temperature increases the rate of biochemical reactions by about 10%.

5. **Hormones.** In a stress situation, the sympathetic nervous system is stimulated and the nerves secrete norepinephrine (NE) which can increase the MR. Thyroxine produced by the thyroid and epinephrine secreted by the adrenal glands increase the metabolic rate.

## Heat Exchange by Passive and Active Mechanisms

As previously indicated, the organism must attain an equilibrium between heat production and heat loss in order to maintain a constant internal body temperature. Figure 2.5 illustrates the potential avenues of heat exchange. The avenues of heat exchange include radiation, convection, conduction, and evaporation, which will be discussed below.

### Radiation (R)

The flow of heat by radiation refers to the heat transfer that occurs when particular electromagnetic waves are emitted by one object and absorbed by another. The organism radiates heat to other objects and receives heat from other warmer objects (thus R is + or − ). The length of these waves, called infrared, is greater than that of visible light, from about 0.75 micron to 50 or more (a micron is one millionth—i.e., $1 \times 10^{-6}$—of a meter). The radiant energy from mammalian skin varies in wavelength from 5 to 20 $\mu$m. Exchanges of heat by radiation between the environment and the body depend on an individual's effective radiating surface area. Other factors influencing radiation include mean surface temperature, skin and clothing emissivity, and average radiant temperature (infrared) of the environment.

**Color.** In the infrared range (5 to 20 $\mu$m), skin color does not have an effect on reflectance; for this reason it is said that the human skin acts as a black body radiator with a power of emissivity close to 1. On the other hand, white clothing has always been considered more suitable than black, both in the tropics because less heat is absorbed and in the polar regions because less heat is lost (6).

**Posture.** The amount of heat gained or lost by radiation also depends on the posture of the organism. According to current estimates, the effective radiating surface area of a person standing with arms and legs spread is approximately 85% of the total skin area (6), whereas for a sitting person it is approximately 70% to 75% of the total body surface area (7). Indeed, when a person faces the sun rising on the horizon, 24.7% of his body surface receives direct radiation, but when the sun is directly overhead, a standing person receives direct radiation on only 4.4% of the body surface (8). In other words,

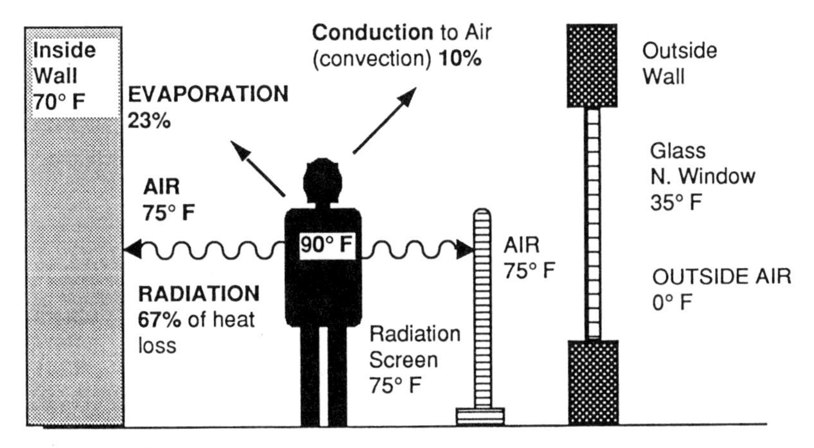


**Fig. 2.5.** Partitioning of heat exchange avenues from man. At air temperature of 24°C (75.2°F) most heat is lost by radiation and the rest by conduction and evaporation. (From G. E. Folk. 1974. Textbook of environmental physiology. Philadelphia: Lea and Febiger.)

vertical posture increases the area available for warming when the sun is low in the morning or evening, and minimizes heat load during the middle of the day (fig. 2.6). The remaining skin surfaces, such as the inner surfaces of legs, arms, and corresponding sites on the trunk of a standing or sitting person, radiate to other skin surfaces and are therefore not a consideration in the effective radiating surface area.

**Surface Area.** Just as variations in posture affect the degree of heat gained or lost by radiation, variations in body surface area have a potential influence on heat exchange by radiation. Body surface area may influence heat exchange in two directions. Under conditions in which the ambient temperature is lower than that of the skin, a large surface area per unit of body weight can be advantageous for facilitating heat loss. However, when the ambient temperature is higher, as is true in the desert, such a ratio can be a disadvantage because it increases the amount of heat that can be gained from the larger surface area exposed.

**Insulation.** Heat gained or lost by radiation is also affected by the degree of insulation. Various investigations have shown that clothed subjects sitting in the sun gain only about half as much heat as when sitting unclothed. However, it must be noted that radiant-heat exchange in persons is complex in natural conditions because of the spectrum of wave-lengths emitted at varying intensities and absorbed by the skin in varying degrees. An additional complicating factor is the fact that the spectrum and intensity of radiation from the terrain are different from those of the sun and sky.

## Convection (C)

Convection refers to the transport of heat by a stream of molecules from a warm object toward a cooler object. The most common exchange of heat by

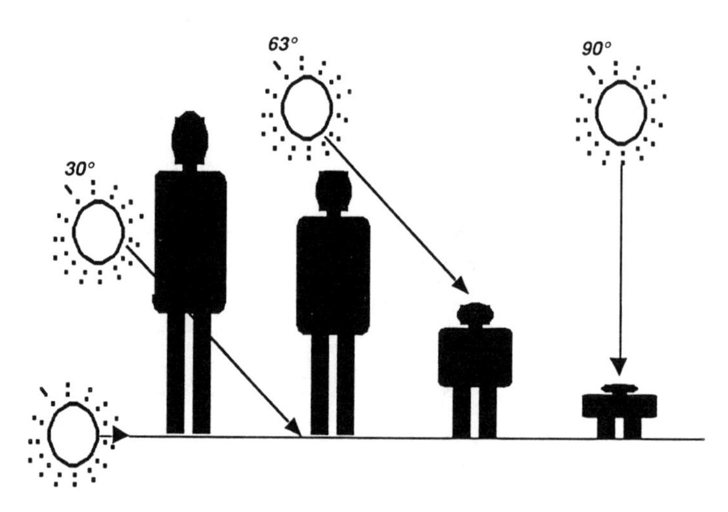


**Fig. 2.6.** Influence of body posture and surface area on effect of direct sunlight on standing man facing the sun at different solar elevations. Vertical posture increases surface area available for warming when sun is low in morning or evening and minimizes heat load during middle of day when sun is high. (Modified from C. R. Underwood and E. J. Ward. 1966. Ergonomics. 9:155–68.)

convection begins with heat conduction from a warm body to surrounding air molecules. The heated air expands, becomes less dense, and rises, taking the heat with it, to be replaced by cooler, denser air. If the ambient air temperature is lower than that of the body, heat is lost, but if it is higher than that of the body, heat is gained through the reverse process (thus C is + or −). Heat exchange by convection depends on (1) the difference in temperature between the body surface and the air, which in turn determines the amount of heat gained or lost by a unit mass of air contacting the skin, and (2) the air movements present, which in turn determine the mass of air that will come in contact with the skin surface. In general there are two different forms of convection: (1) natural convection (rising of warm air) and (2) forced convection, caused by the actions of an outside force (e.g., a fan). Natural convection depends on the natural bouyancy of heated material, and forced convection depends on the force of the air current and the movement of heated material within a medium.

## Conduction (CD)

Conduction refers to transfer of heat by direct physical contact. Heat exchange by conduction takes place internally and externally. Internal conduction occurs from tissue to tissue, especially in the blood, and represents an important force in the distribution of body heat. External conduction occurs through physical contact of the skin with external objects. Conductive heat-loss experiments have usually demonstrated that external conduction represents only a small percentage of total heat exchange. Thus, in studies of adaptation to heat stress, conduction is not usually considered separately, but is discussed with radiation and convection. Furthermore, the areas of skin in contact with surrounding objects are usually small, and direct contact with highly conductive materials is avoided. However, in a clothed individual, conduction of metabolic heat from skin to clothing does take place. When heat reaches the clothing, it is dissipated from the outer clothing surfaces by evaporation and/or convection and radiation, depending on the amount of air movement, the ambient temperature, and the vapor pressure gradients between clothing and environment.

### Evaporation (E)

Among the major avenues of heat exchange, only evaporation of water from the skin (in the form of sweat) and respiratory passages will always result in heat loss and never in heat gain. Heat is lost by evaporation because heat is required in the endothermic conversion of water to vapor. This heat is called the *latent heat of evaporation*. It is estimated that when water evaporates from the body surface 580 kilocalories (kcal) of heat are lost for each liter of water that evaporates. Therefore, evaporative heat lost is determined by the rate of water evaporation from the skin and respiratory system multiplied by the latent heat of water evaporation (580 kcal).

As shown in table 2.2 the relative role played by each of the four avenues of heat loss depends on the interaction of ambient temperature and humidity. Thus, in a comfortable climate with a temperature of 25°C (77°F), an unclothed seated person loses metabolic heat mostly through radiation (67%), and very little heat is lost through evaporation (23%) or convection (10%). At a warm temperature of 30°C (86°F), heat is lost in about equal proportions by radiation (41%), convection (33%), and evaporation (26%). At temperatures higher than 35°C (95°F), 90% of the heat lost is by evaporation, and very little (4% and 6%, respectively) is lost by radiation and convection (9). As the ambient temperature rises, sweat production is increased to maintain thermal homeostasis. Indeed, the sweat rate increases 20 ml for every 1°C rise in air temperature. Once sweating begins, vasodilation must continue to

**TABLE 2.2. Mechanisms of heat loss from the body (nude) at different room temperatures (at constant low air movement)**

| Room Temperature | Radiation (%) | Convection (%) | Evaporation (%) |
|---|---|---|---|
| Comfortable = 25°C (77°F) | 67 | 10 | 23 |
| Warm = 30°C (86°F) | 41 | 33 | 26 |
| Hot = 35°C (95°F) | 4 | 6 | 90 |

*Source:* From G. E. Folk, Jr. 1974. Textbook of environmental physiology. Philadelphia: Lea and Febiger.

increase as the heat load increases to transfer more heat from the internal body core toward the periphery, where it will be dissipated to the environment by the evaporation of sweat. When evaporation is incomplete because of increased humidity, sweat production is a far less efficient means of heat dissipation.

## Evaporation and its Regulation

Evaporation of water from the body occurs in two forms: insensible evaporation and thermal sweating.

### Insensible and Emotional Evaporation

Insensible evaporation, or perspiration (diffusion of water), leaves the body at all times unless the ambient atmosphere is too humid (100% saturated). This moisture diffuses through the pores of the sweat glands at a rate of approximately 900 ml/24 hr. Similarly, diffusion of water from the lungs occurs at about 400 ml/24 hr. Thus on the average, in temperate zones, insensible perspiration results in a heat loss of 522 kcal from the sweat glands (900 ml × 0.58 kcal/ml = 522 kcal) and 232 kcal from the lungs (400 ml × 0.58 kcal/ml = 232 kcal). This water loss through the skin and lungs may vary with altitude and the general quantity of moisture in the air—the higher the altitude, the greater the insensible moisture loss through the skin.

Emotional evaporation occurs through the skin on the forehead, the eccrine glands (soles of the feet and palms of the hands), and the apocrine glands (axillary and pubic). This nonthermal sweating usually occurs when the individual is under emotional stress rather than physical or thermal stress. In general, however, the contributions of insensible water loss and emotional perspiration to thermoregulation of heat stress are minimal.

### Thermal Sweating

**Sweat Glands.** Humans can sweat profusely, and it is this capacity that enables them to adapt to a wide range of heat stresses. Thermal sweating occurs through the *eccrine glands.* The *eccrine glands* are coiled tubular glands that connect the dermis and epidermis of the skin. The secretory portion of the sweat glands is located in the subcutaneous layer, and the excretory duct projects upward through the dermis and epidermis to terminate in the pore at the surface of the epidermis (see fig. 2.7). They are distributed throughout the skin, except for the margins of the limbs, glans penis, glans clitoris, labia minora, and eardrums. They are most numerous on the head, face, trunk, arms, and legs. Humans have between 2 and 5 million sweat

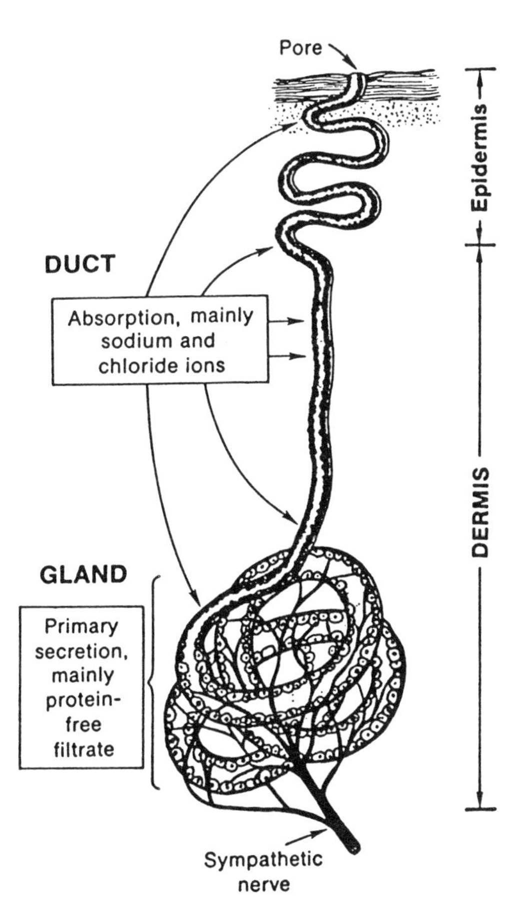


**Fig. 2.7.** A sweat gland innervated by a sympathetic nerve. A primary secretion is formed by the glandular position, but many if not most of the electrolytes are reabsorbed in the duct, leaving a diluted, watery secretion. (Adapted from A. C. Guyton. 1986. Textbook of medical physiology. 7th ed. Philadelphia: W. B. Saunders Company.)

glands with an average distribution of 150 to 340 per square centimeter. They are most numerous on the palms, soles, and then, in decreasing order, on the head, trunk, and extremities. According to various studies there are no ethnic differences in the number of sweat glands or in the gradient of distribution over the body (10–11). However, as indicated in chapter 3, the extent to which the sweat glands become active or dormant appears to depend on the development period of exposure to heat stress (12).

These glands are activated by hypothalamic impulses along sympathetic motor nerve fibers, also termed *cholinergic fibers* because they release acetylcholine. This chemical, usually associated with the parasympathetic nervous system (see section on nervous control of temperature), is a powerful stimulator of the sweat glands.

As long as the skin temperature is greater than the ambient temperature, heat is lost by radiation and convection, but when the ambient temperature is greater than that of the skin, the body gains heat by radiation and convection, which adds to the heat produced by the organism (metabolic rate). Under these conditions, the only means by which the body can rid itself of heat is by evaporation. Therefore, any factor that facilitates or prevents evaporation will cause changes in internal body temperature. Thus, water output from sweat glands increases in proportion to the need for evaporative cooling. For these reasons, water lost through the kidneys, skin, and respiratory tract may range from 10 to 12 l/24 hr for individuals working in a hot environment. The regulation of water intake to balance water output in hot climates depends on thirst.

## Thirst

At present it is not well defined whether the sensation of thirst originates locally in receptors in the mouth and pharynx or is of diffuse origin and results from general dehydration. What is known is that under natural conditions drinking stops before water has been absorbed from the stomach and is capable of diluting the blood. Even with full access to water, a human being undergoes a voluntary dehydration that often reaches 2% to 4% of body weight after heavy sweating. Thirst is thus satisfied before water intake equals loss (13). In fact, when volunteers working under heat stress were asked to drink sufficient quantities of water to replace sweat loss, none were able to do it without reporting abdominal discomfort and nausea (13, 14). Because of this paradoxical drinking behavior, serious dehydration, circulatory strain, and elevated rectal temperature may result when working persons ''voluntarily'' abstain from water despite a plentiful supply (13, 14). If sodium loss is not replaced, thirst will demand replacement only to restore the isotonic sodium concentrations in the extracellular fluid. If water loss is fully replaced without the corresponding sodium replacement, the plasma sodium concentration will decrease, and diuresis will proceed until the isotonic balance is restored. Thus dehydration may result from excessive hydration.

## Sodium Adjustments

As indicated below the sweat glands are stimulated by sympathetic cholinergic nerve fibers as well as by epinephrine or norepinephrine circulating in the blood. The sweat glands are tubular structures consisting of a deep subdermal *coiled portion* that secretes the sweat and a *duct portion* that goes through the dermis and epidermis of the skin (see fig. 2.7). The sweat glands secrete a fluid called *primary secretion* or *precursor secretion.* Mild stimulation of the sweat glands produces secretion of the precursor fluid that contains high concentrations of sodium (about 142 milli equal moles per liter) and chloride (about 104 mEq per liter). As the precursor fluid passes through the duct, sodium and chloride are reabsorbed so that the concentration of these ions is decreased to as low as 5 mEq per liter. On the other hand, when sweat glands are strongly stimulated by the sympathetic nervous system, as occurs during heat stress, the reabsorption of ions is not great, so the concentration of sodium and chloride in the precursor fluid and, hence, the sweat may be as high as 65 mEq per liter. However, with acclimation the concentration of sodium in the sweat decreases. For example, in an unacclimated person the sweat contains high concentrations of sodium that may range from 40 to 65 mEq per liter. After a 2-week period of acclimation, the salt concentration in sweat decreases to between 10 and 20 mEq per liter.

## HEAT TRANSFER FROM THE INTERNAL ORGANS TO THE PERIPHERY

### Heat Conductance by Vasoconstriction and Vasodilation

The circulatory system, in which blood circulates throughout the body, consists of the heart and a network of blood vessels. The blood vessels consist of: (1) the arteries, large-diameter vessels that bring blood from the heart to all parts of the body including the heart; (2) the capillaries, consisting of microscopically thin tubes present in every tissue of the body, forming a dense network that functions as a nutritional channel supplying the tissues with oxygen and other needed substances as well as transporting waste products; and (3) the veins, which consist of a fine network of very thin tubes that return the blood to the heart.

In addition, the muscle and overlying tissues are irrigated by a special variety of capillaries that function merely as intermediaries between the arterial and venous systems. These are called arteriovenous capillaries and they form a network called arteriovenous anastomoses. These capillaries play an important role in the distribution of the body's heat which is produced in the deeper portions of the organism such as the liver. By changing the diameter of the capillaries through *constriction* or *dilation* the rate of blood flow from the internal organs to the periphery can change by as much as 30%, resulting in a corresponding change in heat conductance. As shown in figures 2.8–9, with full vasodilation the heat conductance increases eightfold when compared to the vasoconstricted state.

### Countercurrent

The rate of heat conductance is influenced by the ''countercurrent'' system of blood flow. Countercurrent heat exchange occurs because each outgoing

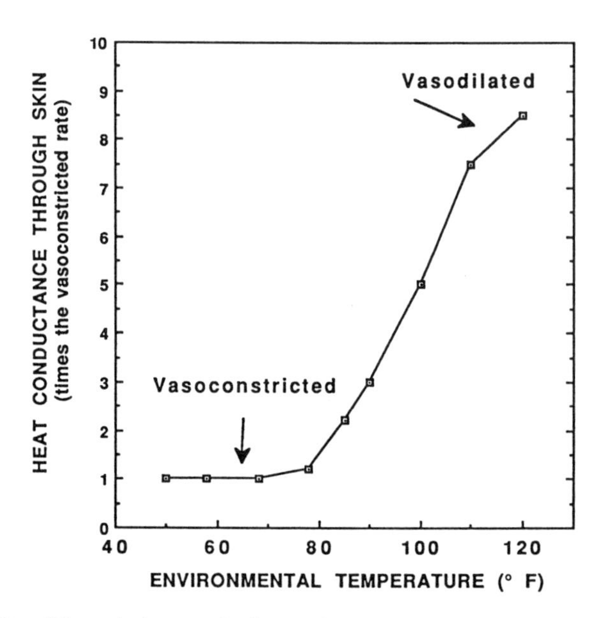


**Figure. 2.8.** Effect of changes in the environmental temperature on the heat conductance from the body core to the skin surface. (Adapted from A. C. Guyton. 1986. Textbook of medical physiology. 7th ed. Philadelphia: W. B. Saunders Company.)

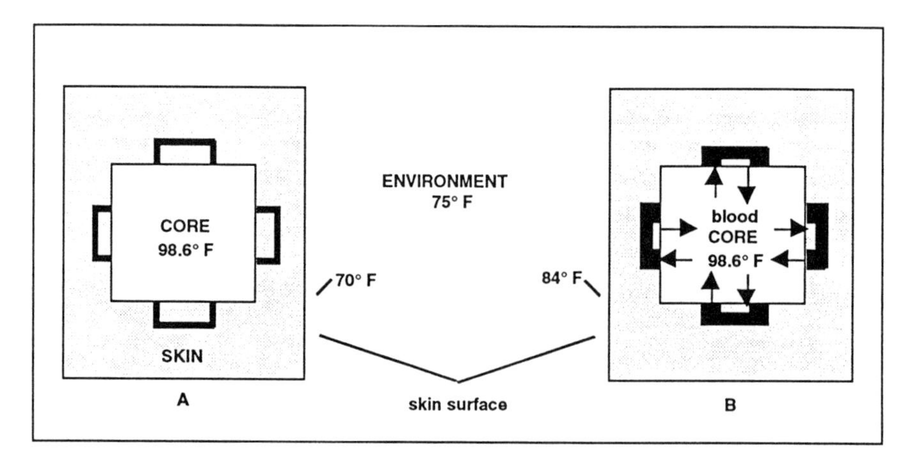


**Fig. 2.9.** Schematization of relationship of skin's insulating capacity to its blood flow. **A**, Skin acts as good insulator, that is, with minimal blood flow temperature of skin surface approaching that of external environment. When skin blood vessels dilate, **B**, the increased blood flow carries heat to body surface, that is, reduces insulating capacity of skin, and surface temperature becomes intermediate between that of core and external environment. (From A. R. Frisancho. 1981. Human adaptation: A functional interpretation. Ann Arbor: University of Michigan Press.)

artery is always surrounded by two incoming and returning veins (fig. 2.10). Because the outgoing arterial blood is always warmer than the incoming venous blood, there is continuous heat exchange. Thus, under cold conditions the arterial blood's temperature leaving the heart is about 37°C, but by the time the blood reaches the wrist, its temperature may decrease to about 23°C. On the other hand, under warm conditions both the arterial and venous blood may be 37°C, hence facilitating heat loss.

In a cold environment most of the venous return from arms and legs is through the deep venae comitantes that receive heat from blood flowing through the arteries and thereby minimize heat loss. Thus heat conductance to the periphery is low, yet actual blood flow to the limbs may be high, protecting tissues of the limbs from cold injury and hypoxia. In a hot environment most of the venous blood flow returns through the peripheral veins, and because these are close to the surface, heat loss to the environment is increased (15, 16). By this pathway external heat loss is maximized with high conductance. Furthermore, blood temperature is directly related to its distance from the body core, so under conditions of heat stress, the temperature of blood in the superficial veins of the fingers is higher than that at the elbow. As a result of these changes, and in order to compensate for the increased cutaneous blood flow needed for heat dissipation, the organism resorts to a compensatory vasoconstriction of internal vascular beds and to an increase of total circulating blood volume, providing more blood for cutaneous flow. With acclimation evaporative cooling improves, and the blood volume returns to pre-exposure levels.

## ROLE OF THE CENTRAL NERVOUS SYSTEM IN TEMPERATURE REGULATION

### Somatic and Autonomic Nervous Systems

As illustrated in figure 2.11 the nervous system is usually divided into the central nervous system and peripheral nervous system.

The central nervous system (CNS) connects the brain and the spinal cord. The body sensations are relayed from receptors to the central nervous system for its interpretation and response. Similarly, most of the nerve impulses that stimulate blood vessels and muscles to contract and glands to secrete are controlled by the central nervous system.

The peripheral nervous system (PNS) consists of the various nerve processes that connect the central nervous system (brain and spinal cord) with receptors, muscles, and glands. The peripheral nervous system is usually divided into two systems: the *afferent* system and the *efferent* system. The afferent system consists of nerve cells that convey information from receptors

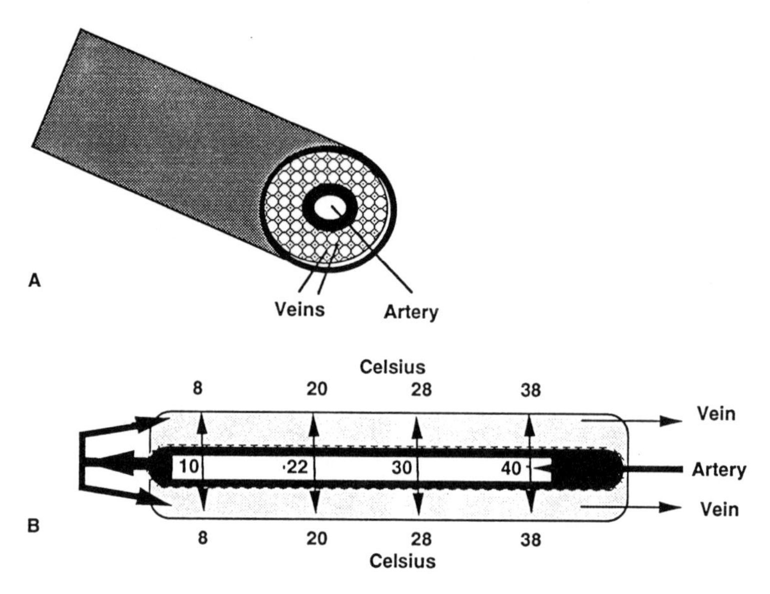


**Fig. 2.10.** Schematization of the anatomical basis for occurrence of countercurrent heat exchange. **A**, In all mammals the arteries that carry warm blood from the internal organs are surrounded by veins that return blood from the periphery. **B** shows a hypothetical temperature gradient occurrence in this concentric countercurrent system. (From G. E. Folk. 1974. Textbook of environmental physiology. Philadelphia: Lea and Febiger.)

in the periphery of the body to the central nervous system. The efferent system consists of nerve cells that transmit information from the central nervous system to blood vessels, muscles, and glands. The efferent system is, in turn, subdivided into somatic and autonomic nervous systems.

The somatic nervous system (SNS) consists of efferent neurons that conduct impulses from the central nervous system to skeletal muscle tissue. The somatic nervous system is under conscious control and therefore is referred to as voluntary. The neuron fibers of the somatic system release the neurotransmitter called acetylcholine (Ach), which stimulates the autonomic nervous system (ANS). These fibers structurally consist of visceral efferent neurons organized into nerves, ganglia, and plexus (network of nerves). These convey impulses from the central nervous system to smooth muscle tissue, cardiac muscle tissue, and glands. Functionally it operates without conscious control, and because it produces responses only in involuntary muscles and glands it is usually considered to be involuntary. The system was originally named autonomic because it was thought that it functioned with no control from the central nervous system, that is autonomous or self-governing. It is now known that the autonomic nervous system is regulated by centers in the brain, in particular by the cerebral cortex, the hypothalamus, and medulla oblongata. However, the old terminology has been retained.

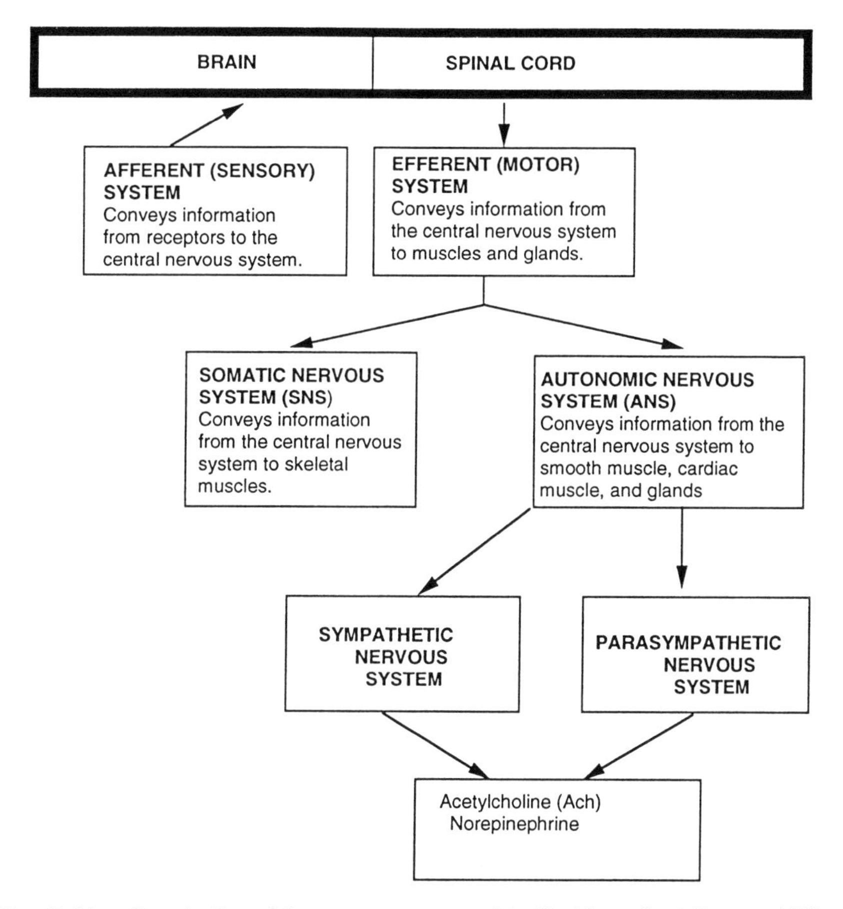


**Fig. 2.11.** Organization of the nervous system. (Modified from G. J. Tortora. 1989. Principles of human anatomy, 5th ed. New York: Harper and Row, Publishers.)

The autonomic nervous system (visceral efferent nervous system) regulates visceral activities such as change in the size of the pupil, accommodation for near vision, dilation and constriction of blood vessels, adjustments of the rate and force of the heartbeat, movements of the gastrointestinal tract, and the secretion of most glands. The efferent neurons of the autonomic system release either acetylcholine or norepinephrine (NE) as their neurotransmitter. The autonomic nervous system consists of the sympathetic and parasympathetic division. The organs innervated by the autonomic nervous system receive visceral efferent neurons from both the *sympathetic* and *parasympathetic* divisions. In general, nerve impulses transmitted by the fibers of one division stimulate the organ to start or increase activity, while nerve impulses from the other division decrease the organ's activity. Organs that receive nerve impulses from both sympathetic and parasympathetic fibers have a *dual innervation*. For this reason, autonomic innervation results in either inhibition or decreased activity. On the other hand, when a somatic neuron stimulates a skeletal muscle the muscle becomes active, and when the neurons cease to stimulate the muscle constriction stops altogether.

### Hypothalamic Regulation of Temperature

All the mechanisms that enable the organism to maintain thermal equilibrium are achieved through a series of temperature regulating centers located in the hypothalamus. Experimental studies have demonstrated that the anterior portion of the hypothalamus is concerned with mechanisms that facilitate heat dissipation, while the posterior portion of the hypothalamus controls the mechanism of heat conservation. For the heat dissipation and heat conservation mechanisms to operate they must receive signals from the temperature receptors located in the skin and deep tissues of the body.

When the ambient temperature is low (below 25°C or 77°F), in an unclothed person the immediate reflex effects are invoked. These reflexes are oriented toward decreasing heat loss through vasoconstriction when internal body temperature falls below the normal range (98°F ± 1). The deep body temperature receptors found in the spinal cord, abdominal viscera, and around the arterial blood send signals to the hypothalamus, which in turn induces shivering of the skeletal muscle. On the other hand, when the ambient temperature exceeds 28°C (80°F) the skin receptors send signals to the hypothalamus to induce vasodilation and thereby facilitate heat loss, and if this response fails to restore the internal body temperature, the deep body temperature receptors signal the hypothalamus to initiate sweating.

The reflex control of sweating seems to be partially related to the stimulation occurring from skin receptors and partially from the temperature of the hypothalamus. It has been postulated that the reflex act of sweating is initiated by thermal receptors in the skin, and the degree of activity may be stimulated by a temperature rise in blood passing through the hypothalamus (6). Another hypothesis maintains that sweating is solely affected by receptors located deep within the skin (6). Increased temperature of the hypothalamic center appears related to the center's increased responsiveness to incoming impulses from the cutaneous thermoreceptors or to deeper nerve signals (17). In any event, signals from the cutaneous thermoreceptors are the first line of defense against environmental heat stress. Heat warms the skin, stimulating these receptors to initiate the reflex arc that elicits sweating and cutaneous vasodilation, thus preventing significant body temperature elevation. However, sweating is initiated not only by a change in skin temperature, but also by an increase in mean body temperature. Even in a cold environment with low skin temperature, a hardworking person sweats profusely. During steady states of work, sweat rate is directly proportional to work rate, metabolic heat production, and the corresponding increase of central body temperature, even without corresponding skin temperature changes.

## CONCEPTS OF TEMPERATURE CONTROL

### Critical Air Temperature

Within an *ambient temperature* range of 25° to 27°C (77° to 80°F), referred to as the "critical air temperature," an unclothed person at rest can maintain body temperature at the basal metabolic rate as heat production from basic metabolic processes is balanced by heat loss to the air (18). Beyond this range the individual must resort to both passive and active mechanisms of heat loss to balance heat gain with heat loss and maintain thermal homeostasis.

### Set Point

Irrespective of the ambient temperature, when the *internal body temperature* rises above 37.1°C (98.8°F) the avenues of heat loss are increased until the temperature is reset to between 36.7° and 37.0°C. On the other hand, when the internal body temperature falls below 37.1°C the activity of the avenues of heat loss are decreased while the mechanisms of heat production such as shivering are increased until the temperature is reset to 37.1°C. In this manner, all the temperature control mechanisms continually attempt to bring the internal body temperature back to the "set point" of 37.1°C.

**Changes in the Set Point.** The extent to which the set point of the hypothalamus is established is determined to a certain extent by the skin temperature. As shown in figure 2.12 when the skin temperature had fallen to 29°C the set point increased from 36.7°C to 37.3°C when the skin temperature was higher than 33°C. By lowering the set point sweating is initiated when the skin temperature is high but it is inhibited when the skin temperature is low. Otherwise, the combined effect of low temperature and sweating at the same time could cause excessive heat loss.

Figure 2.13 shows that when the skin temperature is at a warm level of 31°C the temperature of the hypothalamus needs to fall below 36.5°C before it initiates shivering. On the other hand, when the skin temperature is 20°C the set point at which shivering occurs is at 37.1°C. This means that when the skin becomes cold it drives the hypothalamic thermostat to initiate shivering despite the fact that the hypothalamic temperature itself is still quite hot. Otherwise a cold skin temperature would eventually result in depressed internal body temperature unless heat production is increased. In other words, the feedback between the skin and hypothalamic temperature results in an anticipation of a possible decrease in internal body temperature and actually prevents its occurrence.

## DISORDERS OF TEMPERATURE REGULATION

The average internal body temperature at which the organism functions optimally ranges from 36.5° to 38.0°C (97.8° to 100°F). These temperatures

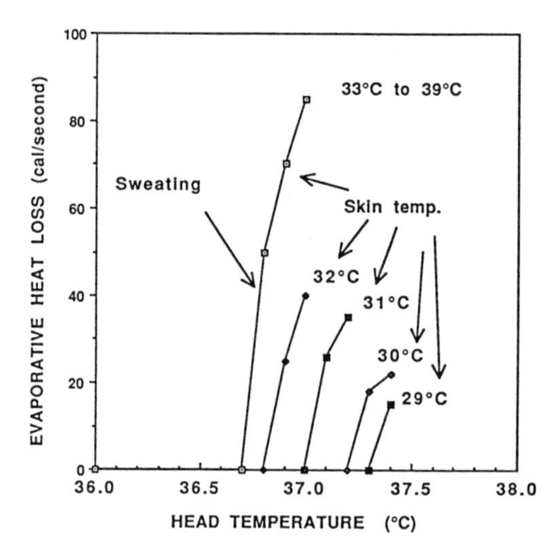


**Fig. 2.12.** Effect of changes in the internal head temperature on the rate of evaporative heat loss from the body. Note also that the skin temperature determines the exact "set point" level at which sweating begins. (Adapted from A. C. Guyton. 1986. Textbook of medical physiology. 7th ed. Philadelphia: W. B. Saunders Company.)

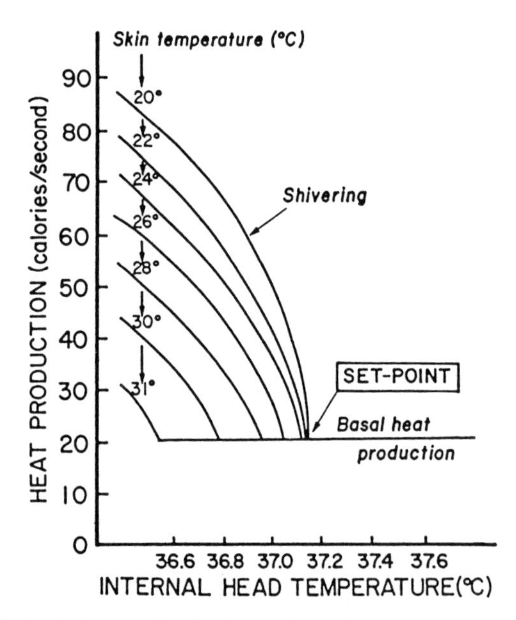


**Fig. 2.13.** Effect of changes in the internal head temperature on the rate of heat production by the body. Note also that the skin temperature determines the exact "set point" level at which shivering begins. (Adapted from A. C. Guyton. 1986. Textbook of medical physiology. 7th ed. Philadelphia: W. B. Saunders Company.)

may change even more with excessive and with extreme temperatures, because the temperature regulatory mechanisms are not immediately effective. With strenuous exercise the internal body temperature can rise as high as 104°F (40°C), and when the body is exposed to severe cold the internal body temperature may fall as low as 96°F (35.5°C), but as shown previously the organism usually succeeds in restoring internal body temperature. However, there are conditions when the organism may be required to function, albeit for short period of time, beyond the normal temperature range, and when this occurs it usually indicates disorder of the thermoregulatory system. Within these categories are fever, heat stroke, heat exhaustion, and frostbite.

## Fever

Fever is a primary disorder of the thermoregulatory system and usually means that the internal temperature is above the usual range of normal. This increase may be caused by bacterial diseases, brain tumor, or environmental conditions that may terminate in heat stroke.

### Role of Pyrogens

Some substances such as proteins or breakdown of proteins and certain other substances such as lipopolysaccharide toxins secreted by bacteria can cause an increase in the set point of the hypothalamic thermostat. Substances that have these characteristics are usually referred to as pyrogens. When the pyrogens cause an increase in the set point of the hypothalamic thermostat all the mechanisms of heat conservation as well as heat production are brought into action. As a result the internal body temperature is increased to approach the set point established at the thermostat within a few hours.

When bacteria or breakdown products of bacteria are present in the tissues or in the blood, these are phagocytized and digested by the blood leukocytes and the tissue macrophages. After digestion these bacterial products are released into the body fluid in a substance called *leukocyte pyrogen* or *endogenous pyrogen*. Experimental studies as reviewed by Stitt (19) indicate that upon reaching the hypothalamus this endogenous pyrogen immediately produces fever, increasing the body temperature in 8 to 10 minutes. According to these studies, endogenous pyrogens cause fever by first inducing the formation of prostaglandin $E_1$ in the local cells of the hypothalamus. The prostaglandin $E_1$, in turn, produces the fever reaction. When prostaglandin formation is inhibited by drugs, the fever is either stopped or reduced. Because aspirin blocks the formation of prostaglandin, it causes the reduction in the degree of fever. The prostaglandin-inhibition effect of aspirin explains why in a normal person aspirin does not decrease body temperature because a normal person does not have pyrogens that induce the formation of prostaglandins which change the set point of the hypothalamic thermostat. For this reason aspirin is known as an antipyretic.

### Resetting Normal Temperature

When the set point of the hypothalamic thermostat is raised, because the blood temperature is less than the set point of the hypothalamic thermostat the individual experiences chills and feels extremely cold, even though his or her body temperature may be already above normal. The skin is cold because of vasoconstriction and the person shivers in order to produce heat. Chills and shivering continue until the body temperature reaches the hypothalamic set point, then when the body temperature reaches this level, the person feels neither cold nor hot. As long as the factor that causes the increase in the set point of the hypothalamic thermostat is present the organism is in equilibrium. However, if the factor that is causing the high temperature is suddenly removed, the set point of the hypothalamic thermostat is also reduced to a lower level. In response to this difference, the hypothalamus attempts to bring back internal homeostasis so that the temperature in the hypothalamus and the body temperature are at the same level. In order to achieve this the hypothalamus induces vasodilation and sweating. As a result of this increased vasodilation and profuse sweating the internal body temperature is brought back to a normal level (fig. 2.14).

## Heat Stroke

Heat stroke is usually referred to as the failure of the organism to dissipate heat. When the internal temperature rises above 106°F the person is likely to develop heat stroke. This is because the hypothalamus at this temperature becomes excessively heated and its heat-regulating ability is drastically depressed so that sweating and vasodilation are impaired. As a result high temperature will be maintained until some means of reducing heat is found. Usual procedures include spraying the skin with cool water or alcohol to facilitate heat dissipation.

## Heat Exhaustion and Dehydration

Heat exhaustion refers to the inability of the organism to maintain internal body temperature due to dehydration and not due to failure of the thermoregulatory system. It is usually associated with excessive sweating and lack of access to water. Return to normal temperature can be achieved by rehydration.

As the ambient temperature and work load increase, nearly all the extra heat produced by the body is lost by evaporation. The production of thermal sweat, however, is a conspicuous and costly physiological event. On a hot day in the desert most individuals can produce sweat at a rate ranging from 0.5

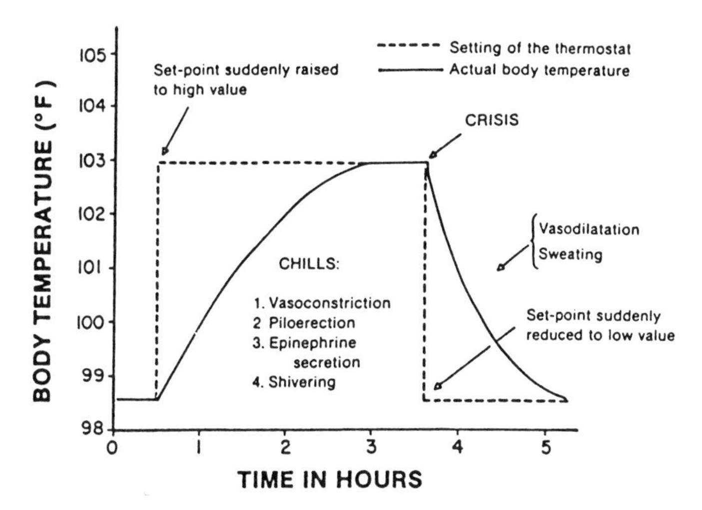


**Fig. 2.14.** Effects of changing the setting of the "hypothalamic thermostat." (Adapted from A. C. Guyton. 1986. Textbook of medical physiology. 7th ed. Philadelphia: W. B. Saunders Company.)

to 4.2 l/hr. Because of this marked increase in sweat loss, even under conditions of adlibitum water intake, there is always some amount of dehydration, since a person's thirst is not sufficient to stimulate replacement of all the lost body water.

## Frostbite

Frostbite is a thermoregulatory disorder that occurs when the body is exposed to extremely low temperature and the skin surface actually freezes. The freezing is called frostbite. Frostbite usually occurs at the earlobes, the digits of the hands and feet, and any other areas exposed to cold. If the freeze has been severe so as to cause extensive formation of ice crystals in the cell, permanent tissue damage may occur. Sometimes gangrene may occur after thawing of the damaged tissues and cause eventual loss of the frostbitten tissue areas.

## Raynaud's Disease

Raynaud's disease is a disorder characterized by spasms of arteries, especially those in the fingers and toes, due to overactivity of the sympathetic nervous system. As a consequence of the spasmodic contractions, the tissues of the fingers receive insufficient blood supply causing hypoxia and severe pain. The disease is bilateral and provoked by exposure to cold. In extreme cases the damaged tissues may become gangrenous.

## OVERVIEW

Humans, like most mammals, are able to maintain a relatively constant internal body temperature independent of environmental temperature. This balance is achieved through the integration of mechanisms that produce heat and regulate heat transfer from the internal organs to the shell or periphery, and mechanisms that facilitate the gain or loss of heat to the environment.

The metabolic rate (MR) represents the energy required to maintain all body functions at rest and active conditions. The basal metabolic rate (BMR) is the minimum level of energy required to sustain the body's vital functions in the waking state. It is usually measured in the postabsorptive and fasting state where food has not been eaten for least 12 hours before the measurement. If the measurements are done under resting conditions but 3 or 4 hours after a meal it is referred to as resting metabolic rate (RMR), representing the minimum level of energy required to sustain the body's vital functions under resting conditions. The RMR is slightly higher than the BMR. However, for most purposes the terms of basal and resting metabolic rate are used interchangeably. The metabolic rate is measured in kcalories or kilojoules.

The metabolic rate measured in kcalories or kilojoules can be obtained by direct and indirect calorimetry. Direct calorimetry measures the total quantity of heat liberated from the body in a given time by measuring the increase in temperature of water circulated in pipes and placed in the chamber where the subject is observed. On the other hand, indirect calorimetry measures the oxygen consumption of the subject from which the rate of heat liberation in the body can be determined. At present there are two methods for measuring oxygen consumption: closed-circuit and open-circuit spirometry. In the closed-circuit method the subject breathes and rebreathes (and hence called closed system) from a prefilled container or spirometer of oxygen. In the open-circuit method the subject breathes ambient air. By measuring the amount of oxygen present in the expired air one can calculate the amount of oxygen consumed.

The source of differences in basal metabolic rate are related to differences in age, physical activity, amount of lean body mass, diet-induced thermogenesis, ambient temperature, and hormone levels.

Within an ambient temperature range of 25° to 27°C (77° to 80°F), referred to as the ''critical air temperature,'' an unclothed person at rest can maintain body temperature at the basal metabolic rate as heat production from basic metabolic processes is balanced by heat loss to the air. To maintain thermal homeostasis the organism must resort to both passive and active mechanisms of heat loss and heat production. The avenues of heat exchange include radiation, convection, conduction, and evaporation. Depending on the ambient temperature, radiation, convection, and conduction can result in loss or gain of heat, but evaporation of sweat always leads to loss of heat.

The evaporation of water from the body results in heat loss through emotional and thermal sweating. Water lost through the kidneys, skin, and respiratory tract may range from 10 to 12 liters per day for individuals working in a hot environment.

When not replaced by water intake excessive sweating may result in dehydration. Under natural conditions drinking stops before water has been absorbed from the stomach and is capable of diluting the blood. Furthermore, if water loss is fully replaced without the corresponding sodium replacement, the plasma sodium concentration will decrease, and diuresis will proceed until the isotonic balance is restored. Thus dehydration may result from excessive hydration.

The concentration of sodium in the sweat decreases from about 60 to 65 mEq per liter before acclimation to heat stress to between 10 and 20 mEq per liter after full acclimation to heat stress.

Both the mechanisms of heat loss and heat production are under direct control of the central nervous system. The hypothalamus and the autonomic nervous system control and initiate heat production and heat loss.

Fever, heat stroke, heat exhaustion, frostbite, and Raynaud's disease represent disorders of the thermoregulatory system.

## References

1. Harris, J. A., and F. G. Benedict. 1919. A biometric study of basal metabolism in man. Washington, DC: Carnegie Institution of Washington.
2. Daly, J. M., S. B. Heymsfield, C. A. Head, L. P. Harvey, D. W. Nixon, H. Katzeff, and G. D. Grossman. 1985. Human energy requirements: overestimation by widely used prediction equation. Am. J. Clin. Nutr. 42:1170–74.
3. Owen, O. E., E. Kavle, and R. S. Owen. 1986. A reappraisal of caloric requirements in healthy women. Am. J. Clin. Nutr. 44:1–19.
4. Owen, O. E., J. L. Holup, and D. A. D'Alessio. 1987. A reappraisal of the caloric requirements of men. Am. J. Clin. Nutr. 46:875–85.
5. Mifflin, M. D., S. T. St Jeor, L. A. Hill, B. J. Scott, S. A. Daugherty, and O. K. Young. 1990. A new predictive equation for resting energy expenditure in healthy individuals. Am. J. Clin. Nutr. 51:241–47.
6. Folk, G. E., Jr. 1974. Textbook of environmental physiology. Philadelphia: Lea and Febiger.
7. Windslow, C. E. A., L. P. Herrington, and A. P. Gagge. 1936. The determination of radiation and convection exchanges by partitional calorimetry. Am. J. Physiol. 116:669–84.
8. Underwood, C. R., and E. J. Ward. 1966. The solar radiation area of man. Ergonomics 9:155–68.
9. Hardy, J. D. 1961. The physiology of temperature regulation. Physiol. Rev. 41:521–606.
10. Knip, A. S. 1969. Measurement and regional distribution of functioning eccrine sweat glands in male and female caucasians. Hum. Biol. 41:380–87.

11. Roberts, D. F., F. M. Salzano, and J. O. C. Wilson. 1970. Active sweat gland distribution in Caingang Indians. Am. J. Phys. Anthrop. 32:395–400.
12. Kuno, Y. 1956. Human perspiration. Springfield, IL: C. C. Thomas.
13. Adolph, E. F. 1947. Physiology of man in the desert. New York: Interscience.
14. Pitt, G. C., R. E. Johnson, and C. F. Consolazio. 1944. Work in the heat as affected by intake of water, salt and glucose. Am. J. Physiol. 142:253–59.
15. Bazett, H. C., and B. McGlone. 1927. Temperature gradients in the tissue of man. Am. J. Physiol. 82:415–51.
16. Gisolfi, C., and S. Robinson. 1970. Central and peripheral stimuli regulating sweating during intermittent work in men. J. Appl. Physiol. 29:761–68.
17. Bazett, H. C. 1949. The regulation of body temperatures. In L. H. Newburgh, ed., Physiology of heat regulation, and the science of clothing. Philadelphia: W. B. Saunders Co.
18. Erickson, H., and J. Krog. 1956. Critical temperature for naked man. Acta Physiol. Scand. 37:35–39.
19. Stitt, J. T. 1981. Neurophysiology of fever. Fed. Proc. 40:2835–42.

CHAPTER 3

# Acclimation and Acclimatization to Hot Climates: Native and Nonnative Populations

---

**Types of Heat Stress**

**Acclimation to Heat Stress**

General Trends

Acclimation of Women

**Individual Factors and Tolerance to Heat Stress**

Age

Body Size and Shape

Fatness

Physical Fitness and Cardiovascular Function

**Acclimatization to Hot Climates**

Nonnative and Native Populations

**Overview**

The fundamental problem for humans and all other homeotherms exposed to heat stress is heat dissipation. Therefore, most of the physiological responses to heat stress are aimed at facilitating heat loss. Successful tolerance of heat stress requires the development of synchronized responses that permit the organism to lose heat in an efficient manner and maintain homeostasis. In this chapter we will focus on the immediate physiological responses to heat stress, the individual factors that affect heat tolerance, and the process of acclimation. These will set the stage for a discussion of the process of acclimatization to heat stress of native and nonnative populations.

## TYPES OF HEAT STRESS

Humans encounter heat stress not only in tropical equatorial areas, but also during the summer in many of the vast land areas of the temperate zones. In general, hot climates are classified as either hot-dry or hot-wet.

**Hot-Wet**. The hot-wet or hot-humid climates are typical of the tropical rain forests usually located within the latitudes of 10° to 20° above or below the equator. Hot-humid climates have the following characteristics: (1) the air temperature does not exceed 35°C (95°F), usually ranging between 26.7° and 32.2°C (80° to 90°F); (2) the average relative humidity exceeds 50%, usually reaching as high as 95%; and (3) there is marked seasonal precipitation. As a result of the high precipitation and hot climate, vegetation is quite abundant and provides ample shade. Because of the combination of moisture and vegetation, much of the solar energy is used to convert liquid water to vapor, which exits in the atmosphere as insensible heat. There is little day-night or seasonal variation in temperature and dew point.

**Hot-Dry**. The hot-dry climates are usually found in desert regions such as those in the southwestern United States, the Kalahari, the Sahara of Africa, and other areas of the world. A hot-dry climate is characterized by (1) high air temperatures, which during the day range from 32.2° to 51.7°C (90° to 125°F); (2) low humidity, usually from 0% to 10%; (3) intense solar radiation; (4) very little precipitation; (5) little or no vegetation; and (6) marked day-night variation in temperature, often exceeding 50°. Because of the lack of vegetation the ground absorbs considerable solar energy and may heat to 32.2°C (90°F). When the ground temperature exceeds that of the surrounding environment, the soil acts as a radiator for long-wave (infrared) radiation. Moreover, the terrain may reflect up to 30% of the incident sunlight. The ambient day air temperature is almost always higher than the skin and clothing; therefore this hotter air heats the individual's body instead of cooling it. In other words, in the hot-dry climate, because of the low moisture content, the solar energy either directly or indirectly heats surfaces as well as the ambient air. In this manner, solar energy exists as a sensible heat in contrast to that of the hot-wet climate. Finally, pervasive desert winds, while facilitating evaporative cooling, at the same time may increase the body's heat by boosting the rate of heat exchange between the hot air and the cooler skin.

## ACCLIMATION TO HEAT STRESS

Studies on acclimation involve exposure of physically well-conditioned subjects to work in a wide range of temperatures and humidities under laboratory conditions. This usually includes 60 to 120 minutes of daily exposure to treadmill or bicycle ergometric work in humid heat (28° to 30°C wet bulb) or dry heat (50°C dry bulb) for periods of 2 or 4 weeks. During these tests data on rectal temperature, skin temperature, pulse rate, ventilation, oxygen consumption, sweat loss, and other variables are monitored at appropriate intervals. These studies indicate that acclimation to heat stress is attained through adjustments of the cardiovascular system. The mechanism that enables the organism to attain thermal homeostasis during the period of initial acclimation is different from those for full acclimation.

### General Trends

**Process of Acclimation.** When exposed to heat stress, the nonacclimated person, because of a poor peripheral heat conductance and a low sweating capacity, exhibits an increased heat and circulatory strain (fig. 3.1). During the first 4 days of exposure to heat stress, an increase in blood flow from the internal core to the shell may increase the peripheral heat conductance from five to six times its normal value. Simultaneously, there is an excessive increase in sweat rate and sodium loss, which may amount to more than 50% of preacclimation levels. Only a small proportion of this sweat is evaporated; a 10% increase in evaporation is accompanied by a 200% increase in non-evaporated sweat (1). This means that during acclimation to heat stress the skin and rectal temperatures are lowered through a wasteful overproduction of sweat. During this stage the circulatory strain is decreased as shown by the return of the heart rate to preexposure levels.

**Attainment of Full Acclimation.** With repeated exposure full acclimation to heat stress is attained through the continued maintenance of high peripheral heat conductance. The increase in vasodilation by increased perfusion of the cutaneous interstitial space also results in an increase in plasma volume and total circulating protein (1, 2). The enhanced peripheral heat conductance is also accompanied by a more complete and even distribution of sweat over the skin (3, 5). Hence with full acclimation evaporation of sweat

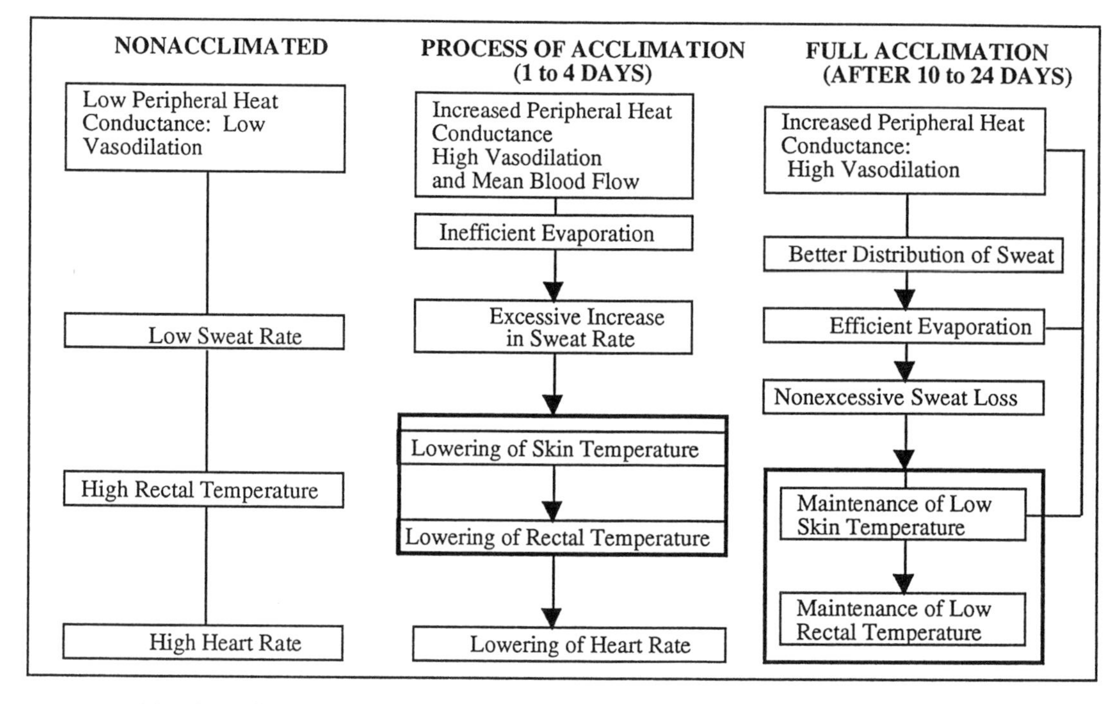


**Fig. 3.1** Schematization of mechanism of acclimation to heat stress. During initial acclimation thermal homeostasis is attained through an increased peripheral conductance, but evaporation is inefficient so that sweat rate becomes excessive. In full acclimation thermal homeostasis is achieved through an increased peripheral heat conductance, better distribution of sweat, and efficient evaporation.

is more efficient and the sweat output is lower than during the first week of acclimation.

Furthermore, the skin temperature threshold for the onset of sweating is decreased, which means that an equivalent rate of sweating is achieved at a lower skin temperature (5–7). As a result of the decrease in skin temperature, the rectal temperature decreases simultaneously and circulatory stability is maintained. During acclimation the renal sodium output is drastically reduced and sweat-sodium concentration is also decreased. However, when considering the total greater sweat loss, even after full acclimation the total loss of sodium exceeds preacclimation levels. Along with an improved physiological response, the disagreeable sensations associated with heat exposure are progressively reduced until individuals are able to work without much discomfort (3).

## Acclimation of Women

All studies indicate that women, like men, can be artificially acclimated to heat, manifesting similar physiological adjustments. During exposure to work in heat stress acclimated women show a reduction in heart rate, a lowering of rectal temperature, a decreased skin temperature, an onset of sweating at a lower skin temperature, and an increased sweating capacity (8–12). However, it appears that the limits of endurance under extreme heat stress for women are less than those for men. This difference appears related to a woman's excessively high skin temperature relative to men that results in a lower thermal gradient for removing metabolic heat and less reserve capacity for moving blood to the skin (3).

## INDIVIDUAL FACTORS AND TOLERANCE TO HEAT STRESS

### Age

Age is an important biological factor that affects tolerance to work in heat stress. In general, subjects over the age of 45 are less able to tolerate heat stress than younger subjects (13, 14). However, the relationship between age and tolerance to heat stress is not linear. As shown by studies on men and women (15, 16), preadolescent subjects (younger than 11 years old) and older subjects (46 to 67 years) have a lower capacity to sweat and a lesser work capacity than adolescents (15 to 16 years) and adults (20 to 29 years). It is evident, then, that at both extremes of age, the capacity to adjust to heat stress is impaired, probably because of incomplete physiological development in young subjects and onset of thermoregulatory mechanism deterioration among older subjects.

### Body Size and Shape

According to Fourier's Law of Heat Flow, the rate of heat loss per minute is directly proportional to body surface and to the difference between temperature of the body core and that of the environment, and inversely proportional to the thickness of the body shell. In other words, the greater the surface area, the greater the rate at which heat leaves the core toward the surrounding environment. Conversely, the thinner the thickness of the barrier between the core and the exterior, the greater the rate of heat flow. However, the relation of physique to heat tolerance is not a simple one and must be considered in reference to different conditions of heat stress and energy production.

Table 3.1 illustrates variability in surface area and heat production in a sample of adult males. From these data the following conclusions are evident. First, when body size increases, the increase in surface area does not keep pace with the increase in body weight. Therefore, a tall, heavy subject has a lower surface area:weight ratio than a short, light subject. In the same manner a tall, lean subject has a greater surface area:weight ratio than a short, stocky subject. That is, for a given body weight, a physique that emphasizes linearity increases the surface area:weight ratio. Second, heat production under resting and active conditions is nearly proportional to body weight so that the heavier the subject, the greater the heat production.

In theory a tall, heavy or short, stocky subject working under heat stress has greater heat production and smaller ranges of heat loss than a short, light or tall, lean subject. Accordingly, under conditions where the ambient temperature is lower than the skin temperature, a short, light or tall, lean subject with a high surface area:weight ratio would produce less heat and have a greater area for sweat evaporation and for heat loss by radiation and convection than a tall, heavy or short, stocky individual. In contrast, under conditions in which the ambient temperature or effective temperature (either dry or humid heat) is higher than the skin temperature, the subjects with high surface area:weight ratio would have little advantage, since the relatively large surface area providing extensive evaporation of sweat would also result in significant heat gains by radiation and convection, negating the positive effects of the evaporative process. These expectations were evaluated through mathematical simulations (17), whereby reference models representing varying body sizes were compared under simulated laboratory heat stress and simulated natural environment heat stress. The results of these analyses indicated that under all but extreme heat load a high surface area and low weight ratio is advantageous for heat loss. This advantage, however, was nullified under conditions of extreme heat and high humidity.

Experimental studies have confirmed in part these theoretical expectations (18). These investigations have shown that in humid heat (30°C, 28°C wet

**TABLE 3.1. Comparison of heat production under resting and active conditions for 35-year-old males who differ in body size**

| Anthropometric Dimension[a] | | | | Resting[b] | Active[c] |
|---|---|---|---|---|---|
| Ht | Wt | SA[3] | SA/Wt[4] | Kcal/day | Kcal/day |
| 150 | 50 | 1.44 | 2.88 | 1612 | 3498 |
| 150 | 60 | 1.55 | 2.58 | 1368 | 2968 |
| 160 | 60 | 1.62 | 2.70 | 1430 | 3103 |
| 160 | 70 | 1.73 | 2.47 | 1530 | 3320 |
| 170 | 72 | 1.84 | 2.56 | 1613 | 3500 |
| 170 | 82 | 1.93 | 2.35 | 1713 | 3717 |
| 180 | 72 | 1.89 | 2.63 | 1675 | 3634 |
| 180 | 92 | 2.11 | 2.29 | 1875 | 4069 |
| 190 | 70 | 2.00 | 2.86 | 2061 | 4472 |
| 190 | 90 | 2.20 | 2.44 | 2263 | 4911 |

[a] Ht = height in cm; Wt = weight in kg; SA = surface area in m derived from figure 2.4; SA/Wt = (SA/Wt × 100)

[b] Calculated with the equation of Mifflin et al., 1990. [Males, REE (calories/day = (10 × weight, kg) + (6.25 × height, cm) − (5 × age) + 5.].

[c] Equals 2.17 the resting energy expenditure.

bulb) subjects with high surface area:weight ratios produced less heat and lost more heat through radiation and convection and therefore had less heat storage than subjects with low surface area:weight ratios (fig. 3.2). On the other hand, as the ambient temperature increased to 35°, 45°, or 50°C, the subjects with a high surface area:weight ratio produced less heat, but because they gained more heat from the ambient temperature through radiation and convection, they had similar heat storage to the subjects with low surface area:weight ratios. Computer simulations indicate that a high surface area:weight ratio is advantageous when evaporative heat loss and radiation-convection heat loss are high (19). On the other hand, when radiation-convection heat losses are low, a high surface area:weight ratio is not advantageous because rather than increasing heat loss, it may increase heat load. Thus it would appear that the role of body size and shape is dependent on the ambient temperature to which the subjects are exposed.

### Fatness

The other component of Fourier's Law is that the rate of heat flow from the core to the exterior is greater as the shell thickness decreases. Various investigations have indicated that fat or obese subjects of all ages and both sexes are under greater heat strain while working under heat stress than their lean counterparts as shown by their higher rectal temperature, higher heart rate, and lower sweat rate (20, 21). These differences, in addition to being related to the insulative effects of subcutaneous fat, are probably related to the fact that overweight or obese subjects have a lower surface area:weight ratio than their lean counterparts. Therefore, the heavier subjects produce more heat and also have fewer avenues for heat loss by radiation and convection.

### Physical Fitness and Cardiovascular Function

Tolerance and acclimation to heat stress are affected by the individual's state of physical fitness, and, in general, the more fit one is, the faster acclimation takes place. It must also be noted that acclimation to heat stress develops when the subject exercises during heat exposure, but will not develop when the subject is exposed to heat stress without any exercise (22).

As measured by maximum oxygen intake (aerobic capacity), tolerance to work in heat stress is related to cardiovascular fitness. Experimental studies have shown that subjects with a high maximum oxygen intake have a greater tolerance to heat stress than those with low oxygen intake (23). In addition, tolerance to heat stress is affected by the capacity to expand the vascular volume. It has been found that heat-intolerant subjects, even when working

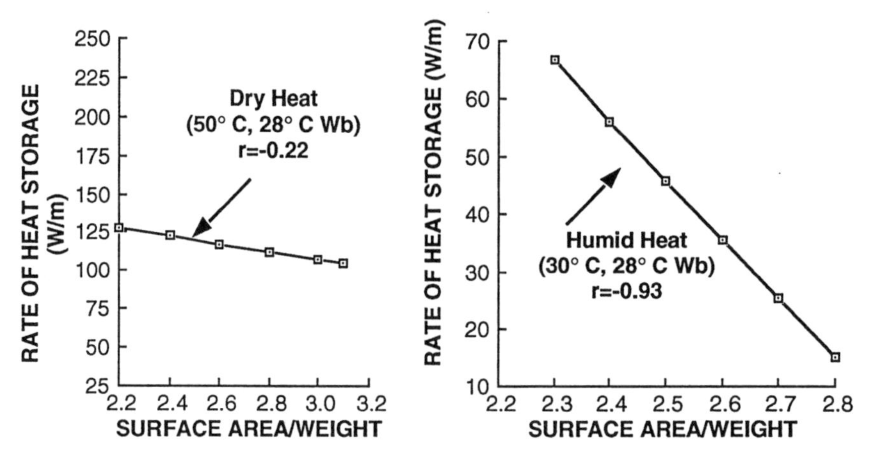


**Fig. 3.2** Relationship between surface area:weight ratio and rate of heat storage in twenty-five subjects performing moderate work at dry heat (50°C) and eight subjects performing mild work in humid heat (30°C). Subjects with high surface area:weight ratio in humid heat (30°C) have lower heat storage because of lower heat production and greater area for heat loss than subjects with low surface area:weight ratios. As temperature increases or in dry heat, the high surface area increases heat load and heat storage. (Based on data from E. Shvartz, E. Saavand, and D. Benor. 1973. J. Appl. Physiol. 34:799–803.)

at a lower absolute work load with a smaller water deficit, are not able to expand their vascular volume to the same degree as the tolerant subjects, probably because the absolute amount of protein in their extravascular space is markedly smaller than for the tolerant subjects (24).

## ACCLIMATIZATION TO HOT CLIMATES

### Nonnative and Native Populations

Since the 1950s several reports have appeared on indigenous adaptations in Mauritania (25), Nigeria (26), and Singapore (25). However, the only detailed studies have been those conducted among whites and Bantus prior to and after acclimation to work in heat stress (27). In addition, the thermoregulatory responses to work in heat stress of Bushmen from the Kalahari Desert (28), Arabs from the Sahara Desert (29), and aborigines from central Australia (30) have also been evaluated. Because these investigations were all conducted under the same thermal and work conditions, they provide an excellent opportunity for evaluating the quality and functional significance of population differences in adaptation to work in heat stress. The following discussion is based on the results of these studies.

**Test Procedures.** One of the difficulties in evaluating population differences in heat acclimatization is the diversity of techniques and reference populations. For this reason, researchers of the Human Science Laboratory of South Africa (27–31) have evaluated the thermoregulatory responses to work in heat stress of whites and Bantus before and after acclimation, and Bushmen, Arabs, and Australian aborigines all under the same standard conditions of temperature and workload. The experimental routine consisted of having the subjects step up and down on a bench with the height adjusted to give a work load of 216 kg-m/min (1560 foot-pounds/min) at a rate of 12 steps per minute. The external work was set to require an oxygen consumption of about 1 l/min. The experiment lasted for 4 hours and was conducted at 33.9°C (93°F) dry bulb and 32.2°C (90°F) wet bulb temperatures, with 24 m/min (80 feet/min) air movement.

### Bantus

**Environment.** The Bantus of South Africa, like those of central Africa, are both agriculturalists and pastoralists, living in a moderately hot environment. The mean daily maximum temperature for January (summer) ranges from 25°C (77°F) to 31.1°C (88°F). The relative humidity ranges from 44% to 68%; the wet bulb temperature ranges from 18.1°C (65°F) to 25.1°C (77°F). The effective general temperature, then, falls within the range of 22°C (71°F) to 25.1°C (79°F) (31).

**Acclimated vs. Nonacclimated.** Since the Bantus are probably more acclimatized to heat stress than whites for an appropriate comparison of the physiological characteristics, the physical conditioning of these groups needs to be standardized. With this purpose Wyndham and his colleagues (27) placed a sample of twenty male Bantus and twenty white male subjects in a program of acclimation to work in heat stress. The white subjects were second-year medical students. The Bantus were all recruits for work in the South African mines. The acclimation procedure included 12 days of work, 4 hours per day, at an air temperature of 36°C (96°F) dry bulb and 33.8°C (93°F) wet bulb, with 24 m/min air movement. The testing procedure was the same as indicated before.

**Anthropometric Characteristics.** As shown in table 3.2 the Bantus (nonacclimated and acclimated) were shorter, lighter in weight, and leaner than the whites. They also had a greater surface area per unit of body weight than whites.

**Thermoregulation.** All twenty nonacclimated Bantus completed the 4-hour experiment, but only ten of the twenty nonacclimated whites were able to do so.

The Bantus in both the nonacclimated and acclimated condition during most of the test sweated at a lower rate while maintaining a lower rectal temperature and a lower heart rate than the whites (fig. 3.3). Furthermore, as illustrated in figure 3.3 the acclimated Bantus, during the last 3 hours of the test, had a systematically lower rectal temperature and lower heart and sweat rates than the whites.

## Bantus: Intertribal Comparisons

An evaluation of a total of 120 nonacclimated male Bantu subjects belonging to six Bantu tribes was made when they arrived (within 1 week) in Johannesburg (31, 32). The Bantus were evaluated under conditions of heat stress following the testing procedures indicated above.

**Anthropometric Characteristics.** As shown in table 3.3 no significant intertribal differences in terms of anthropometric dimensions and work performance or thermoregulatory responses were revealed.

**TABLE 3.2. Comparison of the anthropometric dimensions and thermoregulatory characteristics of South African whites and Bantus before acclimation and after acclimation to heat stress**

| | Nonacclimated | | Acclimated | |
|---|---|---|---|---|
| | Bantus | Whites | Bantus | Whites |
| | $N=22$ | $N=20$ | $N=20$ | $N=10$ |
| Variables | Mean ± SD | Mean ± SD | Mean ± SD | Mean ± SD |
| Height (cm) | 165.9 ± 6.0 | 175.9 ± 7.7 | 166.8 ± 5.9 | 174.6 ± 6.5 |
| Weight (kg) | 59.1 ± 6.1 | 70.2 ± 7.0 | 60.5 ± 5.9 | 69.3 ± 7.0 |
| Surface Area ($m^2$) | 1.66 ± 0.11 | 1.86 ± 0.13 | 1.68 ± 0.10 | 1.84 ± 0.12 |
| SA/W ($cm^2$/kg) | 2.8 ± 0.4 | 2.7 ± 0.1 | 2.8 ± 0.1 | 2.7 ± 0.3 |
| Skinfolds (mm) | 5.5 ± 1.0 | 8.1 ± 3.0 | 6.0 ± 1.6 | 8.3 ± 2.6 |
| **Rectal temp. (°F)** | | | | |
| Rest | 98.7 ± 0.9 | 99.6 ± 0.6 | 98.5 ± 0.4 | 98.5 ± 0.3 |
| 1st hr | 101.4 ± 0.5 | 102.0 ± 0.6 | 100.3 ± 0.4 | 100.1 ± 0.3 |
| 2nd hr | 102.0 ± 0.7 | 102.9 ± 0.8 | 100.5 ± 0.5 | 100.7 ± 0.3 |
| 3rd hr | 102.4 ± 0.7 | 103.2 ± 0.6 | 100.5 ± 0.5 | 100.9 ± 0.4 |
| 4th hr | 103.2 ± 0.9 | 103.7 ± 1.0 | 100.7 ± 0.5 | 101.0 ± 0.4 |
| **Heart rate (b/m)** | | | | |
| Rest | 78.2 ± 9.6 | 91.0 ± 28.8 | 83.6 ± 14.0 | 65.0 ± 11.3 |
| 1st hr | 149.3 ± 11.0 | 152.4 ± 22.6 | 121.0 ± 11.5 | 115.0 ± 15.3 |
| 2nd hr | 150.0 ± 14.6 | 162.2 ± 17.5 | 118.7 ± 13.4 | 124.6 ± 13.8 |
| 3rd hr | 157.4 ± 15.4 | 168.0 ± 13.6 | 118.9 ± 11.1 | 126.0 ± 14.7 |
| 4th hr | 161.9 ± 14.6 | 173.6 ± 16.6 | 122.3 ± 13.9 | 131.4 ± 16.7 |
| **Sweat rate (ml/hr)** | | | | |
| 1st hr | 405.3 ± 160.6 | 592.7 ± 221.6 | 674.8 ± 164.6 | 634.0 ± 157.4 |
| 2nd hr | 438.5 ± 106.2 | 547.1 ± 195.6 | 937.6 ± 71.7 | 1101.6 ± 159.7 |
| 3rd hr | 361.4 ± 117.3 | 417.8 ± 188.5 | 720.2 ± 116.1 | 958.4 ± 227.3 |
| 4th hr | 289.0 ± 131.5 | 446.1 ± 201.4 | 508.4 ± 81.6 | 867.5 ± 231.8 |
| **Sweat Rate (ml/$m^2$/hr)** | 224 ± 79.2 | 276 ± 112.0 | 428 ± 68.7 | 478 ± 107.6 |

*Source:* Modified from C. H. Wyndham. 1966. In P. T. Baker and J. S. Weiner, eds., The biology of human adaptability. Oxford: Clarendon Press.

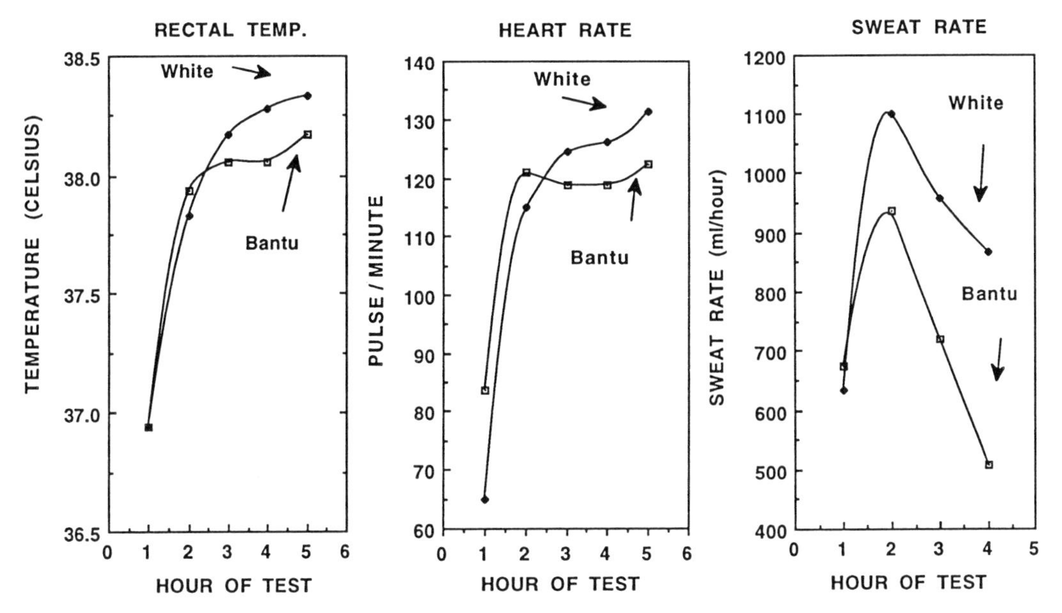


**Fig. 3.3.** Physiological responses to standardized heat stress test of acclimated Bantus and acclimated whites. The Bantus, as those in the nonacclimated state, are able to maintain thermal homeostasis with lower sweat rate and lower heart rate than the acclimated whites. (Based on data from C. H. Wyndham. 1966. Southern African ethnic adaptation to temperature and exercise. In P. T. Baker and J. S. Weiner, eds., The biology of human adaptability. Oxford: Clarendon Press.)

**Thermoregulation.** An outstanding observation of this study was that only three of the 120 Bantu subjects developed rectal temperatures above 40°C (104°F), and only one individual collapsed, with a rectal temperature of 38.9°C (102°F). This finding contrasts markedly with the fact that 50% of the 20 nonacclimated young whites in the aforementioned study failed to complete the 4-hour test, 5 with a rectal temperature of 40°C (104°F), and 5 either collapsed or were exhausted. These findings clearly document the fact that, when exposed to heat, the nonacclimated Bantus are less liable to heat collapse and have greater cardiovascular stability than do Caucasians. These Bantu subjects also had lower sweat rates than the nonacclimated whites.

Thus, it is evident that the Bantus, whether nonacclimated or acclimated, maintain thermal homeostasis with a generally lower sweat rate while working in heat stress than both the nonacclimated and acclimated whites. The fact that the rectal temperatures of the acclimated Bantus were not higher than those of the acclimated Caucasians indicates that the Bantus were losing heat through channels other than sweat evaporation alone.

### Bushmen

**Environment**. The Bushmen are divided into several tribal groups, some of which live along with the Bantus in pastoral tribes. Ethnographic sources

**TABLE 3.3. Comparison of the anthropometric dimensions and thermoregulatory characteristics of South African Bantu**

| | Barotsi | Mpedi | Nyasa | Shangaan | Tswana | Zulu | Basutu | Angola | Xhosa |
|---|---|---|---|---|---|---|---|---|---|
| | *N* = 22 | *N* = 17 | *N* = 20 | *N* = 20 | *N* = 20 | *N* = 20 | *N* = 20 | *N* = 20 | *N* = 21 |
| Variables | Mean ± SD | Mean ± SD | Mean ± SD | Mean ± SD | Mean ± SD | Mean ± SD | Mean ± SD | Mean ± SD | Mean ± SD |
| Height (cm) | 168.1 ± 8.4 | 166.4 ± 5.3 | 163.8 ± 6.2 | 166.4 ± 6.2 | 165.1 ± 5.8 | 168.3 ± 6.8 | 167.6 ± 5.3 | 167.6 ± 5.4 | 167.4 ± 4.4 |
| Weight (kg) | 56.8 ± 2.0 | 56.7 ± 5.8 | 56.9 ± 3.9 | 57.9 ± 6.1 | 53.8 ± 4.8 | 60.4 ± 4.9 | 56.8 ± 4.8 | 60.4 ± 5.8 | 59.0 ± 6.7 |
| Surface Area ($m^2$) | 1.63 ± 0.24 | 1.63 ± 0.27 | 1.62 ± 0.26 | 1.65 ± 0.11 | 1.58 ± 0.27 | 1.68 ± 0.25 | 1.64— | 1.69— | 1.68— |
| SA/W ($m^2$/kg) | 2.9 — | 2.9 — | 2.9 — | 2.9 — | 2.9 — | 2.8 — | 2.9 — | 2.8 — | 2.9 — |
| Skinfolds (mm) | — | — | — | — | — | — | — | — | — |
| **Rectal temp. (°F)** | | | | | | | | | |
| Rest | 98.9 ± 0.7 | 98.7 ± 0.7 | 99.0 ± 0.9 | 98.8 ± 0.7 | 98.9 ± 0.6 | 98.9 ± 0.5 | 98.8 ± 0.9 | 98.6 ± 1.0 | 98.8 ± 0.4 |
| 1st hr | 100.8 ± 0.6 | 100.9 ± 0.5 | 101.0 ± 0.6 | 101.0 ± 0.4 | 101.0 ± 0.5 | 100.7 ± 0.6 | 101.0 ± 0.5 | 100.8 ± 0.9 | 100.9 ± 0.6 |
| 2nd hr | 101.6 ± 0.8 | 101.4 ± 0.7 | 101.9 ± 0.5 | 101.7 ± 0.6 | 101.5 ± 0.7 | 101.4 ± 0.5 | 101.6 ± 4.8 | 101.6 ± 0.9 | 101.6 ± 0.6 |
| 3rd hr | 102.0 ± 1.0 | 102.0 ± 0.9 | 102.4 ± 0.8 | 102.1 ± 0.7 | 101.9 ± 0.8 | 101.9 ± 0.7 | 102.3 ± 0.8 | 102.1 ± 1.0 | 102.1 ± 0.7 |
| 4th hr | 102.4 ± 1.0 | 102.4 ± 1.0 | 102.8 ± 0.8 | 102.5 ± 0.8 | 102.2 ± 0.8 | 102.5 ± 0.8 | 102.8 ± 0.8 | 102.5 ± 1.0 | 102.6 ± 1.0 |
| **Heart rate (b/m)** | | | | | | | | | |
| Rest | 83.1 ± 15.5 | 75.9 ± 14.7 | 85.1 ± 21.2 | 73.4 ± 9.9 | 81.6 ± 18.0 | 82.3 ± 16.4 | 79.6 ± 17.6 | 76.8 ± 15.2 | 82.2 ± 17.2 |
| 1st hr | 137.0 ± 15.5 | 136.1 ± 10.0 | 137.9 ± 18.3 | 141.5 ± 15.1 | 140.3 ± 17.7 | 130.0 ± 17.9 | 134.2 ± 18.1 | 136.7 ± 20.2 | 135.4 ± 13.6 |
| 2nd hr | 143.1 ± 16.6 | 146.1 ± 14.2 | 148.8 ± 18.3 | 152.9 ± 18.1 | 149.7 ± 17.6 | 138.7 ± 18.5 | 141.4 ± 20.3 | 147.5 ± 19.2 | 145.3 ± 15.3 |
| 3rd hr | 149.1 ± 15.9 | 154.2 ± 15.9 | 157.2 ± 18.3 | 158.0 ± 16.5 | 155.7 ± 18.3 | 145.8 ± 18.2 | 149.8 ± 17.6 | 153.6 ± 19.9 | 154.0 ± 17.3 |
| 4th hr | 154.5 ± 15.5 | 161.9 ± 11.0 | 162.6 ± 15.8 | 165.1 ± 16.7 | 166.0 ± 13.0 | 154.9 ± 16.4 | 155.7 ± 16.1 | 154.6 ± 17.8 | 162.0 ± 20.3 |
| **Sweat rate (ml/hr)** | | | | | | | | | |
| 1st hr | 319.0 ± 94.7 | 405.5 ± 104.0 | 272.2 ± 109.8 | 401.2 ± 130.3 | 407.5 ± 159.1 | 361.4 ± 113.0 | 361.4 ± 148.1 | 348.4 ± 159.2 | 303.9 ± 120.2 |
| 2nd hr | 445.1 ± 131.0 | 578.6 ± 177.1 | 375.0 ± 139.6 | 612.3 ± 205.5 | 583.9 ± 179.8 | 512.5 ± 152.0 | 495.9 ± 192.4 | 530.9 ± 132.2 | 428.6 ± 125.5 |
| 3rd hr | 353.2 ± 93.9 | 427.9 ± 153.2 | 369.0 ± 104.1 | 462.4 ± 130.5 | 445.4 ± 104.2 | 411.2 ± 108.3 | 410.2 ± 136.9 | 412.4 ± 126.0 | 383.5 ± 99.5 |
| 4th hr | 307.0 ± 60.0 | 367.6 ± 120.3 | 289.0 ± 91.4 | 389.6 ± 82.2 | 379.9 ± 54.9 | 360.0 ± 80.0 | 347.5 ± 84.3 | 361.8 ± 79.8 | 306.6 ± 71.2 |
| **Sweat Rate (ml/m²/hr)** | | | | | | | | | |
| | 232 ± 61.0 | 273 ± 82.2 | 202 ± 69.5 | 294 ± 84.7 | 297 ± 79.9 | 246 ± 67.4 | 246 ± 82.4 | 245 ± 78.2 | 211 ± 62.4 |

*Source:* Modified from C. H. Wyndham. 1966. In P. T. Baker and J. S. Weiner, eds., The biology of human adaptability. Oxford: Clarendon Press.

indicate that Bushmen have lived in the Kalahari Desert for at least a hundred years (33, 34). The Kalahari Desert is characterized by three seasons: a hot summer with a 5-month rainy season from November to March; a cool, dry winter from April to August; and a hot, dry spring in September and October. The daily mean maximum temperature for the midwinter month of July ranges from 20°C (68°F) to 22.5°C (72.5°F). The minimum daily temperature is about 3.2°C (37.8°F). Dry bulb temperatures, for the summer only, rise above 32°C (89.6°F), and black globe thermometer readings in direct sunlight sometimes rise to 60°C (140°F) (28, 31-34). In the summer the mean maximum and minimum air temperatures for January equaled 30°C (86°F) to 32.5°C (90.5°F) and 18°C (64.4°F), respectively. The annual rainfall varies from 400 to 600 mm.

Despite the high temperatures the loose, sandy soil supports abundant vegetation, and underlying limestone strata allow for the formation of water holes. Because of these geological factors the distribution of water sources is by far the most important ecological determinant of Bushman subsistence. Their camps are anchored to water sources, and they exploit only vegetable foods and game within a reasonable walking distance from water; yet their food resources are varied and abundant. Lee (33) indicates that the Bushmen know over 200 plant and 220 animal species; of these, 85% of the plants and 54% of the animals are edible. The predominant staple is the mongongo (mangetti) nut, which is found in abundant quantities and gives an energy yield of 600 cal/100 g. The major constituents of the diet by weight are 33% mongongo nuts, 37% meat, and 30% vegetable matter (33).

In addition to desert environments, Bushmen also inhabit river areas; these are the so-called river Bushmen, whose economy is a mixture of horticulture, hunting, and gathering. They have a more settled existence than the desert Bushmen, living in fixed villages, sowing crops, and raising domestic animals. These Bushmen are said to be somewhat hybridized with the neighboring Bantu tribes, which would then be reflected in their genetic constitution. During the summer of 1962 the anthropometric dimensions of a sample of eight and fifteen desert Bushmen of the Maxqonbi tribe and ten river Bushmen of the Makanchwe tribe were obtained (28, 31). The physiological measurements of work under conditions of heat stress included evaluations of eleven desert Bushmen and ten river Bushmen (28, 31). The thermoregulatory responses of these Bushmen were compared to those of nonacclimated and acclimated whites studied using the same procedures.

**Anthropometric Characteristics.** As indicated in table 3.4 the desert Bushmen were shorter and lighter in weight, but they had a greater surface area:weight ratio than the river Bushmen. The mean surface area:weight ratio for the desert Bushmen equaled 3.11 and 3.05 $cm^2/kg$ and 2.85 for the river Bushmen.

**Thermoregulation.** These data also show that the rectal temperatures and sweat rates of both Bushmen samples were similar, but the heart rates of the desert Bushmen were lower than those of the river Bushmen. The rectal temperatures and heart rates of the two Bushmen groups were intermediate between the nonacclimated and acclimated whites, but their sweat rates were lower than the acclimated whites. From these data the investigators concluded that the Bushmen are only partially acclimatized to work in heat stress (28, 31). This conclusion is surprising in view of the fact that in their everyday activities the Bushmen are continuously exposed to environmental heat stress and strenuous physical work. Therefore, one would expect a full adaptation.

Furthermore, it should be noted that although the Bushmen had higher

**TABLE 3.4. Comparison of the anthropometric dimensions and thermoregulatory characteristics of river and desert Bushmen from Kalahari Desert of South Africa**

| | Desert—Maxqonbi | | River—Makanchwe |
|---|---|---|---|
| | Takaschane | Motokwe | Okavango |
| | $N=8$ | $N=15$ | $N=10$ |
| Variables | Mean ± SD | Mean ± SD | Mean ± SD |
| Height (cm) | 160.5 ± 6.1 | 157.8 ± 6.4 | 171.3 ± 6.7 |
| Weight (kg) | 47.6 ± 5.2 | 47.7 ± 6.1 | 61.2 ± 4.9 |
| Surface Area (m²) | 1.47 ± 0.11 | 1.48 ± 0.36 | 1.74 ± 0.09 |
| SA/W (cm²/kg) | 3.1 ± 0.0 | 3.1 ± 0.1 | 2.9 ± 0.1 |
| Skinfolds (mm) | 4.8 ± 2.0 | 4.5 ± 0.7 | 4.7 ± 0.0 |
| **Rectal temp. (°F)** | | | |
| Rest | 98.7 ± 0.7 | — | 98.8 ± 0.6 |
| 1st hr | 101.6 ± 0.6 | — | 101.5 ± 0.7 |
| 2nd hr | 102.1 ± 0.8 | — | 101.9 ± 0.6 |
| 3rd hr | 102.1 ± 0.6 | — | 102.2 ± 0.7 |
| 4th hr | 102.2 ± 0.6 | — | 102.4 ± 0.8 |
| **Heart rate (b/m)** | | | |
| Rest | 72.6 ± 8.8 | — | 82.5 ± 12.5 |
| 1st hr | 125.8 ± 12.7 | — | 147.8 ± 18.6 |
| 2nd hr | 132.4 ± 16.4 | — | 150.3 ± 19.2 |
| 3rd hr | 134.0 ± 15.1 | — | 154.5 ± 18.3 |
| 4th hr | 142.9 ± 14.2 | — | 159.5 ± 14.9 |
| **Sweat rate (ml/hr)** | | | |
| 1st hr | 592.7 ± 139.7 | — | 484.9 ± 137.9 |
| 2nd hr | 643.4 ± 181.8 | — | 772.6 ± 258.3 |
| 3rd hr | 591.9 ± 128.8 | — | 650.0 ± 136.2 |
| 4th hr | 481.0 ± 90.5 | — | 505.5 ± 143.2 |
| **Sweat Rate (ml/m²/hr)** | 392 ± 90.8 | — | 347 ± 96.5 |

*Source:* Modified from C. H. Wyndham. 1966. In P. T. Baker and J. S. Weiner, eds., The biology of human adaptability. Oxford: Clarendon Press.

rectal temperatures and heart rates than the acclimated whites, their performance (all completed the 4-hour experiment, and none had rectal temperatures above 40°C) was equal to that of the acclimated whites. This exceptional performance was attained despite their lower sweat rates, suggesting that the Bushmen, like the Bantus, are less susceptible to heat collapse and have greater cardiovascular stability than the whites when exposed to work in heat stress (28, 31).

## Australian Aborigines

**Environment**. Central Australia is also characterized by a hot-dry climate. The mean maximum daily temperature for the summer months of January

**TABLE 3.5. Comparison of the anthropometric dimensions and thermoregulatory characteristics of Australian aborigines from the Weipa and Aurukun Mission Station of central Australia**

| | *N* = 31 |
|---|---|
| Variables | Mean ± SD |
| Height (cm) | 172.0 ± 5.3 |
| Weight (kg) | 56.5 ± 6.5 |
| Surface area ($m^2$) | 1.67 ± 0.10 |
| SA/W ($cm^2$/kg) | 3.0 ± 0.2 |
| Skinfolds (mm) | 6.6 ± 2.1 |
| **Rectal temp. (°F)** | |
| Rest | 99.6 ± 0.4 |
| 1st hr | 101.4 ± 0.5 |
| 2nd hr | 101.7 ± 0.6 |
| 3rd hr | 102.0 ± 0.7 |
| 4th hr | 102.5 ± 0.8 |
| **Heart rate (b/m)** | |
| Rest | 92.6 ± 13.0 |
| 1st hr | 147.4 ± 16.5 |
| 2nd hr | 153.6 ± 12.2 |
| 3rd hr | 160.0 ± 11.9 |
| 4th hr | 167.4 ± 13.5 |
| **Sweat rate (ml/hr)** | |
| 1st hr | 444.3 ± 168.8 |
| 2nd hr | 468.0 ± 139.7 |
| 3rd hr | 390.8 ± 100.6 |
| 4th hr | 322.6 ± 74.8 |
| **Sweat rate (ml/m²/hr)** | 237 ± 69.2 |

*Source:* Modified from C. H. Wyndham. 1966. In P. T. Baker and J. S. Weiner, eds., The biology of human adaptability. Oxford: Clarendon Press.

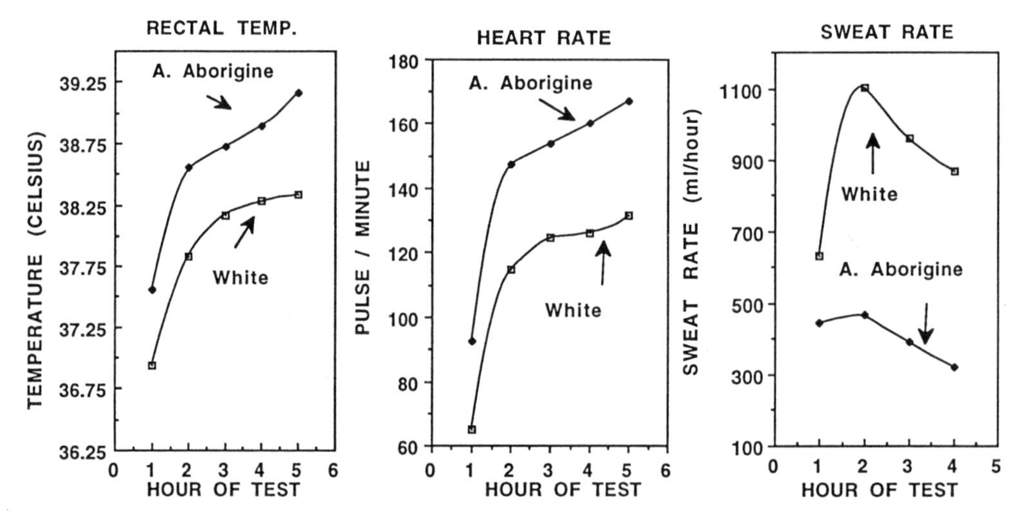


**Fig. 3.4.** Physiological responses to standardized heat stress test of acclimated whites and Australian aborigines. The Australian aborigines show higher increase in rectal temperatures and heart rates than acclimated whites, but have lower sweat rates. (Based on data from C. H. Wyndham. 1966. Southern African ethnic adaptation to temperature and exercise. In P. T. Baker and J. S. Weiner, eds., The biology of human adaptability. Oxford: Clarendon Press.)

through March is 37°C (98.6°F), and the daily minimum temperature is 22°C (71.6°F). In the winter month of July, the mean maximum daily temperature is 21°C (69.8°F), and the minimum is 4°C (39.2°F), with an average annual rainfall of 260 mm. The native flora and fauna of central Australia are quite limited. The subsistence economy of the aborigines is therefore based on the capture of small, slow game and the gathering of wild roots and fruit plants. Compared to the Bushmen, the Australian aborigines live in a more stressful environment in terms of natural nutritional resources, availability of water holes, and presence of high temperatures (31). Two groups of aborigines were studied, fourteen men from Weipa Mission Station and another seventeen men from Aurukun Mission Station (30, 31).

**Anthropometric Characteristics.** Table 3.5 gives the anthropometric dimensions and thermoregulatory characteristics for 31 Australian aborigines. From these data it is evident that the Australian aborigines are as tall as the whites and are taller than the Bantus, and Bushmen. On the other hand, they are lighter in weight than the whites, Bantus, and river Bushmen. As a result of these differences, except for the desert Bushmen, the Australian aborigines have the highest ratio of surface area to weight (2.97 vs. 2.85 for the Bantus and 2.80 to 3.10 for the Bushmen).

**Thermoregulation.** As illustrated in figure 3.4 throughout the heat stress test the rectal temperatures and heart rates of the Australian aborigines were higher than those of the acclimated whites, but their sweat rates were significantly lower than the acclimated whites. In other words, the thermoregulatory responses to heat stress of the aborigines are similar to the Bantus and Bushmen.

### Cultural Adaptation by the Chaamba Arabs

The climate of the Sahara Desert is characterized by higher temperatures and lower humidity than that of the Kalahari. The mean maximum temperature for the midsummer month of July is between 35° and 37.5°C (95° and 99.5°F), and the mean minimum temperature is about 26°C (78.8°F). The annual rainfall is less than 100 mm. Because of the higher temperatures and lower moisture of the Sahara, the direct and reflected solar radiation from bare sand is greater than in the Kalahari.

Following the same procedures as described in the previous studies the thermoregulatory reactions to work in heat stress of fifteen Arabs from the Chaamba tribe, as well as fifteen whites were evaluated (29, 31). The Chaamba tribesmen are of Arab origin; in the past they were pastoral nomads, but at the time of the study they were working as laborers in Hassi-Messaoud. The white subjects were French soldiers who had come from metropolitan France to serve at the garrison in Hassi-Messaoud.

As shown in table 3.6 even though the Chaamba Arabs have a high ratio of surface area to weight (2.91 $cm^2/kg$ vs. 2.60 $cm^2/kg$ for the white French soldiers) their performances in heat were comparable to the white French soldiers. Compared to the South African acclimated whites (given in table 3.2), the Arabs and Frenchmen had higher rectal temperatures and higher heart rates. Although the Arabs had, during the last 2 hours of the experiment, lower sweat rates than the French soldiers, the fact that their rectal temperatures and heart rates were higher than those of the South African whites indicates that the thermoregulatory responses to heat stress of the Arabs were inferior to those of the whites.

This paradoxical finding may be caused by several factors. First, the increased heart rates of the Arabs are indicative of their poor physical conditioning. Second, without exception the Arabs wear clothing that in a hot-dry environment minimizes conductive and radiant energy gains from the environment. Arab clothing reduces heat stress by trapping a boundary of stale air between the skin and the cloth, which serves to insulate the skin from the ambient temperature and solar radiation (35). Thus it is likely that the Arabs have succeeded in protecting themselves so effectively that in their normal state they are seldom exposed to temperatures conducive to heat stress or prolonged dehydration. Therefore, their adaptation to heat stress is probably more cultural than physiological.

## OVERVIEW

Acclimation to heat results in an enhanced tolerance to work under heat stress. This is attained through increased peripheral heat conductance in which the rate of heat transfer from the internal body core to the periphery is increased. Along with increased heat conductance, the capacity to sweat increases to as much as four times the preacclimation capacity, and this increase occurs at a lower skin temperature. Although the sweat-sodium concentration decreases with acclimation because of the increase in total sweat output, the sodium loss exceeds preacclimation levels. Concomitant with increased heat conductance and as a result of the improved cooling efficiency of sweat, the rectal temperature and heart rate return to preacclimation levels, and the disagreeable sensations associated with heat exposure are progressively reduced. Although the limits of endurance under extreme heat stress for women are less than those for men, women with acclimation to heat stress show the same thermoregulation responses as men.

Tolerance to heat stress is influenced by age; very young and very old subjects are less tolerant than adolescents and adults. Body size, shape, and subcutaneous fat influence the rate of heat production and heat loss, and their

**TABLE 3.6. Comparison of the anthropometric dimensions and thermoregulatory characteristics of Chaamba tribesmen of Arab origin and white French soldiers living at Hassi-Messaoud Garrison of Sahara**

| | Chaamba Arabs | White French |
|---|---|---|
| | *N* = 15 | *N* = 15 |
| Variables | Mean ± SD | Mean ± SD |
| Height (cm) | 165.4 ± 5.5 | 170.1 ± 5.0 |
| Weight (kg) | 55.2 ± 5.5 | 70.0 ± 9.3 |
| Surface Area (m$^2$) | 1.60 ± 0.09 | 1.81 ± 0.13 |
| SA/W (cm$^2$/kg) | 2.9 ± 0.2 | 2.6 ± 0.1 |
| Skinfolds (mm) | 6.2 ± 3.4 | 9.8 ± 2.9 |
| **Rectal temp. (°F)** | | |
| Rest | 98.4 ± 0.3 | 99.4 ± 0.3 |
| 1st hr | 101.3 ± 0.5 | 101.0 ± 0.5 |
| 2nd hr | 102.0 ± 0.6 | 101.5 ± 0.4 |
| 3rd hr | 102.4 ± 0.7 | 102.0 ± 0.6 |
| 4th hr | 102.5 ± 0.6 | 102.4 ± 0.6 |
| **Heart rate (b/m)** | | |
| Rest | 80.5 ± 14.1 | 80.7 ± 12.4 |
| 1st hr | 143.9 ± 17.6 | 128.1 ± 11.3 |
| 2nd hr | 156.3 ± 20.3 | 140.5 ± 17.7 |
| 3rd hr | 162.6 ± 18.6 | 153.7 ± 18.3 |
| 4th hr | 167.4 ± 18.2 | 164.6 ± 14.1 |
| **Sweat rate (ml/hr)** | | |
| 1st hr | 576.8 ± 185.2 | 653.9 ± 144.4 |
| 2nd hr | 740.7 ± 235.9 | 826.8 ± 230.8 |
| 3rd hr | 571.2 ± 212.3 | 704.6 ± 246.9 |
| 4th hr | 486.7 ± 147.7 | 594.2 ± 247.1 |
| **Sweat Rate (ml/m$^2$/hr)** | 371 ± 122 | 384 ± 121.7 |

*Source:* Modified from C. H. Wyndham. 1966. In P. T. Baker and J. S. Weiner, eds., The biology of human adaptability. Oxford: Clarendon Press.

significance depends on the degree of heat intensity and work. In the same manner, cardiovascular function influences both tolerance and acclimation to heat stress; subjects in good health and physical condition are able to tolerate heat stress better than subjects who are not as healthy or physically fit.

In general the Bantus, Bushmen, and Australian aborigines are by nature of their place of residence acclimatized to heat stress, and when compared to both acclimated and nonacclimated whites, experience lower sweat rates when working in heat stress. All three groups sweat at a lower rate, yet they maintain a similar core temperature to that of the nonacclimated whites and a slightly higher core temperature than the acclimated whites. Furthermore, their individual performances in heat were better than the nonacclimated whites, since nearly all these men completed the 4-hour experiment, and less than 3% had a rectal temperature above 40°C (104°F). This finding means that these populations have been able to maintain thermal homeostasis in heat stress with a lower body fluid loss. The implication is that in these populations heat dissipation is not simply a function of sweat evaporation, but probably of other physiologically less expensive avenues of heat loss as well, such as radiation and convection. These populations, when compared to the whites, are leaner and have a higher ratio of surface area to weight, thus having the appropriate morphology to facilitate heat loss through radiation and convection.

As shown by experimental studies and computer simulations (chap. 2), heat loss through radiation and convection in subjects with high surface area:weight ratio is maximized in moderate heat stress. The temperature in which the Bantus, Bushmen, and Australian aborigines were tested (32.9°C dry bulb and 32.2°C wet bulb) is appropriate for a maximization of heat loss through radiation, convection, and evaporation. In these circumstances these populations would not need to produce as much sweat as the whites. As indicated by recent investigations, maintenance of high sweat rate during acclimation to heat stress is not an efficient mechanism for maintaining thermal homeostasis. Indeed, with continuous exposure to heat stress the sweat rate does decline, and the skin temperature is maintained at the level necessary to evaporate just the amount of sweat sufficient to maintain thermal balance. For this reason, and given the fact that these populations have been exposed continuously to heat stress during daily activities, they have probably developed a much more sensitively adjusted thermoregulatory control channel between rectal temperature and sweat rate than the whites. Thus, with an increase in core temperature, the Bantus, Bushmen, and aborigines sweat just enough to maintain thermal equilibrium, whereas the white person sweats excessively. This lower sweat rate could then be regarded as advantageous for surviving longer in hot environments where no water is available, since the rate of dehydration would be slower. In view of the limited natural nutritional resources that these populations have, a lower sweat rate would also have an additional adaptive significance in that they would lose less salt, iron, and other minerals than the Caucasians. The sweat loss of Indians from India, Nigerians, and Papago Indians has also been found to be less than that of either residentially or artificially acclimated Europeans exposed to the same standardized conditions (36–38).

Presently, however, we do not know how or by what mechanisms populations that live in tropical, stressful climates acquire the capacity to maintain thermal homeostasis with a lower sweat rate. There are, however, several possible explanations.

First, this capacity may be acquired through long-term residence in tropical climates. It has been shown that the water intake of persons acclimatized to heat is lower than that of nonacclimatized persons (39). Furthermore, this investigation indicates that restricting the water intake of nonacclimatized persons hastens their acclimation process and that long-term residence and work in a hot climate is not necessarily accompanied by a dramatic increase in sweat loss so commonly observed when nonacclimatized persons are first exposed to heat stress. However, persons acclimatized by living and working in a hot climate for as long as 9 months do not have lower water intakes or sweat rates than similar, nonacclimatized persons (40). Therefore, it is doubtful that a low sweat rate can be acquired simply by long-term residence in a hot climate.

Second, this capacity may be acquired through developmental morphological changes oriented toward the maximization of sweat evaporation. In general, the extremities, because of enhanced potential for convective heat loss from their relatively larger surface areas, are the most effective areas for sweat evaporation from the body. As revealed by a growth study of American white children of good nutritional status, acclimatization to the tropical climate of Rio de Janeiro, Brazil, results in development of smaller calf circumference as compared to non-heat-stressed controls (41). Similarly, as illustrated in figure 3.5, Quechua children acclimatized to the tropical climate of the Peruvian lowlands have a proportionally smaller trunk (low sitting height:stature ratio) and greater arm length (high arm length:sitting height ratio) than their Quechua counterparts of the same genetic composition living in the cold climate of the Peruvian highlands (42). Since the extremities have a greater density of functional sweat glands than the trunk and the distal parts of the limbs have a greater density of functional sweat glands than the proximal parts (43–45), the development of leaner extremities would contribute to better sweat evaporation. Furthermore, the reduction in trunk size and increase in relative arm length would result in an increase in the surface area:weight ratio, which would facilitate radiative and convective heat loss. It is likely, then, that populations who have been raised in tropical climates

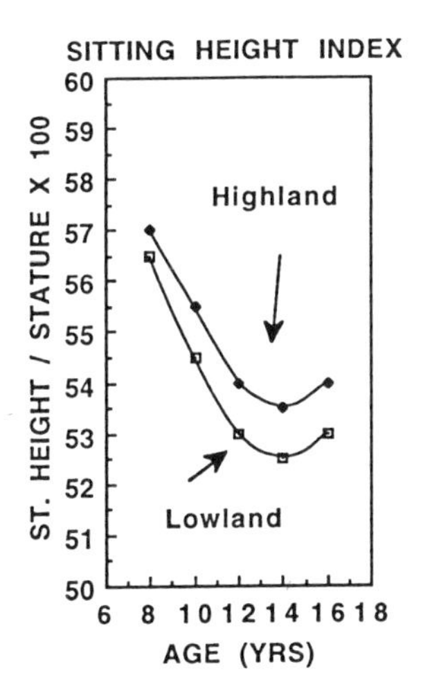


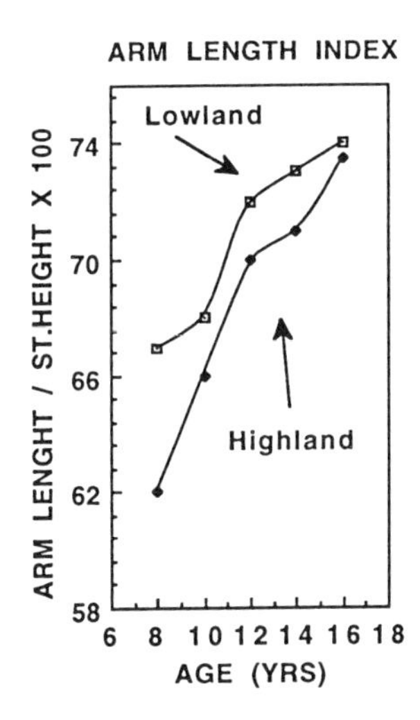


**Fig. 3.5.** Comparison of sitting height index (sitting height/stature × 100) and arm length index (arm length/stature × 100) of Peruvian highland and lowland tropical Quechua male samples. (Modified from S. Stinson and A. R. Frisancho. 1978. Hum. Biol. 50:57–68.)

do have a more efficient sweat evaporation and, therefore, do not have to sweat as excessively as those non-heat-acclimatized populations. Support for this hypothesis is found in studies of sweat gland concentration. Kuno (46) demonstrated that Japanese born and raised in the tropics had a significantly greater number of active sweat glands than their counterparts who migrated to the tropics after childhood. These findings suggest that the extent to which the sweat glands become active or dormant depends on the developmental period of exposure to heat stress.

Finally, the lower sweat rate may be a developmental response to heat stress mediated by nutritional factors. Almost all minerals and vitamins are excreted in the sweat, the principal components being iron, calcium, zinc, and especially sodium chloride. It has been demonstrated that the concentration of sodium chloride in sweat decreases with continuous exposure and acclimatization to heat stress. However, it has also been shown that even in acclimatized subjects there is a positive relationship between sweat output and total sweat-sodium loss (47). This means that when the sweat losses are high, total sodium losses will also tend to be high. In addition, when the sweat rate and the state of acclimatization are held constant, sweat-salt loss is mediated by dietary salt intake. For example, when men worked in a room at 38°C (100.4°F) and 80% relative humidity, the sweat loss in 162 minutes was 25 g, which amounts to approximately 15% of the total estimated body salt (47). Thus, if a high sweat rate is maintained, the salt output must be replaced through the diet if an adequate electrolyte balance is to continue. In view of the limited availability of salt to most tropical, indigenous populations (48), it is likely that the low sweat rate is a developmental response that prevents excessive loss of body fluids and salt.

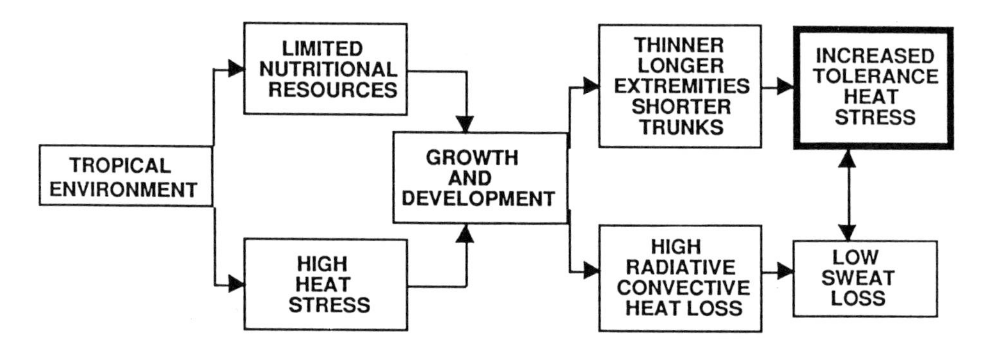


**Fig. 3.6.** Schematization of interaction of tropical environmental stresses influencing adaptation to heat stress. Joint influences of limited nutritional stress and heat stress growth and development in tropical environment result in development of adaptive morphology and increase in heat tolerance capacity.

In summary, major population differences in adaptation to heat stress are reflected in high tolerance to heat stress and low sweat loss of indigenous populations. At the present stage of knowledge, these differences can be explained as a result of interaction of limited nutritional resources and high heat stress associated with a tropical environment operating during the period of growth and development (fig. 3.6). Thus, populations inhabiting tropical climates during growth develop proportionally thinner, longer extremities and shorter trunks than those raised in temperate climates. In turn, this development affects the rate of radiative and convective heat loss. As a result, tropical native populations demonstrate an enhanced tolerance to heat stress and a low sweat rate.

Similarly, viewed in this context, the relationship between body weight and mean annual temperature observed among world populations (49, 50) would reflect a developmental response to the joint effects of heat and nutritional stress associated with tropical climates. Whether the same conclusion is applicable to the observed relationship between head form and climate remains to be determined (51).

## References

1. Mitchell, D., L. C. Senay, C. H. Wyndham, A. J. Van Rensburg, G. G. Rogers, and N. B. Strydom. 1976. Acclimatization in a hot, humid environment: energy exchange, body temperature, and sweating. J. Appl. Physiol. 40:768–78.
2. Senay, L. C., D. Mitchell, and C. H. Wyndham. 1976. Acclimatization in a hot, humid environment: body fluid adjustments. J. Appl. Physiol. 40:786–96.
3. Folk, G. E., Jr. 1974. Textbook of environmental physiology. Philadelphia, PA: Lea and Febiger.
4. Gisolfi, C., and S. Robinson. 1969. Relations between physical training, acclimatization and heat tolerance. J. Appl. Physiol. 26:530–34.
5. Colin, J., and Y. Houdas. 1965. Initiation of sweating in man after abrupt rise in environmental temperature. J. Appl. Physiol. 20:984–90.

6. Houdas, Y., J. Colin, J. Timbal, C. Bontelier, and J. D. Guien. 1972. Skin temperatures in warm environments and the control of sweat evaporation. J. Appl. Physiol. 33:99–104.
7. Wyndham, C. H., G. G. Rogers, L. C. Senay, and D. Mitchell. 1976. Acclimatization in a hot, humid environment: cardiovascular adjustments. J. Appl. Physiol. 40:779–85.
8. Hertig, B. A., and F. Sargent, II. 1963. Acclimatization of women during work in hot environments. Fed. Proc. 22:810–13.
9. Wyndham, C. H., J. F. Morrison, and C. G. Williams. 1965. Heat reactions of male and female Caucasians. J. Appl. Physiol. 20:357–64.
10. Morimoto, T., Z. Slabochova, R. K. Naman, and R. Sargent, II. 1967. Sex differences in physiological reactions to thermal stress. J. Appl. Physiol. 22:526–32.
11. Kamon, E., and B. Avellini. 1976. Physiologic limits to work in the heat and evaporative coefficient for women. J. Appl. Physiol. 41:71–76.
12. Dill, D. B., M. K. Yousef, and J. D. Nelson. 1973. Responses of men and women to two-hour walks in desert heat. J. Appl. Physiol. 35:231–35.
13. Hellon, R. F., and A. R. Lind. 1958. The influence of age on peripheral vasodilation in a hot environment. J. Physiol. (Lond.) 141:262–72.
14. Lind, A. R., P. W. Humphreys, K. J. Collins, K. Foster, and K. F. Sweetland. 1970. Influence of age and daily duration of exposure on responses of men to work in the heat. J. Appl. Physiol. 28:50–56.
15. Wagner, J. A., S. Robinson, S. P. Tzankoff, and R. P. Marino. 1972. Heat tolerance and acclimatization to work in the heat in relation to age. J. Appl. Physiol. 33:616–22.
16. Lofstedt, B. 1966. Human heat tolerance. Lund, Sweden: Dept. of Hygiene, University of Lund.
17. Austin, D. M., and M. W. Lansing. 1986. Body size and heat tolerance: a computer simulation. Hum. Biol. 58:153–69.
18. Shvartz, E., E. Saavand, and D. Benor. 1973. Physique and heat tolerance in hot-dry and hot-humid environments. J. Appl. Physiol. 34:799–803.
19. Austin, D. M. 1977. Body size and heat tolerance: a computer simulation approach. Am. J. Phys. Anthropol. (Abstract). 47:116.
20. Bar-or, O., H. M. Lundegren, and E. R. Buskirk. 1969. Heat tolerance of exercising obese and lean women. J. Appl. Physiol. 26:403–9.
21. Haymes, E. H., R. J. McCormick, and E. R. Buskirk. 1975. Heat tolerance of exercising lean and obese boys. J. Appl. Physiol. 39:457–61.
22. Gisolfi, C. 1973. Work-heat tolerance derived from interval training. J. Appl. Physiol. 35:349–54.
23. Wyndham, C. H., N. B. Strydom, A. J. Van Rensburg, A. J. S. Benade, and A. J. Heyns. 1970. Relation between $VO_2$ max and body temperature in hot humid air conditions. J. Appl. Physiol. 29:45–50.
24. Senay, L. C., Jr. 1975. Plasma volumes and constituents of heat-exposed men before and after acclimatization. J. Appl. Physiol. 38:570–75.
25. Pales, L. 1950. Les sels alimentaires: sels minéraux. O.F., Dakar: Gouvern. Gen. de L'A.
26. Ladell, W. S. S. 1957. Disorders due to the heat. Trans. R. Soc. Trop. Med. Hyg. 51:189–207.
27. Wyndham, C. H., J. F. Morrison, C. G. Williams, G. A. G. Bredell, M. J. E. Von Raliden, L. D. Holdsworth, C. H. Van Graan, A. J. Van Rensburg, and A. Munro.

1964. Heat reactions of caucasians and Bantu in South Africa. J. Appl. Physiol. 19:598–606.
28. Wyndham, C. H., N. B. Strydom, J. S. Ward, J. F. Morrison, C. G. Williams, G. A. G. Bredell, M. J. E. Von Raliden, L. D. Holdsworth, C. H. Van Graan, A. J. Van Rensburg, and A. Munro. 1964. Physiological reactions to heat of bushmen and of unacclimatized and acclimatized Bantu. J. Appl. Physiol. 19:885–88.
29. Wyndham, C. H., B. Metz, and A. Munro. 1964. Reactions to heat of Arabs and caucasians. J. Appl. Physiol. 19:1951–54.
30. Wyndham, C. H., R. K. McPherson, and A. Munro. 1964. Reactions to heat of aborigines and caucasians. J. Appl. Physiol. 19:1055–58.
31. Wyndham, C. H. 1966. Southern African ethnic adaptation to temperature and exercise. In P. T. Baker and J. S. Weiner, eds., The biology of human adaptability. Oxford: Clarendon Press.
32. Wyndham, C. H., N. B. Strydom, C. G. Williams, J. F. Morrison, G. A. G. Bredell, J. Peter, C. H. Van Graan, L. D. Holdsworth, A. J. Van Rensburg, and A. Munro. 1964. Heat reactions of some Bantu tribesmen in southern Africa. J. Appl. Physiol. 19:881–84.
33. Lee, R. B. 1969. !Kung bushman subsistence: an input-output analysis. In A. P. Vayda, ed., Environment and cultural behavior: ecological studies in cultural anthropology. New York, NY: Natural History Press.
34. Truswell, A. S., and J. D. L. Hansen. 1976. Medical research among the Kung. In R. B. Lee and I. DeVore, eds., Kalahari hunters-gatherers. Cambridge, MA: Harvard University Press.
35. Briggs, L. C. 1975. Environment and human adaptation in the Sahara. In A. Damon, ed., Physiological anthropology. Oxford: Oxford University Press.
36. Edholm, O. G. 1966. Acclimatization to heat in a group of Indian subjects. In M. S. Malhatra, ed., Human adaptability to environments and physical fitness. Madras, India: Madras Defense Institute of Physiology and Allied Sciences.
37. Ladell, W. S. S. 1955. Physiological observations on men working in supposedly limiting environments in a West African gold-mine. Br. Med. J. 12:111–25.
38. Hanna, J. M. 1970. Responses of native and migrant desert residents to arid heat. Am. J. Phys. Anthropol. 32:187–96.
39. Yunusov, A. Y. 1970. In Proceedings of a symposium on adaptation of man and of animals to extreme natural environments. Novosibirsk: Siberian Branch of the Academy of the USSR. 12–17.
40. Edholm, O. G. 1972. The effect in man of acclimatization to heat on water intake, sweat rate and water balance. In S. Itoh, K. Ogata, and H. Yoshimura, eds., Advances in climatic physiology. Heidelberg, NY: Springer-Verlag New York.
41. Eveleth, P. B. 1965. The effects of climate on growth. Ann. N.Y. Acad. Sci. 134:750–55.
42. Stinson, S., and A. R. Frisancho. 1978. Body proportions of highland and lowland Peruvian Quechua children. Hum. Biol. 50:57–68.
43. Randall, W. C., and A. B. Hertzman. 1953. Dermal recruitment of sweating. J. Appl. Physiol. 5:399–409.
44. Hofler, W. 1968. Changes in the regional distribution of sweating during acclimatization to heat. J. Appl. Physiol. 25:503–6.
45. Ogawa, T. 1972. Local determinants of sweat gland activity. In S. Itoh, K. Ogata, and H. Yoshimura, eds., Advances in climatic physiology. Heidelberg, NY: Springer-Verlag New York.

46. Kuno, Y. 1956. Human perspiration. Springfield, IL: C. C. Thomas.
47. Ladell, W. S. S. 1964. Terrestrial animals in humid heat: man. In D. B. Dill, E. F. Adolph, and C. G. Wilber, eds., Handbook of physiology, vol. 4. Adaptation to the environment. Washington, DC: American Physiological Society.
48. Gleibermann, L. 1973. Blood pressure and dietary salt in human populations. Ecology of Food and Nutrition 2:83–90.
49. Roberts, D. F. 1953. Body weight, race and climate. Am. J. Phys. Anthropol. 11:533–58.
50. Schreider, E. 1964. Ecological rules, body heat regulation, and human evolution. Evolution 18:1–9.
51. Beals, K. 1972. Head form and climatic stress. Am. J. Phys. Anthropol. 37:85–92.

# CHAPTER 4 Thermoregulation and Acclimation to Cold Stress

Biological responses to cold stress involve mechanisms of heat production and conservation. These responses to cold stress are more complex than those of heat stress adaptation. Successful responses to cold stress require the synchronization of cardiovascular and circulatory systems and, most important, the increased activation of the metabolic process. In this chapter the basic physiological responses with which organisms counteract cold stress, the individual factors that modify and affect these responses, and the process of acclimation to cold stress will be discussed. In subsequent chapters the process of acclimatization of native and nonnative populations is discussed to illustrate the variety of characteristics of human adaptation to cold stress.

## ENVIRONMENTAL FACTORS

Besides low temperatures, there are several other environmental factors that must be taken into account when considering human responses to cold stress. Among these, the most important are wind velocity, humidity, and duration of exposure to cold. In general, it is assumed that a low temperature, usually around 0°C (32°F), with high humidity results in greater cold sensation than with low humidity. However, these assumptions have not been confirmed by experimental studies.

Heat loss is strongly affected by wind velocity; a given temperature and a rapid wind result in greater cold stress than the same temperature with slow wind. Accordingly, an index for deriving the equivalent temperature that results from the actual (or total) temperature and wind velocity has been devised and is shown in table 4.1. For example, if the local temperature is

**TABLE 4.1. Wind chill effect: equivalent effective temperature**

| Wind | Actual Air Temperature (°F) | | | | | | | | | | | |
|---|---|---|---|---|---|---|---|---|---|---|---|---|
| speed | 50 | 40 | 30 | 20 | 10 | 0 | –10 | –20 | –30 | –40 | –50 | –60 |
| (mph) | Equivalent Temperature (°F) | | | | | | | | | | | |
| 5 | 48 | 36 | 27 | 17 | –5 | –5 | –15 | –25 | –35 | –46 | –56 | –66 |
| 10 | 40 | 29 | 18 | 5 | –8 | –20 | –30 | –43 | –55 | –68 | –80 | –93 |
| 15 | 35 | 23 | 10 | –5 | –18 | –29 | –42 | –55 | –70 | –83 | –97 | –112 |
| 20 | 32 | 18 | 4 | –10 | –23 | –34 | –50 | –64 | –79 | –94 | –108 | –121 |
| 25 | 30 | 15 | –1 | –15 | –28 | –38 | –55 | –72 | –88 | –105 | –118 | –130 |
| 30 | 28 | 13 | –5 | –18 | –33 | –44 | –60 | –76 | –92 | –109 | –124 | –134 |
| 35 | 27 | 11 | –6 | –20 | –35 | –48 | –65 | –80 | –96 | –113 | –130 | –137 |
| 40 | 26 | 10 | –7 | –21 | –37 | –52 | –68 | –83 | –100 | –117 | –135 | –140 |
| 45 | 25 | 9 | –8 | –22 | –39 | –54 | –70 | –86 | –103 | –120 | –139 | –143 |
| 50 | 25 | 8 | –9 | –23 | –40 | –55 | –72 | –88 | –105 | –123 | –142 | –145 |

*Source:* Modified from M. Ward. 1975. Mountain medicine: A clinical study of cold and high altitude. London: Crosby Lockwood Staples.

0°C (32°F) and wind velocity is 25 mph, the person loses heat as if the dry bulb temperature were −38.9°C (−38°F). Therefore, the chilling power of the wind can produce an almost supercooling effect on exposed skin.

In general, the degree of cold stress to which a person is exposed is classified as either *acute* or *chronic* cold. *Acute cold stress* refers to severe cold stress for short periods of time. *Chronic cold stress* refers to moderate cold stress experienced for prolonged periods of time, either seasonally or throughout the year. Obviously, the degree of cold stress depends on the amount of insulation. To quantify the amount of thermal insulation and heat exchange, the index called *clo* has been devised (1). One clo of thermal insulation will maintain a resting-sitting person with a metabolic rate of 50 kcal/m$^2$/hr comfortable in an environment of 21°C (70°F) with relative humidity less than 50% and air movement of 6 m/min (20 feet/min). In these basal conditions 1 clo is equivalent to a business suit or 0.64 cm (¼ inch) of clothing. A heavy article of arctic clothing provides about 5 clo units.

## MEASUREMENT OF BODY TEMPERATURE

The body temperature is the sum of internal and skin temperature.

### Internal Body Temperature

The classical site for measurement of internal body temperature is the rectum. Because the rectal temperature changes in the same magnitude as the temperature in the brain due to physical exertion or exposure to heat, it is a good indicator of changes in internal body temperature. The internal body temperature can also be measured under the tongue, in the axilla, in the groin, in the eardrum, or in the esophagus. It should be noted, however, that readings at these sites are either lower or more likely to be affected by external functions than those obtained from the rectum.

### Skin Temperature

The skin temperature is measured with thermocouples or thermistors on the skin at certain locations. The mean skin temperature is calculated by assigning certain factors to each of the measurements in proportion to the fraction of the body total surface area represented by each specific site:

head = 0.07
arm = 0.14
feet = 0.07

| | | |
|---|---|---|
| legs | = | 0.13 |
| thighs | = | 0.19 |
| trunk | = | 0.35 |
| Total | | 1.00 |

### Mean Body Temperature

The mean body temperature (also referred to as heat content of the body) is the product of rectal and mean skin temperature calculated with the following equation:

Mean body temperature = (0.33 × skin temperature) + (0.65 × rectal temperature)

## GENERAL RESPONSES TO COLD STRESS

As previously indicated, within the thermoneutral temperature range of 25° to 27°C (77° to 80°F) the individual is in thermal equilibrium, but below this range the nude individual responds immediately through mechanisms that permit both the conservation of heat and an increase in heat production. The major mechanism concerned with heat conservation is vasoconstriction, alternated with vasodilation and synchronized with the countercurrent system of heat exchange. The major mechanism concerned with heat production is shivering. As a side effect of cold stress, changes occur in the activities of the sympathetic nervous system and vagal reflex. Since each mechanism is multifaceted and in order to have a broader view of the general responses to cold stress, each response will be discussed and illustrated with representative research data dealing with various temperature conditions in cold water and cold air.

### Vasoconstriction

On exposure to cold stress, such as a temperature of 0°C (32°F) or even 15°C (60°F), the cold receptors in the skin of a nude individual are activated to initiate those reflexes involved in conserving heat. This is accomplished through a constriction of the subcutaneous blood vessels (vasoconstriction), which limits the flow of warm blood from the core to the shell (skin). The result of this lower blood flow is a decrease in skin temperature, a reduction of the temperature gradient between the skin surface and the environment, and, consequently, a reduction in the rate of heat loss (fig. 4.1). As a result, during full vasoconstriction heat conductivity of the blood is reduced by as

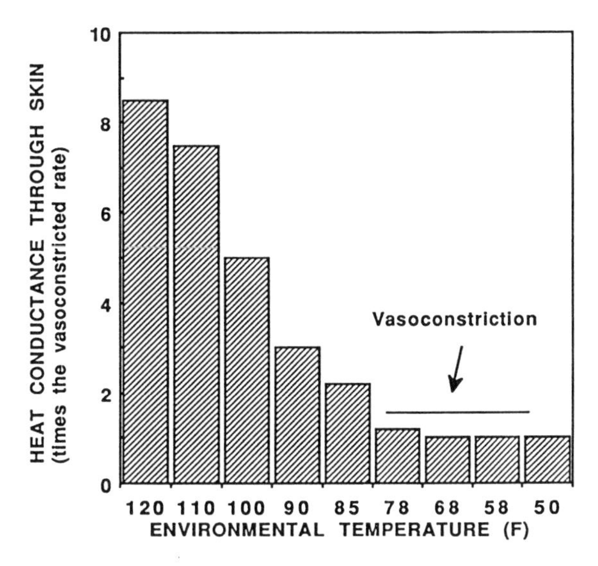


**Fig. 4.1.** Decrease of heat conductance from body core to skin surface in response to decreased ambient temperature.

much as eight times. Concomitant with decreased blood flow to the skin, the blood flow to the viscera and internal organs is augmented. This shift in blood flow is probably responsible for increased blood pressure and heart rate under severe cold stress.

### Countercurrent

The reduction in conductance of the blood is also caused by a deviation of the blood in the veins of the extremities from the superficial to the deep veins. That is, in a cold environment most of the venous return from arms and legs is through the deep venae comitantes that receive heat from blood flowing through the arteries, thereby minimizing heat loss.

As shown in figure 4.2, owing to this system of countercurrent in a person exposed to an ambient temperature of 10°C (50°F), blood leaving the heart will have a temperature of 37°C (98.6°F). As it reaches the hand this blood will have a temperature of 20°C (68°F). In contrast, the returning venous blood at the hand has a temperature of 20°C (68°F) but by the time it reaches the heart it will have a temperature of 28°C (82°F). In other words, as the venous blood returns from the periphery it absorbs nearly 8°C (46.4°F) from the outgoing arterial blood. This countercurrent heat exchange occurs because each outgoing artery is always surrounded by two incoming and returning veins. The outgoing arterial blood is always warmer than the incoming venous blood, consequently there is continuous heat exchange (see also fig. 2.10). In this manner, heat conductance to the periphery is low, yet actual blood flow to the limbs may be high, protecting tissues of the limbs from cold injury and hypoxia.

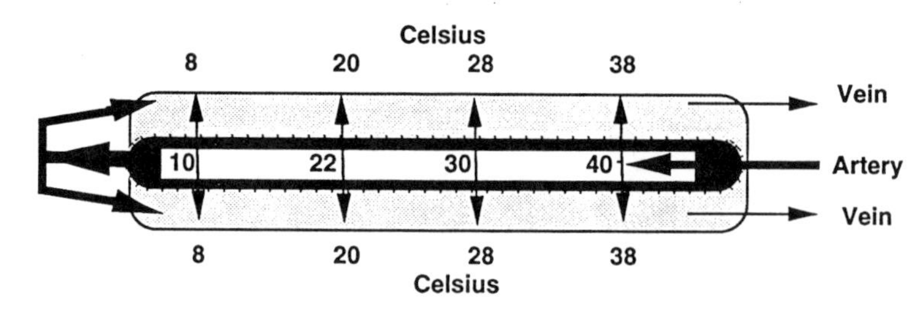

**Fig. 4.2.** Schematization of the system of countercurrent heat exchange. Note that as the arterial blood leaves it gives off heat to the returning venous blood. (From G. E. Folk. 1974. Textbook of environmental physiology. Philadelphia: Leu and Febiger.)

## Hunting Response or Cold-Induced Vasodilation

A typical response to immersion of the finger in cold water is the spontaneous, semirhythmic changes in skin temperature. As shown in figure 4.3, upon immersion digital temperature drops for about 15 minutes until the temperature is about 2.5°C above the ice water temperature. After about 16 minutes, digital temperature rises by some 6° to 8°C and fluctuates thereafter between 4° and 6°C. These fluctuations in temperature are caused by fluctuations in the blood flow and have been termed the Lewis ''hunting'' phenomenon (2). Further studies indicated that this cold-induced vasodilation results from an increase in blood flow because of the sudden opening of the arteriovenous anastomoses (3). This in turn produces the marked variation in temperature. Similar responses to cold were noted in the skin of the ear, cheek, nose, chin, and toes. This mechanism may have importance in cold water conditions for protecting exposed parts of the skin from excessive cooling and injury (2). However, current research has not demonstrated a relationship between this type of response and individual susceptibility to cold injury.

Another effect of cold stress is that pain sensation increases with falling skin temperatures; with rewarming of the skin by cold-induced vasodilation (CIVD), this sensation disappears, giving the sensation that the hand is being immersed in lukewarm water. Recurrence of vasoconstriction coincides with an elevation in the pain sensation. However, pain is observed not only when skin temperature cools, but also when the skin warms up after cooling. Indeed, rewarming of the hand or face at room temperature produces a marked pain sensation. This increase in pain occurs when skin temperature increases rapidly.

## Role of the Central Nervous System in Response to Cold Stress

**Sympathetic Action and Vagal Reflex.** As indicated in chapter 2 the autonomic nervous system (visceral efferent nervous system) regulates visceral

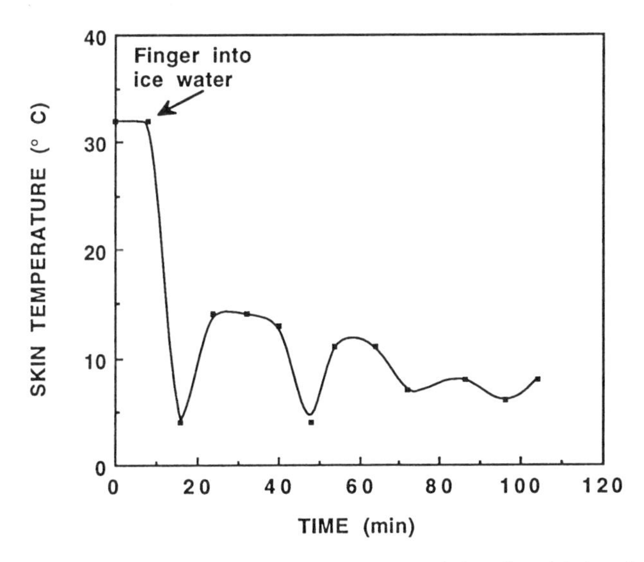


**Fig. 4.3.** Alternation of vasodilation and vasoconstriction (Lewis's hunting phenomenon) in response to immersion of finger in crushed ice. Curve shows large, prolonged temperature oscillations finally giving way to smaller, more rapid ones. (Modified from T. Lewis. 1930. Heart. 15:177–81.)

activities such as change in the size of the pupil, accommodation for near vision, *dilation* and *constriction of blood vessels*, adjustments of the rate and force of the heartbeat, movements of the gastrointestinal tract, and the secretion of most glands. The efferent neurons of the autonomic system release either *acetylcholine* or *norepinephrine* (NE) as their neurotransmitter. The autonomic nervous system consists of the sympathetic and parasympathetic division. For example, when the hand is exposed to cold stress (water at 4°C) there is an increase in blood pressure and an increase in heart rate that results from the vasoconstriction (see fig. 4.4). This response indicates the activation of the sympathetic nervous system. On the other hand, exposure of the face but not of the hand to the same stress causes an increase in blood pressure but a decrease in heart rate (fig. 4.5). The heart is innervated by the autonomic nervous system, specifically by a group of neurons located within the medulla of the brain. Within the medulla there are two groups of neurons: (1) one called the *cardioacceleratory* center that gives rise to sympathetic fibers whose function is to increase the rate of heartbeat and the strength of contraction through secretions of norepinephrine; (2) the other group of neurons is called the *cardioinhibitory* center, which gives rise to parasympathetic fibers that reach the heart via the *vagus nerve*, whose function is to decrease the rate of heartbeat and the strength of contraction through secretions of acetylcholine. The fact that exposure of the face to cold stress results in the decrease of the heart rate indicates an increased activity of the vagal reflex while the increase in blood pressure indicates the activity of the parasympathetic system. Exposure of the hand, however, results only in activation of the sympathetic system (4). It has been suggested

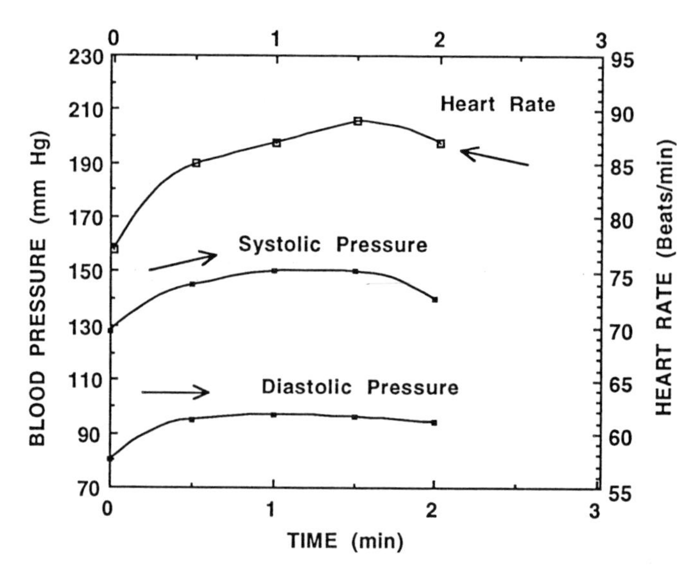


**Fig. 4.4.** Increase in blood pressure and heart rate during immersion of one hand into cold water (4°C) for 2 minutes. The increase in heart rate and blood pressure associated with vasoconstriction suggests activation of sympathetic nervous system. (Modified from J. LeBlanc. 1975. Man in the cold. Springfield, IL: Charles C. Thomas.)

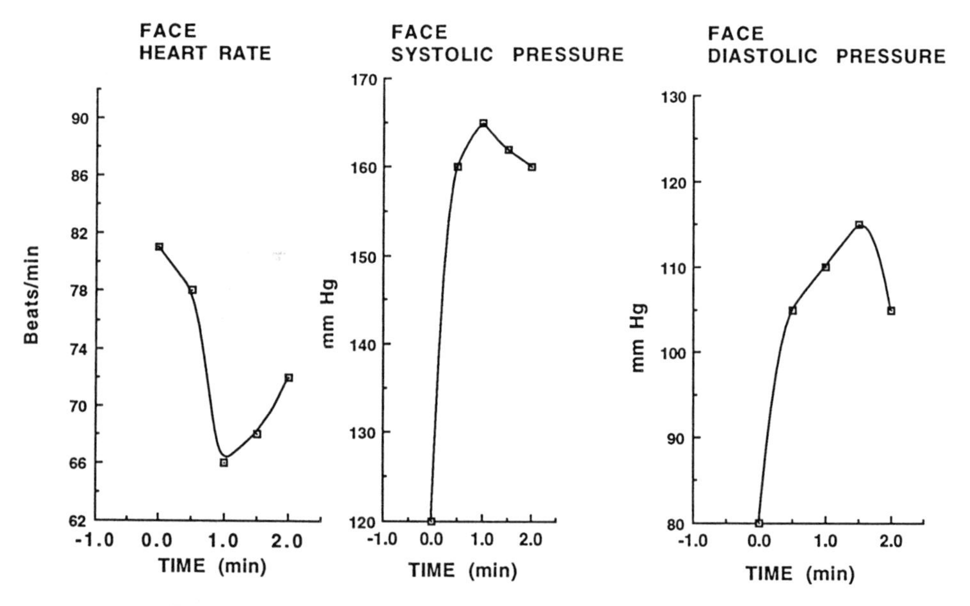


**Fig. 4.5.** Effects of immersion of face in water at 4° C for 2 minutes on heart rate and systolic and diastolic pressure. Decreased heart rate but increased blood pressure during immersion in cold water suggests activation of both vagal reflex and sympathetic nervous system. (Modified from J. LeBlanc. 1975. Man in the cold. Springfield, IL: Charles C. Thomas.)

that pain experienced by angina pectoris patients when their faces are exposed to cold winds may be the result of a decreased heart rate (4). Furthermore, the cold-induced increase in blood pressure along with bradycardia may impose additional stress on the hearts of the subjects. Perhaps, then, the higher cardiovascular mortality that occurs in winter in the United States (5) may have a thermoregulatory cause.

## Metabolic Rate and Shivering

When the vasoregulatory mechanisms of heat conservation are not sufficient to counteract heat loss, the organism adjusts by increasing the rate of heat production. For an unclothed man an increase in heat production usually occurs when the ambient temperature falls below 25°C. The most rapid and efficient way to increase heat production is by voluntary exercise, such as running, which may increase the metabolic rate from a basal value of 1.17 cal/min to 37.94 cal/min (6). However, such high rates of activity cannot be maintained for prolonged periods. Thus, in the absence of voluntary exercise, shivering of the skeletal muscle is the main source of increased heat production. As shown by experimental studies, shivering requires the activation of skin receptors through the lowering of skin temperature (7). The major function of shivering is to increase the rate of heat production (but it also adds to heat loss). In addition, shivering provides improved protection of core heat by augmenting the thermogenesis of the muscle mass. Through this mechanism the temperature of muscle is raised to approach that of the core, thus eliminating the temperature gradient heat loss.

As a result of shivering, the metabolic rate may be increased two to three times the basal value. This increase in heat production is progressive throughout the cold stress. For example, during intense cold, heat production increases from a basal level at pre-exposure of 35.4 cal/m$^2$/hr to 54 cal/m$^2$/hr by the end of the first hour. By the end of the second, third, and fourth hours production had risen to 72, 92, and 96 cal/m$^2$/hr, respectively (8).

Along with shivering, subcutaneous blood vessels dilate (vasodilation) to keep the skin warm and prevent tissue injury from frostbite. Thus, adjustment to cold stress is an interplay between mechanisms to conserve heat and mechanisms to produce and dissipate heat. In terms of energy expenditure, defense against cold is achieved more economically by increasing body heat conservation than by increasing heat production. However, the extent to which these mechanisms are operative in humans depends on the degree of cold stress experienced and on technological and cultural adaptation, as will be shown in later chapters.

## ACCLIMATION TO COLD STRESS

### Experimental Animals

Information on the process of acclimation to cold stress has been derived mainly from studies with animals, specifically rats. These studies have been concerned with the mechanisms that enable an animal to survive severe cold stress. Studies have emphasized energy sources for heat production; this approach involves evaluations and identification of thermogenic hormones, morphological changes, and the role of lipids.

#### Nonshivering Thermogenesis

When rats are continuously exposed to temperatures of 5°C (41°F) for 2 or 3 weeks, they are able to maintain a high metabolism (elevated about 80%) and normal temperature without resorting to shivering. This increase in heat production without muscular movement is referred to as nonshivering thermogenesis (4, 9). The fact that increased heat production associated with continuous cold exposure is *not* the result of shivering is indicated by experiments showing that cold-acclimated rats whose shivering was blocked by curare increased their metabolic rates and their rectal and skin temperatures more effectively when exposed to cold stress than did nonacclimated rats (fig. 4.6). The increased metabolism of the cold-acclimated rats persisted even after the animals had been transferred to warmer climates (30°C).

#### Hormonal Effects

An important factor associated with nonshivering thermogenesis appears to be noradrenaline, which is released from sympathetic nerve endings. Continuous exposure to cold, like any other stress, causes a marked increase in activity of the sympathetic nervous system, as shown by the enhanced secretion of adrenaline and noradrenaline in the urine. It has been demonstrated that nonshivering thermogenesis results from an increased sensitivity to noradrenaline (4, 9, 10-12). Animals injected for 3 weeks with relatively small daily doses of noradrenaline became more sensitive to this hormone and at the same time were made more resistant to the cold (4). This change in the calorigenic effect of noradrenaline has been confirmed in studies of men who were cold-acclimated (4, 13). For this reason, noradrenaline (or norepinephrine) is considered the hormone of cold acclimation. Daily injections for 3 weeks of isoproterenol also produced a thermogenic sensitivity that appeared as important as noradrenaline's effect in both heat production and cold resistance (4, 14). However, noradrenaline and isoproterenol are not the only factors in cold acclimation because repeated injections of these hormones,

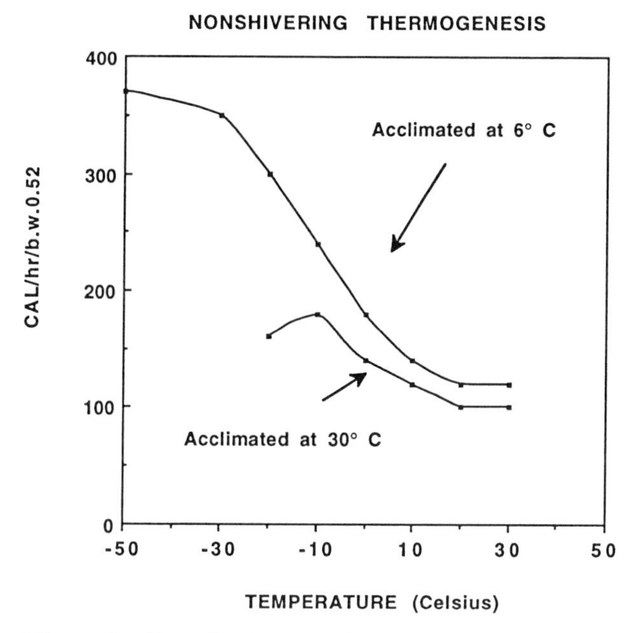


**Fig. 4.6.** Effects of cold acclimation at 6°C on metabolic rate of rats. (Modified from F. Depocas. 1960. Fed. Proc. 19 [Suppl. 5]: 19–24.)

although they produce as much sensitization to catecholamines as cold exposure does, do not elevate the threshold of tolerance as much as actual exposure of the animal to a cold environment (4, 9, 14).

Thyroxine, because of its importance in calorigenesis, is another hormone that plays a significant role in cold acclimation. Various studies have shown that exposure for several weeks to cold stress increased the output of thyroxine by as much as 100%, an amount that can increase heat production of an animal as much as 20% to 30%. Furthermore, continuous treatment of normal animals with relatively small doses of thyroxine (10 ml/24 hr) enhances cold tolerance (4). For these reasons, it is assumed that the most important control components of nonshivering thermogenesis are the combined effects of noradrenaline and thyroxine. However, these hormones are not the only factors in cold acclimation. Experiments have demonstrated that thyroxine- and noradrenaline-treated rats exposed for 3 hours to severe cold stress (−25°C) maintained high colonic temperatures but experienced frostbite of the tail and extremities, whereas naturally cold-acclimated rats maintained a warm and flexible tail even after 6 hours of exposure to the same cold stress (4, 9, 14). All this evidence together suggests, again, that acclimation to cold stress involves more than just sensitization to and activation of calorigenic hormones.

## Morphological Changes

Along with increases in nonshivering thermogenesis, total daily food consumption increases. Similarly, continuous exposure of rats to cold stress is associated with hypertrophy of the thyroid, adrenal cortex, heart, kidney, liver, and digestive tract as compared to non-cold-exposed controls (15). When oxygen consumption of each of these tissues is measured, total oxygen consumption, expressed per unit tissue weight, is greater for the visceral regions than those of the shell, such as the muscle and skin. Thus, with acclimation heat production by the core region becomes more important than that of the surrounding shell (15).

## Lipid Changes

Nonshivering thermogenesis requires rapid mobilization of energy. Because of its abundance, fat appears to be the most important source of fuel. With continuous exposure to cold stress, the triglycerides of depot fat, influenced by noradrenaline, are mobilized and split into free fatty acids and glycerol. Subsequently, the free fatty acids are broken down into two carbon fragments called acetyl-CoA, which in turn are used to liberate energy in the tricarboxylic acid cycle. Also, continuous exposure to cold stress causes the lipoprotein lipase activity to decrease in white adipose tissue, while markedly increasing it in brown adipose tissue and cardiac and skeletal muscles (16). Based on this information, it has been postulated that with acclimation to cold the triglycerides of the very low-density lipoproteins are directed as an energy source to those tissues involved in thermogenesis, such as brown adipose tissue and muscle, but are withheld from white adipose tissue, which under nonacclimated conditions is the usual site of storage (17). Recent investigations indicate that, like the free fatty acids, the ketone bodies found in urine are an important source of energy for rats acclimated to cold (18).

## Brown Adipose Tissue

Animals exposed to cold stress exhibit a marked increase in brown adipose tissue. This tissue's importance in cold acclimation is inferred from evidence that (1) the fall in body temperature of animals exposed to -25°C for 3 hours was inversely related to the weight of brown adipose tissue (14); (2) hypertrophy of brown adipose tissue through treatments with thyroxine and noradrenaline was associated with cold tolerance (4); and (3) removal of brown adipose tissue in cold-acclimated rats significantly reduced the animals' response to the sensitization of noradrenaline and their tolerance to cold (17, 19). The degree of chemical thermogenesis that occurs in animals is nearly proportional to the amount of brown fat present in the animal tissues.

Several hypotheses have been postulated to explain the mechanisms of brown adipose tissue and its actual contribution to nonshivering thermogenesis. First, brown adipose tissue's thermogenic capacity and its special vascularization (the positions of arteries and veins are closely juxtaposed in both cervical brown fat pads and in the interscapular pad) return metabolically warmed blood to the thorax through its venous drainage. This arrangement results in a bathing of thoracic and cervical spinal cord areas with warmed blood and adequate heat to supply the heart and sympathetic chain (15, 20). This "metabolic warming blanket" protects the central body core from the peripheral cooling effects of a cold environment (4, 9). Second, brown adipose tissue is an important source of noradrenaline for the organism. The brown fat of cold-acclimated animals has been found to contain three times more noradrenaline than that of nonacclimated controls (4). Third, because of its special anatomical location combined with its high calorigenic properties, brown adipose tissue controls shivering (21). This is supported by evidence that local heating of cervical, but not lumbar vertebral canals suppresses shivering and diminishes heat production. Fourth, brown adipose tissue contributes to nonshivering thermogenesis by releasing into the circulation a hormonal factor such as noradrenaline (17). Removal of the interscapular brown fat reduces noradrenaline sensitivity by 40% and significantly reduces cold tolerance even in cold-acclimated animals (17, 19).

### Humans

In contrast to many animal studies, investigations of human acclimation have been limited both in scope and quantity. Studies that have been done with humans have been concerned mostly with the occurrence of nonshivering thermogenesis and the maintenance of core and skin temperatures.

In one experiment five subjects were placed in a 15°C room for 2 weeks, with the subjects nude 24 hours a day. By the second week their metabolic rates were higher than the controls, whereas skin temperature showed no decline at all (22). In another experiment thirty-six subjects dressed in shorts only were exposed for 4 weeks for daily 8-hour periods to temperatures of 5° to 11°C (23). In this experiment there was a decrease in shivering, an increase in nonshivering thermogenesis, and later initiation of shivering (fig. 4.7). Another study (24) evaluated thermoregulatory responses of 10 male subjects who were acclimated for 2 months by water immersion (4 per week) to temperatures of 10° and 15°C. Results of this study indicate that after acclimation to cold stress three changes occur: (1) an increase in metabolic rate; (2) a decrease in heat conductance brought about by vasoconstriction; and (3) a decrease in total heat loss.

The above findings together suggest that when sufficiently exposed to cold stress, humans behave like other species and can acclimate to cold through similar metabolic changes. However, a recent study of the thermoregulatory responses of seventeen black American soldiers who underwent

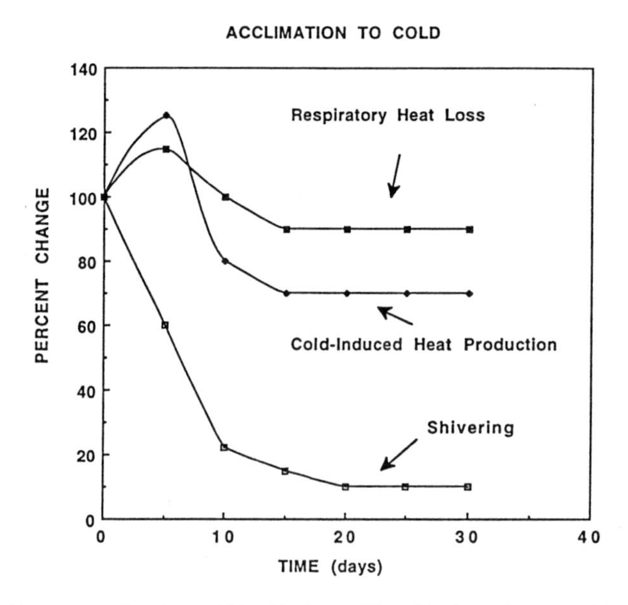


**Fig. 4.7.** Human acclimation within 20 days. After 20 days (except nights) of exposure to cold chamber (12°C) shivering declines whereas heat production continues to increase. (Modified from T. R. A. Davis. 1961. J. Appl. Physiol. 16: 1011–15.)

4-hour daily exposures at 5°C (41°F) for 8 weeks (5-day work weeks) wearing minimal clothing indicates that after this period of acclimation, although there was a significant reduction in shivering activity and warmer skin temperatures, the rectal temperatures and heat production were not affected by these daily exposures (25). Furthermore, a recent study evaluated the acclimation process through short intermittent exposure to severe cold stress among a group of subjects during 16 days in the Arctic (26). Results of this study indicate that intermittent exposure to severe cold stress can result in habituation that involves facultative body cooling and a decreased metabolic and hormonal response. A corroboration of this finding can be found in the evaluations of the thermoregulatory characteristics of J. Etienne after his journey of 63 days to the North Pole where the ambient temperature varied from − 12° to − 52°C (10.4° to − 61.6°F) (27). After his journey on the sea ice to the North Pole, Etienne experienced a decrease in metabolic rate and rectal temperature but an increase in skin temperature of the extremities during the cold test when compared to a similar test before the expedition. It would appear then that Etienne's acclimatization to cold stress involved a redistribution of heat from the internal to the peripheral organs despite the low metabolic rate.

## INDIVIDUAL FACTORS AND TOLERANCE TO COLD

As indicated in chapter 3, heat flow follows Fourier's Law which indicates that heat loss per minute is directly proportional to body surface, the difference between the temperature of the body, and of the environment. It is also inversely proportional to thickness of the body shell. When considering the sources of individual variability in cold tolerance, the most important factors include surface area and insulation.

### Surface Area

Tolerance to cold stress is affected by the size and shape of the individual. When expressed as per unit of body weight, the heat required to maintain a constant internal body temperature will be greater in a small rather than a large individual (1, 4, 9, 28). This is because surface area exposed to the environment, all other factors being constant, is greater per unit of body weight if the total body is smaller. For this reason, in a cold environment, all factors being constant, a small individual would be expected to produce relatively more heat than a large individual in order to maintain homeostasis. The same principle applies to children who, when compared to adults, have a high surface area:weight ratio (SA/weight). Analyses of the thermoregulatory characteristics of highland Peruvian Quechua Indians (29) revealed a significant negative correlation between rectal temperatures and the measurements of hand and foot size when hands and feet were exposed for 2 hours at 10°C (fig. 4.8). In the same manner, among Japanese and European white subjects born and raised in Hawaii, a positive correlation between the measurements of trunk size and mean finger temperatures was found when the hands were exposed to 0°C moving air temperatures (30).

### Insulation of Fat

One of the most important factors affecting heat loss is the degree of artificial or natural insulation. In humans, subcutaneous fat represents the most important form of natural insulation. Because the subcutaneous fat layer is not very well vascularized, the thermal conductivity of fat is much less than that of muscle. For this reason, the greater the fat layer, the lower the skin temperature, and, consequently, the smaller the gradient between the body's surface temperature and that of the environment (4, 9, 31, 32). As a result of these interactions, the greater the fat layer, the lower the total heat loss. As shown in figure 4.9, when men were exposed to water at 15°C for 30 minutes, the higher the skinfold thickness, the smaller the drop in rectal temperature (31). For the same reason, the frequency and intensity of shivering is more pronounced in thin persons than in fat ones (4). Increased heat production as a result of cold immersion is also much greater in thin subjects because of less tissue insulation (4, 31, 33).

For human survival in cold water, the amount of subcutaneous fat, along with the amount of artificial insulation such as clothing, is the most important consideration (4, 31, 33). However, the insulative effectiveness of subcutaneous fat during immersion in cold water is affected by the degree of physical

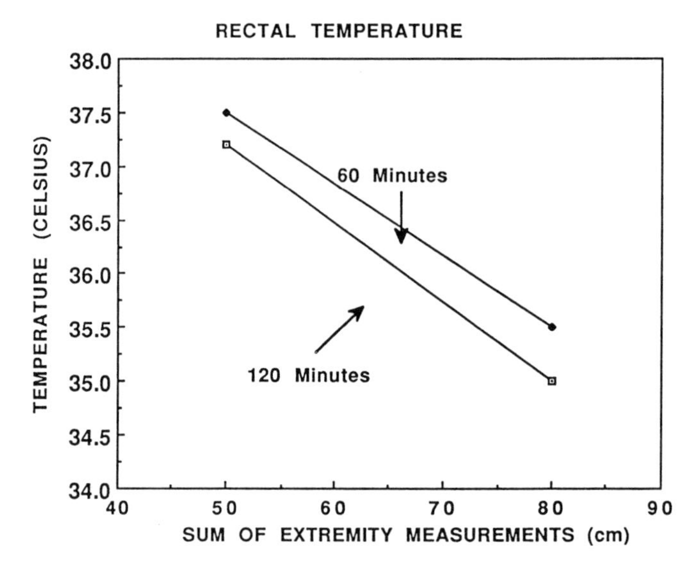


**Fig. 4.8.** Relationship between rectal temperature and sum of hand and foot size measurements on Quechua Indians during exposure to 10°C. The higher the available surface area of extremities, the lower the rectal temperature. (From C. A. Weitz. 1969. Morphological factors affecting responses to total body cooling among three human populations at high altitude. M.A. Thesis. University Park, PA: Pennsylvania State University.)

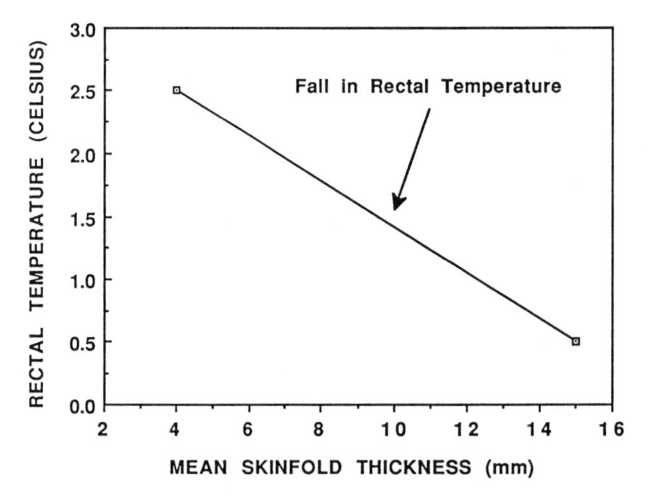


**Fig. 4.9.** Relationship of subcutaneous fat thickness and heat loss in cold water. The lower the mean skinfold thickness, the greater the fall in rectal temperature. (Modified from W. R. Keatinge. 1969. Survival in cold water: The physiology and treatment of immersion hypothermia and of drowning. Oxford: Blackwell Scientific Publications)

activity. At a given fat thickness the subjects who were at rest experienced a lesser fall in rectal temperature than those who were swimming (31). Thus, because cold water facilitates heat loss by convection, movements associated with swimming, while increasing heat production, also accentuate heat loss.

For human survival in cold water this heat loss from activity could be detrimental even for subjects with a high fat layer.

### Gender

Since women on the average are fatter than men, one would expect that under conditions of cold stress their internal body temperature would be more stable. However because women usually have lower body weight than men, they have a high ratio of surface area to weight and as such have greater avenues for heat loss and a lower heat-producing capability due to their lower body weight.

**Rest.** In a recent study the thermoregulatory response of 10 men and 8 women who were exposed at rest in air and during one hour immersion in water at 20°, 24° and 28°C was evaluated (34). Results of this study indicate that women, despite their greater amount of fatness, experience a greater cold stress than men. The relative greater cold stress is related to the high surface-low weight that characterizes women. Thus it would appear that extra insulation derived from the greater amount of fat possessed by women is not sufficient to counteract the low heat production and high heat loss associated with high surface-low weight ratio.

**Exercise.** In a recent study the thermoregulatory response of 10 men and 8 women who exercised for one hour in air and water at 20°, 24° and 28°C was evaluated (35). Results of this study indicate that, contrary to that found under resting conditions, during exercise both men and women maintain similar thermoregulatory responses at each water temperature. For both men and women the thermoregulatory benefits of exercise were due largely to the added heat production from physical activity.

### Physical Fitness

**Work Capacity and Maximal Oxygen Intake.** Some of the major adaptive effects of physical training include increased vascularization, increased size of the striated and cardiac muscles, and increased maximum aerobic capacity. It is generally agreed that the maximum oxygen intake per unit of body weight (or aerobic capacity) during maximal work is a measure of the individual's work capacity because it reflects the capacity of the working muscles to use oxygen and the ability of the cardiovascular system to transport and deliver oxygen to the tissues. The rate of oxygen consumption increases linearly with the magnitude of work. As an exercising subject approaches the point of exhaustion or fatigue, his or her oxygen consumption will reach a maximum and remain at that level even with further increase in work. This peak value is referred to as the individual's *maximal oxygen intake.* Maximum oxygen intake may be obtained directly or indirectly. In a recent study

(36) where 17 healthy male volunteers were exposed to cold stresses ranging from 1°, 5° and 10°C it was found that heat production was directly related to the maximal oxygen intake. This finding suggests that the sensitivity of the thermoregulatory system as shown by the increased heat production is increased with physical fitness. In other words, fit subjects have more efficient thermoregulatory abilities against cold stress than unfit individuals.

### Age

In human newborns all the thermoregulatory responses (enhanced thermogenesis, vasomotor responses, sweat secretion, etc.) can be triggered immediately after birth, even in premature infants weighing about 1,000 grams (37). Thus, their heat production can be raised 100–200% above the resting metabolic rate without shivering. However, because newborns have a surface area to volume ratio that is 2–3 times larger than those of adults, they are at a disadvantage when exposed to cold temperature (for example, a 3.5 kg [7.7 lbs] and 56 cm [22 in.] newborn has a surface area of 0.23 $m^2$ or SA/wt = 6.57, while an adult man of 70 kg [154 lbs] and 170 cm [67 in.] tall has a surface area of 1.80 $m^2$ or SA/wt = 2.57). Furthermore, because the insulating layer of fat as measured by skinfold thickness in newborns is very low, vasoconstriction is not as effective as in adults in reducing heat loss. For this reason, if thermal balance is to be maintained at the minimal metabolic rate in the newborn the ambient temperature must be higher (32–34°C) than that of adults.

As measured by peripheral responses of the hand, thermoregulatory responses are better in young adulthood than during old age. During immersion in 10°C water (with normal room temperatures) cold-induced vasodilation of the fingers was more frequent and the ''hunting waves'' more rapid during shorter periods in the nonadult and adult subjects than in the older adult subjects (over 70 years) (38). In a recent study, men ranging in age from 20 to 73 years were exposed at rest for two hours to 10°, 15°, 20° and 28°C room temperatures (39). This study indicates that older men have a greater increase in metabolic rate than their younger counterparts. This finding suggests that older individuals are more susceptible to colder stress than young individuals. Furthermore, older women, despite their greater amount of fat, were as susceptible to cold as young women. These differences are probably caused by the effects of aging—decreased vascularization and peripheral blood flow and diminished heat conductivity. Obviously, there is a critical need for further research in this area.

## OVERVIEW

Initial responses to cold stress are oriented at conserving heat through vasoconstriction alternated with vasodilation. These responses appear to be

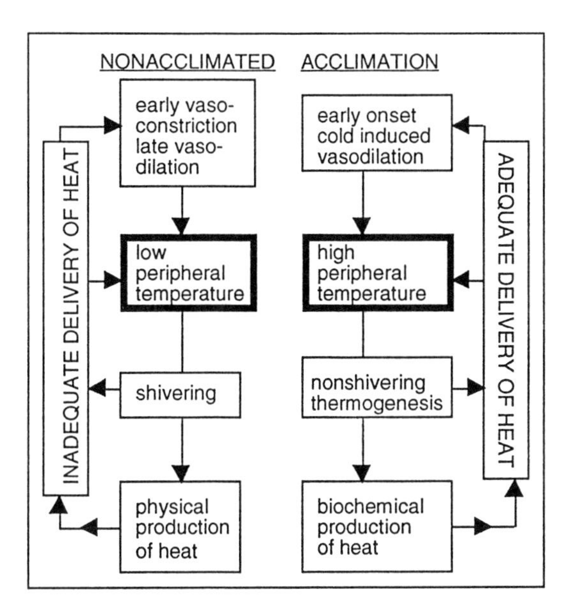


**Fig. 4.10.** Schematization of thermoregulatory responses to cold before and after acclimation. Before acclimation onset of cold-induced vasodilation is delayed and increase in heat production is attained through shivering, but delivery of heat is still inadequate, resulting in low peripheral temperature. After acclimation cold-induced vasodilation occurs earlier, heat production is increased through nonshivering thermogenesis, and peripheral temperature is higher.

mediated through the activity of the central nervous system as demonstrated by the action of the sympathetic system and vagal reflex.

As schematized in figure 4.10, when exposure to cold stress is severe, the organism increases its rate of heat production through shivering before acclimation. As a result of shivering the metabolic rate increases two to three times the normal rate; thus the rate of heat production is increased. However, because of inefficient delivery of heat the peripheral temperature is cold. On the other hand, acclimation to cold stress appears linked to maintenance of warm skin temperatures, made possible by an increased ability to produce heat. Increased heat production is achieved without shivering. The source of energy for nonshivering thermogenesis appears to be white and brown adipose tissue deposits converted into free fatty acids. Lipid sources of energy appear to be mediated through the increased activity of calorigenic hormones such as noradrenaline (norepinephrine) and thyroxine.

Limited studies with humans suggest that acclimation to cold stress is associated with increased ability to tolerate low temperatures. This is attained through a maintenance of warmer skin temperatures and increased heat production. However, at present it is not clear to what extent acclimation leads to an increased metabolic rate without shivering.

Tolerance to cold stress appears to be influenced by an individual's age and physical fitness, degree of insulation derived from subcutaneous fat, and the ratio of surface area to weight. Females, despite their higher amount of subcutaneous fat, are less tolerant to cold stress than males, and this difference appears to be related to their high surface area and low weight which facilitates heat loss and limits heat production.

## References


1. Burton, A. C., and O. G. Edholm. 1955. Man in a cold environment. London: Edward Arnold (Publishers).
2. Lewis, T. 1930. Vasodilation in response to strong cooling. Heart 15:177–81.
3. Grant, R. T., and E. F. Bland. 1931. Observations on arterio-venous anastomoses in human skin and in bird's foot with special reference to reaction to cold. Heart 15:385–407.
4. LeBlanc, J. 1975. Man in the cold. Springfield, IL: Charles C. Thomas, Publisher.
5. Smolensky, M., F. Halberg, and F. Sargent. 1972. Chronobiology of the life sequence. In S. Itoh, K. Ogata, and H. Yoshimura, eds., Advances in climatic physiology. Heidelberg, NY: Springer-Verlag New York.
6. Consolazio, C. F., R. E. Johnson, and L. J. Pecora. 1963. Physiological measurements of metabolic functions in man. New York, NY: McGraw-Hill.
7. Benzinger, T. H. 1959. On physical heat regulation and the sense of temperature in man. Proc. Natl. Acad. Sci. U.S.A. 45:645–59.
8. Glickman, N., H. Mitchell, R. Keeton, and E. Lambert. 1967. Shivering and heat production in men exposed to intense cold. J. Appl. Physiol. 22:1–8.
9. Folk, G.E. 1974. Textbook of environmental physiology. Philadelphia, PA: Lea and Febiger.
10. Hsieh, A. C. L., and L. D. Carlson. 1957. Role of the thyroid in metabolic response to low temperature. Am. J. Physiol. 188:40–44.
11. Depocas, F. 1960. Calorigenesis from various organ systems in the whole animal. Fed. Proc. 19 (Suppl. 5):19–24.
12. Jansky, L. 1966. Body organ thermogenesis of the rat during exposure to cold and at maximal metabolic rate. Fed. Proc. 25:1297–1302.
13. Carlson, L. D., and A. C. L. Hsieh. 1965. Cold. In O. G. Edholm, and A. L. Bacharach, eds., The physiology of human survival. New York, NY: Academic Press.
14. LeBlanc, J., and A. Villemaire. 1970. Thyroxine and noradrenaline or noradrenaline sensitivity, cold resistance, and brown fat. Am. J. Physiol. 218:1742–45.
15. Smith, R. E., and D. J. Hoijer. 1962. Metabolism and cellular function in cold acclimatization. Physiol. Rev. 42:60–142.
16. Radomski, M. W., and T. Orme. 1971. Responses of liproprotein lipase in various tissues to cold exposure. Am. J. Physiol. 220:1852–56.
17. Himms-Hagen, J. 1969. The role of brown adipose tissue in the calorigenic effect of adrenaline and noradrenaline in cold-acclimated rats. J. Physiol. (Lond.) 205:393–403.
18. Hiroshige, T., K. Yoshimura, and S. Itoh. 1972. Mechanisms involved in thermoregulatory heat production in brown adipose tissue. In S. Itoh, K. Ogata, and H. Yoshimura, eds., Advances in climatic physiology. Heidelberg, N.Y.: Springer-Verlag New York.

19. Leduc, J., and P. Rivest. 1969. Effets de l'ablation de la graisse brune interscapulaire sur l'acclimatation au froid chez le rat. Rev. Can. Biol. 28:49–66.
20. Smith, R. E., and J. C. Roberts. 1964. Thermogenesis of brown adipose tissue in cold-acclimated rats. Am. J. Physiol. 206:143–48.
21. Bruck, K. 1970. Brown adipose tissue. New York, NY: L.O. Lindberg.
22. Iampietro, P. F., D. E. Bass, and E. R. Buskirk. 1957. Diurnal oxygen consumption and rectal temperature of men during cold exposure. J. Appl. Physiol. 10:398–400.
23. Davis, T. R. A. 1961. Chamber cold acclimatization in man. J. Appl. Physiol. 16:1011–15.
24. Bittel, J. H. M. 1987. Heat debt as an index for cold adaptation in men. J. Appl. Physiol. 62:1627–34.
25. Newman, R. W. 1969. Cold acclimation in Negro Americans. J. Appl. Physiol. 37:316–19.
26. Radomski, M. W., and C. Boutelier. 1982. Hormone response of normal and intermittent cold-preadapted humans to continuous cold. J. Appl. Physiol. 53:610–16.
27. Bittel, J. H. M., G. H. Livecchi-Gonnot, A. M. Hanniquet, C. Poulain, and J.-L. Etienne. 1989. Thermal changes observed before and after J.-L. Etienne's journey to the North Pole. J. Appl. Physiol. 58:646–51.
28. Pugh, L. G. C., and O. G. Edholm. 1955. The physiology of channel swimmers. Lancet 2:761–68.
29. Weitz, C. A. 1969. Morphological factors affecting responses to total body cooling among three human populations tested at high altitude. M.A. Thesis. University Park, PA: Pennsylvania State University.
30. Steegman, A. T. 1974. Ethnic and anthropometric factors in finger cooling: Japanese and Europeans of Hawaii. Hum. Biol. 46:621–31.
31. Keatinge, W. R. 1969. Survival in cold water: the physiology and treatment of immersion hypothermia and of drowning. Oxford: Blackwell Scientific Publications.
32. Daniels, F., and P. T. Baker. 1961. Relationship between body fat and shivering in air at 15°. C. J. Appl. Physiol. 16:421–25.
33. Buskirk, E. R., R. H. Thompson, and G. D. Whedon. 1963. Metabolic response to cold air in men and women in relation to total body fat content. J. Appl. Physiol. 18:603–12.
34. McArdle, W. D., J. R. Magel, T. J. Gergley, R. J. Spina, and M. M. Toner. 1984. Thermal adjustment to cold-water exposure in resting men and women. J. Appl. Physiol. 56:1565–71.
35. McArdle, W. D., J. R. Magel, R. J. Spina, T. J. Gergley, and M. M. Toner. 1984. Thermal adjustment to cold-water exposure in exercising men and women. J. Appl. Physiol. 56:1572–77.
36. Bittel, J. H. M., C. Nonotte-Varly, G. H. Livecchi-Gonnot, G. L. M. J. Savourey, and A. M. Hanniquet. 1988. Physical fitness and thermoregulatory reactions in a cold environment in men. J. Appl. Physiol. 65:1984–89.
37. Bruck, K. 1978. Heat production and temperature regulation. In U. Stave, ed., Perinatal physiology. New York, NY: Plenum Publishing.
38. Spurr, G. B., B. K. Hutt, and S. M. Horratt. 1955. The effects of age on finger temperature responses to local cooling. Am. Heart J. 50:551–55.
39. Wagner, J. A., and S. M. Horvath. 1985. Influences of age and gender on human thermoregulatory responses to cold exposures. J. Appl. Physiol. 58:180–86.

CHAPTER 5

# Acclimatization to Cold Environments: Native and Nonnative Populations

---

**Australian Aborigines**
- Environment
- Physiological Adaptation
- Cultural Adaptation

**Kalahari Bushmen**
- Environment
- Physiological Adaptation
- Cultural Adaptation

**Alacaluf Indians**
- Environment
- Physiological Adaptation

**Norwegian Lapps**
- Environment
- Physiological Adaptation

**Eskimos**
- Environment
- Technological Adaptation
- Physiological Adaptation

**Athapascan Indians**
- Environment

Technological Adaptation
Physiological Adaptation

**Algonkians**

Environment
Technological Adaptation
Physiological Adaptation
Behavioral Responses

**Peruvian Quechuas**

Environment
Observation in the Natural State
Technological Adaptation
Observation Under Laboratory Conditions

**Ama Women Divers From Korea**

Acclimatization to Cold Water Immersion
Deacclimatization to Cold Water Immersion

**Whites**

Canadian Soldiers Acclimatized to Northern Manitoba
Acclimatization to Antarctica and North Pole
Gaspé Fishermen

**Overview**

Investigators concerned with human acclimatization and adaptation to cold have centered their investigations on the study of indigenous populations who live and work in the cold. In general, the methodology employed in these studies can be classified into three major types: (1) the night-long cold-bag technique that has been used to study the thermoregulatory characteristics of Australian aborigines, Kalahari bushmen, Alacaluf Indians, Norwegian Lapps, Eskimos, Athapascan Indians, and Peruvian Quechuas; (2) the short-period laboratory whole-body-cooling technique that has been used to evaluate the thermoregulatory characteristics of Eskimos, Asiatic divers, Peruvian Quechuas, and European whites; and (3) the short-period laboratory extremity-cooling technique that complements the whole-body-cooling technique. These approaches, although their results are not strictly comparable, provide valuable information about human adaptation.

In this chapter the thermoregulatory responses to cold of Australian aborigines, Kalahari bushmen, Alacaluf Indians, Norwegian Lapps, Eskimos, Algonkians, Athapascan Indians, and Peruvian Quechuas will be discussed separately. The aim is not to determine populational differences in cold adaptation, but to ascertain the mechanisms that enable a given population to overcome cold stress. Most studies of adaptation to cold have been concerned with measurements of metabolic rates and rectal and skin temperatures; little attention has been given to technological adaptations. The exceptions are those studies conducted among Eskimos and Peruvian Quechua Indians, both of which have been extensively studied for technological and physiological adaptation to cold. For this reason, the Eskimos and, because of my research experience, especially the Quechuas, will be discussed in more detail than the other populations. Furthermore, the climatic conditions in which a given population lives will always be presented before summarizing their adaptations to cold.

Because of complex environmental conditions and the role of cultural and technological responses with which populations respond, it is difficult to determine the extent to which cold stress can modify human thermoregulatory characteristics. To do so one needs to study individuals or populations that are periodically and voluntarily exposed to cold stress. For this reason, this chapter also discusses the thermoregulatory characteristics of Asiatic divers who are periodically exposed to cold stress, and of European whites who, because of their economic activities or for investigative purposes, were exposed to cold stress.

## AUSTRALIAN ABORIGINES

### Environment

The aborigines of central Australia are exposed during the day to heat stress and during the night to moderate cold stress, as a result of a great diurnal-nocturnal variation in temperature. In the summer months of January through March the mean daily maximum temperature is 37°C (98.6° F), and the minimum is 22°C (71.6°F). In the winter months, such as July, the mean daily maximum and minimum temperatures are 21°C and 4°C (69.8° and 39.2°F) (1). Given the fact that Australian aborigines do not wear clothing and have no adequate housing during either the winter or summer nights, they are exposed to moderate cold stress.

Although by gathering wild foods and hunting small game (kangaroos, opossum, lizards, etc.) the aborigines of central Australia are able to attain an adequate dietary intake (2), they are subjected to frequent seasonal food shortages.

### Physiological Adaptation

The thermoregulatory characteristics of the aborigines from central and northern Australia have been studied both in winter and summer (3, 4). The test consisted of the subjects sleeping overnight (8 hours) in an air temperature ranging from 3° to 5°C (37.4° to 41°F), with insulation ranging from 2.9 to 3.4 clo units (1 clo = 38 cal/m$^2$/hr).

As shown in figure 5.1, the aborigines compared to white control subjects studied under the same conditions tolerated a greater lowering of the skin and rectal temperatures, resulting in a 30% reduction in heat conductance from core to shell. Hence, metabolic heat production was also lower. The aborigines appear to oppose cold stress by both increasing insulation of the body shell through vasoconstriction and by tolerating moderate hypothermia without metabolic compensation. The same pattern of body cooling without metabolic compensation during moderate cold exposure was found among seven aborigines studied in the summer (4). Furthermore, the aborigines living in the northern, tropical, less cold-stressed region of Australia exhibited a metabolic and thermoregulatory response that was intermediate between the aborigines of central Australia and the control whites (1).

### Cultural Adaptation

The Australian aborigines do not wear clothing except for a genital covering. During the night they do not shield themselves from the cold air, and they

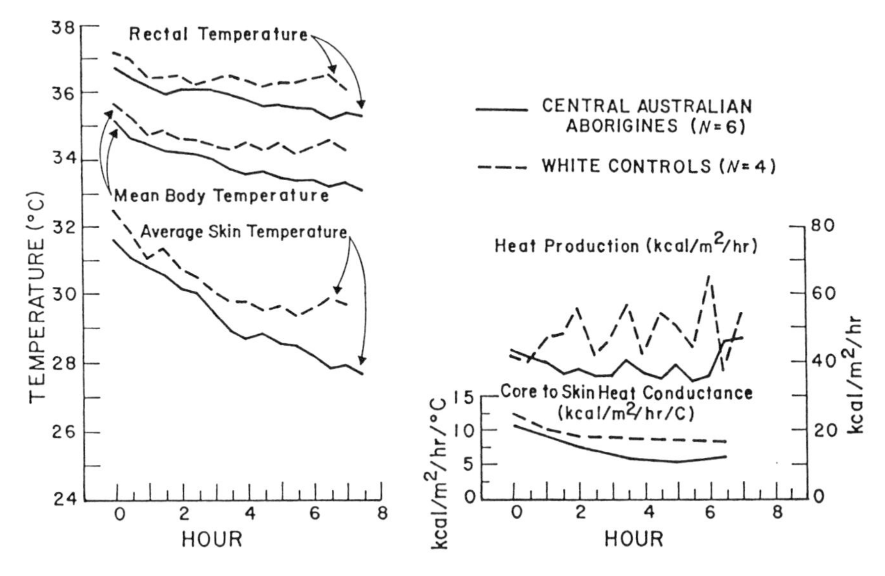


**Fig. 5.1.** Thermal and metabolic responses of six central Australian aborigines and four control whites during a night of moderate cold exposure in winter. For the Australian aborigines, adaptation to the moderate night cold temperatures (about 3°C) involves a decrease in skin temperature and metabolic rate. (Modified from P. F. Scholander, H. T. Hammel, J. S. Hart, D. H. LeMessurier, and J. Steen. 1958. J. Appl. Physiol. 13:211–18.)

sleep on the leeward side of a windbreak hastily made from brush and between small fires. Field studies indicate that the glowing embers of the fire, if attended frequently, would provide sufficient radiant heat to the body through the exposed side to balance heat loss, but the Australian aborigines do not do so. Thus, the degree of cold exposure in the sleeping microenvironment is below the thermoneutral temperature. Despite this cold stress the Australian aborigines were able to sleep comfortably without shivering, whereas the white controls studied under the same conditions shivered continuously and were unable to sleep.

## KALAHARI BUSHMEN

### Environment

The Bushmen of the Kalahari Desert, like the Australian aborigines, are exposed during the night to moderate chronic cold stress. The average daily minimum temperatures for the month of July in the Kalahari Desert range from 3° to 55°C, the same as found in central Australia. Also like the aborigines, the Bushmen do not wear clothing except for a genital covering. Most field reports agree that the Bushmen have an abundant variety of flora and fauna and, given their excellent skills in gathering and hunting, are able to attain a good dietary intake (5). Indeed, the evaluations of nutritional status indicate that the Bushmen maintain good health and nutrition. However, as shown by the changes in body weight and subcutaneous fat (6), the Bushmen are periodically subjected to restrictions in their dietary intakes, which at times reach levels of chronic undernutrition (6).

## Physiological Adaptation

A sample of Kalahari Bushmen, using the same night-long cold exposure tests as those used with the Australian aborigines, was evaluated by Hammel et al. (7). As shown in figure 5.2, after the second hour, the Bushmen had lower rectal temperatures and lower metabolic rates than the white controls, but skin temperatures and heat conductivity were higher. The increased heat conductivity of the Bushmen has been attributed to their small amount of subcutaneous fat (1). Although rectal temperatures and metabolic rates were lower than those of the whites, the Bushmen, like the Australian aborigines, experienced a higher grade of sleep throughout the entire night of the test. However, other investigations indicate that Bushmen subjects, when exposed to air temperatures of 5°, 10°, 15°, and 20°C for 2 hours, each maintained a lower skin temperature with marked increase in metabolic rate (8, 9). In other words, according to this study the thermoregulatory response of the Bushmen was different from that of the aborigines in that their lower skin temperatures were caused by a smaller skinfold thickness, which resulted in an enhanced rate of heat loss, in turn eliciting an increase in metabolic rates. It must be noted that this cold test is not comparable to that given the Australian aborigines, since it was given in four sequential 2-hour night exposures rather than 8-hour night-long exposures. In these studies the Bushmen were awakened and moved from their shelters near the fires to the test site and there exposed for 1½ to 2 hours on a stretcher, naked except for the genital covers. It is quite possible that these conditions could have caused an increase in the subjects' anxiety levels, which may have increased their metabolic rates.

## Cultural Adaptation

Field investigations indicate that through efficient use of fires and skin cloaks during cold nights the Bushmen have been able to create a microclimate around their bodies that is close to the thermoneutral temperature of 25°C (10). These investigations indicate that Bushmen employ the following techniques. First, in the early part of the evening the Bushmen build large fires and sit around them, huddled in their skin cloaks. Second, they sleep in groups of three or four, in families or in single-sex groups. Third, when they

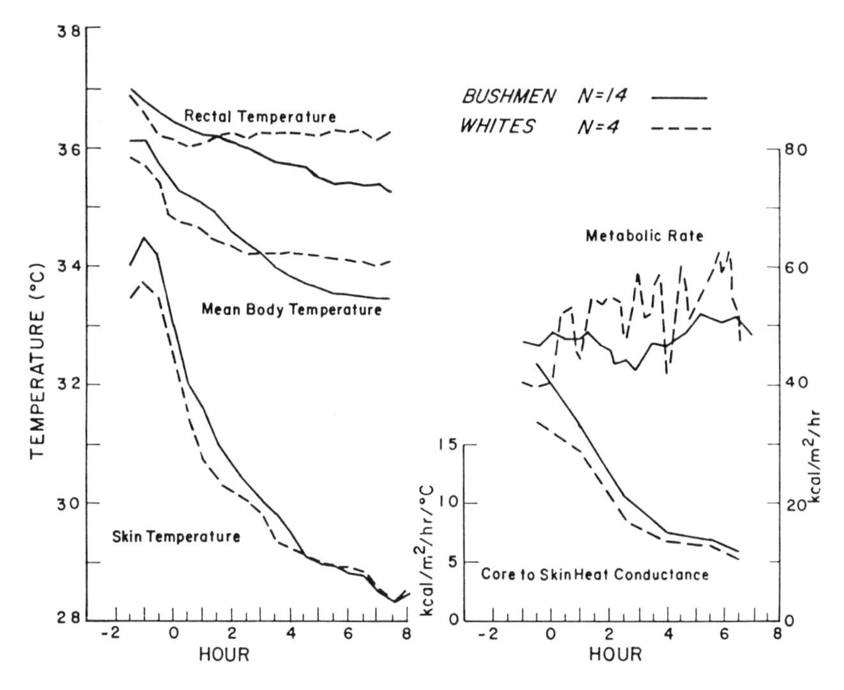


**Fig. 5.2.** Thermal and metabolic responses of fourteen Kalahari Bushmen and four control whites during a night of moderate cold exposure in winter. Adaptation to moderate night cold temperatures (3° to 8° C) for Bushmen involves greater body cooling than for whites, but unlike the Australian aborigines their excessive heat loss (conductance) is compensated for by increased metabolic rate. (Modified from H. T. Hammel. 1964. Terrestrial animals in cold: Recent studies of primitive man. In D. B. Dill, E. F. Adolph, and C. G. Wilber, eds., Handbook of physiology, vol. 4. Adaptation to the environment. Washington, DC: American Physiological Society.)

retire, they lie with their feet to the fire and have their blankets (skin cloaks) tucked in around their bodies and over their heads. When the fires die down on cold nights, they wake up and stoke the fires. Fifth, their huts of grass and boughs are used as a windbreak, and for this purpose the huts are placed in a half-circle.

## ALACALUF INDIANS

### Environment

The Alacaluf Indians of Tierra del Fuego in southern Chile numbered only fifty as of 1960. These Indians have little protection from the wind and rain other than crude huts, and they are poorly clothed. As shown by temperature recordings at Puerto Eden on Wellington Island, the Indians are exposed to moderate, chronic cold stress. The mean maximum temperature for the summer month of January is 13.1°C (55.6°F), and the mean minimum temperature is 7°C (44.6°F). In the winter month of July, the mean maximum and minimum temperatures are 5.3°C (41.5°F) and – 0.3°C (31.5°F), respectively (1). The area of Tierra del Fuego is also characterized by high precipitation and high winds.

### Physiological Adaptation

Using the night-long cold-exposure tests, like those employed in the aborigine and Bushmen studies, a sample of Alacaluf Indians was studied (11). As shown in figure 5.3 the thermoregulatory responses of the Alacaluf Indians are similar to those of whites studied under equal conditions of humidity and temperature. However, during the entire test the foot temperatures of the Indians were maintained at about 2° to 3°C higher than those of the whites. Furthermore, evaluations of foot temperatures after immersion in water at 5°C for 30 minutes showed that the Alacaluf Indians sustained the stress without any signs of pain from cold, whereas the whites tested under the same conditions were in great agony (11).

In summary, the Alacaluf Indians respond to the humid cold stress of Tierra del Fuego with the maintenance of high metabolic rates and high extremity temperatures. Furthermore, they appear to exhibit the ability to tolerate cold stress without pain.

## NORWEGIAN LAPPS

### Environment

The Lapp shepherds from Norway, because of their nomadic life-style, have not developed formal housing and thus are continuously exposed to cold stress. They are reindeer hunters and herders who spend most of the daylight hours outdoors and sleep in poorly heated tents at night. During the night the outdoor temperature usually falls from – 25°C (– 13°F) to – 30°C (– 22°F) (12). The inside-tent temperature, initially at a comfortable level, by morning is the same as outdoors because of the dying fire. The underblanket consists of single reindeer hide, and the top blanket is made of skins of long-haired sheep. In contrast to the nomadic shepherds, there are also village Lapps who are farmers living in formal modern houses that provide adequate protection against the cold.

### Physiological Adaptation

Figure 5.4 shows that the metabolic responses and rectal temperatures of the Lapp villagers and white controls during night-long cold exposure were comparable (13). In contrast, the shepherds experienced a substantially smaller

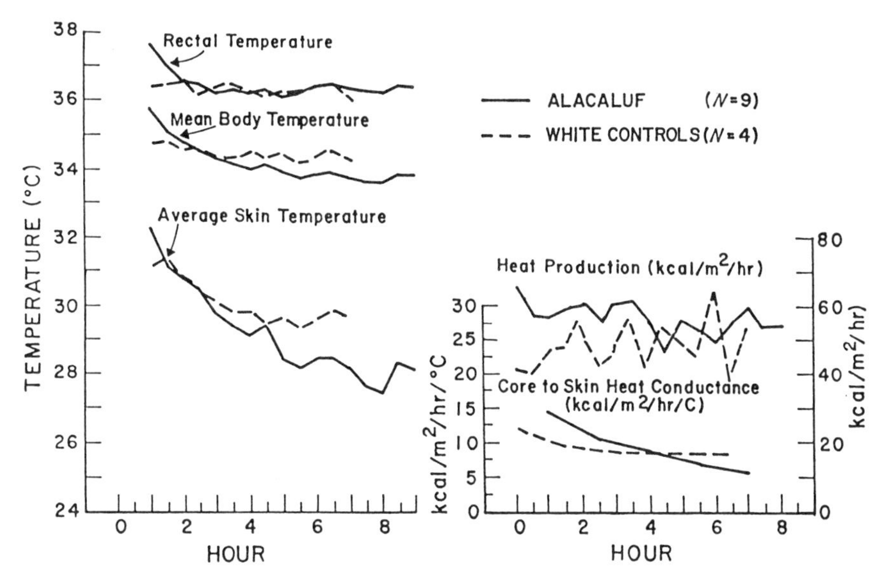


**Fig. 5.3.** Thermal and metabolic responses of nine Alacaluf Indians during a night of moderate cold exposure and four control whites studied in central Australia under similar conditions in winter. The Alacaluf respond to night cold temperatures (3°C) in a manner similar to control whites, but they begin with high metabolic rates, which are maintained throughout night. (Modified from H. T. Hammel. 1964. Terrestrial animals in cold: Recent studies of primitive man. In D. B. Dill, E. F. Adolph, and C. G. Wilber, eds., Handbook of physiology, vol. 4. Adaptation to the environment. Washington, DC: American Physiological Society.)

metabolic response accompanied by a 1°C greater drop in rectal temperature (down to 35.8°C by the fifth hour of testing). However, the foot temperatures of the shepherds were maintained at a higher level than those of the villagers or controls. The shepherds also slept more comfortably than the other two groups. Furthermore, Norwegian and Finnish Lapps exhibited an earlier onset of cold-induced vasodilation and less pain than the white controls when immersing their hands in ice water for 15 minutes (14, 15).

In summary, the Norwegian Lapp shepherds respond to cold stress with a lower metabolic rate increase and higher peripheral temperatures than white controls.

## ESKIMOS

### Environment

The world's population of Eskimos, numbering 50,000 to 60,000, occupies the northwestern coast of North America and across the Bering Strait into Asia. The Eskimo culture is most characteristically found around the shores of Baffin Island and the northern parts of Hudson Bay, in latitudes 69° to

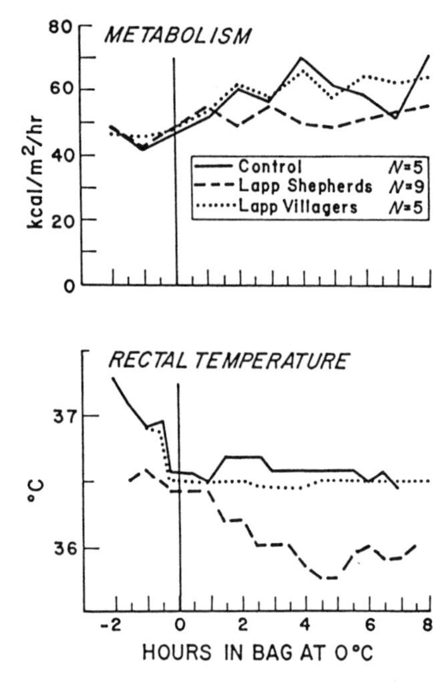


**Fig. 5.4.** Thermal and metabolic responses of nine nomadic Lapp shepherds, five Lapp villagers, and five control whites during night of moderate cold exposure in winter. The Lapp shepherds during the cold night (0°C) sleep had a lower metabolic rate and rectal temperature than control whites and Lapp villagers. (Modified from K. L. Andersen, Y. Loyning, J. D. Nelms, O. Wilson, R. H. Fox, and A. Bolstad. 1960. J. Appl. Physiol. 15:649–53.)

70° N. In this central area, as throughout the arctic zone, the principal factors that distinguish the seasons are length of daylight and temperature. In the Arctic archipelago the average winter temperature ranges from −46°C (−50°F) to −31°C (−35°F); in the summer the temperature rarely goes above 46°F. Equally important is the drastic change in sunlight hours. During the brief arctic summer the sun shines almost continually for 2 months; in the fall there is a maximum of 12 hours of sunshine per day, followed by a complete absence of sun in midwinter. With the coming of spring the sun again shines for increasing periods each day until the annual cycle is complete. It is quite evident that such an inhospitable environment requires thorough technological and physiological adaptation.

## Technological Adaptation

**Housing.** The native Eskimo housing, whether permanent, temporary, or "igloo" type is well-insulated. The walls made of whale rib rafters are covered with a double layer of seal skin alternated with moss. This design permits considerable air to be trapped that, in turn, acts as further insulation. In addition, through efficient use of underground tunnels and by placing the source of heat, usually an oil, blubber, or coal lamp, at a lower level than the main floor, the Eskimos have produced a useful system of heat exchange whereby cold air is warmed before it reaches the area where the people live (16, 18). Thus, despite subzero outside temperatures the indoor night and day temperatures average between 10° and 21°C for the coastal Eskimos and around 0°C for the Baffin Island Eskimos (12–17).

**Clothing.** In most areas the Eskimo clothing is made of caribou. This is the preferred skin because of its high-quality insulation, its light weight, and its suppleness compared to seal skin. When caribou is not available, as among the polar Eskimos, polar bear skins are the replacement (19). It has been determined that Eskimo clothing, which consists of caribou fur in thicknesses of 1½ to 3 inches, provides insulation equivalent to 7 to 12 clo units (20). That is, the insulative efficiency of Eskimo clothing ranges from 266 to 456 cal/m$^2$/hr.

Although it is evident that this type of clothing has succeeded in creating a comfortable microclimate, it has by no means removed cold stress, especially for the extremities. Although Eskimos wear snowshoes and short skin mittens at times, during their daily activities such as fishing their hands and feet are continuously subject to cold stress. As indicated by ethnographic accounts (18, 19, 21), during hunting and fishing the Eskimos continually dip their hands in cold water and expose their feet to severe cold; occasionally they experience whole-body chilling while waiting motionless at a fishing hole. In fact, some hunters have been known to sit over the breathing holes of seals for up to 72 hours at a time in subzero temperatures (22). Thus, despite their efficient technological adaptations, the Eskimos are not always "tropical men in arctic clothing" since the subsistence activities of the coastal and inland Eskimos induce conditions of prolonged extremity chilling and, at times, body chilling. In response to this cold stress, the Eskimos have developed specialized thermoregulatory characteristics.

## Physiological Adaptation

**Metabolic Rate.** Most studies agree that the Eskimos' metabolic rate during warm and cold conditions is between 13% and 45% higher than that of white controls or predicted from standards (fig. 5.5) (23–35). However, there is disagreement as to the source of increased heat production. Investigations of thyroid function have not shown any major relationship between thyroid levels and increased metabolic rates in Eskimos. Nutritional studies have indicated, on the other hand, that the higher metabolic rates may be related in part to the Eskimos' dietary habits. Earlier investigations in East Greenland described the Eskimos as "the most exquisitely carnivorous people on earth." (23) The Eskimos' daily diet was reported to contain an excessive amount of

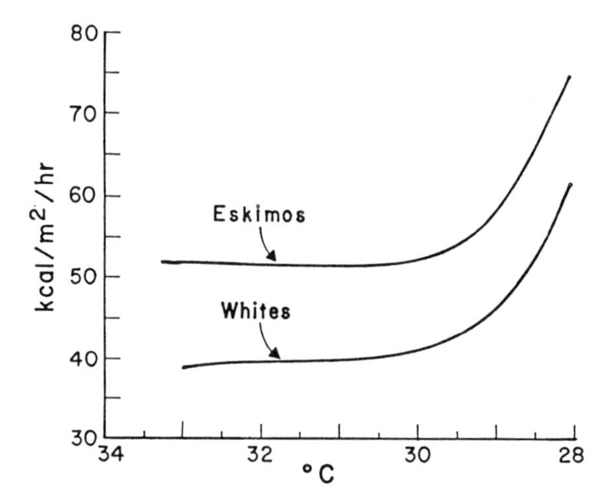


**Fig. 5.5.** Relationship between skin temperature and metabolic rate during cold exposure of Eskimos and whites. At given temperature Eskimos respond with greater metabolic rate. (Modified from T. Adams and B. G. Covino. 1958. J. Appl. Physiol. 12:9–12.)

animal protein (280 g), a great deal of fat (135 g), and a small amount of carbohydrates (54 g), of which more than one half was glycogen derived from meat. Controlled dietary studies and metabolic measurements of Eskimos from four localities in northern Alaska during the summer and winter concluded that the dynamic action of a protein diet was the main source of high metabolism (17). On the other hand, other studies indicate that the Eskimos' elevated metabolism during sleep was not correlated with either protein or fat energy fractions in the diet (36). Furthermore, the more cold-stressed inland Eskimos have been found to have a greater metabolic rate than the less cold-stressed coastal Eskimos, who were studied under similar dietary conditions (37, 38). Thus it would appear that the Eskimos' elevated metabolism is not caused only by high protein intake.

**Peripheral Temperature.** All studies indicate that when either the feet or hands of Eskimos are cooled in water or air at any temperature, peripheral temperatures and degree of spontaneous fluctuations in blood flow are greater among Eskimos than among white controls (fig. 5.6) (27, 30, 39, 40). This response results in improved performance and reduced cold sensation for the Eskimo, which is reflected in the Eskimos' remarkable ability to work with bare hands and to continue to perform fine movements in the winter cold (27). Furthermore, when the subjects' hands were exposed to a 10°C water bath for 2 hours, the Eskimos experienced only a mild cold sensation during immersion but did not exhibit signs of pain, whereas the white controls experienced first the sensation of severe cold in the immersed hand and then a deep aching pain. Similarly, all the Eskimos sustained exposure of fingers to water baths at 0°C, but 75% of the white controls could not finish the test either because of frostbite or intense pain (41). The Eskimos' higher

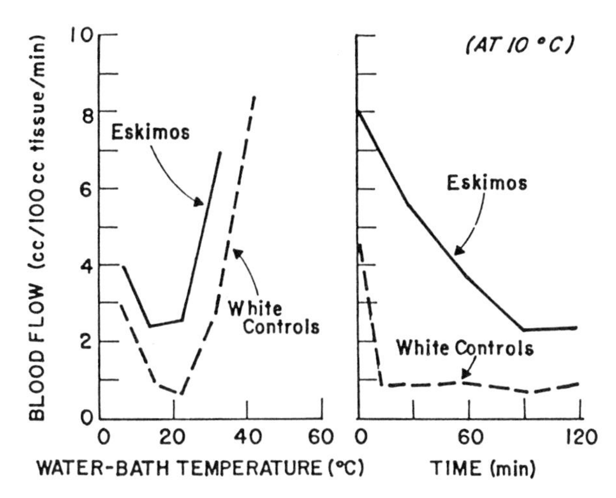


**Fig. 5.6.** Hand blood flow during different water-bath temperatures and immersion of hand in water bath at 10° C. Eskimos respond to cold temperatures with greater blood flow than white controls. (Modified from G. M. Brown and J. Page. 1953. J. Appl. Physiol. 5:221–27.)

peripheral temperature is also associated with greater amounts of red blood cells, higher plasma volumes, and more globulins (28, 42, 43). Since the increase in red blood cells requires a greater vascular bed, it is very plausible that the Eskimos' high peripheral temperature is maintained through an enhanced vascularization.

In summary, the climatic conditions in which Eskimos live provide severe acute and chronic cold stress. The Eskimos' material culture, in the form of housing and clothing, represents one of the most important adaptations to cold stress. However, because of their subsistence economy, the Eskimos are still exposed to prolonged extremity and, at times, whole-body chilling. The Eskimo diet is based on a high fat and protein intake that provides calorigenic effects and, in view of their natural resources, is an adaptive response to cold stress by its contribution to the maintenance of high metabolic rates. All evidence confirms the Eskimos' increased metabolic rates, high peripheral temperatures, and remarkable tolerance to cold exposure of the extremities. As shown in figure 5.7, the minimum finger temperature of Eskimo children and women when exposed to air temperatures between $-3°$ and $-7°C$, despite their smaller hand volumes, are as high or higher than those of white men accustomed to outdoor cold (44). Thus the Eskimos' ability to maintain high extremity temperatures and their ability to tolerate cold appear to be acquired during growth. That is, the thermoregulatory characteristics of Eskimos reflect the influence of developmental adaptation (or developmental acclimatization). The thermoregulatory characteristics of Eskimos from the village of Igloolik were recently studied (45). The population of this community is in a period of rapid acculturation and is adopting some of the eating and cultural habits of whites. Furthermore, the women and children who

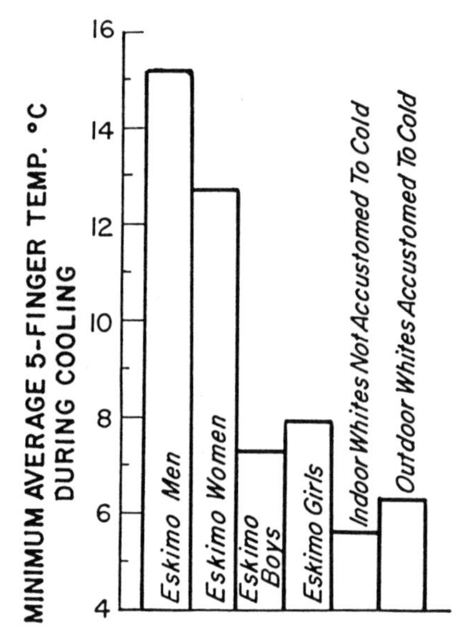


**Fig. 5.7.** Peripheral temperature of left middle finger of Eskimo men, boys, outdoor whites, and indoor whites during cold air exposure (−3° to −7° C). Eskimo men maintain higher skin temperatures than both outdoor and indoor whites, and Eskimo children, in spite of smaller hand volumes, maintain temperatures equal to adult indoor whites. (Modified from L. K. Miller and L. Irving. 1962. J. Appl. Physiol. 17:449–65.)

attend comfortably heated schools are less cold-exposed than the adult males who still make their living by hunting. As shown in figure 5.8, the Eskimo women and children during immersion of the hand in cold water are as tolerant as the Eskimo adult male hunters and more tolerant than the whites. These findings suggest that the Eskimos' developmental response to cold is probably mediated by genetic factors.

## ATHAPASCAN INDIANS

### Environment

The Athapascan Indians neighbor the Eskimos in northern Canada. The climatic conditions of this area are very similar to those for the Eskimos; however, the cold is not as intense as that encountered in northern Alaska or on Baffin Island. Among Athapascan-speaking people, the Indians from Old Crow have been studied most extensively by physiologists (39, 46, 47). Based on climatological data collected in Fort Yukon in 1958, the mean temperature for January was −28°C, and for September the mean maximum temperature was 21°C. These temperatures are made more severe by high winds, especially in winter. The short summer has both warm and chilly days, but frost is so frequent that vegetables are not grown.

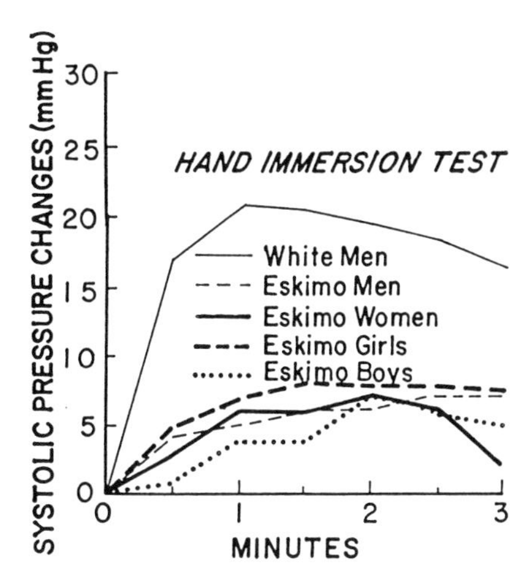


**Fig. 5.8.** Changes in systolic blood pressure in white male subjects, Eskimo men and women, and Eskimo boys and girls during immersion of hand in water at 4° C for 3 minutes. Eskimo women and children, despite the fact that they live in comfortably heated houses, tolerate cold stress as well as adult Eskimo hunters and better than whites. (Modified from J. LeBlanc. 1975. Man in the cold. Springfield, IL: Charles C. Thomas.)

The subsistence economy of the Athapascan people is based on hunting, trapping, and fishing. As with the Eskimos, the subsistence activities of the Athapascan people involve severe cold stress for the extremities as well as severe whole-body cooling.

## Technological Adaptation

**Housing.** Aboriginal housing consists of excavated caves or inverted V-shaped houses whose sides and roofs are covered with moss to retain heat. A fire is also maintained in the center of the dwelling. Other housing forms are simple bivouacs and cabins that are not as well insulated as those of the Eskimos.

**Clothing.** The clothing of the Old Crow people consists of a duffed parka (blanket cloth), wool underclothing, flannel shirt, wool or cotton trousers, moosehide mitts, and wool-duffed socks under smoke-tanned moosehide moccasins.

## Physiological Adaptation

**Whole-Body Cooling.** The basal metabolic rates of the Athapascan Indians, when expressed per kilogram of body weight, were about 20% to 14% higher than expected on the basis of white standards (46). Evaluations using the night-long exposure technique indicated that these Indians exhibit the same metabolic responses as white controls (fig. 5.9) (39, 46, 47). However, even

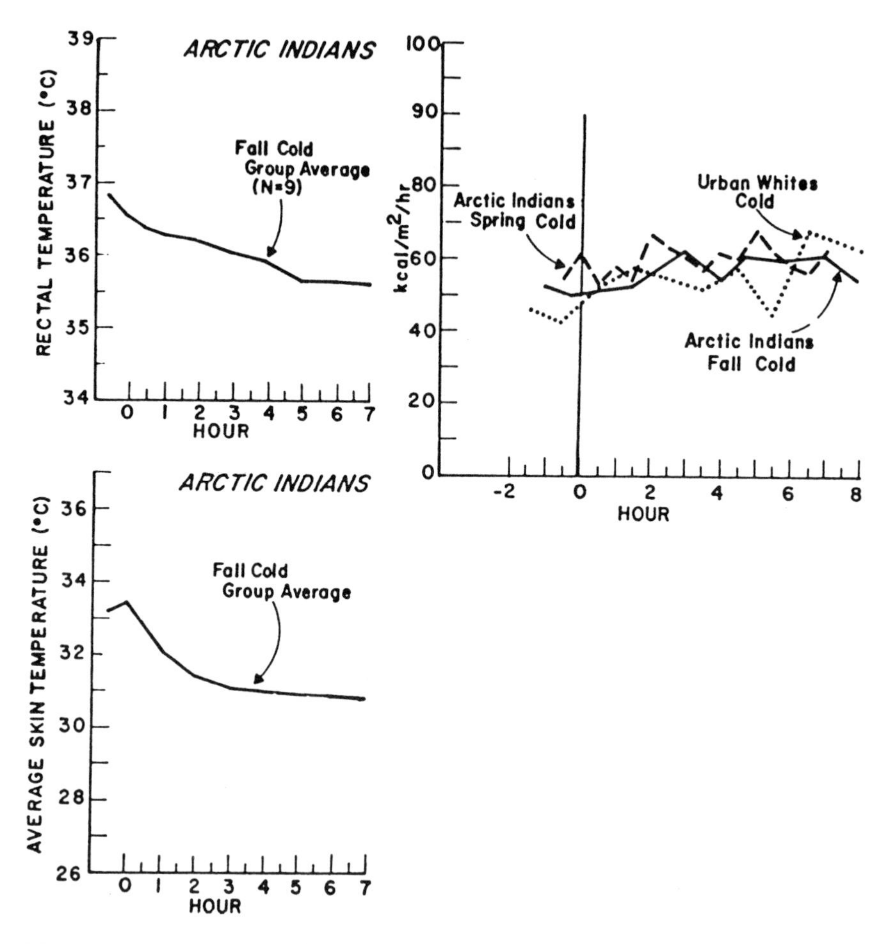


**Fig. 5.9.** Thermal and metabolic responses of Arctic Indians and control whites. Arctic Indians exposed to moderately cold nights have an increased metabolic rate and skin temperature similar to white controls. (Based on data from H. T. Hammel. 1964. Terrestrial animals in cold: Recent studies of primitive man. In D. B. Dill, E. F. Adolph, and C. G. Wilber, eds., Handbook of physiology, vol. 4. Adaptation to the environment. Washington, DC: American Physiological Society.)

though the Athapascan Indians' rectal temperatures declined at a faster rate than those of the white controls, their foot temperatures were always maintained at a higher level. Myographic records indicate that although the Athapascan Indians exhibited the same frequency of shivering bouts as the white controls, they were able to sleep better and for longer periods than whites tested under the same conditions (46, 47).

**Extremity Cooling.** Temperature evaluations of hands exposed to either cold water (5°C) or ice water indicate that the Athapascan Indians had warmer hand temperatures than white controls (39, 46). Furthermore, the Indians had a more rapid rewarming and suffered less pain than the whites.

In summary, the Athapascan Indians differ from whites in their ability to distribute heat toward the extremities and tolerate cold stress. However, these differences are less well defined than those observed between Eskimos and Europeans, suggesting that the adaptation to cold stress of Athapascan Indians and other subarctic populations is based on both physiological mechanisms and behavioral and cultural responses as well (48, 49).

## ALGONKIANS

### Environment

The Algonkian-speaking populations include the Blackfoot, Cree, and Ojibwa, who live in northern Ontario. Studies of the Algonkian adaptation to cold stress have focused on the Ojibwa. As documented by Winterhalder (50), the Ojibwa live in an environment characterized by low temperatures, high snowfall, and abundant forest.

As inferred from climatic data of Big Trout Lake in northern Ontario, the minimum daily temperatures rise from a monthly mean of −29°C (−20°F) in January to 11°C (52°F) in July, and the maximum daily temperature ranges from a low of −19°C (−2°F) in January to 21°C (70°F) in July. The total annual precipitation in northern Ontario is about 60 cm, and the major precipitation occurs during the summer months. Snow accumulation begins in October and continues to a peak in February, declines slightly in March and falls quickly in April. Contrary to popular views, the Algonkians and, specifically, the Ojibwa live in an environment characterized by abundant forest. The boreal forest of northern Ontario contains a mosaic of vegetation habitats rich in its variety of trees and shrubs. In terms of fauna, the Ojibwa have access to moose, hare, caribou, beaver, muskrat, ruffled grouse, spruce grouse, and fish.

In summary, the Ojibwa of northern Ontario live in an environment conducive to chronic cold exposure. They have been living in this extreme climate for at least 6,000 to 7,000 years, and therefore have developed a complex set of adaptive responses that enable them to survive in an otherwise harsh environment.

### Technological Adaptation

**Shelter and Clothing**. In the past, the standard winter field dwelling consisted of a wedge-shaped or conical tepee with walls composed of vertically laid sticks chinked against the cold and wind with moss. Snow was banked around the lower perimeter for insulation. The interior was kept warm by a central fire. At present, trapping camps have been replaced by permanent cabins or wall tents, and small "air-tight" wood stoves are used. But these cabins are still banked with snow and insulated with sheets of pasteboard from boxes or with thick clear plastic.

During hunting trips, in the absence of tents and when caught unprepared, the Algonkians use the moose skin as a makeshift sleeping robe. To sleep they find shelter in a snow bank and cover themselves with evergreen foliage and snow. According to Marano (51), under these conditions the hunter apparently can sleep well enough.

Although the Algonkians have abandoned their nomadic hunting pattern they do still maintain the same practices of thermoregulation with regard to young children. Newborns and young children, unlike Quechua children, do not sleep close to the parents, but on top of a raised platform or in a hammock constructed of rope, in their own well-insulated sleeping bags. The cradleboard is still in use and is a valuable aid in protecting the child from exposure to cold. A child in a cradleboard is warm and safe, and can be carried in such a way as to leave the hands of the parent free for other tasks.

As indicated by measurements of 24-hour room temperature during January and early February at Weagamow (52), the insulative efficiency of the Algonkian cabin is not good. Through the night and day the temperature is below the critical neutral zone. The highest mean temperature was 23°C (73°F) at 8 P.M., and the 6 A.M. low was 1.7°C (35°F). The lowest temperature recorded was at 8 A.M. at −5°C (22°F) (see fig. 5.10). In other words, it is quite evident that the technological adaptation has decreased but not removed the presence of cold stress, and without special thermal protection both infants and adults would suffer hypothermia.

The standard clothing in the past included several layers of clothes made of caribou fur. Face masks made of fox and rabbit skin were used. Today, commercially available parkas and gloves are worn, but fox and rabbit skins are still used to protect and warm the hands after cold exposure. Rabbit skins stuffed down trousers around the groin are used for rapidly rewarming chilled hands. Moccasins made of moose hide are still widely used. Typically the body of the Algonkian moccasin, although it rises only to ankle level, is extended upward several inches by an added open-fronted piece which wraps around, overlapping right and left, and is lashed in place. This design permits adjustment of the moccasins' insulation to variable temperatures and facilitates putting them on and taking them off, as well as permitting them to dry rapidly. Today moccasins are still worn for general use, especially in winter travel where long distances are to be covered. Due to the constant low temperatures, waterproofing is not needed. Compared to commercial boots, these moccasins are easier to use with native snowshoes and bindings because they are lighter and more flexible.

## Physiological Adaptation

Steegmann (48) and Hurlich and Steegmann (49) studied the thermoregulatory responses of the Algonkians of Weagamow, using samples of northern and

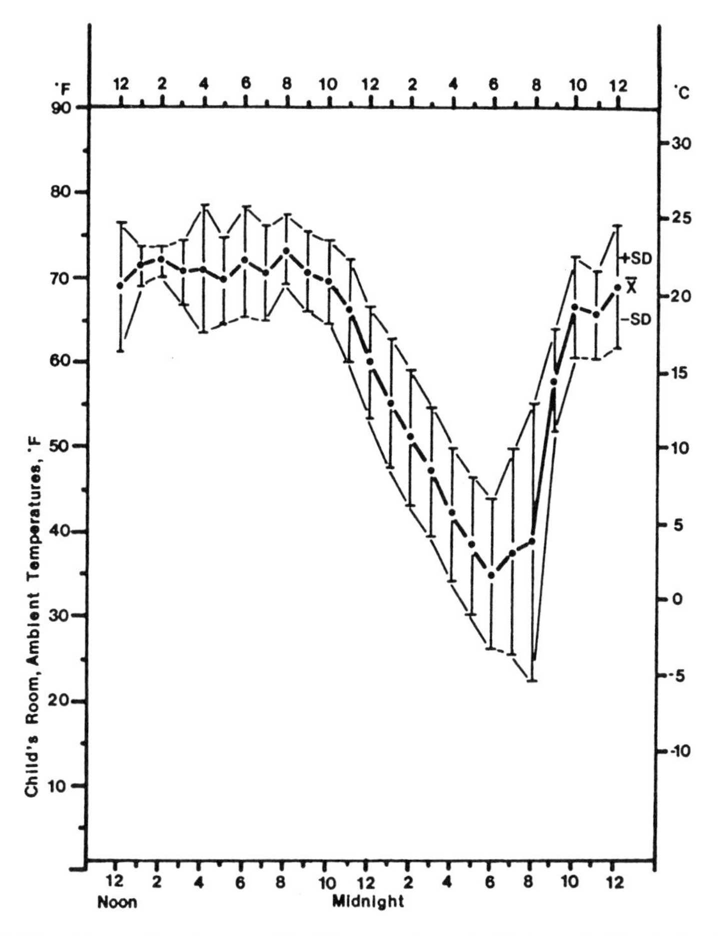


**Fig. 5.10.** Twenty-four hour ambient temperatures in 10 households during the winter of 1974 in Weagamow, Ontario. The temperatures were measured near the sleeping places of the youngest children. Note that by 5 o'clock in the morning, the temperature has gone down below freezing. (Adapted from A. T. Steegmann, Jr., M. G. Hurlich, and B. Winterhalder. 1983. Coping with cold and other challenges of the Boreal Forest: an overview. In A. T. Steegmann, Jr., ed., Boreal Forest adaptations: The Northern Algonkians. New York: Plenum Press.)

southern Chinese for comparison (53). Measurements of finger temperatures were taken during 30 minute immersion of the hand in 5°C water. Results of these studies summarized by Steegman et al. (52) indicate that the Algonkians: (1) maintained higher finger temperatures than both Chinese samples; (2) exhibited faster rewarming; and (3) experienced less cold pain than the Chinese samples.

## Behavioral Responses

Along with the physiological responses, the Algonkians have developed a complex set of behavioral responses that protect them from frostbite and permit them to work under otherwise severe cold conditions. These include:

(1) **Use of Tight, Well-Insulated Clothing and Replenishment of Body Fluid Lost.** Contrary to the practice of the Eskimos (Inuits), the Algonkians wear their clothing tightly and do not adjust it for ventilation while doing heavy outdoor work. As a result, the Algonkians maintain a warm core temperature even though they may sweat. To compensate for the sweat loss the Algonkians rest frequently while doing outdoor work and drink tea that replenishes the loss of body fluid.

(2) **Fires.** Because of the ready availability of firewood, Algonkians build fires at the ''slightest pretext.'' Once a fire is started, no matter how low the temperature, they strip and use the fire to dry their sweat-dampened clothes.

(3) **Pacing of Physical Work.** Daily walks were paced in such a way as to avoid excessive sweating.

(4) **Protection of Hands and Feet Through the Use of Mittens.** Due to the demands of the Algonkians' subsistence activities, their hands and feet are frequently exposed to cold water and air for sustained periods. To avoid frostbite the Algonkians work with bare hands only when necessary and unavoidable. To minimize the exposure of the hands to cold, manual activities are assisted by the use of teeth as a ''third hand.'' Steegman et al. (52) report that when fish are stuck in gill nets, older men will sometimes grip the fish snout with their incisors and rapidly work the tangled fish out of the net with both hands. They rapidly rewarm their hands by stuffing them into highly insulated mittens and rabbit skins. In addition, to avoid frostbite they dip their hands in cold water (which by nature is above the freezing point of flesh).

(5) **Posture.** When not working the Algonkians always sit either on their heels or on insulated surfaces. At all costs they avoid sitting on the bare ground, and if possible they sit on moose hides or evergreen foliage.

In summary, the climatic conditions in which the Algonkians live are as severe as those of the Eskimo. The Algonkians' material culture, as indicated by the low indoor temperatures of their shelters, is not as efficient as that of the Eskimos. On the other hand, the Algonkians have developed a unique and complex set of behaviors that enables them to work under otherwise intolerably cold conditions. The use of heavy, tightly worn insulation guarantees the maintenance of core temperature, and the frequent rests and use of fire make it possible to maintain both thermal and fluid homeostasis. The Algonkian physiological responses, in terms of maintenance of warm peripheral temperatures, fit the pattern of chronic cold-stressed populations. The Algonkians may be considered another example of developmental acclimatization to cold. As indicated by Marano (51), the everyday activities of Algonkian children provide stimulus for the development of adaptive physiological responses to cold stress.

## PERUVIAN QUECHUAS

The highland Quechua populations from the Peruvian Andes and other mountain areas of South America are exposed to a variety of stresses including hypoxia, cold, low humidity, and high levels of solar radiation. These climatic stresses interact with other stresses such as limited energy and food resources and disease. The pattern of adaptation to each of these stresses depends on the interaction of environmental, cultural, and human biological parameters. For this reason, interpretation of the cold adaptation of the highland populations requires a synergic interpretation of all stresses, since any adaptive pattern developed by a population represents a compromise to often conflicting and antagonistic stresses.

This discussion is based on the multidisciplinary research centered in the altiplano population of the district of Nuñoa situated at a mean altitude of 4150 m in southern Peru (54).

### Environment

The economic focus of the highland region is herding sheep, llamas, and alpacas, the major domestic animals of the altiplano. However, the subsistence pattern of much of the indigenous population between 3000 and 4000 m is based on a mixed economy of corn, potatoes, barley, wheat, and native chnopodium (canihua and quinua) cultivation and the herding of sheep and llamas. On the other hand, at elevations above 4000 m, such as those of the altiplano region, agriculture is not practiced because of severe cold and frost, and the subsistence pattern is primarily pastoral. The sheep, llama, and alpaca skins are used for bedding, the sheep and alpaca wool is used for clothing, the dung is used for fuel and fertilizer, and the meat is used for private consumption as well as for trade. The hides and animals are also often traded for cash, although the greater part of the family income is derived from the sale of wool. With this income cereal foods, additional clothing, yarn dyes, coca leaves, alcohol, and other small luxuries can be purchased. All the meat and the wool for basic clothing is provided by herding. Every Indian owns some of these domestic animals, and many own highland ponies as well. Llamas and horses are used as pack animals, whereas sheep and alpacas are raised solely for their wool and meat.

By most standards the altiplano of the Nuñoa region is cold. The mean annual temperature is 8.3°C (47°F), which is well below the thermoneutral zone for humans. The mean monthly temperature ranges from 10°C in the warmest month of November to 5.5°C in the coldest month of June. Daily temperatures throughout the year are lower than those considered comfortable for most populations. Seasonal variation in temperature in the highlands is primarily the result of the monsoon pattern in the Pacific, which creates dry and wet seasons.

The dry season extends from April to September. About 50% of the days are free of cloud cover. As a result, daytime temperatures, especially outdoors, are pleasant with high solar radiation, but the afternoons and nights are very cold because the clear sky increases heat loss through long-wave radiation. Most of the nights have temperatures below freezing, and a low of −8°C has been recorded for this period. The wet season extends from October to March and is characterized by almost daily thunderstorm activity producing heavy local precipitation. The precipitation most often occurs as sleet, rain, hail, or snow and is accompanied by high winds. The daytime cloud cover usually reduces solar radiation and causes a drop in temperature to about 5°C. Snow may accumulate up to 4 cm but is frequently melted by sunshine. Because the heavy cloud cover reduces solar radiation, the daytime outdoor climate is less pleasant than during the dry season, but since sunlight hits at a greater angle at this time of year, the temperatures are actually higher. Furthermore, because the cloud cover reduces heat loss through long-wave radiation, the nights are warmer than in the dry season. Both in the dry and wet seasons the varying temperatures are accompanied by gusty winds that add to the cold stress when temperatures are low. The coolness is accentuated by the lack of internal heating in houses.

## Observation in the Natural State

Evaluations of the thermoregulatory responses of individuals in their natural state were obtained during both the dry and wet seasons (55).

**Cold Stress in the Dry Season.** The greatest cold stress in the dry season is night time (55–57). As shown in figure 5.11 even though the interior house temperatures averaged only 3.6°C (38.5°F), the subjects were not greatly cold stressed. Even though the severity of the ambient cold stress is ameliorated by the fact that the subjects usually sleep clothed and in groups of two to four as shown in figure 5.11, the children are the most cold stressed. The greater cold stress of children is related to the fact that traditionally they do not wear trousers until about the age of 6 years and until then wear poorly insulated clothing. Furthermore, young children assist in the herding of alpacas and llamas, an activity that leads to considerable cold stress especially in the afternoons. Thus, the everyday activities of Quechua children provide stimulus for the development of adaptive physiological responses to cold stress.

**Cold Stress in the Wet Season.** During the wet season the potential for cold stress is rather high because rain, snow, sleet, and hail fall daily and are driven by high winds. Since this season coincides with the time of planting

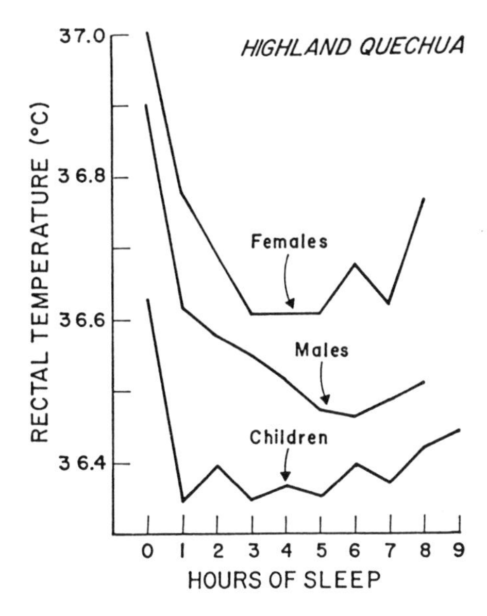


**Fig. 5.11.** Rectal temperature of Quechua females, males, and children while sleeping in their own houses. Rectal temperatures did not decrease to initial low levels even though interior house temperatures were only 3.6° C. The children show a rapid fall in rectal temperature for the first hour and then attain a certain equilibrium for remainder of night. (Modified from J. M. Hanna. 1976. Natural exposure to cold. In P. T. Baker and M. A. Little, eds., Man in the Andes: A multidisciplinary study of high-altitude Quechua. Stroudsburg, PA: Dowden, Hutchinson and Ross.)

and cultivation as well as herding, exposure of the entire body, and extremities in particular, to cold stress cannot be avoided. However, the body surface temperatures of 58 individuals ranging in age from 3 to 80 years were not very low and in fact approximated the thermoneutral zone, despite the fact that ambient temperatures were 11°C (56). This adequate temperature regulation was maintained even during the coldest days of the rainy season.

In summary, although the climatic conditions are severe and provide potential cold stress, evaluations of peripheral and rectal temperatures indicate that both at night during the dry season and by day during the wet season the Quechua Indians live within tolerable limits.

## Technological Adaptation

The success of the Quechua populations in preventing severe body cold stress reflects the effectiveness of their technological adaptations, the most important of which are housing, bedding, and clothing.

**Housing.** The housing of the highland natives differs with variations in altitude and subsistence pattern. A distinct advantage of the more sedentary, mixed economy populations living below 4000 m is individual or community ownership of land. In higher regions of the altiplano the large herds make individual ownership of pasture land impractical. Personal ownership provides for a greater economic investment in the land itself. Thus, at elevations of 4000 m the houses are built of adobe and are permanent. These adobe houses seem quite effective in protecting against cold stress in that they maintained the indoor temperatures more than 10°C above outdoor temperatures (58).

On the other hand, because of a pastoral economy requiring high mobility, the housing at those elevations above 4300 m is more temporary. It usually consists of two or three circular dwellings, about 30 square feet in area, constructed of piled stones and roofed with straw. These houses have the advantage of minimal economic investment and may be abandoned without great economic loss each time the family moves on to new grazing lands (fig. 5.12). The insulative effectiveness of this type of housing, however, is very inadequate, as shown by the fact that the average indoor-outdoor differential temperature of twenty-one houses equaled only 3.7°C (57). In many cases, especially above 4500 m, the inside temperature nearly equalled the outside one.

In summary, housing at around 3500 to 4000 m provides adequate protection against cold stress; however, at higher elevations, because the housing is very temporary, it has only a minimal effect on the severity of cold stress. Nevertheless, it does provide protection against wind and rain and consequently helps reduce heat loss through conduction and radiation.

**Bedding.** The adaptive significance of Nuñoa Quechua bedding has been studied by comparing thermoregulatory characteristics while the subjects were sleeping overnight (8 hours) under three different conditions: (1) in light woolen sleeping bags at 23°C, (2) in light woolen sleeping bags at 4°C, and (3) in their own bedding at 4°C. The results of this study indicate that when native bedding was used at 4°C, the skin temperatures of Quechua men throughout the night were comparable to those attained when sleeping in the thermoneutral conditions of 23°C (56). However, the metabolic rate showed cyclic variations indicating bouts of shivering but not so severe as to disturb the sleeping subjects. The authors point out that if clothing were used, as is the native practice, a shiver-free sleep would be maintained. Thus, it would appear that native bedding provides adequate thermal protection against cold stress.

**Clothing.** The effectiveness of the Nuñoa Quechua clothing has been studied by comparing the thermoregulatory responses of men and women to a standard cold stress of 10°C for 2 hours with and without clothing (59). This study indicated that the insulative value of the men's clothing without poncho and hat equaled 1.21 clo units and 1.43 clo units for women without shawls and hats. With clothing the metabolic heat production was lower than

**Fig. 5.12.** Natural cold exposure in circular house of pastoral Quechuas living above 4500 m in district of Nuñoa in southern Peruvian highlands.

without. Furthermore, the skin temperatures, including hands and feet uncovered, were maintained at higher levels than without clothing. Use of clothing resulted in a 4°C increase in microclimate temperature (temperature of the skin under clothing), and the resulting caloric saving equaled 139 kcal for men and 107 kcal for women. Although there was no sex difference in skin temperature when exposed nude, when clothing was used men maintained lower hand temperatures than women, but higher foot temperatures. This sex difference reflects differences in the pattern of local acclimatization (59). Men experience greater cold exposure of the foot than women because of their daily herding and agricultural activities, whereas women experience more daily cold exposure to the hand from cooking, washing, and weaving. These daily exposure differences become evident when the stress is milder (10°C wearing clothing) and body heat content is not a major problem, so local vascular adaptations are more operative.

In summary, the most singly important technological adaptation to cold stress is clothing. However, the use of gloves and socks for protection of the hands and feet of adults and children is not normally practiced.

## Observation Under Laboratory Conditions

**Whole-Body Cold Exposure.** Studies of whole-body cooling, including nude exposures of 2 hours at 10°C, were conducted in a laboratory built especially

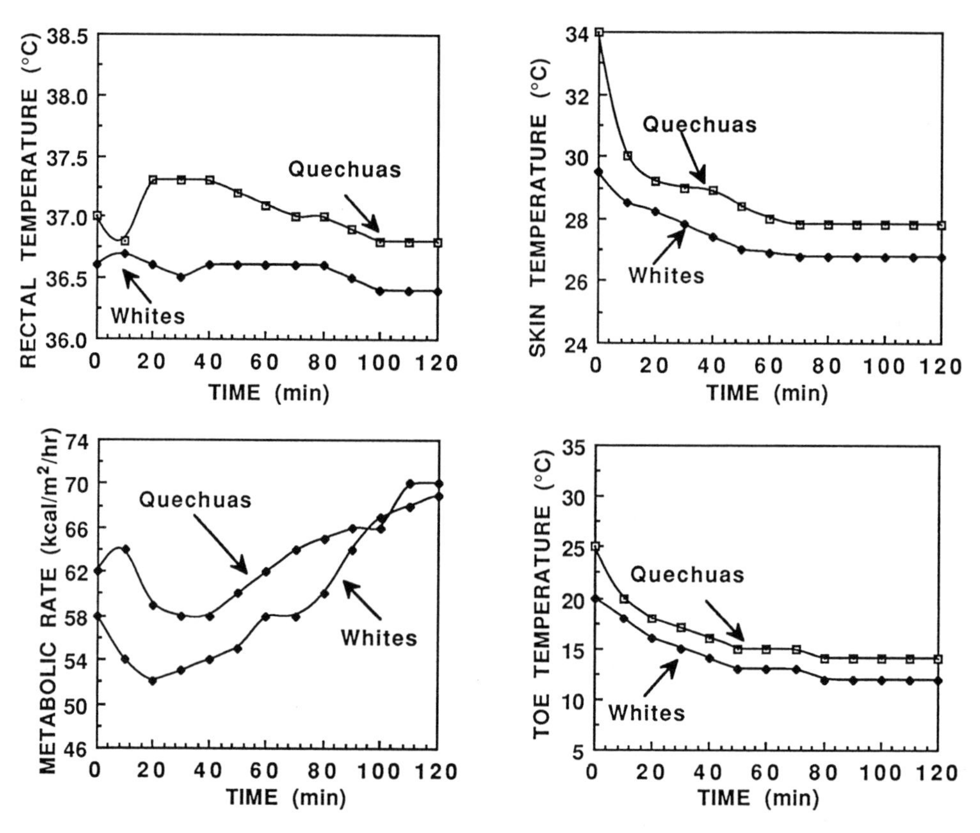


**Fig. 5.13.** Thermal and metabolic responses of highland Quechua Indians from Nuñoa and control whites during 2-hour exposure at 10° C at 4150 m altitude. Highland Quechuas maintain higher rectal and peripheral temperatures, especially of the toe, than white controls, whereas metabolic rate during the first 1½ hours is higher than that of whites. (Modified from J. M. Hanna. 1976. Natural exposure to cold. In P. T. Baker and M. A. Little, eds., Man in the Andes: A multidisciplinary study of high-altitude Quechua. Stroudsburg, PA: Dowden, Hutchinson and Ross.)

for this purpose at 4150 m in the town of Nuñoa (54). As illustrated in figure 5.13, the highland Nuñoa Quechua maintained higher foot temperatures and hence higher mean-weighted skin temperatures than the sea-level United States whites. Associated with increased peripheral extremity temperatures, the Quechuas had higher rectal temperatures. Furthermore, although metabolic heat production during the first hour of the test was higher for the Quechuas than for sea-level whites, the rest of the test results were comparable for the two groups. In other words, the central highland Quechuas, when compared to sea-level controls, showed a metabolic compensation that enabled them to maintain warmer skin temperatures (60).

**Extremity Cooling.** Tests of extremity cooling were done under a variety of conditions including exposure of the hand and foot to air temperatures of 0°C; exposure of the foot in water at 4°, 10°, and 15°C; and exposure of the hand in water at 4°C (61, 62). These studies indicate that in each test the

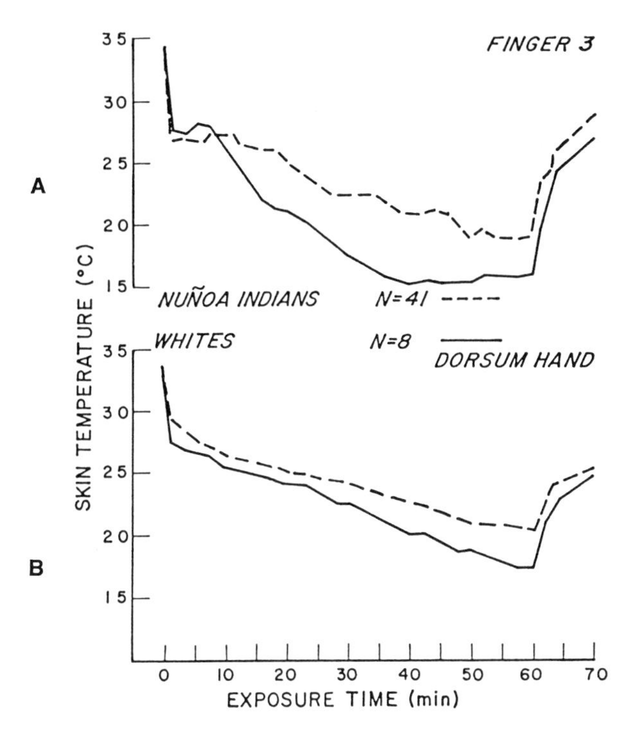


**Fig. 5.14.** Skin temperature of the hand of Nuñoa highland Quechuas and white controls during 1-hour exposure to 0° C air at 4150 m. **A**, Comparison of temperature of subjects' third finger. **B**, Comparison of temperature of subjects' dorsum hand. The highland Nuñoa Quechuas had warmer peripheral temperatures than white controls who resided 14 months at high altitude. (Modified from M. A. Little. 1976. Physiological responses to cold. In P. T. Baker and M. A. Little, eds., Man in the Andes: A multidisciplinary study of high-altitude Quechua. Stroudsburg, PA: Dowden, Hutchinson and Ross.)

highland Nuñoa Quechuas displayed warmer skin temperatures and hence greater blood flow to the surface of the extremities than white subjects who were tested at sea level and after a residency of 14 months at high altitude (fig. 5.14). It must be noted that the greatest highland Quechua-white differences occurred at a water temperature of 10° to 15°C. This finding suggests that the peripheral vasomotor system of the highland Quechuas operates more effectively in moderate cold stress (61, 62).

Figure 5.15 compares the responses of 1-hour exposure of the foot to 0°C air temperatures of thirty Nuñoa Quechua adults, twenty-nine young Quechuas aged 7 to 19 years, twenty-six adult whites, and twenty-eight young whites aged 7 to 18 years. These data demonstrate that the adult and nonadult whites maintained the same low peripheral temperatures. In contrast, the adult Quechuas maintained higher temperatures than the nonadults, but both groups had systematically warmer foot temperatures than the whites. The fact that there were no young adult differences among the whites suggests

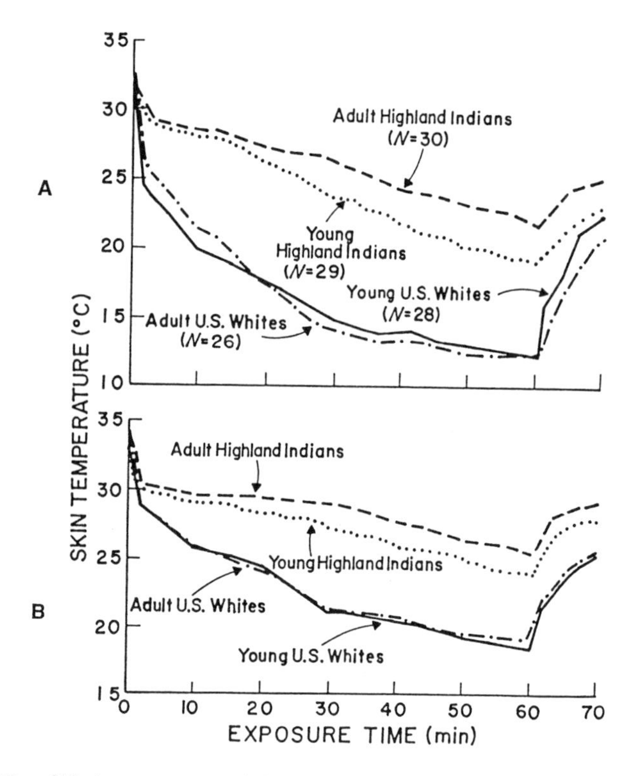


**Fig. 5.15.** Skin temperatures of the foot of adult Nuñoa highland Quechuas, young highland Quechuas, adult lowland whites, and young lowland whites during 1-hour exposure to 0° C air tested at 4150 m. **A**, Temperature comparison of subjects' first toe. **B**, Temperature comparison of dorsum foot of subjects. Both adult and young highland Quechuas had higher peripheral temperature than white controls, suggesting that developmental acclimatization influences high peripheral temperatures of Andean Indians. (Modified from M. A. Little. 1976. Physiological responses to cold. In P. T. Baker and M. A. Little, eds., Man in the Andes: A multidisciplinary study of high-altitude Quechua. Stroudsburg, PA: Dowden, Hutchinson and Ross.)

that developmental acclimatization is one factor contributing to the elevated extremity temperatures of Andean Indians (61, 62). In other words, the existence of a relationship between age and foot temperatures of the Quechuas indicates that cold stress is present during the developmental period and that some acclimatization to this stress has taken place.

In summary, the environmental conditions in which the highland Quechua Indians live provide potential cold stress. The Quechua Indians, through the use of clothing, sleeping patterns, and housing, have successfully modified and ameliorated the severity of cold stress. Evaluations of the thermoregulatory characteristics in natural and laboratory settings demonstrate that the Quechua Indians respond to cold stress with metabolic compensation and great heat flow to the extremities. The fact that maintenance of high peripheral temperatures characterizes both children and adult Quechua Indians suggests that the thermoregulatory characteristics of the Indian native are acquired through developmental acclimatization to cold (61, 62). The everyday activities of Quechua children provide stimulus for the development of adaptive physiological responses to cold stress

## AMA WOMEN DIVERS FROM KOREA

For centuries, Korean women from the Korean peninsula of Ama made their living by diving for plant and animal food. They have dived wearing only lightweight *cotton swimsuits* and face masks for underwater vision. They are initiated into their profession by the age of 12 years and continue it for most of their lives. The diving water temperature falls to 10°C (46°F) in the winter and rises to 27°C (80.6°F) in the summer, yet the women continue to dive throughout the year (63). They dive as deep as 20 m while tied to a heavy ballast belt and holding on to another weight of up to 15 kg (64). Depending on the temperature, they engage in diving for periods of 1 to 5 hours daily, rewarming themselves at open fires during the winter (63–66). In the course of a dive, their oral temperatures routinely fall between 33° and 35°C (91.4° and 95°F). This degree of chronic, intermittent cold exposure is perhaps the most severe form of cold stress to which humans voluntarily submit (63).

### Acclimatization to Cold Water Immersion

During 1960–1961 Hong and colleagues (63–67) studied the thermoregulatory characteristics of the Ama Korean women divers.

**Metabolic Rate.** As shown in figure 5.16, the basal metabolic rate of the Ama women was greater in the winter (63–67), whereas that of the nondiving controls continued at a constant level. Furthermore, the basal metabolic rate (BMR) of the Ama varied as an inverse function of the water temperature at diving; BMRs of 35% above normal were observed in the winter when the water temperature fell to 10°C (63). Daily urine samples indicated that excretion of nitrogen was the same in the Ama and in the control subjects. For this reason, the increased metabolic rate of the Ama must reflect the influence of the acclimatization response to cold and cannot be caused by differences in diet (63, 67).

**Shivering Threshold.** Using as a criterion the diving water temperature at which 50% of the women shivered, it was found that among Ama women exposed to sequentially colder water baths, the divers start shivering at a lower temperature (28.2°C) than the controls (29.9°C) (fig. 5.17). In other words, among the Ama the shivering threshold is lower, allowing them to tolerate a reduced core temperature without shivering (63, 65, 66). Although continuous exposure to cold enhances the mechanism of heat production, it

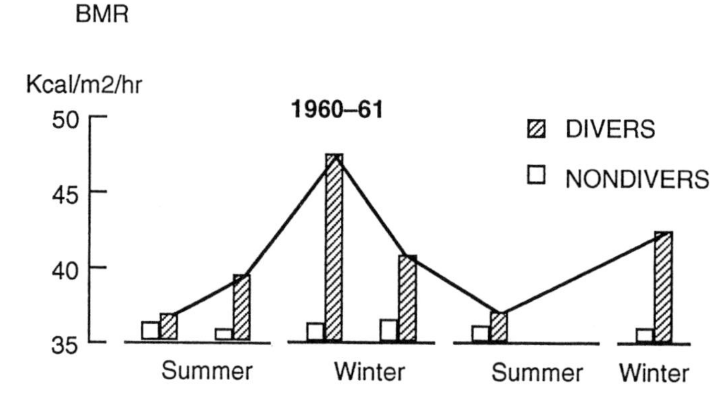


**Fig. 5.16.** Metabolic rate among Ama women divers and nondivers from Korea. The metabolic rate for the divers is higher in the winter, but in the nondivers there are no seasonal differences. (Adapted from data of T. Sasaki. 1966. Fed. Proc. 25:1163–68.)

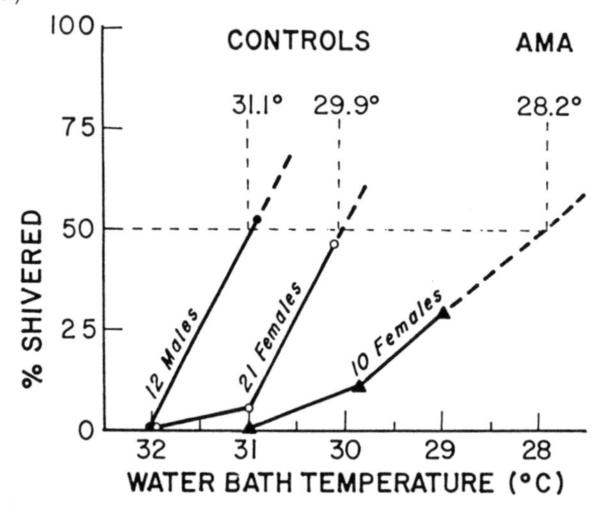


**Fig. 5.17.** Relationship between water temperature and shivering threshold among Korean Ama diving women and controls. The winter cold-acclimatized Ama women shiver at lower temperature than controls. (Modified from S. K. Hong. 1963. Fed. Proc. 22:831–33.)

also leads to a lower threshold at which heat production starts. The effects of acclimatization also are reflected in the elevated tissue insulation of Ama and other Asiatic divers (65, 66).

**Vascular Response to Hand Immersion in Cold Water.** The finger skin temperature and blood flow during hand immersion in 6°C (42°F) among the Ama divers was lower than in nondivers (68).

## Deacclimatization to Cold Water Immersion

Since 1977 the Ama women have been diving wearing *wet suits* instead of the traditional cotton swimsuits. As a result of this change the Ama diving women are no longer exposed to the severe water stress to which previous cotton-suit divers were subject on a daily basis (69). This change provided a unique opportunity to test whether the thermoregulatory characteristics of the Ama women were developed through repeated exposures to severe cold water stress or reflected a genetic adaptation. With this purpose Park and colleagues (69, 70) during 1980–1982 studied the thermoregulatory functions of the Ama women.

**Metabolic Rate.** As shown in figure 5.18 and table 5.1, contrary to results for women wearing cotton suits, the basal metabolic rate of the Ama women who wore wet suits did not show any seasonal changes. Furthermore, the basal metabolic rate (BMR) of the divers and nondivers were similar through all seasons.

**Body Fat.** Measurements of skinfold thickness as shown in table 5.2 indicate that the divers and nondivers in 1980–1982 were significantly fatter than their counterparts studied in 1960–1961.

**Shivering Threshold.** In 1982 divers and nondivers shivered at the same water temperature. In other words, the low shivering threshold observed in 1960s divers returned to the control values (in this case the nondivers), despite the fact that in the 1980s divers were fatter than in the 1960s.

**Vascular Response to Hand Immersion in Cold Water.** The finger skin temperature and blood flow during hand immersion in 6°C (42°F) among the Ama divers was similar to that of nondivers. This finding suggests that the local vascular acclimatization observed among the cold stressed divers studied in the 1960s disappeared in contemporary divers. It also suggests that the low finger skin temperature observed in previous divers was a result of generalized cold body stress rather than cold stress to the hands as it occurs with the Gaspé fishermen.

## WHITES

Studies on acclimatization and habituation of whites have been conducted among populations who live in cold areas in northern Canada, the antarctic region, and in the fishing areas of the Gaspé Peninsula of Canada.

### Canadian Soldiers Acclimatized to Northern Manitoba

Studies of the thermoregulatory responses of Canadian soldiers who were transferred from Winnipeg to northern Manitoba in November and remained there until the following spring have provided valuable information on habituation to cold (71). Because of their duties, these soldiers spent from 8 to 10 hours a day outdoors, 6 days a week throughout the winter. The average

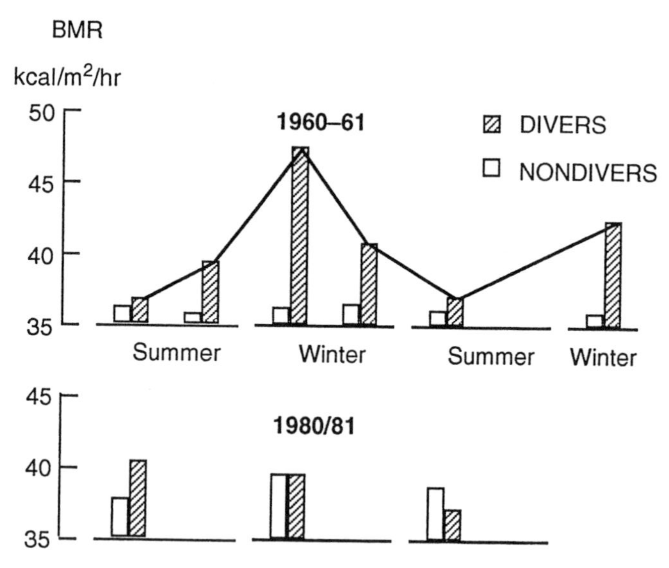


**Fig. 5.18.** Metabolic rate among Ama women divers and nondivers from Korea studied in 1960–1961 and 1980–1981. In 1960–1961, as result of diving with cotton swimsuits, the Ama women were cold stressed and developed a high metabolic rate in the winter, but in 1980–1981, with the use of wet suits, the divers were not cold stressed and, therefore, there are no seasonal differences in metabolic rate. (Adapted from data of Park et al. 1983. J. Appl. Physiol.: Respirat. Environ. Exercise Physiol. 54:1708–16.)

air temperature was −24°C (−11.2°F). These subjects were exposed for 1 hour at 9.5°C (49°F) immediately after their arrival in Manitoba and then again in January and March. Compared to the values measured in November and to basal levels, the metabolic rates showed a marked reduction throughout the winter while the skin temperatures remained constant (fig. 5.19).

## Acclimatization to Antarctica and North Pole

The thermoregulatory characteristics of eight subjects who spent 5 weeks in Antarctica in the autumn and again in the winter have provided valuable information about the development of habituation to cold stress. The results of this study indicate that residence in Antarctica during the winter was associated with a reduction in metabolic rates, but the rectal and skin temperatures were not changed (72). The most consistent change from autumn to spring was a continuous increase in body weight. On the other hand, when five men who resided for 12 months in Antarctica were compared to the controls (their metabolic rates were measured before the move to Antarctica), their response to cold (5°C) was characterized by lower metabolic rates and lower rectal temperatures. At a given skin temperature the increase in metabolic rate was lower after 6 months in Antarctica. The skin temperatures were also lowered, except for the toes which were higher. The subjective reactions indicated that with continued residence in the cold, most of the subjects found they were comfortable with ever-decreasing amounts of heavy arctic clothing. A corroboration of this finding can be found on the evaluations of the thermoregulatory characteristics of J. Etienne after his journey of 63 days to the North Pole where the ambient temperature varied from – 12° to – 52°C (10.4° to – 61.6°F) (73). After his journey on the sea ice to the North Pole Etienne experienced a decrease in metabolic rate and rectal temperature but an increase in skin temperature of the extremities during the cold test when compared to a similar test before the expedition. From these studies it can be inferred that prolonged residence of European whites in cold climates is associated with a reduction in metabolic rate and a redistribution of heat from the internal to the peripheral organs, so that the skin temperature of the trunk is lowered (but not of the hands), exemplifing habituation to cold.

**TABLE 5.1. Seasonal changes in basal metabolic rate (BMR)**

| | BMR, kcal/m²/hr | | % Deviation from Standard | | |
|---|---|---|---|---|---|
| | Nondivers | Divers | Nondivers | Divers | |
| Season | Mean ± SE | Mean ± SE | Mean ± SE | Mean ± SE | P value |
| | | | 1960–1961 | | |
| Spring | 36.6 ± 1.7(20) | 41.0 ± 5.2(20) | + 4.1 ± 4.9 | + 16.9 ± 3.4 | <0.05 |
| Summer | 36.4 ± 0.9(20) | 36.6 ± 0.9(20) | + 3.7 ± 2.2 | + 5.0 ± 7.5 | >0.1 |
| Autumn | 35.7 ± 1.0(20) | 39.4 ± 0.9(20) | + 1.8 ± 1.9 | + 12.7 ± 7.5 | <0.005 |
| Winter | 96.4 ± 0.7(20) | 47.0 ± 2.6(20) | + 3.5 ± 1.5 | + 35.1 ± 7.5 | <0.005 |
| | | | 1980–1981 | | |
| Spring | 38.5 ± 1.6(18) | 37.0 ± 1.8(18) | + 5.4 ± 4.3 | – 0.4 + 43 | >0.2 |
| Summer | 38.1 ± 1.1(16) | 40.6 ± 2.0(17) | + 4.3 ± 3.1 | + 11.6 ± 5.4 | >0.2 |
| Winter | 39.3 ± 1.2(16) | 39.4 ± 1.4(16) | + 7.3 ± 3.4 | + 8.8 ± 4.1 | >0.5 |

*Note:* The number of divers is given in parentheses.
*Source:* Adapted from Y. S. Park, D. W. Rennie, I. S. Lee, Y. D. Park, K. S. Paik, D. H. Kang, D. J. Suh, S. H. Lee, S. Y. Hong, and S. K. Hong. 1983. Time course of deacclimatization to cold water immersion in Korean women divers. J. Appl. Physiol.: Respirat. Environ. Exercise Physiol. 54:1708–16 (70).

**TABLE 5.2. Body fat contents and mean skinfold thickness of Ama divers and nondivers from Korea**

| | | | Age | Wt | Body Fat | Mean Skinfold |
|---|---|---|---|---|---|---|
| | | | (yrs) | (kg) | (%) | (mm) |
| Year | Group | *N* | Mean ± SE | Mean ± SE | Mean ± SE | Mean ± SE |
| '61 | Nondivers | 17 | 33.0 ± 2.0 | 55.0 ± 1.0 | 11.5 ± 0.8 | 2.31 ± 0.29 |
| | Divers | 9 | 42.0 ± 3.0 | 56.0 ± 2.0 | 11.4 ± 1.0 | 2.24 ± 0.27 |
| '80–'81 | Nondivers | 39 | 38.2 ± 1.1 | 55.5 ± 1.4 | 25.8 ± 0.8 | 9.03 ± 0.54* |
| | Divers | 18 | 41.4 ± 1.5 | 56.0 ± 2.0 | 25.8 ± 0.9 | 8.85 ± 0.63* |

*Source:* Adapted from Y. S. Park, D. W. Rennie, I. S. Lee, Y. D. Park, K. S. Paik, D. H. Kang, D. J. Suh, S. H. Lee, S. Y. Hong, and S. K. Hong. 1983. Time course of deacclimatization to cold water immersion in Korean women divers. J. Appl. Physiol.: Respirat. Environ. Exercise Physiol. 54:1708–16 (70).
*Significantly ($p$<0.05) greater than respective value in 1961.

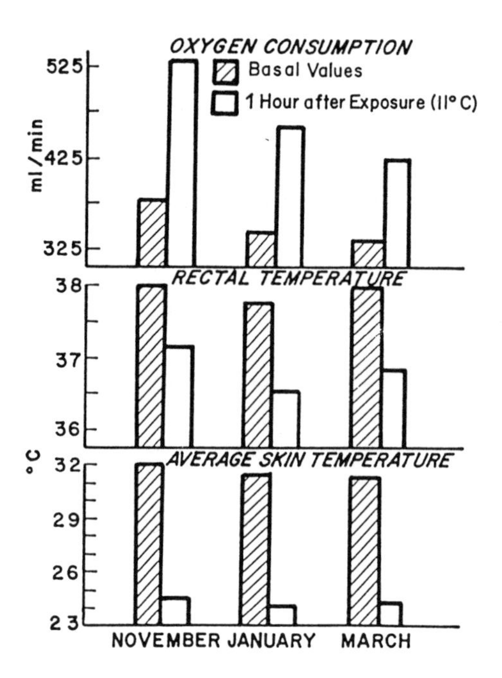


**Fig. 5.19.** Thermoregulatory responses to cold stress among Canadian soldiers who lived outdoors all winter (January, February, and March). Habituation to cold resulted in reduced oxygen consumption (metabolic rate) and rectal temperature but no change in skin temperature. (Modified from J. LeBlanc. 1975. Man in the cold. Springfield, IL: Charles C. Thomas.)

### Gaspé Fishermen

Studies of Gaspé fishermen accustomed to fishing in cold water (approximately 9.4°C) have provided valuable information about the development of local acclimatization to cold stress (74, 75).

The fourteen fishermen included in the study have been fishing for 2 to 25 years. They routinely expose their bare hands to cold sea water for several hours a day. The thermoregulatory characteristics of this group were compared to fourteen non-cold-stressed control subjects. The cold test involved immersing the hand in cold water with a temperature of 2.5°C (35°F) for 10 minutes followed by a further 10 minutes in room air. As shown in figures 5.20 and 5.21 the fishermen maintained higher finger skin temperatures than did white controls. Furthermore, as shown in figure 5.21 the total heat flow from the fishermen's hands was significantly higher than from the control subjects. Thus the fishermen's response is an example of local acclimatization to cold.

## OVERVIEW

The major conclusion derived from these studies is that humans have developed a great capacity to tolerate varying degrees of cold stress. In general,

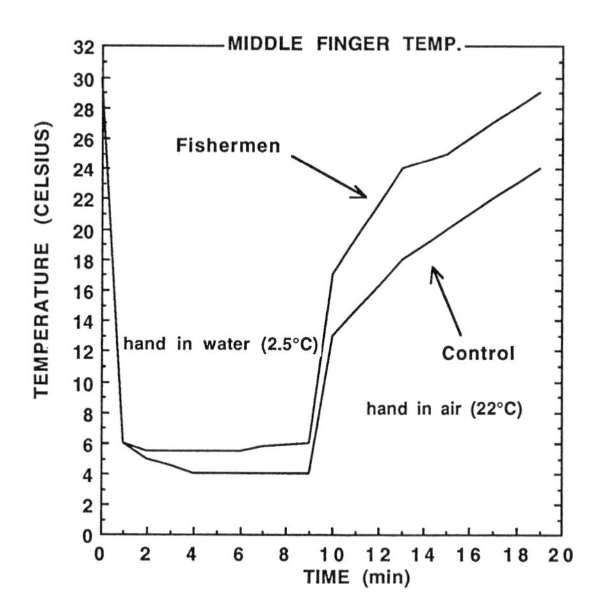


**Fig. 5.20.** Variations in skin temperature of middle finger of Gaspé fishermen and control subjects during immersion of hand into cold water. Note that the fishermen maintain higher temperature than the controls. (Adapted from J. LeBlanc, J. A. Hildes, and O. Heroux. 1960. Tolerance of Gaspé fishermen to cold water. J. Appl. Physiol. 15:1031–34.)

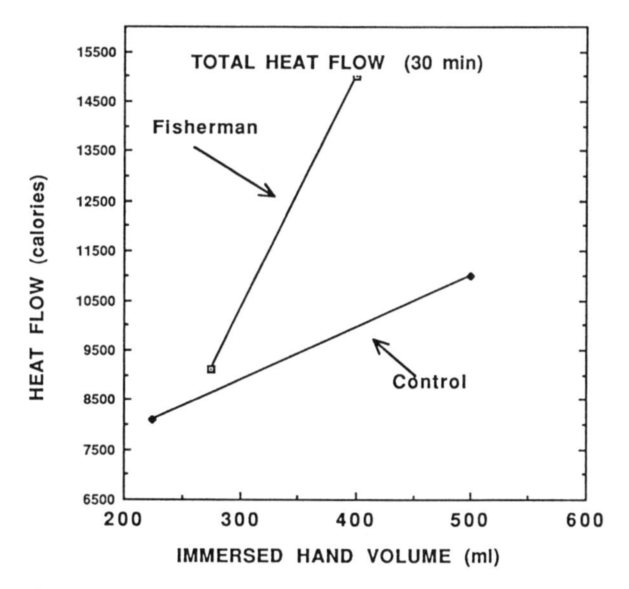


**Fig. 5.21.** Relationship between immersed hand volume and heat flow in Gaspé fishermen and control subjects on immersion of the hand for 30 minutes in a water bath initially set at 5° C. (Adapted from J. LeBlanc, J. A. Hildes, and O. Heroux. 1960. Tolerance of Gaspé fishermen to cold water. J. Appl. Physiol. 15:1031–34.)

populations inhabiting regions of moderate cold stress, such as those of the desert, have developed physiological mechanisms oriented toward conserving body heat. On the other hand, populations living in the regions of severe cold stress, such as the Arctic and Andes, have developed efficient technological and cultural adaptations that ameliorate but do not replace the cold stress in response to which they have developed physiological responses. The various adaptive responses that enable populations to overcome cold stress are schematized in figure 5.22.

Using the ability to sleep comfortably in the cold as an index of adaptation, indigenous populations exhibit a high degree of adjustment to cold stress, but they differ in manner of attaining the adaptation. The Australian aborigines inhabiting the deserts of central Australia sleep well by tolerating colder peripheral temperatures. In contrast, populations inhabiting the colder regions of the world appear to sleep well because they can keep their peripheral and especially their extremity temperatures warm. Considering that the desert, both in duration and intensity of cold, provides conditions of moderate cold stress and adds no risk of frostbite, letting the body shell cool is an excellent response. In contrast, the arctic, subarctic, and Andean regions are characterized by severe cold stress and letting the body shell cool would lead to considerable body injury and frostbite. Furthermore, maintaining a low metabolic rate and low peripheral temperature is very economical from the energy standpoint and may be adaptive under conditions of limited nutritional resources as occurs among the Australian aborigines. On the other hand, populations inhabiting regions of severe cold stress do have access to adequate nutritional resources, and therefore maintaining a high metabolic rate and high peripheral temperature is an affordable response.

As expected, along with these environmental and thermoregulatory differences, the technological adaptations to cold are more efficient and advanced in the populations inhabiting the cold regions than in those in the deserts of Australia or the Kalahari. However, despite the high degree of technological adaptation that is exhibited by the Eskimos, and to a lesser extent by the Peruvian Quechuas, and because of the requirements of their subsistence economies, their environments provide conditions of moderate body cold stress and severe extremity cold stress.

At least three physiological mechanisms of cold adaptation set the Eskimos and Quechuas apart from white controls or tropical populations, such as the Australian aborigines. First, they have a high metabolic rate that is not explainable solely on the basis of nutritional factors. Second, a test of whole-body exposure as well as local extremity exposure to cold indicates that they maintain high levels of blood flow to the extremities, resulting in warm hands and feet and a correspondingly greater heat loss than for whites. Third, the high peripheral temperatures of the extremities and high tolerance to cold

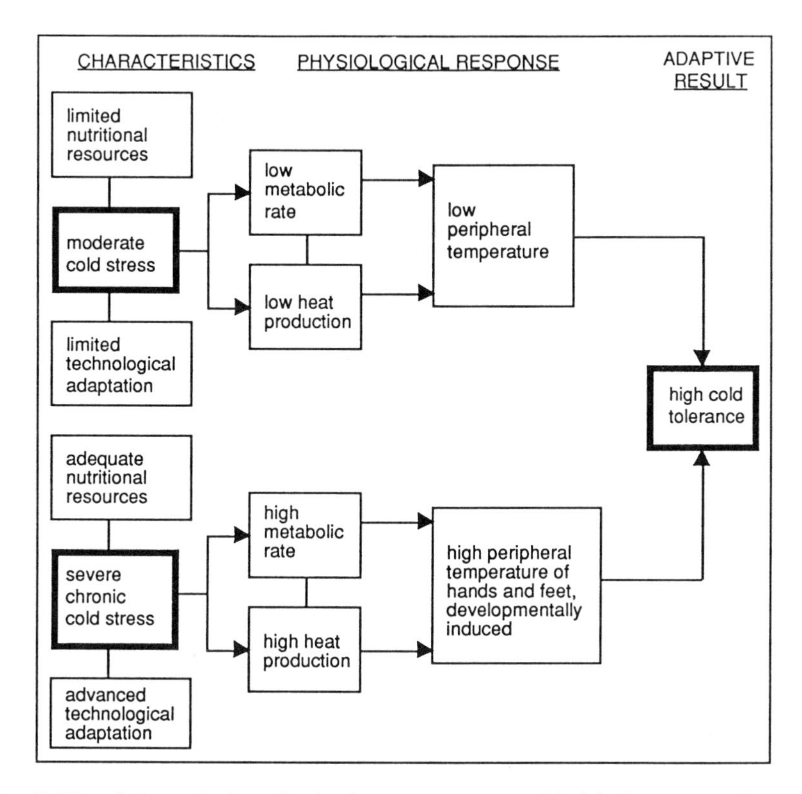


**Fig. 5.22.** Schematization of adaptive responses to cold of indigenous and nonindigenous populations acclimatized to environments of moderate and severe cold stress. Adaptation to cold stress has resulted in the development of specialized thermoregulatory responses, which are intimately related to the severity of cold stress and degree of access to nutritional resources and technological adaptation.

of Eskimos and highland Quechuas appear to reflect the influence of developmental acclimatization. Among the Eskimos, but not among the Quechuas, these characteristics appear to be based on specific genetic characteristics.

Studies of Korean divers provide conclusive evidence that chronic exposure to cold stress increases the metabolic rate. These findings are in marked contrast to those of European whites living in cold climates, which indicate a lowering of metabolic rates and concomitant decreases in rectal and skin temperatures. The cause of these differences can be explained by the severity of cold stress experienced by each population. The divers are exposed to severe cold stress for prolonged periods throughout the year and for most of their lives. In contrast, the whites, although they have lived for prolonged periods in cold climates, do not have comparable severity and length of exposure to cold because of their use of clothing. In fact, whites who exposed themselves to cold in the Norwegian mountains as much as possible did show increased metabolic rates (76). For this reason, it can be concluded that when sufficiently exposed to cold, all humans adapt to cold through increased metabolic rates and with an attendant increase in peripheral temperature. Investigations conducted among women divers from Hokkaido Island of Japan found that in the cold-acclimatized Ainu a small dose of noradrenaline produced a greater elevation in metabolic rate and plasma levels of free fatty acids and ketone bodies than in the non-cold-acclimatized Japanese (77). Since more adrenaline elicits an increase in oxygen consumption by enhancing the oxidation of fatty acids, these findings suggest that, as a result of cold acclimatization, the calorigenic effects of noradrenaline are increased and free fatty acids are oxidized faster in the Ainu than in nonacclimatized Japanese. These findings suggest that the maintenance of high metabolic rates that characterize cold-acclimatized populations is attained through a faster oxidation of free fatty acids. The high metabolic rates observed among Eskimos, the Ama women from Korea, and divers from Hokkaido, and to a lesser extent among Peruvian Quechuas and Alacaluf Indians may be viewed as examples of acclimatization to cold stress. On the other hand, the fact that with the adoption of wet suits the high metabolic rates and low shivering threshold that characterized the Ama divers are no longer present indicates that acclimatization to cold stress is *reversible.*

As learned from the studies of the vascular response to cold water immersion of the Gaspé fishermen continued exposure of the hand to cold stress results in the maintenance of high finger temperatures and decreased pain sensation. In other words, the Gaspé fishermen like the Quechuas and Eskimos and Athapaskans have developed thermoregulatory responses that enable them to carry on with their daily activities.

On the other hand, studies of European whites suggest that when exposed to relatively moderate cold stress, the organism can adapt without metabolic compensation. This kind of adaptation is attained through continuous exposure leading to habituation, whereby the organism becomes accustomed to sustaining some degree of hypothermia. As indicated by LeBlanc (71), "if life is not endangered, the responses of the body seem oriented to retaining individual identity and preserving homeostasis, avoiding unnecessary challenges." In a sense one can say that the thermostat is lowered to a more economical level. The fact that adaptation to moderate cold stress is attained through habituation suggests involvement of the central nervous system (71, 72). Along with reduced cold sensation, habituation to cold brings a reduced activation of the sympathetic nervous system as shown by decreased changes in blood pressure with exposure to cold. The fact that northern Japanese have a higher finger vasodilation and earlier onset of cold-induced vasodilation than those of southern origin (78–80) may be explained on the basis of local acclimatization to cold stress rather than habituation. There is conclusive evidence that blacks have lower finger temperatures when exposed to ice water than whites and that in Korea and Alaska cold injury to the hands occurs with greater frequency (81–84). Based on the similarities between South African blacks and whites (85) and thermoregulatory responses to whole-body exposure to cold between United States blacks and whites (86), it would not be surprising if the observed differences in responses to hand cooling and cold injury were proved to reflect different degrees of acclimatization.

## References


1. Hammel, H. T. 1964. Terrestrial animals in cold: recent studies of primitive man. In D. B. Dill, E. F. Adolph, and C. G. Wilber, eds., Handbook of physiology, vol. 4. Adaptation to the environment. Washington, DC: American Physiological Society.
2. McArthur, M. 1960. Food consumption and dietary levels of groups of aborigines living on naturally ocurring foods. In C. P. Mounttord, ed., Anthropology and nutrition. New York, NY: Cambridge University Press.
3. Scholander, P. F., H. T. Hammel, J. S. Hart, D. H. LeMessurier, and J. Steen. 1958. Cold adaptation in Australian aborigines. J. Appl. Physiol. 13:211–18.
4. Hammel, H. T., R. W. Elsner, D. H. LeMessurier, H. T. Anderson, and F. A. Milan. 1959. Thermal and metabolic responses of the Australian aborigine exposed to moderate cold in summer. J. Appl. Physiol. 14:605–15.
5. Lee, R. B. 1969. !Kung bushman subsistence: an input-output analysis. In A. P. Vayda, ed., Environment and cultural behavior: ecological studies in cultural anthropology. New York, NY: Natural History Press.
6. Wilmsen, E. 1978. Seasonal effects of dietary intake on Kalahari san. Fed. Proc. 37:65–72.
7. Hammel, H. T., J. S. Hildes, D. C. Jackson, and H. T. Anderson. 1962. Thermal and metabolic responses of the Kalahari Bushmen to moderate cold exposure at night. Tech. Rep. 62–44. Fort Wainwright, AL: Arctic Aeromedical Laboratory.
8. Ward, J. S., G. A. C. Bredell, and H. G. Wenzel. 1960. Responses of Bushmen and Europeans on exposure to winter night temperatures in the Kalahari. J. Appl. Physiol. 15:667–70.
9. Wyndham, C. H., and J. F. Morrison. 1958. Adjustment to cold of Bushmen in the Kalahari Desert. J. Appl. Physiol. 13:219–25.
10. Wyndham, C. H. 1964. Southern African ethnic adaptation to temperature and exercise. In P. T. Baker, and J. S. Weiner, eds., The biology of human adaptability. Oxford: Clarendon Press.
11. Elsner, R. W. 1963. Comparison of Australian Aborigines, Alacaluf Indians, and Andean Indians. Fed. Proc. 22:840–43.
12. Milan, F. A. 1962. Racial variations in human response to low temperature. In L. P. Hannon, and E. Viereck, eds., Comparative physiology of temperature regulation. Fort Wainwright, AL: Arctic Aeromedical Laboratory.
13. Andersen, K. L., Y. Loyning, J. D. Nelms, O. Wilson, R. H. Fox, and A. Bolstad. 1960. Metabolic and thermal response to a moderate cold exposure in nomadic Lapps. J. Appl. Physiol. 15:649–53.
14. Krog, J., B. Folkow, R. H. Fox, and K. L. Andersen. 1960. Hand circulation in the cold of Lapps and North Norwegian fishermen. J. Appl. Physiol. 15:654–58.

15. Krog, J., M. Alvik, and K. Lund-Larsen. 1969. Investigations of the circulatory effects of submersion of the hand in the water in Finnish Lapps, the ''Skolts.'' Fed. Proc. 28:1135–37.
16. Jenness, D. 1928. People of the twilight. New York, NY: Macmillan.
17. Rodahl, K. 1952. Basal metabolism of the Eskimo. J. Nutr. 48:359–68.
18. Murdock, G. P. 1964. Our primitive contemporaries. New York, NY: Macmillan.
19. Nelson, R. K. 1966. Literature review of Eskimo knowledge of the sea ice environment. Tech. Rep. AAL-TR-65-7. Fort Wainwright, AL: Arctic Aeromedical Laboratory.
20. Scholander, P. F., V. Walters, R. Hock, and L. Irving. 1950. Body insulation of some arctic and tropical mammals and birds. Biol. Bull. 99:225–36.
21. Forde, C. D. 1963. Habitat, economy and society. New York, NY: E. P. Dutton and Co.
22. Perry, R. 1966. The world of the polar bear. Seattle, WA: University of Washington Press.
23. Krough, A., and M. Krogh. 1913. A study of the diet and metabolism of Eskimos. Copenhagen: Bianco Lund.
24. Heinbecker, P. 1928. Studies on the metabolism of Eskimos. J. Biol. Chem. 80:461–75.
25. Rabinowitch, I. M., and F. C. Smith. 1936. Metabolic studies of Eskimos in Canadian Eastern Arctic. J. Nutr. 12:337–56.
26. Hoygarrd, A. 1941. Studies on the nutrition and physiopathology of Eskimo. Skrifter 9. Oslo: Norske Videnskaps Akademi.
27. Brown, G. M., and J. Page. 1953. The effect of chronic exposure to cold on temperature and blood flow of the hand. J. Appl. Physiol. 5:221–27.
28. Brown, G., J. Malcom, J. D. Hatcher, and J. Page. 1953. Temperature and blood flow in the forearm of the Eskimo. J. Appl. Physiol. 5:410–20.
29. Brown, G. M., G. S. Bird, L. M. Boag, D. J. Delahaye, J. E. Green, J. D. Hatcher, and J. Page. 1954. Blood volume and basal metabolic rate of Eskimos. Metabolism 3:247–54.
30. Adams, T., and B. G. Covino. 1958. Racial variations to a standardized cold stress. J. Appl. Physiol. 12:9–12.
31. Rennie, D. W., and T. Adams. 1957. Comparative thermoregulatory responses of Negroes and white persons to acute cold stress. J. Appl. Physiol. 11:201–4.
32. Rennie, D. W., B. G. Covino, M. R. Blair, and K. Rodahl. 1962. Physical regulation of temperature in Eskimos. J. Appl. Physiol. 17:326–32.
33. Hart, J. S., H. B. Sabean, J. A. Hildes, F. Depocas, H. T. Hammel, K. L. Andersen, L. Irving, and G. Foy. 1962. Thermal and metabolic responses of coastal Eskimos during a cold night. J. Appl. Physiol. 17:953–60.
34. Milan, F. W., and E. Evonuk. 1966. Oxygen consumption and body temperature of Eskimos during sleep. Tech. Rep. AAL-TR-66-10. Fort Wainwright, AL: Arctic Aeromedical Laboratory.
35. Rodahl, K., and B. Issekutz. 1965. Nutritional requirements in the cold. Symposia on Arctic Biology and Medicine. Fort Wainwright, AL: Arctic Aeromedical Laboratory.
36. MacHattie, L., P. Haab, and D. W. Rennie. 1960. Eskimo metabolism as measured by the technique of 24-hour indirect calorimetry and graphic analysis. Tech. Rep. AAL-TR-60-43. Fort Wainwright, AL: Arctic Aeromedical Laboratory.
37. Milan, F. A., J. P. Hannon, and E. Evonuk. 1963. Temperature regulation of

Eskimos, Indians and Caucasians in a bath calorimeter. J. Appl. Physiol. 18:378–82.
38. Milan, F. A., R. W. Elsner, and K. Rodahl. 1961. Thermal and metabolic responses of men in the Antarctic to a standard cold stress. J. Appl. Physiol. 16:401–4.
39. Meehan, J. P. 1955. Individual and racial variations in a vascular response to a cold stimulus. Milit. Med. 116:330–34.
40. Page, J., and G. M. Brown. 1953. Effect of heating and cooling the legs on hand and forearm blood flow in the Eskimo. J. Appl. Physiol. 5:753–58.
41. Eagan, C. J. 1963. Introduction and terminology: habituation and peripheral tissue adaptation. Fed. Proc. 22:930–33.
42. Paul, J. R., J. T. Riordan, and L. M. Kraft. 1951. Serological epidemiology. Antibody patterns in North Alaskan Eskimos. Immunology 66:695–713.
43. Luzzio, A. J. 1966. Comparison of serum proteins in Americans and Eskimos. J. Appl. Physiol. 21:685–88.
44. Miller, L., and L. Irving. 1962. Local reactions to air cooling in an Eskimo population. J. Appl. Physiol. 17:449–55.
45. LeBlanc, J. 1975. Man in the cold. Springfield, IL: Charles C. Thomas, Publisher.
46. Elsner, R. W., J. D. Nelms, and L. Irving. 1960. Circulation of heat to the hands of Arctic Indians. J. Appl. Physiol. 15:662–66.
47. Irving, L., K. L. Andersen, A. Bolstad, R. W. Elsner, J. A. Hildes, Y. Loyning, J. D. Nelms, L. P. Peyton, and R. D. Whaley. 1960. Metabolism and temperature of Arctic Indian men during a cold night. J. Appl. Physiol. 15:635–44.
48. Steegmann, A. T. 1977. Finger temperatures during work in natural cold: the Northern Ojibwa. Hum. Biol. 49:349–74.
49. Hurlich, M. D., and A. T. Steegmann, Jr. 1979. Contrasting laboratory response to cold in two sub-arctic Algonkian villages: An admixture effect? Hum. Biol. 51:255–78.
50. Winterhalder, B. 1983. History and ecology of the Boreal Zone in Ontario. In A. T. Steegmann, Jr., ed., Boreal Forest adaptations: the Northern Algonkians. New York, NY: Plenum Press.
51. Marano, L. 1983. Boreal Forest hazards and adaptations: the present. In A. T. Steegmann, Jr., ed., Boreal Forest adaptations: the Northern Algonkians. New York, NY: Plenum Press.
52. Steegmann, A. T., Jr., M. G. Hurlich, and B. Winterhalder. 1983. Coping with cold and other challenges of the Boreal Forest: an overview. In A. T. Steegmann, Jr., ed., Boreal Forest adaptations: the Northern Algonkians. New York, NY: Plenum Press.
53. So, J. K. 1975. Genetic acclimatizational and anthropometric factors in hand cooling among North and South Chinese. Am. J. Phys. Anthropol. 43:31–38.
54. Baker, P. T., and M. A. Little, eds. 1976. Man in the Andes: a multidisciplinary study of high-altitude Quechua. Stroudsburg, PA.: Dowden, Hutchinson and Ross.
55. Hanna, J. M. 1976. Natural exposure to cold. In P. T. Baker, and M. A. Little, eds., Man in the Andes: a multidisciplinary study of high-altitude Quechua. Stroudsburg, PA.: Dowden, Hutchinson and Ross.
56. Mazess, R. B., and R. Larsen. 1972. Responses of Andean highlanders to night cold. Int. J. Biometeorol. 16:181–92.
57. Baker, P. T. 1966. Micro-environment cold in a high altitude Peruvian population. In H. Yoshimura, and J. S. Weiner, eds., Human adaptability and its methodology. Tokyo: Japanese Society for the Promotion of Sciences.

58. Baker, P. T. 1966. Ecological and physiological adaptation in indigenous South Americans with special reference to the physical environment. In P. T. Baker, and J. S. Weiner, eds., The biology of human adaptability. Oxford: Clarendon Press.
59. Hanna, J. M. 1970. A comparison of laboratory and field studies of cold response. Am. J. Phys. Anthropol. 32:227–32.
60. Blatteis, C. M., and L. O. Lutherer. 1976. Effect of altitude exposure on thermal regulatory response of man to cold. J. Appl. Physiol. 41:848–58.
61. Little, M. A. 1976. Physiological responses to cold. In P. T. Baker and M. A. Little, eds., Man in the Andes: a multidisciplinary study of high-altitude Quechua. Stroudsburg, PA: Dowden, Hutchinson and Ross.
62. Little, M. A., and J. M. Hanna. 1978. The responses of high-altitude populations to cold and other stresses. In P. T. Baker, ed., The biology of high-altitude peoples. New York, NY: Cambridge University Press.
63. Hong, S. K. 1963. Comparison of diving and non-diving women of Korea. Fed. Proc. 22:831–33.
64. Kita, H. 1965. Review of activities: harvest, seasons, and diving patterns. In H. Rahn and T. Yokoyama, eds., Physiology of breath-hold diving and the Ama of Japan. Washington, DC: National Research Council Publication 1341.
65. Rennie, D. W., B. G. Covino, B. J. Howell, S. H. Song, B. S. Kang, and S. K. Hong. 1962. Physical insulation of Korean diving women. J. Appl. Physiol. 17:961–66.
66. Rennie, D. W. 1965. Thermal insulation of Korean diving women and nondivers in water. In H. Rahn and T. Yokoyama, eds., Physiology of breath-hold diving and the Ama of Japan. Washington, DC: National Research Council Publication 1341.
67. Sasaki, T. 1966. Relation of basal metabolism to changes in food composition and body composition. Fed. Proc. 25:1163–68.
68. Paik, K. S., B. S. Kang, D. S. Han, D. W. Rennie, and S. K. Hong. 1972. Vascular responses of Korean ama to hand immersion in cold water. J. Appl. Physiol. 32:446–50.
69. Kang, D. H., Y. S. Park, Y. D. Park, I. S. Lee, D. S. Yeon, S. H. Lee, S. Y. Hong, D. W. Rennie, and S. K. Hong. 1983. Energetics of wet-suit diving in Korean women divers. J. Appl. Physiol.: Respirat. Environ. Exercise Physiol. 54:1702–7.
70. Park, Y. S., D. W. Rennie, I. S. Lee, Y. D. Park, K. S. Paik, D. H. Kang, D. J. Suh, S. H. Lee, S. Y. Hong, and S. K. Hong. 1983. Time course of deacclimatization to cold water immersion in Korean women divers. J. Appl. Physiol.: Respirat. Environ. Exercise Physiol. 54:1708–16.
71. LeBlanc, J. 1975. Man in the cold. Springfield, IL: Charles C. Thomas, Publisher.
72. Wyndham, C. H., R. Plotkin, and A. Munro. 1964. Physiological reactions to cold of men in the Antarctic. J. Appl. Physiol. 19:593–97.
73. Bittel, J. H. M., G. H. Livecchi-Gonnot, A. M. Hanniquet, C. Poulain, and J-L. Etienne. 1989. Thermal changes observed before and after J-L. Etienne's journey to the North Pole. J. Appl. Physiol. 58:646–51.
74. LeBlanc, J., J. A. Hildes, and O. Heroux. 1960. Tolerance of Gaspé fishermen to cold water. J. Appl. Physiol. 15:1031–34.
75. LeBlanc, J. 1962. Local adaptation to cold of Gaspé fishermen. J. Appl. Physiol. 17:950–52.
76. Scholander, P. F., H. T. Hammel, K. L. Andersen, and Y. Loyning. 1958. Metabolic acclimation to cold in man. J. Appl. Physiol. 12:1–8.
77. Itoh, S., and A. Kuroshima. 1972. Lipid metabolism of cold-adapted man. In S. Itoh, K. Ogata, and H. Yoshimura, eds., Advances in climatic physiology. Heidelberg, NY: Springer-Verlag New York.

78. Kondo, S. 1969. A study on the acclimatization of the Ainu and the Japanese with reference to hunting temperature reaction. J. Faculty Science, Tokyo University, Sect. V. 3–4:253–65.
79. Yoshimura, H., and T. Iida. 1952. Studies on the reactivity of skin vessels to extreme cold. Part 2. Factors governing the individual difference of the reactivity or the resistance against frostbite. Jpn. J. Physiol. 2:177–85.
80. Yoshimura, H., and T. Iida. 1950. Studies in the reactivity of skin vessels to extreme cold. Part 1. A point test on the resistance against frostbite. Jpn. J. Physiol. 2:147–59.
81. Schuman, L. M. 1953. Epidemiology of frostbite, Korea, 1951–1952. In Cold injury-Korea, 1951–1952. Rep. 113. Ft. Knox, KY: U.S. Army Medical Research Laboratory. 205–568.
82. Meehan, J. P. 1955. Individual and racial variations in a vascular response to a cold stimulus. Milit. Med. 116: 330–34.
83. Impietro, P. F., R. F. Goldman, E. R. Buskirk, and D. E. Bass. 1959. Response of Negro and white males to cold. J. Appl. Physiol. 14:798–800.
84. Rennie, D. W., and T. Adams. 1957. Comparative thermoregulatory responses of Negro and white persons to acute cold stress. J. Appl. Physiol. 11:201–4.
85. Wyndham, C. H. 1964. Southern African ethnic adaptation to temperature and exercise. In P. T. Baker, and J. S. Weiner, eds., The biology of human adaptability. Oxford: Clarendon Press.
86. Newman, R. W. 1969. Cold acclimation in Negro Americans. J. Appl. Physiol. 37:316–19.

PART 3

# Skin Color and Adaptation to Solar Radiation

CHAPTER 6

# Biological Responses, Acclimatization to Solar Radiation, and Evolution of Population Skin Color Differences

## NATURE OF SOLAR ENERGY

Every second inside the sun thermonuclear reactions convert approximately 564 million tons of hydrogen into 560 million tons of helium. During this process, large quantities of matter are destroyed and released from the sun's surface in the form of radiant energy. This radiant energy spreads in every direction at the speed of light and arrives at the outer limits of the earth's atmosphere at an intensity of 135.30 $mW/cm^2$ or slightly less than 2 $cal \times min^{-1}$. This value is called the *solar constant*. In ascending order the solar spectrum includes (1) short-wave, high-energy X rays, (2) ultraviolet rays, (3) visible light range, (4) infrared rays, and (5) long-wave, low-energy radiowaves. In general, about one half of the energy from the sun that reaches the earth's surface is in the form of infrared radiation, and the remaining half consists mostly of visible and ultraviolet light. The other forms of radiation such as X rays are screened out by the atmosphere (1).

In the process of passing through the atmosphere, ultraviolet, visible, and infrared radiation are absorbed, scattered without loss of energy, and reflected in every direction, depending on the specific wavelength involved and the degree of atmospheric turbidity and cloudiness. In general the rate of scattering is inversely proportional to the wavelength; the shorter the wavelength, the greater the amount of scattering (1, 2).

Wavelengths shorter than 290 nm are shielded from reaching the earth by the formation of ozone ($O_3$). With chlorophyll in plants, the atmospheric oxygen is continually replaced by photosynthesis and converted to ozone by radiation of wavelengths shorter than 240 nm. Ozone absorbs ultraviolet radiation from about 200 to 320 nm and blocks radiation below 290 nm. In this manner, the ozone layer protects the biosphere from biocidal radiations shorter than 290 nm.

In terms of human adaptation to solar radiation, the narrow wavelength band between 290 and 315 nm is the most crucial, since it produces vitamin D and is responsible for killing microorganisms in the atmosphere. The functional influences of solar radiation on humans are several, both direct and indirect. Among these the most important include (1) the provision of light and heat, (2) synthesis of biochemical components such as vitamin D, and (3) influence of hormonal and neuronal changes of pigmentation. In general, these influences take place through the skin and its components.

## SKIN STRUCTURE AND COLOR

### Skin Structure

Human skin consists of the dermis and epidermis, which are connected by a basement membrane and an intricate system of cells and blood vessels that enable the skin to function as one of the important factors in human thermoregulation and as protection against the deleterious effects of solar radiation. As shown in figure 6.1 the epidermis's outer surface has a stratum corneum (horny layer). This layer is extremely tough, chemically resistant, and almost impenetrable. Its average thickness is about 15 $\mu$m. It consists of many flat cells that have lost most of their cytoplasm and nuclei but contain many filaments of keratin in a matrix formation. These cells are held together by an extremely strong cement substance of unknown composition. The thickness of the stratum corneum layer is remarkably consistent. This layer is produced by a cell line called *keratinocytes,* that, because of their appearance, are known as the *prickle cells.* These prickle cells migrate outward during cell duplication and division. As they approach the stratum corneum, or horny layer, they accumulate granules, becoming the granular layer located directly below the stratum corneum.

Below the prickle cells are located the basal cells and the Langerhans' cells; interspersed among these are the melanocytes, or melanoblasts. The melanocytes synthesize the pigment melanin in a process that begins with oxidation of the amino acid tyrosine, which is then acted on by the enzyme tyrosinase. Melanin is deposited along with protein in specific subcellular organelles called *melanosomes.* These melanosomes are then transferred to keratinocytes of the skin and hair through the dendritic processes (3). It has been estimated that each melanocyte "services" about thirty-six keratinocytes. In other words, there is functional as well as structural integration of a melanocyte with its associated keratinocytes.

## Skin Color and its Measurement

The mature melanosome or melanin granule is dispersed into smaller melanin particles as the cells move outward. These cells range in color from brown to black. The melanin granules are about 1 mm in size and scatter themselves to form a cap over the nucleus of the keratinocyte, thus protecting the nuclear deoxyribonucleic acid (DNA) of the epithelium. As shown in recent studies, individual and population differences in skin color appear related to differences in both the amount of melanin synthesis and in the number of melanocytes (2, 4, 6). Among blacks or dark-skinned populations, such as the Australian aborigines and Solomon Island natives, melanocyte cells are greater and melanin granules are larger and more dispersed than those found in whites or Asians (2, 4, 6).

In addition to melanin, oxyhemoglobin, reduced hemoglobin, and carotene contribute to skin color. The most useful method of measuring these influences is by measurement of skin reflectance with either a photoelectric reflectometer or a reflectance spectrophotometer (see fig. 6.2). Both of these

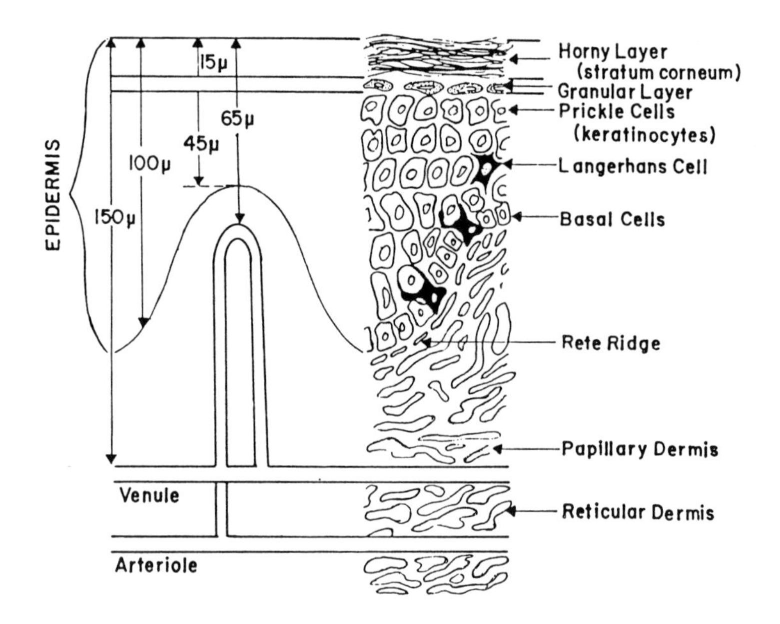


**Fig. 6.1.** Diagram of human skin. (Modified from F. Daniels. 1974. Radiant energy. Part A. Solar radiation. In N. B. Slonim, ed., Environmental physiology. Saint Louis, MO: C. V. Mosby Co.)

instruments work on the same principle of reflectance and have similar parts. Therefore, with appropriate statistical adjustments, their values are comparable (7). The main unit includes a galvanometer, a constant voltage transformer, and controls; the search unit consists of a lamp and a photocell. Light from the lamp passes through exchangeable glass filters of known wavelength to the surface being measured. The light that is diffusely reflected from this surface acts on the photocell. Thus the amount of light reflected from a skin surface, as compared with reflectance by standard white magnesium surface (giving 100% reflectance) is measured by the photocell and recorded on the galvanometer (7–12). Reflectance readings are usually taken with glass filters identified as blue, green, tri-amber, tri-blue, tri-green, and red. These filters have transmission peaks that range from 420 to 670 millimicrons. The filters and their transmission peaks and absorption capacities as used with the reflectometer (7–14) are given below.

| Glass Filter | Millimicron (m$\mu$) | Absorbs |
|---|---|---|
| Blue | 420 | Melanin and oxyhemoglobin |
| Green | 525 | Hemoglobin and carotene |
| Tri-amber | 600 | Melanin |
| Tri-blue | 450 | Melanin |
| Tri-green | 550 | Oxyhemoglobin and melanin |
| Red | 670 | Melanin |

The first important characteristic of spectral reflectance of the human skin is that reflectance is directly proportional to wavelength; the lower the reflectance, the shorter the wavelength and the darker the skin color. Melanin has no specific absorption bands but general areas of absorption range from infrared to ultraviolet (2).

### Body Sites of Measurement

The most common body sites where skin reflectance is measured include: (1) the inner upper arm distal to the axillary region of the arm (see fig. 6.2); (2) the subscapular area; and (3) the forehead. Readings in the inner upper arm are used to evaluate the biological characteristics of skin color, while the forehead is used to evaluate the effects of tanning on skin color.

### Age and Sex Differences

As measured by skin reflectance in the inner arm from infancy to the beginning of adolescence there is rapid darkening in skin color, followed by a lightening through adulthood. Furthermore, females through all ages are lighter than males (12, 13).

### Inheritance of Skin Color

**Population Level.** The heritability of skin color has been very little studied. As shown by the studies of admixture of white Europeans and black Africans (15) and South American blacks (10) population differences in skin color are related to three or four genes. As shown in figure 6.2 at the short wavelength (400 to 500 nm) the mean skin reflectance for the Black hybrid (F1) is closer to the African black (assumed to be homozygous for black skin color) than to the European whites (assumed to be homozygous for white skin color). This finding suggests that genetic contribution to skin color is greater in blacks than in whites, and based on analyses of these data it has been suggested that about 70% of variance in skin color in the offspring of black Africans and white Europeans is due to additive genetic contribution (15).

**Individual Level.** Studies of parent-offspring and sibling resemblance in skin reflectance indicate that skin color is inherited in a Mendelian and polygenic form (13). These studies also indicate a high degree of assortative mating in skin color so that light skin individuals tend to marry other light skinned and vice versa. According to research among Japanese (11) assortative mating and social selection for skin color has favored the concentration of genes for lighter skin color in the upper classes. Thus it appears that attractiveness for light skin color might have contributed to enhance any possible

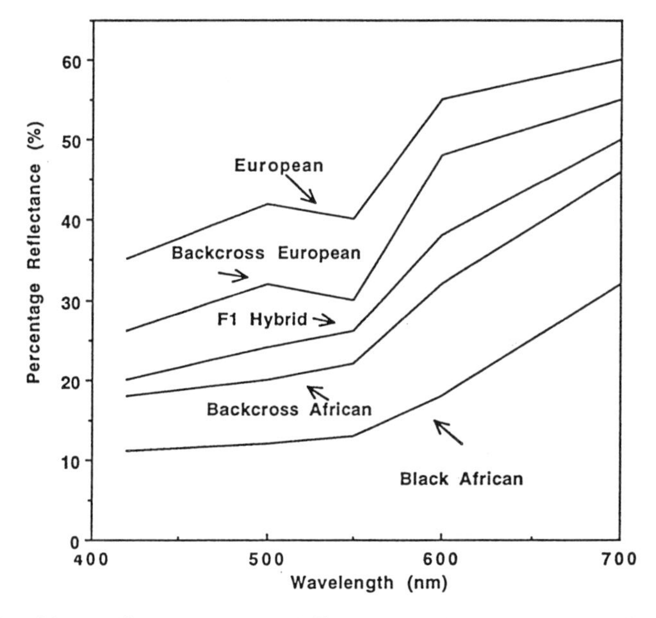


**Fig. 6.2.** Mean reflectance curves of European, African, and various hybrid groups.

genetic distinctions between social groups. Despite the effect of assortative mating as shown in figure 6.3 additive genetic variance accounted for about 56% of the total phenotypic variability in skin color.

## INFLUENCE OF RADIATION ON HUMAN SKIN

### Sunburn

When solar radiation is intense, 290 to 315 nm, and the skin is not tanned, an individual develops sunburn. The general sequence of events includes a latent period of several hours, followed by blood vessel dilation manifested in erythema, which reaches its maximum between 8 and 24 hours after exposure. As a result of sunburn, there is general discomfort, a reduction in the pain threshold, and severe blistering. From sunburn the skin may also develop secondary infection or a suppression of sweating. Susceptible individuals, such as those with low melanization, may develop desquamation and peeling of the sunburned area with continued exposure; this in turn reduces the possibility of maintaining adequate melanization. Therefore, repeated sunburn may lead to degenerative changes in both the dermis and epidermis, and skin cancer is one of the possible consequences (2).

### Melanization

On exposure to ultraviolet radiation the skin becomes melanized or tanned by two distinct processes: (1) immediate tanning and (2) delayed tanning (16).

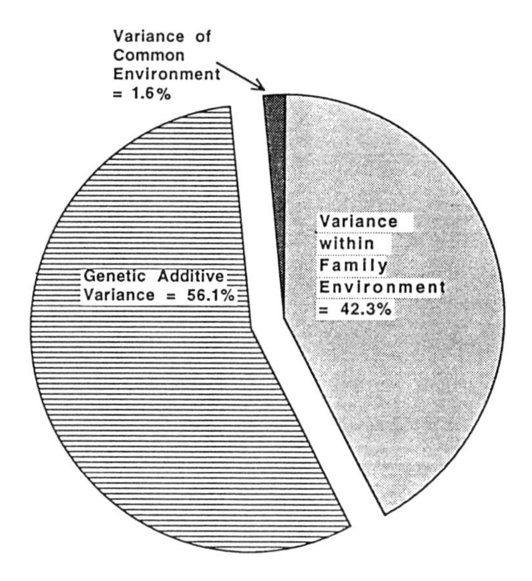


**Fig. 6.3.** Components of phenotypic expression in skin color of Peruvian Mestizos. Note that of the total phenotypic variance (Vp) in skin color, 56% is due to the effect of additive genetic inheritance ($V_A$), 42% is due to the variance within family environment ($V_{EW}$), whereas 2% is due to variance of common environment ($V_{EC}$).

## Immediate Tanning

Immediate tanning is produced by both ultraviolet (320 to 380 nm) and visible light (400 to 700 nm). Within 5 to 10 minutes after exposure to midday summer sun the skin becomes hyperpigmented. When withdrawn from exposure to light for about 4 hours, the tanned area fades almost to nonexposure levels. The rate of depigmentation appears to be related to length of exposure to ultraviolet light; after prolonged exposure (90 to 120 minutes) residual hyperpigmentation may be visible for as long as 24 to 36 hours. The rapid rate of immediate tanning is brought about by (1) rapid darkening of preformed melanin, (2) rapid transfer of preformed melanosomes (melanin) from the basal location to the upper Malpighian layer and stratum corneum, and (3) rapid distribution of melanosomes in melanocytes. The rapid darkening of preformed melanin occurs either because of photo-oxidation of melanosomes or because of the enzyme-mediated oxidation of melanin.

## Delayed Tanning

Delayed tanning is optimally stimulated by exposure to short ultraviolet radiation between 290 and 315 nm and to a lesser extent by exposure to long-wave ultraviolet radiation and visible radiation. Because delayed tanning involves production of new melanosomes, it appears slowly over a period of 48 to 72 hours. Current theories (16) postulate that delayed melanization occurs from (1) an increased number of functional melanocytes resulting from a proliferation of melanocytes and possible activation of dormant or resting melanocytes, (2) hypertrophy of melanocytes and increased ramification of their dendrites, (3) increased melanosomal synthesis, (4) increased rate of melanization in melanosomes, (5) transfer of melanosomes from melanocytes to keratinocytes because of increased turnover in keratinocytes, and (6) activation of tyrosinase because of the direct effect of radiation on the tyrosinase-inhibiting sulfhydryl compounds present in the epidermis (17).

Figure 6.4 illustrates the development of skin tanning in relation to length of exposure to summer sunlight in Idaho. These data show that with constant 40-minute exposures a rapid melanization (darkening) of the skin occurs, which is stabilized by the end of the second week (broken lines). However, these data also show that the greater the length of daily exposure to sunlight (straight lines),the greater the degree of melanization. In other words, the amount of melanization is proportional to amount of daily exposure to solar radiation.

## Synthesis of Vitamin D in Human Skin

### Types of Vitamin D

Humans derive vitamin D either from the diet or from photosynthesis in the skin.

### Dietary Source

Ergocalciferol is the synthetic form of vitamin $D_2$ made by irradiation of ergosterol from plants and fungi, whereas the natural form, cholecalciferol, is found in small amounts in most seafoods. As shown in table 6.1, the dietary sources of vitamin D are limited to seafood and, to a lesser extent, dairy foods. To meet childhood requirements for vitamin D of 5 to 10 mg daily, one would have to consume between 11 and 23 g of swordfish, sardines, or mackerel or 100 g of herring. For littoral people these dietary sources might be easily accessible, but for the majority of populations they are not.

### Photosynthesis

The same wavelengths of solar radiation, 290 to 315 nm, that may cause sunburn, cataracts, and skin cancer, also assist in the synthesis of vitamin $D_3$. During exposure to sunlight the ultraviolet-B (UV-B) portion of the solar spectrum penetrates the skin and synthesizes vitamin D. In 1949 it was found that when the skin was irradiated it produced a photoproduct which was

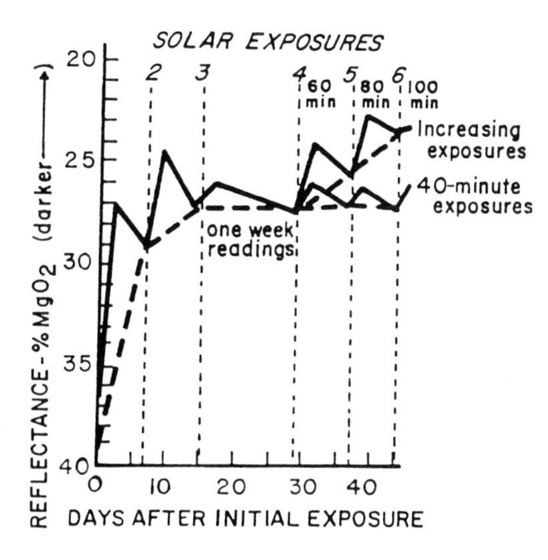


**Fig. 6.4.** Development of tanning in relation to length of exposure to summer light. After weeks of exposure, amount of melanization increases in proportion to length of time of daily exposure to solar radiation. (Modified from F. Daniels. 1974. Radiant energy. Part A. Solar radiation. In N. B. Slonim, ed, Environmental physiology. Saint Louis, MO: C. V. Mosby Co.)

**TABLE 6.1. Dietary sources of vitamin D**

| Food | Vitamin D per Portion |
|---|---|
| Swordfish | 45 μg/100 g |
| Sardines, canned | 29-39 μg/100 g |
| Salmon, raw | 4-14 μg/100 g |
| Salmon, canned | 6-12 μg/100 g |
| Mackerel, raw | 28 μg/100 g |
| Herring, raw | 8 μg/100 g |
| Herring, canned | 8 μg/100 g |
| Halibut | 1 μg/100 g |
| Shrimp | 4 μg/100 g |
| Oysters | 0.1 μg/3–4 medium-sized |
| Liver | |
| Beef | 0.2 μg/100 g |
| Calf | 0.0-0.3 μg/100 g |
| Chicken | 1.2-1.7 μg/100 g |
| Eggs | 0.6 μg/average yolk |
| Cow's milk | <0.2 μg/8 oz |
| Cream | 0.4 μg/1 oz coffee cream |
| Butter | 0.1 μg/2 pats |
| Cheese | 0.3-0.4 μg/100 g |

*Source:* From R. M. Neer. 1975. The evolutionary significance of vitamin D, skin pigmentation, and ultraviolet light. Am. J. Phys. Anthropol. 43:409–16.

called previtamin D (18). According to current research (18–21) sunlight falling on human skin goes through the following process: (1) sunlight interacts with the 5,7-diene of 7-dehydrocholesterol (provitamin $D_3$); (2) the 5,7-diene absorbs these photons and undergoes a photochemical reaction to produce previtamin $D_3$; and (3) once previtamin $D_3$ is formed in the skin it is isomerized to become secosterol vitamin $D_3$. It is estimated that more than 90% of previtamin $D_3$ synthesis occurs in the epidermis of the skin. According to these investigations, if all the available 7-dehydrocholesterol in 8 $cm^2$ of human skin were converted to vitamin $D_3$, this would provide the daily requirement for vitamin D.

As illustrated in figure 6.5 vitamin D synthesized in the skin ($D_3$) or absorbed from the diet ($D_2$) enters the liver where it is hydrolyzed. This process produces 25-hydroxycalciferol (25-HCC). From the liver 25-HCC is transported to the kidney where the compound undergoes further hydroxylation. In the kidney the 25-HCC is converted to 1,25 dihydroxycalciferol (1,25 DHCC) under hypocalcemic conditions, or under normal conditions it is converted to 24,25 dihydroxicholecalciferol (2, 19, 22). Thereafter, vitamin D, in either form, enters the circulation and is subsequently bound to the nuclei of intestinal epithelial cells. It is assumed that in these cells vitamin D acts as a messenger, instructing the DNA to transcribe information to RNA, which in turn transcribes a message to form the enzyme needed for calcium ($Ca^{++}$) transport from gut lumen into the circulation. In this manner, vitamin D presumably regulates calcium absorption through biochemical events, enabling the formation and maintenance of an internal calcified skeleton.

## Function of Vitamin D in Calcium Balance

Although the direct mechanisms whereby vitamin D influences bone mineralization are not well known, there are several functions attributable to vitamin D that have a direct effect on proper bone mineralization. First, it is well known that vitamin D stimulates intestinal calcium and phosphorus absorption. Second, vitamin D stimulates the mobilization of calcium and phosphate from the bone through resorption. By contributing calcium and phosphate to the blood pool, vitamin D helps maintain a plasma concentration of these minerals sufficient for proper bone calcification. Hormonal factors also interact with vitamin D in bone resorption as well as in mineralization. The influence of the parathyroid hormone in mobilizing calcium is decreased when there is a vitamin D deficiency (2). Third, it has been suggested that vitamin D stimulates renal tubular transport of calcium and phosphorus (20, 23).

It is now evident that the observed abnormalities in calcium metabolism are directly related to the availability of vitamin D, which in turn depends

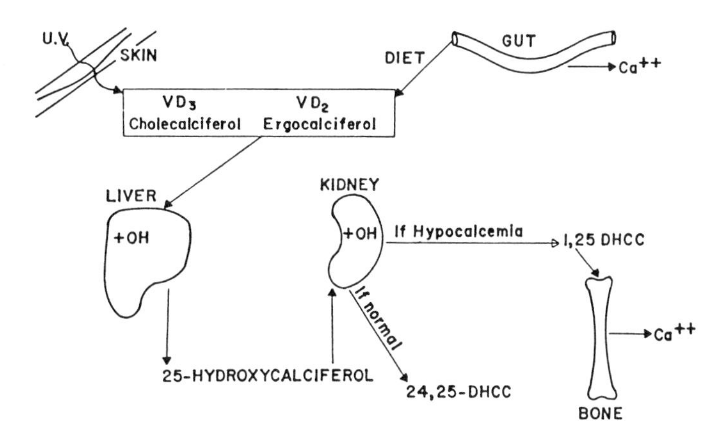


**Fig. 6.5.** Diagram of vitamin D synthesis. Through the action of ultraviolet radiation vitamin D is synthesized in skin or absorbed from diet, after which it is hydrolized in liver, transported to kidney, and from there enters the circulation to be used during calcium metabolism. (Modified from R. M. Neer. 1975. Am. J. Phys. Anthropol. 43:409–416.)

on the availability of solar radiation. Clinical studies clearly indicate that childhood rickets is readily cured with 0.05 to 0.1 mg (2000 to 4000 I.U.) of vitamin D daily for 6 to 12 weeks, prolonged exposure to winter sunlight, or 5 minutes exposure to artificial ultraviolet radiation equivalent to strong summer sunlight at 36° latitude (24). However, bone disorders associated with a vitamin D deficiency may occur in humans even though dietary intake of vitamin D and its precursors is normal, and there has been sufficient exposure to sunlight or ultraviolet radiation. Such bone disorders are usually related to metabolic factors that inhibit activation of vitamin D (24, 25).

## Evidence for the Role of Solar Radiation on Vitamin $D_3$ Synthesis

### *Epidemiological Evidence*

From the seventeenth to the nineteenth century children who lived in the inner cities of Poland, England, and United States had a very high incidence of rickets, whereas children living in the rural areas were free of this disease (26–29). Similarly, in China, India, and Japan, where the children were often malnourished and lived in squalor, rickets was a clinical rarity, while in middle-class London and Manchester, where the children received reasonably good nutrition, rickets abounded (29–30). Rickets usually appeared in the first year of life and was characterized by muscle weakness, deformities of the long bones including bowed legs, knucklelike projections along the rib cage known as the rachitic rosary, and deformities of the pelvis that were often permanent. The consequences of this disease were quite profound, especially for young women, in whom a deformed and narrow pelvis (see fig. 6.6) would cause difficulty with childbirth and result in a high incidence of infant and maternal morbidity and mortality. In the 1800s various investigators suggested that rickets was caused by lack of exposure to sunlight.

*Experimental Evidence*

In 1919 Huldschinsky (16) exposed four rachitic children to radiation from a mercury arc lamp and several months later they were completely cured of the disease. Similarly, in 1921 Hess and Unger (31) exposed eight rachitic children to natural sunlight on the roof of a hospital in New York City over a period of several months and showed by radiologic evaluation that this treatment healed the rachitic lesions. Later on several investigators (32, 33) demonstrated that exposure of the diet to radiation from a mercury vapor lamp could also impact antirachitic activity and cure rickets in experimental animals. Based on these findings milk was initially fortified with provitamin D that had been exposed to UV radiation, and eventually with the simple addition of vitamin D. Today each quart of milk in the United States is fortified with 400 IU (10 $\mu$g, the recommended daily allowance for adults) of either vitamin $D_2$ or vitamin $D_3$.

*Seasonal Effects*

As shown in figure 6.7 despite the availability and consumption of vitamin D-fortified dairy products in Michigan, variations in plasma vitamin D levels paralleled variations in solar radiation between summer and winter. In contrast, in Puerto Rico where there are minimal seasonal differences in solar intensity, there are no seasonal variations in vitamin D levels (34). In the same manner, in London variations in blood 25-OH vitamin D are associated with seasonal differences in solar intensity (35, 36). Reflecting these seasonal differences in vitamin D synthesis is the fact that in Swedish adults intestinal calcium absorption is more efficient in the summer, declining progressively during the winter (37). Another example illustrating the seasonal differences in vitamin D synthesis and its influence on calcium metabolism is that before the generalized use of vitamin D-fortified milk incidence of childhood rickets in the northern United States was greater in the winter than in the summer (25). The incidence of rickets in New Haven, Connecticut, at a latitude of 40° N, was greater than in Puerto Rico at a latitude of 18 N. Similarly, in Scotland the incidence of infantile hypocalcemia and defective teeth is much higher among children who developed in utero during the winter (38, 39). It is evident, then, that the limiting factor in cutaneous vitamin D production is the amount of absorbed solar radiation.

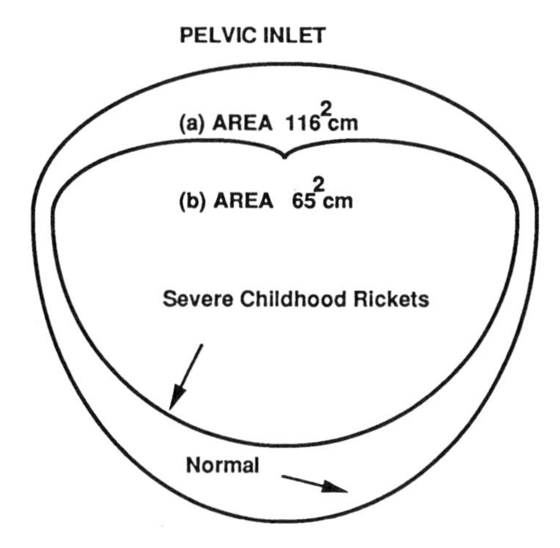


**Fig. 6.6.** Outline of normal pelvis and pelvis deformed by childhood rickets. Deformed pelvis impairs reproductive efficiency. (Modified from N. J. Eastman. 1965. Obstetrics, 11th ed. New York: Appleton-Century-Crofts.)

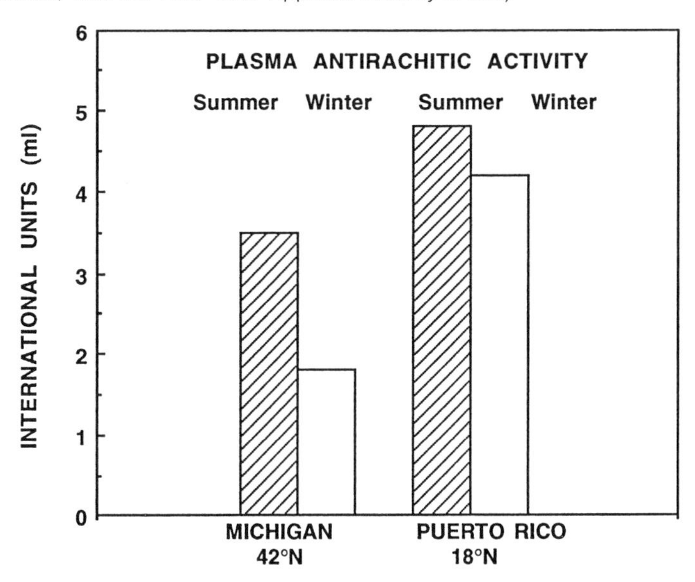


**Fig. 6.7.** Comparison of seasonal variation in plasma antirachitic activity between Michigan and Puerto Rico. In Michigan, with low solar radiation in winter, there are significant differences in plasma antirachitic throughout the year. In Puerto Rico there are no seasonal variations in plasma antirachitic activity. (Modified from R. W. Smith and B. Frame. 1964. Am. J. Clin. Nutr. 14:1498–508.)

## OVERVIEW

Solar radiation is the major source of energy for the terrestrial biosphere. It affects all levels of life from a single photon acting on one electron in a molecule through all levels of molecular organization—organelles, cells, organs, individuals, and populations. In this chapter we have been concerned mostly with middle-length and long-wave ultraviolet radiation. Depending on the degree of acclimatization, the effects of solar radiation on the health and well-being of the organism can be both positive and negative.

When exposed to ultraviolet and visible solar radiation, the unacclimatized organism responds with reddening of the skin or erythema, followed by initial tanning from the darkening of peripheral melanin molecules. With continued exposure the melanocytes of the epidermis increase their synthesis of melanin and granules; the granules are then extruded toward the stratum corneum after which they are taken up into the keratinocytes. Along with this secondary hyperpigmentation, the thickness of the superficial stratum corneum or horny layer of the epidermis increases. The two processes represent a protective response in that they help retard the passage of ultraviolet light into the deeper layers of the skin. Without these two processes, excessive exposure to solar radiation in the unacclimatized organism might result in sunburn and related cellular injuries and, possibly, in skin cancer.

Solar radiation plays an important role in the synthesis of vitamin D. As shown by studies indicating marked seasonal differences in plasma vitamin D levels in regions characterized by seasonal differences in solar radiation and use of fortified vitamin D foods, the majority of vitamin D is synthesized through solar radiation stimulus. Because vitamin D plays an important role in the metabolism of calcium, variations in solar radiation intensity are reflected in marked seasonal differences in the incidence of rickets. Therefore individual and populational differences in melanin-producing capacity have profound influences on the well-being and survival of the organism.

## References

1. Thekaekara, M. P., and A. J. Drummond. 1971. Standard values for the solar constant and its spectral components. Nature Phys. Sci. 229:6–9.
2. Daniels, F., Jr. 1974. Radiant energy. Part A. Solar radiation. In N.B. Slonim, ed., Environmental physiology. St. Louis, MO: C. V. Mosby Co.
3. Fitzpatrick, T. B., W. C. Quevedo, Jr., G. Szabo, and M. Seiji. 1971. The melanocyte system. In T. B. Fitzpatrick, ed., Dermatology in general medicine. New York, NY: McGraw-Hill.
4. Garcia, R. I., R. E. Mitchell, J. Bloom, and G. Szabo. 1977. Number of epidermal melanocytes, hair follicles, and sweat ducts in the skin of Solomon Islanders. Am. J. Phys. Anthropol. 47:427–34.
5. Mitchell, R. E. 1968. The skin of the Australian aborigine: a light and electromicroscopical study. Australas. J. Dermatol. 9:314–28.
6. Szabo, G., A. B. Gerald, M. A. Pathak, and T. B. Fitzpatrick. 1972. The ultrastructure of racial color differences in man. In V. Riley, ed., Pigmentation: its genesis and biologic control. New York, NY: Appleton-Century-Crofts.
7. Garrad, C., G. A. Harrison, and J. J. T. Owen. 1967. Comparative spectrophotometry of skin color with EEL and Photovolt instruments. Am. J. Phys. Anthropol. 27:389–96.

8. Lasker, G. W. 1954. Seasonal changes in skin color. Am. J. Phys. Anthropol. 12: 553–58.
9. Weiner, J. S., G. A. Harrison, R. Singer, R. Harris, and W. Japp. 1964. Skin color in Southern Africa. Hum. Biol. 36:294–307.
10. Harrison, G. A., J. T. T. Owen, F. J. DaRocha, and F. M. Salzano. 1967. Skin color in Southern Brazilian populations. Hum. Biol. 39:21–31.
11. Hulse, F. S. 1967. Selection for skin color among Japanese. Am. J. Phys. Anthropol. 27:143–56.
12. Conway, D., and P. T. Baker. 1972. Skin reflectance of Quechua Indians: the effects of genetic admixture, sex and age. Am. J. Phys. Anthropol. 36:267–82.
13. Frisancho, A. R., R. Wainwright, and A. Way. 1981. Heritability and components of phenotypic expression in skin reflectance of mestizos from the Peruvian low-lands. Am. J. Phys. Anthropol. 55:203–8.
14. Little, M. A., and C. J. Sprangel. 1980. Skin reflectance relationships with temperature and skinfolds. Am. J. Phys. Anthropol. 52:145–51.
15. Harrison, G. A. and J. J. T. Owen. 1964. Studies on the inheritance of human skin colour. Ann. Hum. Genet., Lond. 28:27–37.
16. Quevedo, W. C., Jr., T. B. Fitzpatrick, M. A. Pathak, and K. Jimbow. 1975. Role of light in human skin color variation. Am. J. Phys. Anthropol. 43:393–408.
17. Pathak, M. A., Y. Hori, G. Szabo, and T. B. Fitzpatrick. 1971. The photobiology of melanin pigmentation in human skin. In T. Kawamura, T. B. Fitzpatrick, and M. Seiji, eds., Biology of normal and abnormal melanocytes. Tokyo: University of Tokyo Press.
18. Holick, M. F. 1987. Photosynthesis of vitamin D in the skin: effect of environmental and life-style variables. Fed. Proc. 46:1876–82.
19. DeLuca, H. F., J. W. Blunt, and H. Rikkers. 1971. The vitamins, 2nd ed. W. H. Sebrell, and R. S. Harris, eds., Vol. 3. New York, NY: Academic Press.
20. Omdahl, J. L., and H. F. DeLuca. 1973. Regulation of vitamin D metabolism and function. Physiol. Rev. 53:327–72.
21. McLaughlin, J. A., R. R. Anderson, and M. F. Holick. 1982. Spectral character of sunlight modulates phytosynthesis of previtamin $D_3$ and its photoisomers in human skin. Science. 216:1001–3.
22. Neer, R. M. 1975. The evolutionary significance of vitamin D, skin pigment, and ultraviolet light. Am. J. Phys. Anthropol. 43:409–16.
23. Yendt, E. R., H. F. DeLuca, D. A. Garcia, and M. Gohanim. 1970. Clinical aspects of vitamin D. In H. F. DeLuca, and J. W. Suttie, eds., The fat-soluble vitamins. Madison, WI: University of Wisconsin Press.
24. Loomis, W. F. 1970. Rickets. Sci. Am. 223:76–91.
25. Loomis, W. F. 1967. Skin-pigment regulation of vitamin-D biosynthesis in man. Science. 157:501–6.
26. Griffenhagen, G. 1952. A brief history of nutritional diseases. II. Rickets. Bull. Natl. Inst. Nutr. 2(No. 9).
27. Holick, M. F. 1986. Sunlight and skin: their role in vitamin-D nutrition for humans. In A. D. Roe, ed., Nutrition and the skin. New York, NY: Liss.
28. Mettler, C. C., and F. A. Mettler, eds. 1947. Pediatrics in the seventeenth century: Social circumstances. In C. C. Mettler and F. A. Mettler, eds., History of Medicine; a correlative text, arranged according to subjects. Philadelphia, PA: Blakiston.
29. Sniadecki, J. 1939. Cited by Mozolowski, W. Jedrzej Sniadecki (1768-1838) on the cure of rickets. Nature. (Lond). 143:121.

30. Palm, T. A. 1890. The geographic distribution and etiology of rickets. Practitioner 45:270–79, 321–42.
31. Hess, A. F., and L. J. Unger. 1921. Cure of infantile rickets by sunlight. J. Am. Med. Assoc. 77:39.
32. Hess, A. F., and M. Weinstock. 1924. Antirachitic properties imparted to inert fluids and green vegetables by ultraviolet irradiation. J. Biol. Chem. 62:301–13.
33. Steenbock, H. 1924. The induction of growth-promoting and calcifying properties in a ration exposed to light. Science 60:224–25.
34. Smith, R. W., J. Rizek, and B. Frame. 1964. Determinants of serum antirachitic activity. Special reference to involutional osteoporosis. Am. J. Clin. Nutr. 14: 98–108.
35. Stamp, T. C. B., and J. M. Round. 1974. Seasonal changes in human plasma levels of 25-hydroxy-vota,om D. Nature 247:563–85.
36. Haddad, J. G., and K. J. Chuy. 1971. Competitive protein binding radioassay for 25-hydroxycholecalciferol. J. Clin. Endocrinol. Metab. 33:992–95.
37. Malm, O. J. 1958. Calcium requirement and adaptation in adult men. Scand. J. Clin. Lab. Invest. (suppl.). 10:36.
38. Roberts, S. A., M. D. Cohen, and J. O. Forfar. 1973. Antenatal factors associated with neonatal hypocalcemic convulsions. Lancet 2:809–11.
39. Purvis, R. J., W. J. Barrie, G. S. MacKay, E. M. Wilkinson, F. Cockburn, N. R. Belton, and J. O. Forfar. 1973. Enamel hypoplasia of the teeth associated with neonatal tetany: A manifestation of maternal vitamin-D deficiency. Lancet 2: 811–14.

# CHAPTER 7 Evolution of Population Skin Color Differences

In general, populational differences in skin color are associated with two climatic and environmental factors. First, on each continent skin color is inversely related to latitude and temperature; the closer the population to the equator and the higher the temperature, the darker the skin color (fig. 7.1). Second, the intensity of solar radiation is directly related to latitude and altitude; the closer the location to the equator and the higher the altitude, the greater the radiation intensity. Along with these climatic associations, skin color differences are also related to differences in incidence of skin cancer and malignant melanoma; the lighter the skin and the lower the latitude, the higher the incidence of skin cancer and malignant melanoma. With a view to the long-term consequences of these associations, biologists have postulated different theories to explain populational differences in skin color. Among these the most important are those concerned with skin cancer and vitamin D synthesis.

## SKIN CANCER AND MALIGNANT MELANOMA: ADAPTIVE SIGNIFICANCE OF DARK SKIN

### Epidemiological Evidence

According to United States mortality data, the death rate from skin cancer in whites is approximately three times greater than that for blacks. It has been demonstrated that this difference is related to the amount of melanin pigmentation by the fact that frequency of skin cancer is greater among albino Bantus than among nonalbino Bantus (1). Furthermore, among whites the incidence of skin cancer is higher in the southern than in the northern United States (1–2). Similarly, red-haired, blue-eyed, freckled individuals have a greater probability of developing carcinoma of the skin if engaged in outdoor occupations than do dark-skinned individuals (3).

In early studies the influence of solar radiation on malignant melanoma was not well defined. However, the importance of solar radiation as an etiological factor in the incidence of malignant melanoma in white populations has since been recognized (4). Furthermore, like skin cancer, the incidence of malignant melanoma is greater among those with outdoor occupations and lighter pigmentation than with darker skin color (4). These investigations clearly demonstrate the association of skin cancer and malignant melanoma with solar radiation and degree of pigmentation. It has been suggested that these differences result from genetic differences that evolved through the influence of natural selection. Others have questioned the evolutionary significance of skin cancer differences in view of the fact that mortality from skin cancer usually occurs after completion of the reproductive period. However, recent literature indicates that mortality from skin cancer and malignant

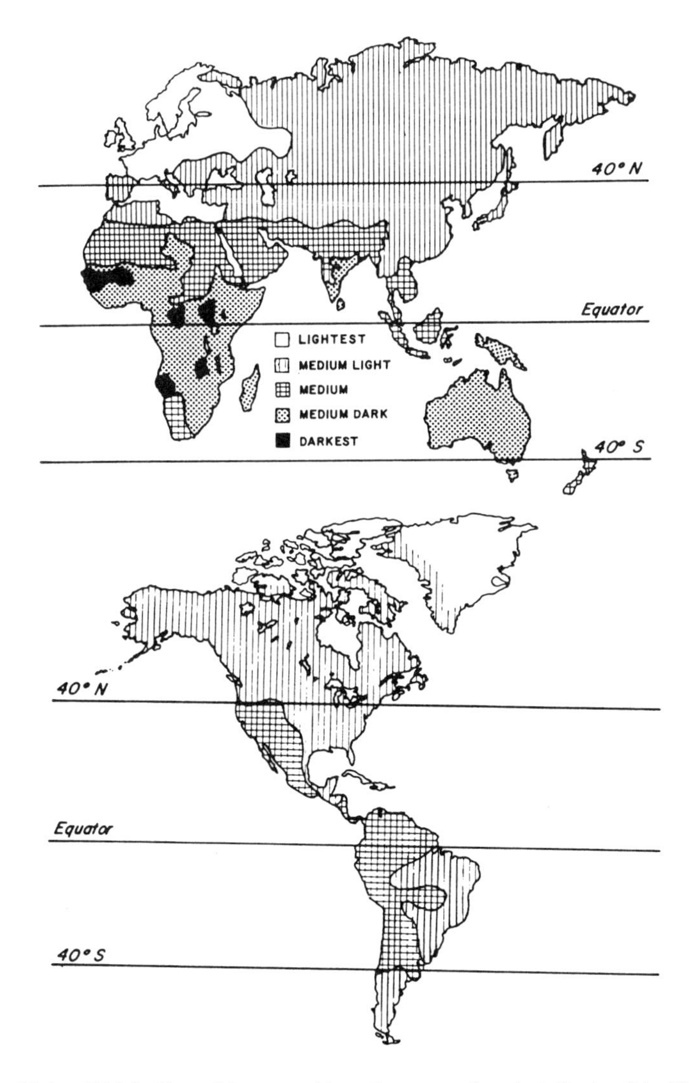


**Fig. 7.1.** Distribution of human skin color according to latitude. (Modified from C. L. Brace and A. Montague. 1965. Man's evolution. New York: Macmillan Co.)

melanoma together, although occurring at lower frequencies, does occur prior to completion of the reproductive period.

## Evolutionary Trends

As shown in epidemiological data, in Australian whites living in a region of high solar radiation, approximately 669 deaths per 100,000 between the ages of 10 and 49 years were recorded to be caused by malignant melanoma, and about 8 deaths per 100,000 in the same age range were caused by other skin cancers (4–6). In the United States and Australia the survival rate for individuals suffering from melanoma under the age of 50 years is approximately

50% because of improved medical care and is approximately 90% for skin cancer (4–6). One would expect that among populations without access to adequate medical care, such as those of early man, the actual mortality rate from melanoma would equal about 1338 deaths per 100,000 and about 89 deaths per 100,000 for other skin cancers. The total combined mortality for these two diseases without modern medical care would equal 1427 per 100,000, or approximately 142/10,000. In view of the lower incidence of malignant melanoma and skin cancer in dark-skinned populations, the selection against light skin color would be high in areas of intense solar radiation. Given the nearly 3 million years of human evolution, it is conceivable that this level of selection could have had a strong influence on the evolution of dark skin color. Through computer simulation, it has been estimated that with optimizing selection and a 6% maximum difference in fitness, the evolution of the range of human skin color differences would have taken about 800 generations with no dominance and about 1500 generations with 80% dominance (7). This would suggest that changes in skin color could have taken place within a range of 24,000 to 45,000 years.

## VITAMIN D SYNTHESIS: ADAPTIVE SIGNIFICANCE OF LIGHT SKIN

### Physiological Evidence

According to spectrophotographic studies the rate at which ultraviolet radiation penetrates the epidermis is inversely related to the amount of melanin, so that the darker the skin, the lower the absorption rate (8–9). Assuming there is a relationship between the absorption of solar radiation and vitamin D synthesis various investigators have postulated that dark skin evolved in order to prevent vitamin D synthesis in populations exposed to high amounts of sunlight, and that variations in human skin pigmentation among the world's populations arose from, and are maintained by, the necessity of regulating the synthesis of vitamin D within certain physiological limits (8–10).

A basic premise of this hypothesis is that the human skin has a high capacity for synthesizing vitamin D. According to Loomis (8) 1 $cm^2$ of human white skin synthesizes up to 18 IU of vitamin D over a 3-hour exposure period to solar radiation. Therefore it is estimated that an individual exposing 1.5 $m^2$ surface area of skin for 6 hours at the equator would produce in excess of 800,000 IU of vitamin D without sufficient melanin in the stratum corneum to filter the intense solar radiation, an amount well within toxic quantities. Recent studies, however, question the concept that prolonged exposure to solar radiation of whites leads to vitamin D intoxication (11). These studies demonstrated that samples of white skin when exposed to simulated sunlight irrespective of length of exposure convert no more than 10 to 20% of the previtamin $D_3$ to vitamin D. Prolonged exposure to sunlight only causes previtamin $D_3$ to isomerize to the two biologically inactive photoproducts known as lumisterol and tachysterol (12). This finding questions the adaptive significance of dark skin color with reference to its capacity for vitamin D synthesis. For this reason the origin of dark skin color may not be explained by the hypothesis of vitamin D synthesis (8). On the other hand, when specimens of white and black skin were exposed to simulated sunlight under the same conditions, the exposure times for the formation of previtamin $D_3$ increased exponentially with the degree of pigmentation (12, 13). Therefore, under conditions of low solar radiation white skin would be more adaptive than dark skin. If this is the case the high frequency of white skin color in the northern latitudes which is characterized by low solar radiation may be viewed as a by-product of adaptive responses to facilitate vitamin D synthesis.

### Epidemiological Evidence

The hypothesis that white skin is adaptive under conditions of low solar radiation is supported by several findings. First, blacks are more susceptible to rickets than whites in northern latitudes (8). Before the widespread use of dietary vitamin D supplements that started in the 1930s, the incidence of deformed pelvis among black women studied in the 1950s was significantly greater (15%) than that of white women (2%) (14) (see fig. 6.5 in previous chapter). The size reduction in the pelvic inlet and the absence of cesarean operations in any population would impair reproductive efficiency. Second, the incidence of vitamin D deficiency rickets and osteomalacia in Saudi Arabia is higher than expected (15). Recent studies indicate the low vitamin $D_3$ status of Saudis is related to their avoidance of sunlight exposure and dress code which involves the complete covering of the skin except for face and hands (16). Similarly, dark-skinned Pakistanis in Boston required longer exposure to ultraviolet radiation than whites to synthesize the same amount of Vitamin D (17). Therefore, the low vitamin D status of Saudis is probably related to their dark skin.

## OVERVIEW

As shown by photochemical and photometric studies, populational differences in skin color appear related to different rates of melanin synthesis and the size of the melanin granules as well as to the number of melanocytes. Various hypotheses have been postulated to explain the origin of populational differences in skin color. These proposals suggest that cutaneous melanin pigmentation is, or has been, associated with (1) resistance to sunburn, solar degeneration, and skin cancer; and (2) regulation of vitamin D biosynthesis by regulating penetration of ultraviolet light into the skin. In addition, earlier hypotheses have postulated that melanin may have been adaptive as camouflage for early man in evading predators and for hunting (18). These theories have directly linked cutaneous melanin with human fitness, thus emphasizing that skin pigmentation has been a major target for natural selection. Analysis of vital statistics from Australia indicates that deaths from skin cancer and malignant melanoma, which are directly associated with intensity of solar radiation and inversely related to pigmentation, occur at a rate of 1:10,000 for individuals under the age of 50 years. Given the nearly 3 million years of human evolution, it is conceivable that selection by skin cancer and malignant melanoma may indeed have been an important factor in the development of populational differences in skin color.

It is evident that populational differences in human skin color developed in response to more than one environmental factor, since no single selective force can account for the distribution of existing skin color variations. Therefore, as schematized in figure 7.2, populational differences in skin color probably evolved as a by-product of the competing selective forces of skin cancer (or malignant melanoma) and vitamin D synthesis.

In addition, several hypotheses have been postulated which at present are either not substantiated by current research or have no adaptive meaning. These include a hypothesis that postulates that the origin of populational differences in skin color is related to thermoregulatory differences between dark and light skin color (19). The hypothesis postulates that brunette and other intermediate-colored skins would be most adaptive under hot-dry conditions (19). This hypothesis is based on the finding that blacks exercising unprotected in the desert had a greater increase in rectal temperatures than did whites (19). It is assumed that this difference is related to an enhanced heat absorption by dark skin and thereby an increase in the gradient between skin and ambient temperature, enhancing the rate of heat radiation to portions of the environment cooler than the skin temperature and receiving less heat from the environment above that of skin temperature. This possibility, however, is not substantiated by current research that indicates that in the infrared range (5 to 20 m), skin color does not have an effect on reflectance; for this reason it is said that the human skin acts as a black body radiator with a power of emissivity close to 1.

Another hypothesis postulates that cold injury or frostbite may be an important factor in the evolution of white skin (20). According to this hypothesis, as humans moved away from the equator toward the colder temperatures of the northern regions, and in the absence of well-developed technologies, white skin was favored by natural selection. This hypothesis is based on evidence that during the two world wars, as well as the Korean

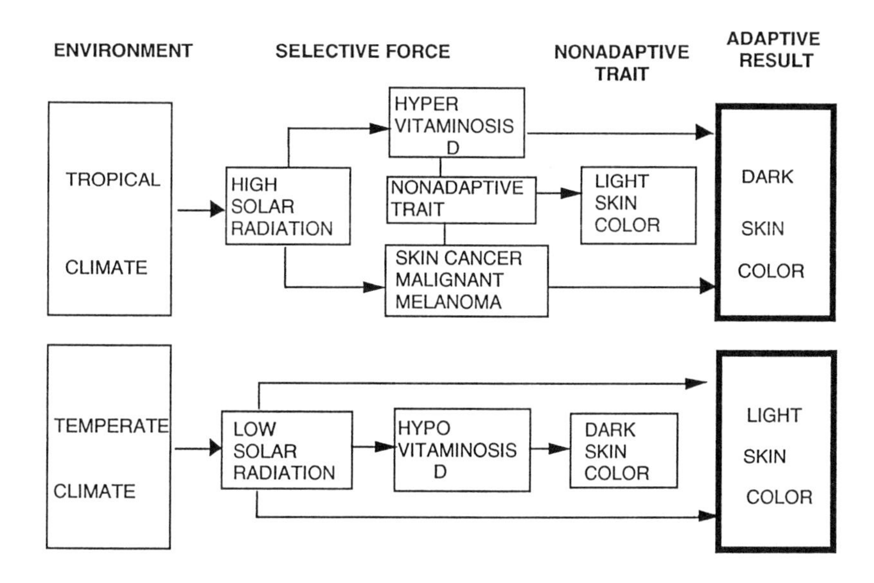


**Fig. 7.2.** Schematization of selective forces in tropical and temperate climates that affect human variation in skin color. Population differences in skin color are viewed as result of evolutionary compromise to selective forces present in world climates.

War, the incidence of frostbite was greater among black soldiers than among whites. Furthermore, studies on tissue tolerance to severe cold indicate that pigmented skins of guinea pigs exposed to freezing temperatures suffered greater tissue damage than nonpigmented skins of the same species (20). At present, however, there are no studies in humans that corroborate these findings.

All the previous hypotheses have considered the pigmentary system as directly involved in responding to solar radiation. The disease hypothesis maintains that pigmentation is an indirect consequence of human adaptation to tropical disease (21). This hypothesis is based on the following facts: (1) the organism's principal line of defense against disease is the reticuloendothelial system (RES), which covers the phagocytic action and production of gamma globulin antibodies; (2) there is an inverse relationship between the activity of RES and adrenocortical activity—the lower the adrenocortical activity, the higher the RES activity; and (3) a deficiency in the enzyme 11B-hydroxylase leads to cortical deficiency. The crux of this hypothesis is that the pituitary gland, in order to reactivate adrenal function, increases production of the adrenocorticotropic hormone (ACTH); parallel with ACTH production, it also secretes the melanocyte-stimulating hormone (MSH). Therefore this hypothesis delegates pigmentation to a secondary phenomenon resulting from an adaptation to tropical disease. Furthermore, it maintains that a single enzyme deficiency in the adrenal cortex may be the primary basis for human variations in skin color. In support of this hypothesis, Wassermann (21) indicates that pigmented populations such as American blacks, Indians, and

Malayans, as shown by anatomical and physiological studies, do have decreased adrenocortical activity. Although these characteristics are accurate, a major flaw in this hypothesis is the assumption that infectious diseases were major selective factors during human evolution. Current anthropological and epidemiological data indicate that infectious disease became prevalent and had significant selective force only after the introduction of agriculture and the consequent increase in population size, which in evolutionary perspective occurred in too short a time to account for skin color variations.

Other theories, although recognizing the role of melanin pigmentation, have identified variations in human skin color as by-products of other events of natural selection. This would include associating skin color with selection for eye color. Because decreasing pigmentation of the fundus of the eye results in an increased sensitivity to longer wavelengths of light (greater than 50 m), it has been postulated that this characteristic might have arisen in the cave dwelling period of European prehistory as an adaptive response to lower levels of light intensity (22). However, experimental studies indicate that variations in retinal and iris pigmentation under varying conditions of light stress are not related to visual acuity (23–24). Another theory states that selection for resistance to tropical disease by hypertrophy of the reticuloendothelial system is accompanied by a passive enhancement of cutaneous pigmentation. Still another theory postulates that dark skin color evolved as a mimic response to the tropical forest (18, 21).

In a recent synthesis Robins (25) argues against the vitamin D hypothesis for the evolution of white skin color. He bases his argument on the fact that skeletal remains in northern zones of North America were free of rickets, and yet, they were of dark skin. His argument also assumes that humans have a high capacity for vitamin storage. For these reasons, he maintains that white skin could not have evolved to facilitate vitamin D synthesis in places of low solar radiation. However, as previously indicated, populations living in Alaska, such as the Eskimos, derive their vitamin D not from the sun, but from their diet. Furthermore, human ability for storing all vitamins is extremely limited. It is evident that these arguments are not sufficient to disprove the importance of the vitamin D hypothesis. Recently Brace (26) has applied the Probable Mutation Effect Hypothesis to explain the evolution of light skin color of European populations. According to this hypothesis, the adaptive significance of dark skin was diminished by the use of clothing. This cultural attribute led to a relaxation of selection for dark skin resulting in accumulation of change mutations leading to light skin color.

## References

1. Oettle, A. G. 1966. Epidemiology of melanoma in South Africa. In G. Della Porta, and O. Muhlbock, eds., The structure and control of the tulanocyte. Heidelberg, NY: Springer-Verlag New York.

2. Burdank, F. 1971. Patterns in cancer mortality in the United States: 1950–1967. NCI Monograph 33. Washington, DC: U.S. Department of Health, Education, and Welfare.
3. Hall, A. F. 1950. Relationships of sunlight, complexion and heredity to skin carcinogenesis. Arch. Dermatol. Syph. 61:589–610.
4. Gellin, G. A., A. W. Kopf, and L. Garfinkel. 1969. Malignant melanoma: a controlled study of possibly associated factors. Arch. Dermatol. 99:43–48.
5. Lee, J. A. H. 1972. Sunlight and the etiology of malignant melanoma. In W. H. McCarthy, ed., Melanoma and skin cancer. Sydney, N.S.W. Australia: Government Printer.
6. Beardmore, G. L. 1972. The epidemiology of malignant melanoma in Australia. In W. H. McCarthy, ed., Melanoma and skin cancer. Sydney, N.S.W. Australia: Government Printer.
7. Livingstone, F. 1969. Polygenic models for the evolution of human skin color differences. Hum. Biol. 41:480–93.
8. Loomis, W. F. 1967. Skin-pigment regulation of vitamin-D biosynthesis in man. Science 157:501–6.
9. Thomson, M. L. 1955. Relative efficiency of pigment and horny layer thickness in protecting the skin of Europeans and Africans against solar ultraviolet radiation. J. Physiol. (Lond.) 127:236–46.
10. Murray, F. G. 1934. Pigmentation, sunlight and nutritional disease. Am. Anthropol. 36:438–45.
11. Holick, M. F. 1987. Photosynthesis of vitamin D in the skin: effect of environmental and life-style variables. Fed. Proc. 46:1876–82.
12. Holick, M. F., J. A. MacLaughlin, and S. H. Doppelt. 1981. Factors that influence the cutaneous photosynthesis of previtamin $D_3$. Science 211:590–93.
13. Holick, M. F. 1986. Sunlight and skin: their role in vitamin-D nutrition for humans. In A. D. Roe, ed., Nutrition and the skin. New York, NY: Liss.
14. Eastman, N. J. 1956. Obstetrics, 11th ed. New York, NY: Appleton-Century-Crofts.
15. Elidrissy, A. T. H., and S. H. Sedrani. 1981. Infantile Vitamin D deficiency rickets in Riyadh. Is material vitamin D deficiency a possible factor? Calcif. Tissue Int. 33:47.
16. Sedrani, S. H., A. T. H. Elidrissy, and K. M. El Arabi. 1983. Sunlight and vitamin-D status in normal Saudi subjects. Am. J. Clin. Nutr. 38:129–32.
17. Lo, C. W., P. W. Paris, and M. F. Holick. 1986. Indian and Pakistani immigrants have the same capacity as Caucasians to produce vitamin D in response to ultraviolet irradiation. Am. J. Clin. Nutr. 44:683–85.
18. Cowles, R. B. 1959. Some ecological factors bearing on the origin and evolution of pigment in the human skin. Am. Naturalist 93:283–93.
19. Baker, P. T. 1958. The biological adaptation of man to hot deserts. Am. Naturalist 92:337–57.
20. Post, P. W., F. Daniels, Jr., and R. T. Binford, Jr. 1975. Cold injury and the evolution of "white" skin. Hum. Biol. 47:65–80.
21. Wasserman, H. P. 1964. Human pigmentation and environmental adaptation. Arch. Environ. Health 11:691–94.
22. Daniels, F., Jr., P. W. Post, and B. E. Johnson. 1972. Theories of the role of pigment in the evolution of human races. In V. Riley, ed., Pigmentation: Its genesis and biologic control. New York, NY: Appleton-Century-Crofts.
23. Hoffman, J. M. 1975. Retinal pigmentation, visual acuity and brightness levels. Am. J. Phys. Anthropol. 43:417–24.

24. Short, G. B. 1975. Iris pigmentation and photopic visual acuity: a preliminary study. Am. J. Phys. Anthropol. 43:425–34.
25. Robins, A. H. 1991. Biological perspectives on human pigmentation. Cambridge: Cambridge University Press.
26. Brace, C. L. 1991. Stages of human evolution. New York: Prentice Hall.

PART 4

# Adaptation to High Altitude

CHAPTER 8

# Respiratory Function and Gas Exchange in the Lungs and Tissues

---

**Pulmonary Structure and Function**
- Lungs

**Pulmonary Volumes and Capacities**
- Static Volumes

**Pulmonary Capacities**
- Dynamic Volumes

**Pulmonary Ventilation and Respiration**
- Pulmonary Ventilation
- Minute Respiratory Volume
- Alveolar Ventilation

**Gaseous Exchange in the Lungs and Tissues**
- Basic Properties of Gases: Partial Pressure of Gases
- Barometric Pressure
- Partial Pressure
- Partial Pressure of Oxygen in Respiratory Passages
- Diffusion of Gases from the Alveoli to the Blood

**Pulmonary Diffusing Capacity**
- Diffusing Capacity for Oxygen
- Diffusing Capacity for Carbon Dioxide
- Measurement of Diffusing Capacity

## PULMONARY STRUCTURE AND FUNCTION

The respiratory system consists of the nose, pharynx, larynx, trachea, bronchi, and lungs. The function of the respiratory system is to exchange gases such as *oxygen* and *carbon dioxide* between the atmosphere and the blood. In this section the basic principles governing the process of gas exchange as it occurs in the lungs, blood, and tissues, as well as the factors that modify them, will be discussed.

### Lungs

While all the organs of the respiratory system participate in respiratory function, the actual gas exchange occurs in the lungs. The lungs are paired, cone-shaped organs situated in the thoracic cavity (see fig. 8.1). They are separated from each other by the heart. The lung is protected by two layers of serous membrane called pleural membrane. The outer layer, called the parietal pleura, is attached to the wall of the thoracic cavity. The inner layer or visceral pleura covers the lungs themselves. Between the visceral and pleural membranes is the pleural cavity, which contains a lubricating fluid secreted by the membranes. Inflammation of the pleural membrane is known as pleurisy, and causes friction during breathing that can be painful when the swollen membranes rub against each other.

The nose is directly connected to the pharynx, larynx, and trachea. The trachea in turn is subdivided into two branches called bronchi. These bronchi are connected directly with the two lobes of the lung. Upon entering the lungs the bronchi divide into smaller branches called secondary bronchi, and the secondary bronchi continue to divide into smaller branches called terminal bronchioles. This continuous branching from the trachea resembles a tree trunk with its branches; as such it is commonly referred to as the bronchial tree (see fig. 8.1).

The bronchioles continue to subdivide into smaller ducts called alveolar ducts (see fig. 8.1b). Around these alveolar ducts the alveolar sacs are formed. Each hollow unit is referred to as an alveolus, or plural alveoli. A collection of alveoli that share a common opening is called an alveolar sac. (There are about 250 million alveolar sacs in the two lungs, each alveolus having an average diameter of about 0.1 mm.) The structures formed by the subdivision of bronchioles and alveolar ducts are called respiratory units. Thus, the respiratory unit consists of a respiratory bronchiole, alveolar ducts, and sacs or alveoli. The epithelium of the alveolar sacs is formed by very thin membranes. Over the alveoli, the blood vessels disperse into a capillary network. The epithelium of the capillaries like that of the alveoli consists of a very thin membrane. The fact that the membranes of the alveoli and capillaries

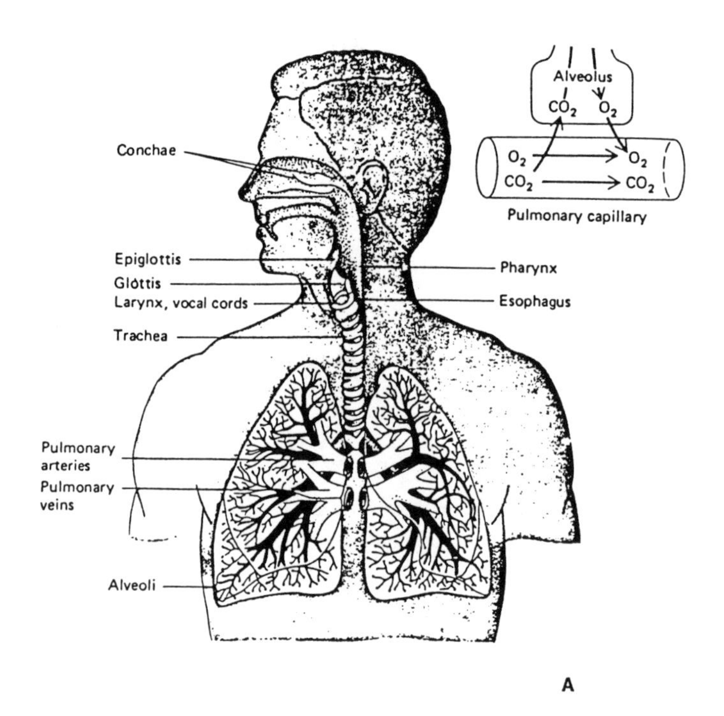


A

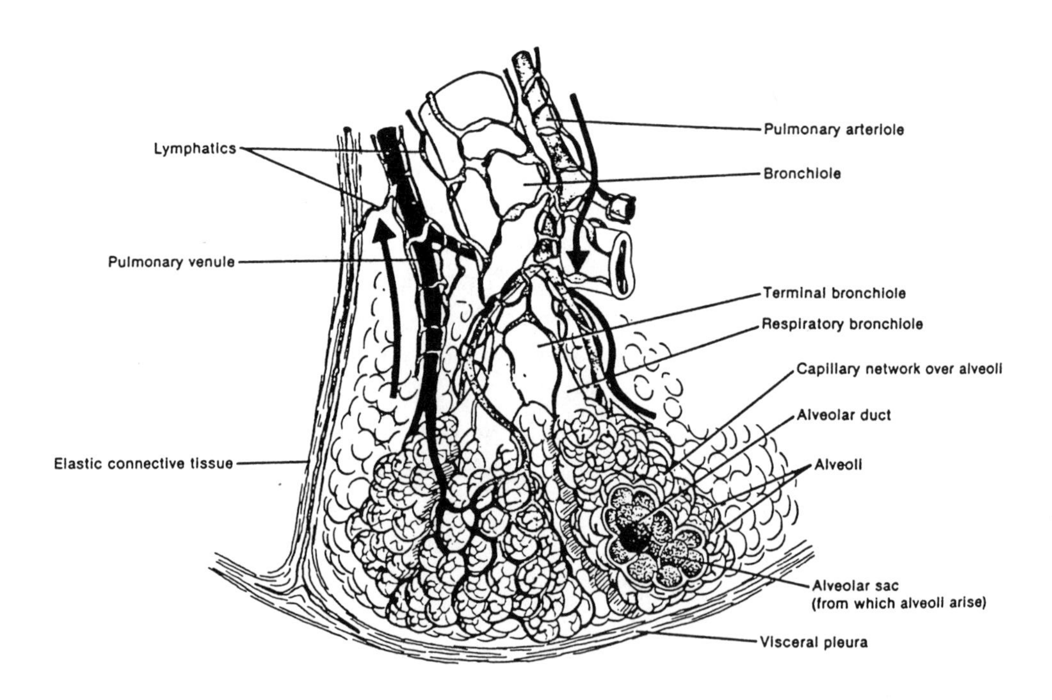


B

**Fig. 8.1.** The respiratory passages. **A,** Diagram of the bronchial tree in relation to the lungs. **B,** Diagram of a lobule of the lung showing the respiratory membrane where the alveolar sacs are located. (Adapted from A. C. Guyton. 1986. Textbook of medical physiology. 7th ed. Philadelphia: W. B. Saunders Company.)

are very thin (0.1 $\mu$m average thickness) facilitates exchange of gases between the air in the alveolar sacs and the pulmonary blood.

### Respiratory Membrane and the Alveoli

The membranes through which gas exchange takes place are collectively called the *respiratory membrane,* also called pulmonary membrane (see fig. 8.1b). The respiratory membrane consists of four thin membranes. Despite the large number of layers, the overall thickness of the respiratory membrane is in some areas less than 0.1 micron. Histological studies indicate that the total surface area of the respiratory membrane in an adult male is about 70 $m^2$ (750 $ft^2$). Taking into account the fact that the total quantity of blood in the capillaries of the lungs at any given instant is only about 60 to 100 ml, to cover the surface area of the respiratory membrane (70 $m^2$) the blood is thinly spread. This factor facilitates the exchange of gases. All gases of respiratory importance are highly soluble in the lipid substances of the cell membranes and for this reason can diffuse through the membranes with great facility. Furthermore, the capillaries are very narrow (7 microns) so that the red blood cells must flow through them in single file. This feature gives each red blood cell maximum exposure to the available oxygen. In general the rate of diffusion of carbon dioxide is 20 times as rapid as diffusion of oxygen, and oxygen in turn diffuses about twice as fast as nitrogen. Thus, any factor that increases the thickness of the respiratory membrane decreases the diffusion, as occurs when excessive fluid accumulates in the interstitial space of the membrane which may lead to pulmonary edema. Similarly, any factor that decreases the surface area of the respiratory membrane reduces the rate of gas diffusion, and conversely any factor that increases the surface area increases the rate of gas diffusion. For example, removal of one lung decreases the surface area to half of normal. Also coalescence of the alveoli as it occurs in emphysema reduces the respiratory membrane despite the fact that each alveolar sac in this condition is larger. Conversely, enlargement of the lung leads to an enlargement of the respiratory membrane and enhances the rate of gas diffusion.

## PULMONARY VOLUMES AND CAPACITIES

### Static Volumes

The volume of air in the lungs depends not only on the rate of ventilation but also on the size of the lungs. The maximum volume to which the lungs can be expanded consists of four volumes: (1) the tidal lung volume, (2) the inspiratory reserve volume, (3) the expiratory reserve volume, and (4) the residual lung volume (see fig. 8.2).

**Tidal Volume (ml).** The volume of air inspired and expired with each normal breath is called the tidal volume. It amounts to about 500 ml in males and 450 ml in females.

**Inspiratory Reserve Volume (ml).** The inspiratory reserve volume is the extra volume of gas that can be inspired over and beyond the normal tidal inspiration. It averages about 3000 ml in males and 2800 ml in females.

**Expiratory Reserve Volume (ml).** The expiratory reserve volume is the extra volume of gas that can be expired by forceful expiration after the end of a normal tidal expiration. It averages about 1100 ml in males and 1000 ml in females.

**Residual Lung Volume (ml).** The residual lung volume is the volume of gas left in the lungs after a maximal forced expiration. The volumes average about 1000 ml in males and 800 ml in females. Under normal circumstances it is determined by body size and age.

## PULMONARY CAPACITIES

### Dynamic Volumes

The pulmonary capacities refer to the sum of two or more of the lung volumes. These include four capacities: (1) the inspiratory capacity, (2) the functional residual capacity, (3) the forced vital capacity, and (4) the total lung capacity.

**Inspiratory Capacity (ml).** The inspiratory capacity is equal to the sum of tidal volume and the inspiratory reserve volume. This is the amount of gas that a person can breathe beginning at the normal expiratory level and expanding the lungs to the maximum amount.

**Functional Residual Capacity (ml).** The functional residual capacity equals the sum of expiratory reserve volume and the residual lung volume. This is the amount of gas remaining in the lungs at the end of a normal expiration.

**Forced Vital Capacity (ml).** The forced vital capacity is equal to the sum of inspiratory reserve volume, the tidal volume, and the expiratory reserve volume. The forced vital capacity is the maximum volume of air that a person can expel from the lungs after first filling the lungs to the maximum extent and then expiring to the maximum extent. The forced vital capacity averages about 4200 ml in males and 4000 ml in females. Under normal circumstances within each gender it is determined by body size and age.

**Total Lung Capacity (ml).** The total lung capacity is equal to the sum of the forced vital capacity and residual lung volume. It averages about 5200 ml

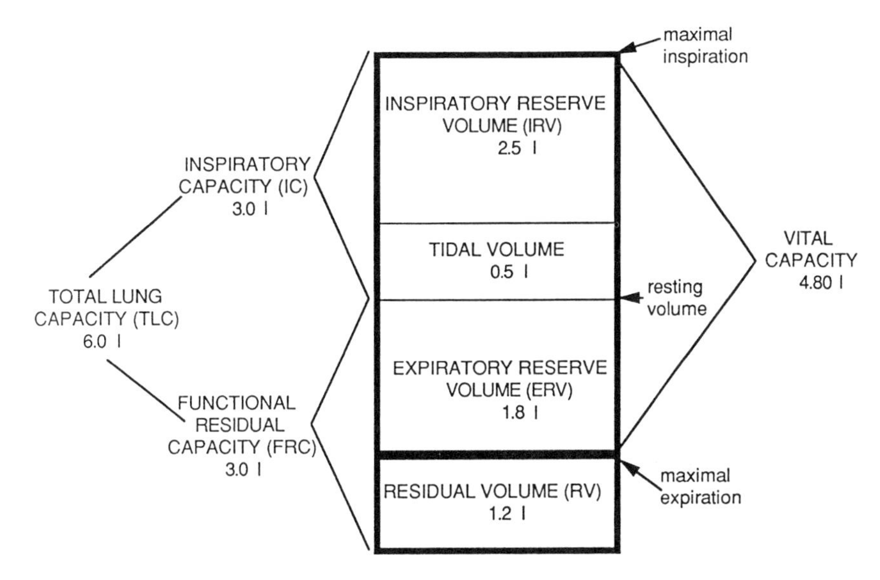


**Fig. 8.2.** Static measures of lung volume.

in males and 4000 ml in females. It is determined by the factors that determine the forced vital capacity and residual lung volume.

## PULMONARY VENTILATION AND RESPIRATION

The process of respiration consists of four major components: (1) pulmonary ventilation, (2) diffusion of oxygen and carbon dioxide between the alveoli and the blood, (3) transport of oxygen and carbon dioxide in the blood and body fluids to and from the cells, and (4) regulation of ventilation.

### Pulmonary Ventilation

Pulmonary ventilation means the inflow and outflow of air between the atmosphere and the lungs. This is accomplished through the process of inspiration and expiration of air. By inspiration the atmospheric air is brought into the lung and by expiration the air from the lung alveoli is brought out into the atmosphere.

### Minute Respiratory Volume

The total amount of new air moved into the respiratory passages each minute is referred to as minute respiratory volume. The volume of air inspired and expired with each normal breath is called the tidal volume, which in a young adult male averages about 500 ml, while the respiratory rate averages about 12 per minute. Thus, the minute respiratory volume averages about 6 liters per minute (500 ml × 12 = 6000 ml). The minute respiratory volume during maximal or strenuous exercise can increase from a resting value of 6 liters/minute to 100 or 200 liters/minute. The increase in pulmonary ventilation can occur due to an increase in tidal volume, respiratory rate, or both.

### Alveolar Ventilation

Alveolar ventilation refers to the volume of air in the alveoli that is renewed by atmospheric air each minute. The rate of alveolar ventilation is always less than the minute respiratory volume, because a large portion of the inspired air goes to fill the passageways (the nasal passages, the pharynx, the trachea, and the bronchi), the membranes of which are not capable of significant gaseous exchange with the blood. This increased space is called the dead space air, which averages about 150 ml in an adult male. Therefore, the volume of alveolar air that enters with each breath is equal to the tidal volume minus the dead space volume. Thus the alveolar ventilation with an average of 500 ml of tidal volume and 150 ml of dead space volume and 12 breaths per minute is 4200 ml (Alveolar Ventilation = [500 ml – 150 ml] × 12 = 4200 ml/min), which is less than the total ventilation of 6000 ml/min (or minute respiratory volume).

When the availability of oxygen is low, as occurs at high altitude, or when the oxygen demands exceed the availability, as occurs during exercise, the alveolar ventilation increases drastically. For example, during strenuous maximal exercise an increase in breath rate from 12 to 20 breaths per minute can raise the alveolar ventilation from a resting value of 4200 ml/min to 7000 to 17,000 ml/min.

## GASEOUS EXCHANGE IN THE LUNGS AND TISSUES

Once the alveoli are ventilated with fresh air, the next step in the respiratory process is the *diffusion* of oxygen from the alveoli into the pulmonary blood and diffusion of carbon dioxide from the pulmonary blood into the alveoli.

### Basic Properties of Gases: Partial Pressure of Gases

All gases are governed by physical laws. First, following Boyle's Law, the volume of gas is inversely related to its pressure; so that the lower the pressure the higher the volume. Second, according to Charles' Law, the volume of a gas, assuming that pressure remains constant, is directly related to its temperature; so that the higher the temperature the greater the volume of the gas. Third, following Dalton's Law, each gas in a mixture of gases exerts its own pressure as if all the other gases were not present.

### Barometric Pressure

Atmospheric air consists of about 78.62% nitrogen ($N_2$), 20.95% oxygen ($O_2$), 0.3% carbon dioxide ($CO_2$), and 0.40% water ($H_2O$) (see table 8.1). These gases together at sea level exert a pressure of 760 mm Hg:

$$\text{Atmospheric Pressure} = PO_2 + PCO_2 + PN_2 + PH_2O$$

This pressure is known as the *barometric pressure.* (In honor of Evangelista Torricelli, the inventor of the barometer, barometric pressure is referred to in torr units.)

### Partial Pressure

The pressure of *each* gas is referred to as its partial pressure and is denoted by a P in front of the symbol for the gas; the partial pressure of oxygen therefore is represented by $PO_2$. The partial pressure exerted by each component of the atmospheric air can be determined by multiplying the percentage of the mixture and the total atmospheric pressure. For example, to find the partial pressure of oxygen in the sea level atmosphere simply multiply the percentage of oxygen present in the air (at 20.95%) by the total barometric pressure (760 mm Hg):

$$PO_2 = 0.2095 \times 760 \text{ mm Hg} = 159.22 \text{ mm Hg}$$

Thus at sea level the oxygen enters the respiratory passages with a force of 159 mm Hg. Since each gas diffuses from the area where its partial pressure is greater to the area where its partial pressure is less, the higher the partial pressure of gas, the higher the forces that enable the gas to diffuse through the pulmonary membrane.

**TABLE 8.1. Components of atmospheric air and its corresponding partial pressures of respiratory gases as they enter the lungs**

| | Atmospheric Air (mm Hg) | In Trachea (mm Hg) | In Alveoli (mm Hg) |
|---|---|---|---|
| $O_2$ | 159.2 | 149.3 | 104.0 |
| $CO_2$ | 0.3 | 0.3 | 40.0 |
| $N_2$ | 597.5 | 563.4 | 569.0 |
| $H_2O$ | 3.0 | 47.0 | 47.0 |
| Total | 760.0 | 760.0 | 760.0 |

### Partial Pressure of Oxygen in Respiratory Passages

As shown in table 8.2, as soon as the atmospheric air enters the respiratory passages it is exposed to the fluids covering the respiratory surfaces and becomes totally humidified and decreases its pressure.

#### Trachea

The water vapor (according to Charles' Law of gas volumes and temperature) expands the volumes of the air entering the respiratory passages and therefore dilutes all the gases in the inspired air. Since the partial pressure of water vapor at normal body temperature is 47 mm Hg, the partial pressure of oxygen is also reduced so by the time it reaches the trachea the partial pressure of oxygen has been reduced from 159 to 149 mm as follows.

$$0.2095 \times (760 - 47) = 149 \text{ mm Hg}$$

#### Alveoli

In the same manner the partial pressure of oxygen is reduced to about 104 mm Hg by the time it reaches the alveoli. Thus, at sea level the oxygen can diffuse through the pulmonary membrane into the pulmonary blood with a force of 104 mm Hg by the time it reaches the alveoli.

### Diffusion of Gases from the Alveoli to the Blood

#### Partial Pressure of Oxygen in Arterial and Venous Blood

The partial pressure of oxygen in the lung alveoli is 104 mm Hg while the partial pressure of oxygen in the venous blood entering the pulmonary capillaries is 40 mm Hg (see fig. 8.3). Since the diffusion of gas in general is

**TABLE 8.2. Partial pressure of oxygen, carbon dioxide, and hemoglobin percent oxygen saturation as it leaves the lung and passes through the venous blood**

| Oxygen in Inspired Air | | | | | |
|---|---|---|---|---|---|
| Trachea | Alveoli | Arterial | | Venous | |
| $PO_2$ (mm Hg) | $PO_2$ (mm Hg) | $PO_2$ (mm Hg) | Saturation (%Hb) | $PO_2$ (mm Hg) | Saturation (%Hb) |
| 149.0 | 104.0 | 100.0 | 98.0 | 40.0 | 75.0 |

| Carbon Dioxide in Expired Air | | | |
|---|---|---|---|
| Trachea $PCO_2$ (mm Hg) | Alveoli $PCO_2$ (mm Hg) | Arterial $PCO_2$ (mm Hg) | Venous $PCO_2$ (mm Hg) |
| 0.3 | 40.0 | 41.0 | 46.0 |

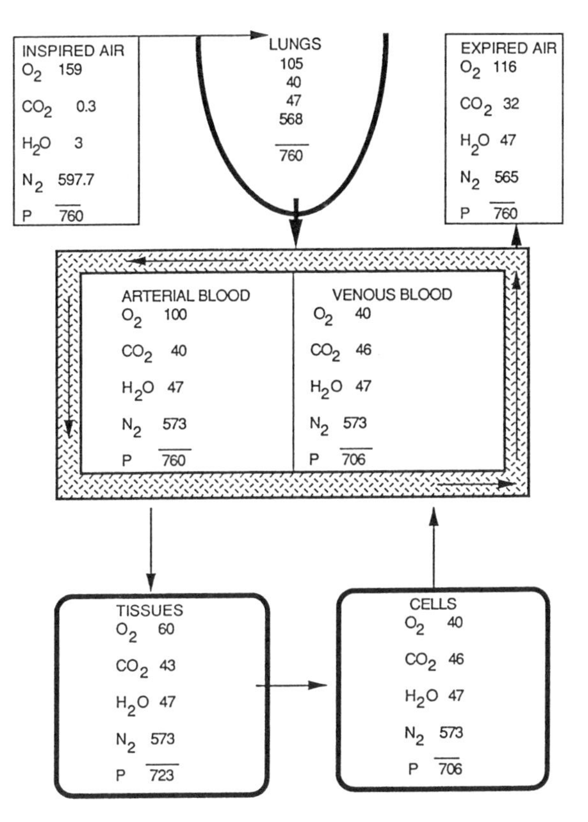


**Fig. 8.3.** Representative values. Typical values of gas tensions in inspired air, lungs, expired air, and blood, at rest. Barometric pressure, 760 mm Hg. As a result of the diffusion of gases from areas of high pressure to areas of low pressure tensions (assume to be free from water) of oxygen and carbon dioxide vary markedly in venous and arterial blood.

directly influenced by the pressure, oxygen gradient diffuses from the alveoli into the deoxygenated venous blood in direct proportion to the pressure difference. Therefore, the $PO_2$ of the venous blood rises to 100 mm Hg as it leaves the lung in the form of arterial blood. From there the oxygenated blood enters the heart, where it is pumped by the left ventricle into the aorta and through the systemic arteries to the capillaries in the peripheral tissues. From the tissue capillaries it diffuses into the tissue cells where the partial pressure ($PO_2$ ranges from 6 to 40 mm Hg) is lower than that of the tissue capillaries, and oxygen diffuses from the capillary blood into the tissues. Thereafter, the deoxygenated blood ($PO_2 = 40$ mm Hg) returns to the heart in the form of venous blood. From here the right ventricle of the heart pumps the deoxygenated blood back through the pulmonary artery to the lungs where it will pick up oxygen once again.

In the same manner, as oxygen diffuses from the alveoli into the pulmonary capillaries and eventually into the tissue cells, the by-product of cell respiration, carbon dioxide, needs to be disposed of. This occurs in an inverse manner so that upon arriving at the lungs the partial pressure of carbon dioxide in the deoxygenated blood is high ($PCO_2 = 45$ mm Hg), while the $PCO_2$ in the alveoli is low ($PCO_2 = 40$ mm Hg). Because of this difference, the carbon dioxide diffuses from the pulmonary deoxygenated venous blood into the alveoli until the $PCO_2$ of the venous blood decreases to the same level as that of the alveoli ($PCO_2 = 40$ mm Hg). From the alveoli the carbon dioxide is exhaled and expelled into the atmosphere where the $PCO_2$ is very low ($PCO_2 = 0.3$ mm Hg).

In summary, the diffusion of gases from the alveoli into the pulmonary blood is facilitated by anatomical factors associated with the pulmonary membrane. These include (1) the thinness of the alveolar-capillary membrane (less than 0.5 microns), (2) the large surface area of the alveoli (about 70 $m^2$), and (3) the large number of capillaries. In addition, the diffusion of gas is directly related to the force exerted by the partial pressure of gases, and in general the greater the pressure difference, the greater is the movement of gases.

## PULMONARY DIFFUSING CAPACITY

The *Pulmonary Diffusing Capacity* is defined as the number of milliliters of gas diffusing across the pulmonary membrane per minute and per millimeter of mercury of partial pressure difference between the alveolar air and the pulmonary capillary blood. That is,

Diffusing Capacity = gas flow/mean driving pressure

### Diffusing Capacity for Oxygen

The average diffusing capacity for oxygen under resting conditions is about 21 ml per minute, and under strenuous exercise can increase to about 65 ml per minute. This three-fold increase is usually due to: (1) opening up of a number of previously dormant pulmonary capillaries, thereby increasing the surface area of the blood into which the oxygen can diffuse; (2) dilation of all the pulmonary capillaries that were already open, therefore also increasing the surface area; and (3) stretching of the alveolar membranes, consequently increasing their surface area and decreasing their thickness.

### Diffusing Capacity for Carbon Dioxide

The diffusing capacity for carbon dioxide under resting conditions has been estimated from measurements of oxygen diffusing capacity to be about 420 ml per minute and 1300 ml per minute under strenuous exercise. These measurements are derived based on the fact that the diffusion of carbon dioxide is 20 times that of oxygen.

### Measurement of Diffusing Capacity

The pulmonary diffusing capacity is usually evaluated through measurements of carbon monoxide diffusing capacity. A small amount of carbon monoxide is breathed into the alveoli, and the partial pressure of the carbon monoxide in the alveoli is measured from an appropriate alveolar sample. Under most conditions the carbon monoxide pressure in the blood is zero because the hemoglobin combines with carbon monoxide so rapidly that the pressure never has time to build up. For this reason, the pressure difference of carbon monoxide across the respiratory membrane is equal to its partial pressure in the alveoli. Therefore, by measuring the volume of carbon monoxide absorbed in a short period of time and dividing this by the carbon monoxide partial pressure, the carbon monoxide diffusing capacity may be determined. Subsequently, the oxygen diffusing capacity is derived by multiplying the carbon monoxide diffusing capacity by a factor of 1.23. This is because the diffusing coefficient for oxygen is 1.23 times that for carbon monoxide. Since the average diffusing capacity for carbon monoxide in adult males is about 17 ml per minute the average diffusing capacity for oxygen is equal to 21 ml per minute ($17 \times 1.23 = 20.91$).

## TRANSPORT OF OXYGEN IN THE BLOOD

Once oxygen has diffused from the alveoli into the pulmonary blood, it is transported by the hemoglobin on the red blood cells to the tissue capillaries where it is released for use by the cells. In a similar manner, the carbon dioxide is brought back by the hemoglobin from the tissue cells to the lung alveoli where it is expelled to the air. This transport is achieved by the hemoglobin in interaction with the partial pressure, pH, temperature, and 2,3-DPG (2,3-Diphosphate Glycerate) of the red blood cells.

### Hemoglobin and Partial Pressure

Oxygen is carried in the arterial blood in two forms: (1) physically dissolved in the blood water and (2) chemically bound to hemoglobin molecules. Normally each liter of arterial blood contains 200 ml of $O_2$, of which 3 ml (or 1.5%) is dissolved in the blood water and 197 ml (or 98.5%) is chemically bound to hemoglobin. Hemoglobin (Hb) is a protein that consists of four iron atoms, each of which is capable of combining with a molecule of oxygen. When hemoglobin (Hb) combines with oxygen ($O_2$), it is called oxyhemoglobin ($HbO_2$). The reaction of hemoglobin and oxygen is usually written:

$$O_2 + Hb \longrightarrow HbO_2$$

Deoxyhemoglobin Oxyhemoglobin

The combination of hemoglobin and oxygen is characterized by an extremely loose bond so that the combination is easily reversible. Thus, when the $PO_2$ is high, as in the pulmonary blood capillaries, oxygen binds with the hemoglobin, but when the $PO_2$ is low, as in the peripheral tissues (tissue capillaries), the oxygen is released from the hemoglobin. Furthermore, the oxygen does not become ionic oxygen but is carried as molecular oxygen to the tissues where, because of the loose, readily reversible combination, it is released into the tissue fluids in the form of dissolved molecule oxygen rather than ionic oxygen. This is the basis for oxygen transfer from the lungs to the tissues.

## The Oxyhemoglobin Dissociation Curve

### The Percent Saturation

One way of expressing the proportion of hemoglobin bound with oxygen is as *percent saturation*. The percent saturation is equal to the content of oxygen in the blood divided by the oxygen-carrying capacity of hemoglobin in the blood times 100:

$$\%\ \text{Saturation in Hemoglobin} = \frac{O_2 \text{ bound to Hb}}{O_2 \text{ capacity of Hb}} \times 100$$

The percent saturation of hemoglobin is usually referred to as the oxyhemoglobin dissociation curve. The oxyhemoglobin dissociation curve is a plot of how the availability of oxygen affects the reversible chemical reaction of oxygen and hemoglobin.

As indicated before (see table 8.2), by the time the oxygen reaches the arterial blood it has a pressure of 100 mm Hg. Because of this high pressure about 98% of the hemoglobin is bound with oxygen, but the same hemoglobin entering into the venous blood is only 75% saturated with oxygen. This is because by the time the blood reaches the tissue capillaries, the $PO_2$ is low ($PO_2 = 40$ mm Hg) and therefore the hemoglobin does not retain as much oxygen. Thus, as illustrated in figure 8.4 the relationship between $PO_2$ and $HbO_2$ is not linear. It is an S-shaped curve, nearly flat when $PO_2$ is above 70 mm Hg and steep when the $PO_2$ is below 70 mm Hg. *In other words, hemoglobin has a high affinity for oxygen under high pressure but low affinity under low pressure.* Physiologists express this affinity in terms of the partial pressure of oxygen in plasma ($PO_2$) associated with 50% oxygen saturation of blood at 37° C and a pH of 7.4, known as the *oxygen-hemoglobin dissociation curve.*

### Oxygen Carrying Capacity of Blood (ml of oxygen per 100 ml of blood)

The oxygen carrying capacity of an individual is dependent on the amount of hemoglobin in that person's blood. Blood contains an average of about 15

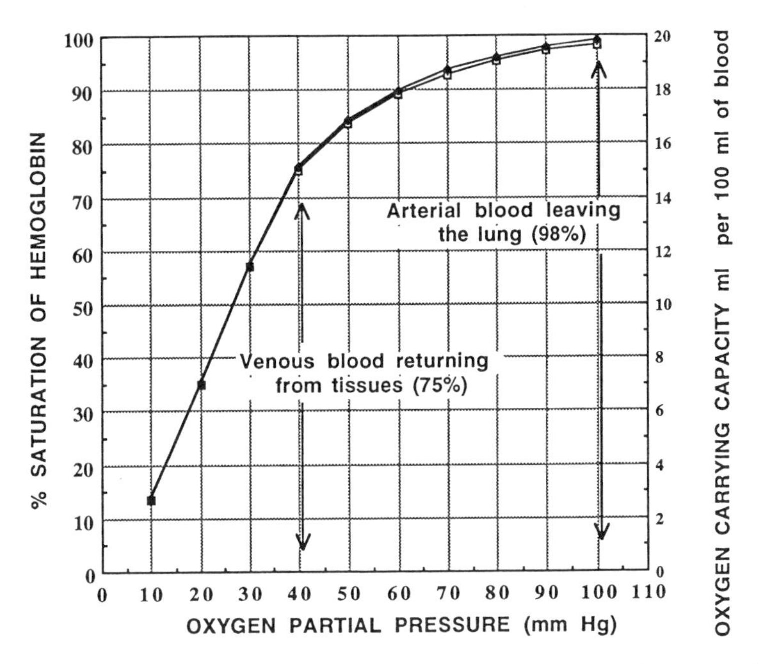


**Fig. 8.4.** Oxygen-hemoglobin dissociation curve. The percent oxygen saturation of hemoglobin is shown on the left while the quantity of oxygen carried on each 100 ml of blood under normal conditions is shown on the right ordinate.

grams of hemoglobin per 100 ml of blood in men and 13 g per 100 ml of blood in women, and each gram of hemoglobin can bind with a maximum of 1.34 ml of oxygen. Therefore, on the average, the hemoglobin in 100 ml of blood combines with a total of about 20 ml of oxygen (15 g × 1.34 ml = 20 ml $O_2$) in males and 17 ml of oxygen in females (13 g × 1.34 ml = 17 ml $O_2$). Therefore, on the average, the hemoglobin in 100 ml of blood has a capacity to combine with about 20 ml of oxygen in males and with 17 ml of oxygen in females when the hemoglobin is saturated as shown in figure 8.4. (These values 20 ml and 17 ml of oxygen are also expressed as volumes percent.)

## Oxygen Uptake in the Lungs

As shown in figure 8.4 in the arterial blood ($PO_2$ = 100 mm Hg) the hemoglobin is 98% saturated with oxygen, therefore, the total quantity of oxygen bound with hemoglobin in the arterial blood is about 19.7 ml/100 ml of blood in males and about 17.1 ml/100 ml in females (0.98 × 1.34 ml $O_2$ × 13 g Hb = 17.1). On the other hand, in the venous blood ($PO_2$ = 40 mm Hg) the hemoglobin is 75% saturated with oxygen and thus the quantity of oxygen bound with hemoglobin in the venous blood in males equals only 15.1 ml/100 ml of blood (0.75 × 1.34 ml $O_2$ × 15 g Hb = 15.1) and 13.1 ml/100 ml in females (0.75 × 1.34 ml $O_2$ × 13 g Hb = 13.1).

This means that, as shown in table 8.3, in passing through the lungs, each ml of blood has loaded about 4.8 ml of oxygen from each 100 ml of blood (19.9 – 15.1 = 4.8 ml/100 ml of blood) in males and about 4.0 ml of oxygen from each 100 ml of blood in females (17.1 – 13.1 = 4.0 ml/100 ml of blood). Assuming a cardiac output of 5 liters/min, the oxygen uptake in the lungs is 240 ml $O_2$/minute in males and 200 ml $O_2$/minute in females.

## Oxygen Delivery to the Tissues

As blood flows from the arteries into the systemic capillaries, it is exposed to lower $PO_2$'s, and because of the pressure gradient oxygen is released by the hemoglobin. The $PO_2$ in the capillaries varies from tissue to tissue, being high in some (e.g., kidney) and relatively lower in others (e.g., myocardium). As can be seen in figure 8.4, the oxyhemoglobin dissociation curve is very steep in the range of 40 to 10 mm Hg. This means that a small decrease in $PO_2$ can result in a substantial further dissociation of oxygen and hemoglobin, and hence unloading more oxygen for use by the tissues. At a $PO_2$ of 40 mm Hg the hemoglobin is about 75% saturated with oxygen, and thus the quantity of oxygen bound with hemoglobin in the venous blood in males equals only 15.1 ml/100 ml of blood (0.75 × 1.34 ml $O_2$ × 15 g Hb = 15.1) and 13.1 ml/100 ml in females (0.75 × 1.34 ml $O_2$ × 13 g Hb = 13.1). At a $PO_2$ of 20 torr, hemoglobin is only 32% saturated with oxygen. The total blood oxygen content in males is 6.4 (0.32 × 1.34 × 15 = 6.4) and 5.57 (.32 × 1.34 × 13 = 5.6) ml $O_2$/100 ml of blood. This means that a decrease of 8.7 in males (15.1 – 6.4 = 8.7) and 7.5 (13.1 – 5.6 = 7.5) in females ml $O_2$/100 ml of blood is achieved with 20 mm Hg decrease in $PO_2$.

**TABLE 8.3. The partial pressure of oxygen in the arterial and mixed venous blood and its relationship to oxygen uptake in the lungs**

| Characteristics | Males | Females |
|---|---|---|
| Hb (g/100 ml) | 15.0 | 13.0 |
| Arterial Blood | | |
| $PO_2$ (mm Hg) | 100.0 | 100.0 |
| %Hb saturation (%) | 98.0 | 98.0 |
| $O_2$ carrying capacity of arterial blood (ml of oxygen per 100 ml of blood) | 19.9 | 17.1 |
| Mixed venous | | |
| $PO_2$ (mm Hg) | 40.0 | 40.0 |
| %Hb saturation (%) | 75.0 | 75.0 |
| $O_2$ carrying capacity of venous blood (ml $O_2$/100 of blood) | 15.1 | 13.1 |
| Loading of $O_2$ in the lung (ml $O_2$/100 of blood) | 4.8 | 4.0 |
| Loading of $O_2$ in the lung (ml $O_2$/min)[1] | 240.0[1] | 200.0[1] |

[1]Assuming a cardiac output of 5.0 l/min

## Oxygen Uptake During Strenuous Exercise

The fact that hemoglobin in the venous blood is still 75% saturated with oxygen indicates that a large reserve of oxygen still exits. This reserve can be utilized when the need arises, as is the case in exercise. In heavy exercise the muscles utilize oxygen at a rapid rate, which causes the $PO_2$ within the cells to lower to as little as 20 mm Hg. This increases the $PO_2$ gradient between arterial and venous blood and the diffusion of oxygen out of the blood. At this pressure ($PO_2 = 20$ mm Hg) the hemoglobin of the venous blood is 35% saturated with oxygen and the amount of oxygen present in the venous blood is only 7.0 ml/100 ml of blood. Therefore, about 12.7ml of $O_2$/100 ml of blood ($19.7 - 7.0 = 12.7$) during exercise as much as three times oxygen is transported in each volume of blood that passes the tissues normally. If one takes into account that with exercise the cardiac output (amount of blood pumped by heart per minute) increases from the resting value of 5 liters/minute as much as 5 times, the oxygen transport to the tissues can be increased drastically.

## Factors Affecting the Oxyhemoglobin Dissociation Curve

### Acidity

The amount of oxygen released from hemoglobin is also affected by the level of acidity in the blood, such that the lower the pH of the blood the lower the oxygen-carrying capacity of the hemoglobin (see fig. 8.5a). That is, an increased acidity shifts the oxyhemoglobin dissociation curve to the right, while a decreased acidity shifts the curve to the left. This is referred to as the Bohr Effect. The Bohr Effect occurs because deoxyhemoglobin is a weaker acid than oxyhemoglobin. For this reason, deoxyhemoglobin readily accepts the hydrogen ion liberated by the dissociation of carbonic acid, therefore allowing more carbonic acid to be transported in the form of bicarbonate ion. This is usually referred as the isohydric shift. Conversely the association of hydric ions with amino acids of hemoglobin decreases the affinity of hemoglobin for oxygen, therefore shifting the oxyhemoglobin dissociation curve to the right when the pH is low or the pressure of carbon dioxide is high. The equation of this reaction is usually written:

$$H^+Hb + O_2 \qquad\qquad H^+ + HbO_2$$

### 2,3-DPG

The amount of oxygen bound to hemoglobin under low pressure as it occurs in the tissue blood is decreased by the concentration of 2,3 dephosphoglycerate (2,3-DPG) in the red blood cells. It has been found that the higher the concentration of 2,3-DPG in the red blood cells, the lower is the hemoglobin

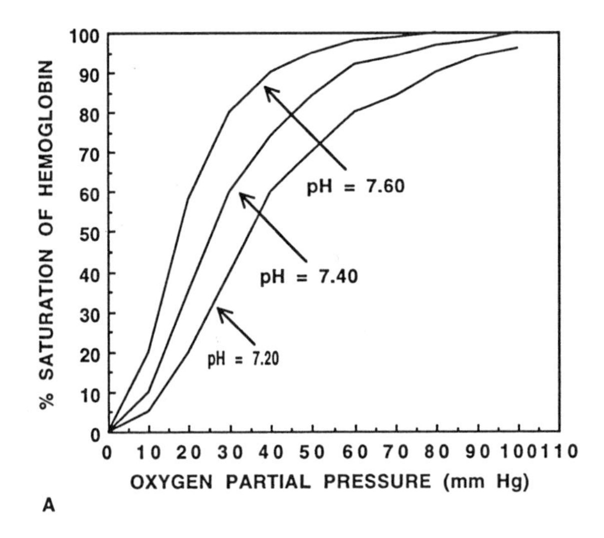


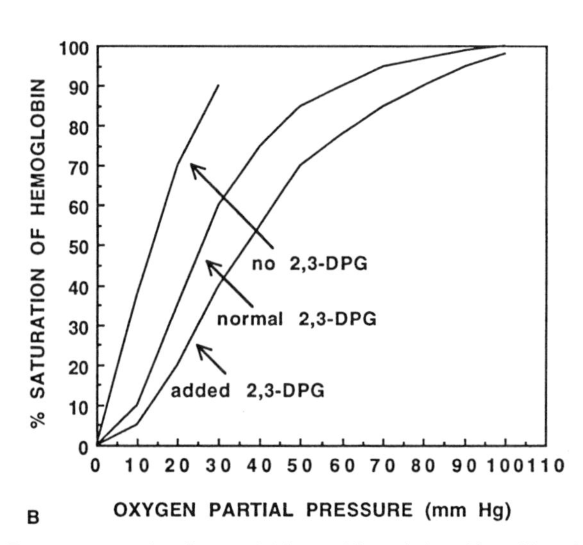


**Fig. 8.5.** Percent saturation hemoglobin and its relationship with acidity and 2,3-DPG. **A,** The lower the pH of the blood (acidity), the lower the oxygen carrying capacity of the hemoglobin. **B,** On the other hand, the amount of oxygen released from hemoglobin is increased when the concentrated 2,3-DPG is high.

affinity for oxygen in the venous blood. Thus, 2,3-DPG enhances oxygen delivery to the tissues and helps maintain the release of oxygen by hemoglobin (see fig. 8.5b).

## Temperature

The amount of oxygen released from hemoglobin is also affected by temperature. High temperatures shift the oxyhemoglobin dissociation curve to the right; low temperatures shift the curve to the left. However, because the internal body temperature is maintained within narrow limits through homeostatic processes, the effect of temperature on hemoglobin affinity under normal conditions is not especially important.

## Oxygen-Buffer Function of Hemoglobin

The regulation of oxygen and carbon dioxide concentrations in the extracellular fluid depends principally on the chemical characteristics of the hemoglobin, which is present in all the red blood cells. Hemoglobin combines with oxygen as the blood passes through the lungs. Then, as the blood passes through the tissue capillaries the hemoglobin will not release oxygen into the tissue fluid if too much oxygen is already there, but if the oxygen concentration is too little, sufficient oxygen will be released to reestablish an adequate tissue-oxygen concentration. Thus, the regulation of oxygen concentration in the tissues is vested principally in the chemical characteristics of hemoglobin itself. This regulation is called the oxygen-buffering function of hemoglobin.

While the main function of hemoglobin is to transport oxygen, it also functions as an "oxygen buffer" system, through which it controls the oxygen pressure in the tissues and prevents it from decreasing to low levels. Were it not for the hemoglobin oxygen buffer system, drastic variations in the $PO_2$ of the tissues would occur during exercise and with every change in metabolism and atmospheric oxygen concentration.

### Value of Hemoglobin for Maintaining Constant $PO_2$ in the Tissue Fluids

As pointed out above, even under basal conditions the tissues require about 4.4 ml of oxygen for every 100 ml of blood passing through the tissue capillaries. To release this much oxygen (4.4 ml) the $PO_2$ must fall to about 40 mm Hg. Therefore, the tissue capillary $PO_2$ must be maintained below 40 mm Hg so as to permit release of oxygen from the hemoglobin. Of course, when the demand for oxygen increases, as in heavy exercise, the $PO_2$ of the capillaries decreases dramatically which causes a proportional increase in the release of oxygen from hemoglobin. This in turn helps to prevent the capillary $PO_2$ from falling below 15 mm Hg.

### Value of Hemoglobin for Adjusting the Tissue $PO_2$ to Different Concentrations of Atmospheric Oxygen

As shown by the oxygen-hemoglobin dissociation curve of figure 8.4, when the $PO_2$ of the arterial blood is 60 mm Hg the hemoglobin is about 89% saturated with oxygen, and when the $PO_2$ in the venous blood is 38 mm Hg the hemoglobin saturation is about 70%. This means that even though the arterial $PO_2$ has been reduced by about 38 mm Hg (98 mm Hg versus 60 mm Hg) the oxygen uptake has been reduced by only 8% (97% hemoglobin saturation with $PO_2$ of 98 mm Hg and 89% hemoglobin saturation with a $PO_2$ of 60 mm Hg).

On the other hand, when the $PO_2$ of the venous blood is 40 mm Hg the hemoglobin saturation is 75%, and when the $PO_2$ in the venous blood is 38 mm Hg the hemoglobin saturation is only 70%. In other words, the release of oxygen from hemoglobin is maximized to adjust to the low pressure of venous blood. Through this process, without change in hemoglobin quantity, and with a slight decrease in $PO_2$ of venous blood, the tissues still remove approximately 4.5 ml of oxygen from every 100 ml of blood passing through the tissues (19.78 – 15.30 = 4.5 ml of oxygen per 100 ml of blood). Thus, the tissue $PO_2$ hardly changes despite the marked fall in the arterial $PO_2$ from 98 to 60 mm Hg.

## TRANSPORT OF CARBON DIOXIDE IN THE BLOOD

Carbon dioxide is one of the major end-products of the oxidative reactions in cells. If all the carbon dioxide formed in the cells should continue to accumulate in the tissue fluids, the mass action of the carbon dioxide itself would soon halt all the energy-giving reactions of the cells. The removal of carbon dioxide involves a nervous mechanism that controls the expiration of carbon dioxide through the lungs and in this way maintains a constant and reasonable concentration of carbon dioxide in the extracellular fluid. That is, a high carbon dioxide concentration excites the respiratory center, causing the person to breathe rapidly and deeply. This increases the expiration of carbon dioxide and therefore increases its removal from the blood and extracellular fluid, and the process continues until the concentration returns to normal. Under normal circumstances about 200 ml of carbon dioxide per minute is produced by the tissue metabolism of a 70-kg person which needs to be transported by the venous blood to the lung for its removal from the body. Carbon dioxide is transported in the venous blood in three forms: (1) in physical solution, (2) chemically combined with amino acids in blood proteins, and (3) as bicarbonate ions. Figure 8.6 schematizes the various forms of carbon dioxide transport in the blood.

### Physically Dissolved

About 5% of the total carbon dioxide remains physically dissolved in the plasma and red blood cells.

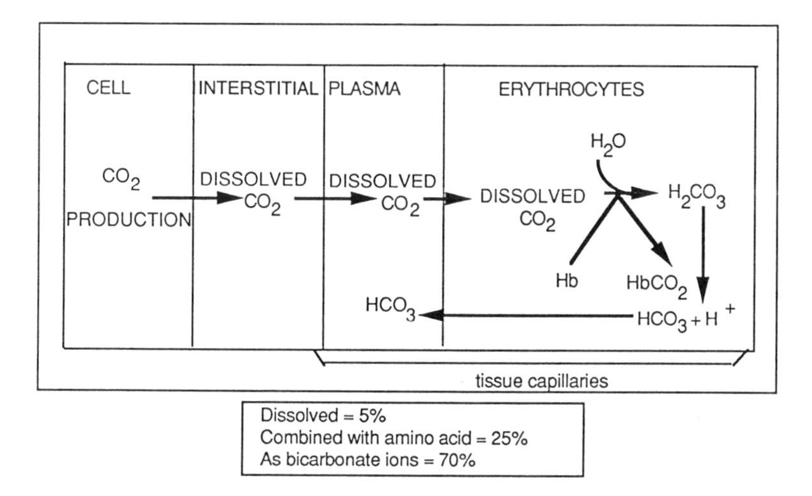


Dissolved = 5%
Combined with amino acid = 25%
As bicarbonate ions = 70%

**Fig. 8.6.** Schematization for the various forms of carbon dioxide transport in the blood.

### Chemically Combined with Amino Acids

About 25% of carbon dioxide combines chemically with the terminal amine groups in blood proteins, forming carbamino compounds. This reaction occurs very rapidly and does not require enzymes. After carbon dioxide is bound with hemoglobin, it is transported by the venous blood to the lungs where the hemoglobin combines with oxygen and releases carbon dioxide from its terminal amino groups.

### As Bicarbonate Ions

The remaining 70% of carbon dioxide transported by the blood is carried as bicarbonate ions. Combining $CO_2$ and $H_2O$ produces carbonic acid:

$$CO_2 + H_2O \xrightarrow{\text{Carbonic Acid}} H_2CO_3$$

Then the carbonic acid dissociates into hydrogen ions and a bicarbonate ion:

$$CO_2 + H_2 \xrightarrow{\text{Carbonic anhydrase}} H_2CO_3 \rightarrow \overset{\text{Bicarbonate}}{H^+ + HCO_3^-}$$

The chemical reaction occurs primarily in the red blood cells because they contain large quantities of the enzyme carbonic anhydrase but the plasma does not. Although the addition of carbon dioxide to a liquid results ultimately in bicarbonate and hydrogen ions, the reaction occurs very slowly unless it is catalyzed by the enzyme carbonic anhydrase.

In summary, all of the reactions of carbon dioxide are reversible. That is, they can proceed in either direction depending on the $PCO_2$. Thus, when the body produces $CO_2$, the partial pressure of $CO_2$ in the tissues is increased creating a gradient which produces net diffusion of $CO_2$ from cells to venous blood and from venous blood to alveoli in the lungs (see fig. 8.7). Conversely, when the concentration of $CO_2$ decreases the gradient is also reduced, and thus $HCO_3^-$ and $H^+$ combine to give $H_2CO_3$, which generates carbon dioxide and water.

## COMBINATION OF HEMOGLOBIN WITH CARBON MONOXIDE

The affinity of hemoglobin for carbon monoxide is about 230 times greater than the affinity for oxygen. As illustrated in figure 8.8, the carbon monoxide-hemoglobin dissociation curve is almost identical with the oxygen-hemoglobin dissociation curve, except that the pressures of the carbon monoxide shown on the abscissa are at a level 1/230 of those in the oxygen-hemoglobin dissociation curve of figure 8.4. For this reason, a carbon monoxide pressure of only 0.4 mm Hg in the alveoli, 1/230 that of the alveolar oxygen, permits the carbon monoxide to compete equally with the oxygen for combination with the hemoglobin. Furthermore, carbon monoxide combines with hemoglobin at the same site on the hemoglobin molecule as does oxygen. It can therefore effectively block the combination of oxygen with hemoglobin because oxygen cannot be bound to iron atoms already combined with carbon monoxide. Carbon monoxide has a second deleterious effect: it shifts the oxyhemoglobin dissociation curve to the left. Thus carbon monoxide can both prevent the loading of oxygen into the blood in the lungs and it can also interfere with the unloading of oxygen at the tissues. This can be seen in figure 8.8.

These two factors cause half the hemoglobin in the blood to become bound with carbon monoxide instead of with oxygen. Carbon monoxide is particularly dangerous for several reasons. Due to the high affinity for hemoglobin a person breathing very low concentrations of carbon monoxide can slowly reach life-threatening levels of carboxyhemoglobin (COHb) in the blood. The effect is cumulative and what is worse, because the gas is colorless, odorless, and tasteless, it does not cause any reflex coughing or sneezing, increase in ventilation, or feeling of difficulty in breathing, and the individual is unaware of the danger. A carbon monoxide pressure of 0.7 mm Hg (a concentration of about 0.1% in the air) can be lethal. Treatment procedures for carbon monoxide poisoning usually include administration of pure oxygen and carbon dioxide, which increases alveolar ventilation and reduces the alveolar carbon monoxide concentration by increasing its release from the blood. Living in urban areas causes small amounts of carboxyhemoglobin to be present in the blood of normal healthy adults. However, smoking, be it in a rural or urban area, may increase the carboxyhemoglobin to 5 to 8 times that of a nonsmoker.

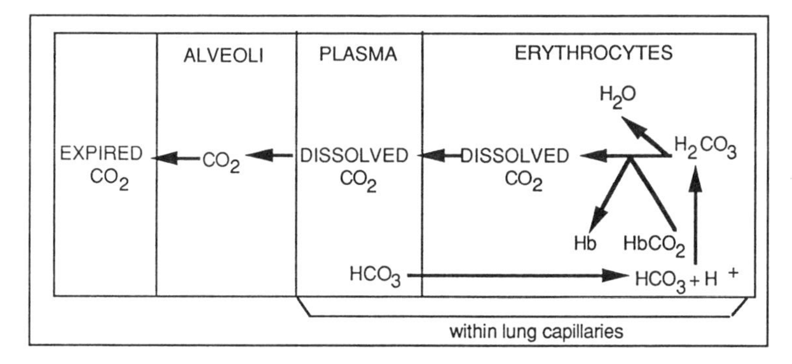


**Fig. 8.7.** Schematization of carbon dioxide movement in the lung capillaries.

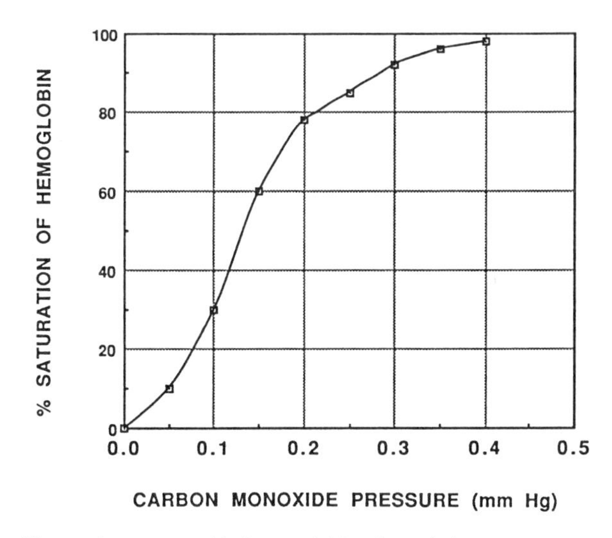


**Fig. 8.8.** The carbon monoxide-hemoglobin dissociation curve.

## OTHER FACTORS AFFECTING OXYGEN TRANSPORT

### Anemia

When the amount of hemoglobin decreases, the total amount of oxygen bound to hemoglobin in arterial blood is also reduced. For example, a reduction of hemoglobin from 15 to 10 g Hb results in reduction of 6.6 ml of oxygen/100 ml in the arterial blood that leaves the lungs. This is because with 15 g of Hb the arterial blood contains 19.7 ml of oxygen/100 ml of blood ($0.98 \times 1.34 \times 15 = 19.7$) while with 10 g Hg it contains only 13.1 ml of oxygen/100 ml ($0.98 \times 1.34 \times 10 = 13.1$).

### Methemoglobin

A high concentration of methemoglobin reduces the hemoglobin affinity for oxygen. Methemoglobin is hemoglobin with iron in the ferric ($Fe^{3+}$) state. It can be caused by nitrite poisoning or by toxic reactions to oxidant drugs, or it can be found congenitally in patients with hemoglobin M. Iron atoms in the $Fe^{3+}$ state will not combine with oxygen.

### Fetal Hemoglobin

Fetal hemoglobin (HbF) in red blood cells has a dissociation curve to the left of that for adult hemoglobin (HbA), which has a high affinity for oxygen at both high and low pressures. Therefore, its presence during postnatal life decreases the oxygen delivery to tissues.

### Abnormal Hemoglobins

Abnormal hemoglobins may have either increased or decreased affinities for oxygen. For example, Hb Seattle and Hb Kansas have lower affinities for oxygen than does HbA; Hb Rainier has a higher affinity for oxygen.

### Myoglobin

Myoglobin increases the storage and transport capacity of oxygen in skeletal muscle. Myoglobin (Mb), a heme protein that occurs naturally in muscle cells, consists of a single polypeptide chain attached to a heme group. As such it combines chemically with a single molecule of oxygen and is similar structurally to a single subunit of hemoglobin. The dissociation curve for myoglobin is shifted to the left of that of normal adult hemoglobin. That is, at lower $PO_2$'s much more oxygen remains bound to myoglobin. As blood passes through the muscle, oxygen leaves hemoglobin and binds to myoglobin. It can be released from the myoglobin when conditions cause lower $PO_2$'s.

## OVERVIEW

The extraction of oxygen from the air is accomplished through the function of the respiratory system. The lungs and corresponding aveolar sacs that constitute the respiratory membrane provide a medium for the gas exchange between the body's internal fluid environment and the gaseous external environment. To accomplish this goal, the lungs have a large volumetric space that is indicated by the measure of residual lung volume and vital capacity. To take advantage of this characteristic at a given time the lungs can move air in and out at a rapid rate. This process is called pulmonary ventilation. In general, pulmonary ventilation is geared to maintain favorable concentrations of alveolar oxygen and carbon dioxide to assure adequate aeration of the blood flowing through the lungs. It is estimated that minute ventilation ranges between 6 and 10 liters at rest and may reach as high as 200 liters during exercise.

Once the alveoli or respiratory membrane is ventilated with fresh air the oxygen diffuses into the pulmonary blood. This diffusion is governed by the barometric pressure and the corresponding pressures of each gas which are referred to as partial pressures. The total barometric pressure at sea level of the air is about 760 mm Hg, and since 21% of the air is composed of oxygen the corresponding partial pressure of oxygen is about 160 mm Hg ($0.21 \times 760 = 160$). Thus, as the oxygen passes through the trachea it becomes moist and loses some of its pressure, so that by the time it reaches the alveolar area it has only 104 mm Hg. This means that at sea level oxygen has a force of 104 mm Hg, with which it will diffuse into the pulmonary blood.

As a result of the high force exerted by the barometric pressure, oxygen ($PO_2$ 104 mm Hg) enters into the pulmonary blood in large quantities where the hemoglobin of the red blood cells traps it and transports it to the tissue capillaries where it is released for use by the cells. When hemoglobin (Hb) combines with oxygen ($O_2$), it is called oxyhemoglobin ($HbO_2$). On the average the hemoglobin in 100 ml of blood has a capacity to combine with about 20 ml of oxygen in males and with 17 ml oxygen in females. Due to the unique quality of hemoglobin, it binds with oxygen under high pressure as occurs in the lungs, but when the pressure is low, as occurs in the tissues, it releases oxygen to the cells. As a result of this metabolic function, the hemoglobin becomes partially deoxygenated. The relationship between the amount of oxygen bound with hemoglobin is usually expressed as the percent saturation, or oxyhemoglobin dissociation curve. In general, the hemoglobin saturation changes very little until the $PO_2$ falls below 60 mm Hg. Because this low pressure occurs in the tissues, the quantity of oxygen bound to hemoglobin falls sharply. Thus, oxygen is released rapidly from capillary blood and flows into the tissues in response to the cells' metabolic demands. At rest, only about 25% of the blood's total oxygen is released to the tissues; the remaining 75% returns to the heart in the venous blood. This is the arteriovenous oxygen difference and indicates that an ''automatic'' reserve of oxygen exists so that cells can rapidly obtain oxygen should the metabolic demands increase suddenly. Increases in acidity, temperature, carbon dioxide concentration, and red blood-cell 2,3-DPG are known to cause alteration in the hemoglobin's ability to transport and release oxygen.

Evidently the organism's ability to extract oxygen from the air and transport it into the cells requires the synchronized action of both cardiovascular and respiratory systems. As such, all factors that modify the size and the structure of these systems play an important role in the organism's ability to survive under conditions of low oxygen availability, as occurs under conditions of high-altitude hypoxia, which will be discussed in the next chapter.

### Suggested References

Tortora, G. J. 1989. Principles of human anatomy. New York, NY: Harper and Row, Publishers.

Guyton, A. C. 1986. Textbook of medical physiology. Philadelphia, PA: W. B. Saunders Co.

Dunn, C. D. R. 1971. The differentiation of haemopoietic stem cells. Baltimore, MD: Williams and Wilkins.

Weatherall, D. J., ed. 1982. Advances in red blood cell biology. New York, NY: Raven Press.

Levitzky, M. G. 1986. Pulmonary physiology. New York, NY: McGraw-Hill, Inc.

Katch, F. I. 1991. Exercise physiology: energy, nutrition, and human performance. Philadelphia, PA: Lea and Febiger.

CHAPTER 9

# Cardiovascular System

---

## COMPONENTS OF THE CARDIOVASCULAR SYSTEM

As illustrated in figure 9.1 the cardiovascular system is a continuous vascular circuit that consists of the heart and a system of blood vessels through which blood is transported to the lungs, where it can obtain oxygen and deliver it to the tissue cells for their metabolic activities. Thus, the heart, blood vessels, and blood constitute the cardiovascular system.

### Heart

The blood can perform its multifaceted function only if it circulates continually through the body. The pump that drives the blood through the vessels is the heart. The heart is a hollow, muscular organ that weighs about 342 grams (11 oz) and beats between 72,000 to 150,000 times a day to pump blood through 1 million meters (60,000 miles) of blood vessels.

#### Anatomical Features

For all general purposes the heart consists of two hollow organs—the right half and the left half—with muscular walls (fig. 9.2). Each half comprises an upper chamber called the atrium and a lower chamber called the ventricle. The right half receives oxygen-depleted blood (called venous blood) from the entire body and sends it to the lungs, where the blood picks up oxygen. The oxygenated blood (called arterial blood) is returned to the left half of the heart and hence distributed to the organs of the body. Thus the right heart pumps only deoxygenated blood, while the left heart pumps only oxygenated blood.

The right atrium receives blood from three major veins: (1) the superior vena cava, which brings blood from parts of the body superior to the heart, (2) the inferior vena cava, which brings blood from the parts of the body inferior to the heart, and (3) the coronary sinus, which drains blood from most of the vessels supplying the walls of the heart. Once the blood enters the right atrium it passes into the right ventricle. From the right ventricle the blood is ejected through the pulmonary trunk. The pulmonary trunk divides into a right and left pulmonary artery, each of which carries deoxygenated blood to the lungs. In the lungs, the blood takes up oxygen and releases its carbon dioxide. Thereafter, the blood returns from the lungs into the left atrium of the heart via four pulmonary veins. The blood then passes into the left ventricle. The ascending aorta then divides into the coronary artery, arch of the aorta, thoracic aorta, and duodenal aorta. These four arteries transport blood to all body parts except the lungs. The ascending aorta gives off two coronary branches to the heart muscle.

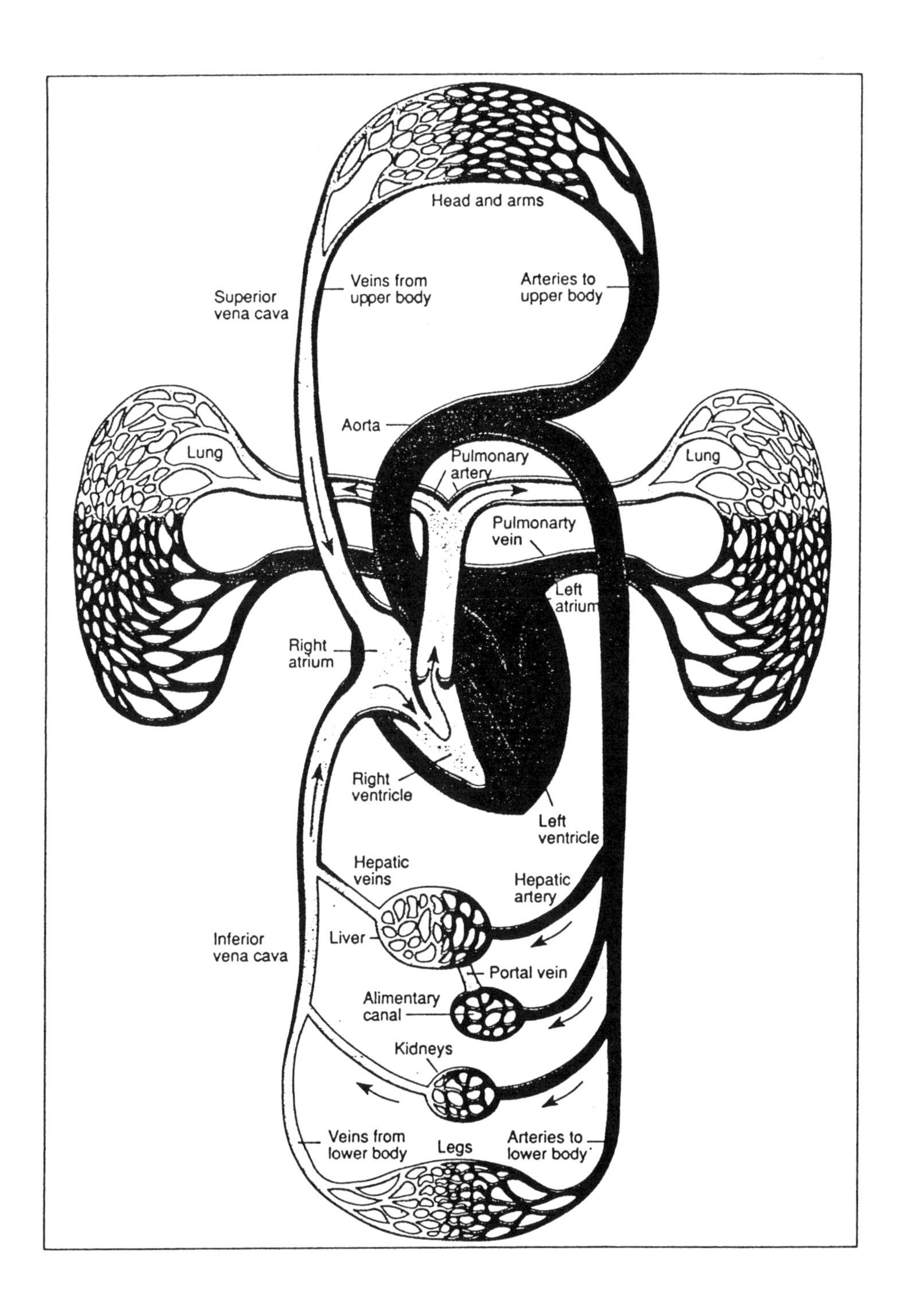


**Fig. 9.1.** Circulatory routes. Schematic view of the cardiovascular system consists of the heart and the pulmonary and systemic vascular circuits. (Adapted from W. D. McArdle, F. L. Katch, and V. L. Katch, 1991. Exercise physiology. Philadelphia: Lea and Febiger.)

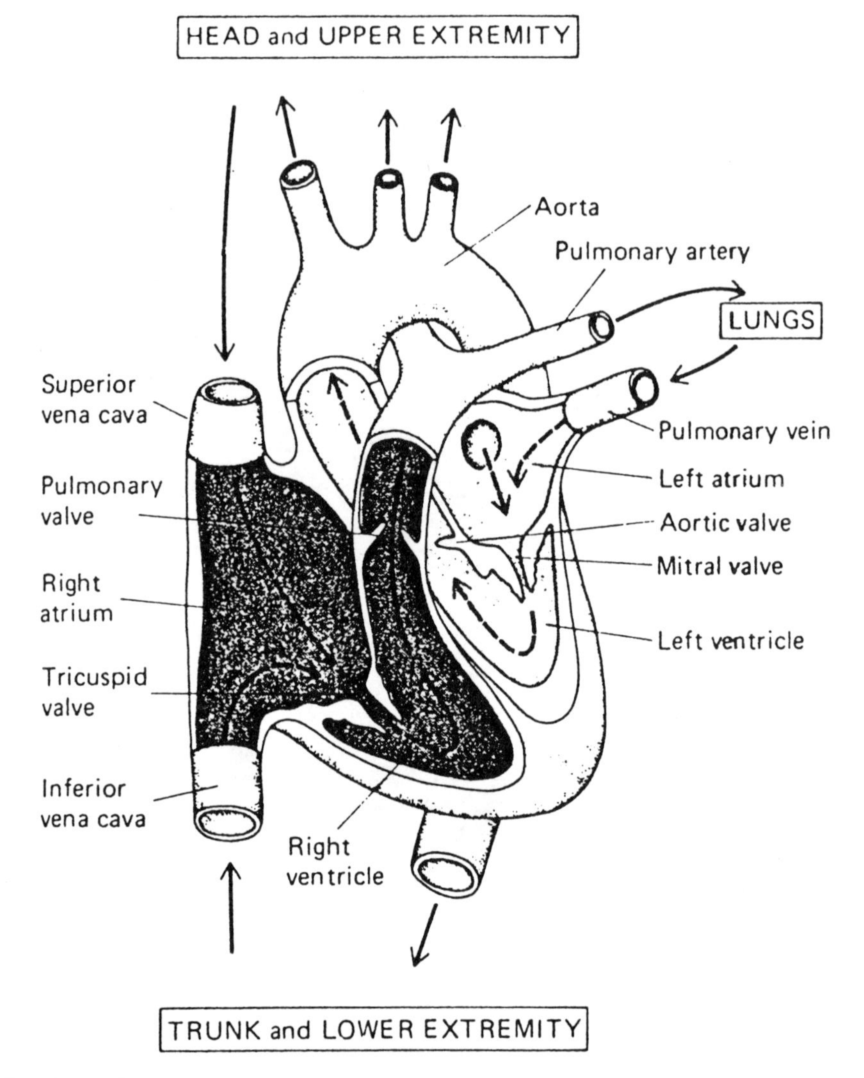


**Fig. 9.2.** Structure of the heart.

## Cardiac Output and Stroke Volume

*Cardiac output* is the volume of blood pumped by each ventricle per minute. On the other hand, the *stroke volume* is the amount of blood ejected by each ventricle during each beat. Therefore, the cardiac output is derived by computation:

Cardiac output (liter/min) = heart rate (beat/min) × stroke volume (liter/beat)

Under the resting condition cardiac output is about 5.0 liter/min, which is derived assuming a heart rate of 72 beats/min, and a stroke volume of 70 ml/beat.

Cardiac output = 72 beats/min × 0.07 liter/beat = 5.0 liters/min

During exercise the cardiac output may increase to 20 and 30 liters/min due to increases in both heart rate and stroke volume. For example, the heart rate may increase to 200 beat/min while the stroke volume may increase to 100 ml/beat, resulting in a cardiac output of 20 liters/min.

### The Blood Vessels

The blood travels away from the heart into the tissues of the body and then returns to the heart through a network of specialized large and small tubes. These include the arteries, arterioles, capillaries, venules, veins, and vasa vasarum.

#### Arteries, Arterioles, Capillaries, and Veins

Arteries are the large tubes that carry blood from the heart to the tissues. As they leave the heart, these arteries divide into medium-sized muscular vessels that go toward the various regions of the body. The medium-sized arteries divide into small arteries, which in turn subdivide into smaller arteries called *arterioles.* Once the arterioles enter into the tissues they subdivide into countless microscopic vessels called *capillaries.* It is through the membranes of the capillaries that all organic and inorganic substances are exchanged between the blood and the body tissues. As the capillaries leave the tissues they reunite to form small veins, called *venules.* These venules, in turn, merge to form progressively larger tubes called *veins.* These *veins* transport blood from tissues back to the heart (fig. 9.3). The blood vessels that supply blood and nutrients to the arteries and veins are called *vasa vasarum.*

## THE CIRCULATORY SYSTEM

### Subdivision of the Circulatory System

The circulation of blood through the vessels of the organism is divided into two major sections arranged one after the other: pulmonary circulation and systemic circulation.

#### Pulmonary Circulation

The flow of deoxygenated blood from the right ventricle to the lungs and the return of oxygenated blood from the lungs to the left atrium is called the pulmonary circulation (fig. 9.3). As indicated above, the pulmonary trunk

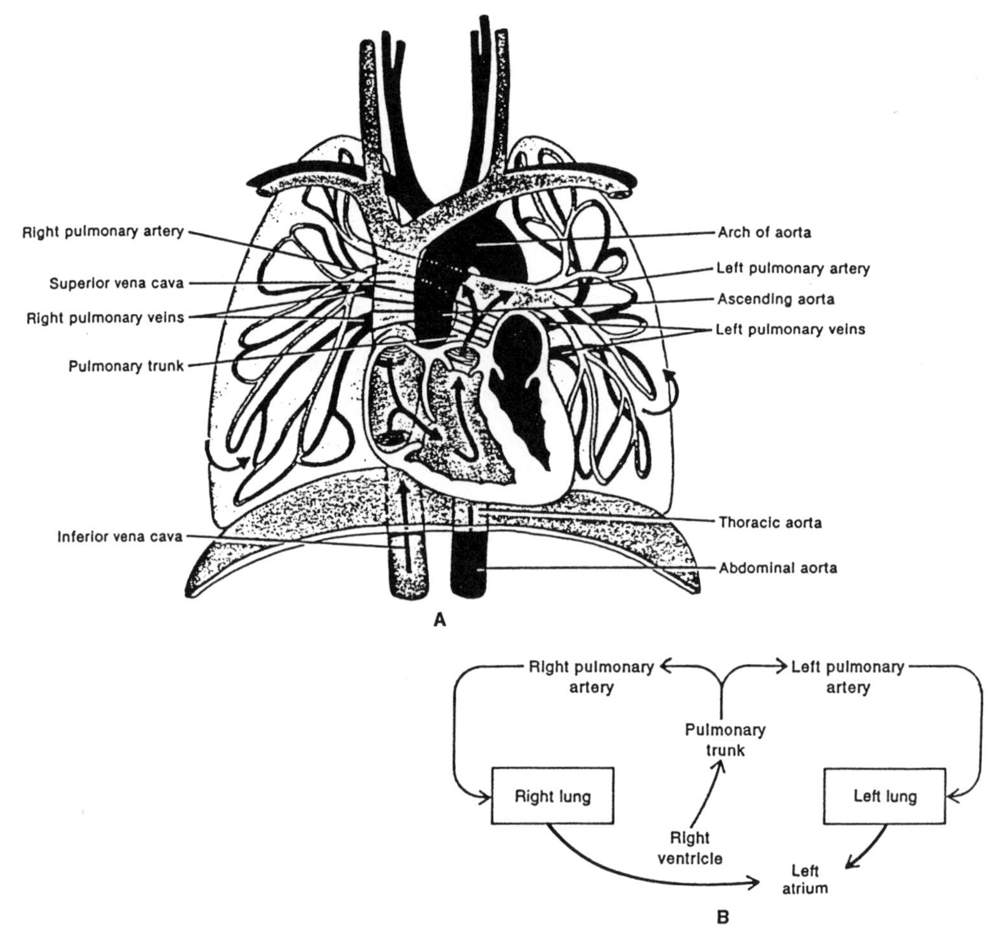


**Fig. 9.3.** Pulmonary circulation. **A,** Diagram. **B,** Schematization of the circulation throughout the right lung, pulmonary artery, left lung, and left atrium of the heart.

arises from the right ventricle and divides itself into two branches called the right and left pulmonary arteries. Each artery carries blood into the corresponding half of the lung (the right artery into the right lung and the left artery into the left lung). On entering the lungs the branches divide and subdivide continuously until they form capillaries around the alveoli in the lungs. Here the oxygen breathed in by the lungs is passed from the alveoli into the blood while carbon dioxide is passed from the blood into the alveoli to be breathed out of the lungs. The capillaries unite, venules and veins are formed, and eventually two pulmonary veins leave from each lung and transport the oxygenated blood to the left atrium, and from there it is pumped into the left ventricle. (Note that the pulmonary veins are the only postnatal veins that carry oxygenated blood.) Contraction of the left ventricle then ejects blood into the systemic circulation.

### Systemic Circulation

The flow of blood from the left ventricle to all parts of the body and back to the right atrium is called the systemic circulation. The purpose of the systemic circulation is to distribute oxygen and nutrients to all body tissues and remove carbon dioxide and other wastes from the tissues. All systemic arteries branch from the aorta, which emerges from the left ventricle (see fig. 9.1).

In the systemic circulation the oxygenated blood is pumped by the left ventricle of the heart into the aorta, and is then distributed into the numerous arteries. The stream of blood is divided among many regional vascular beds in parallel, each of which supplies a particular organ (heart, brain, liver, kidney, skeletal, musculature, skin, etc.). Each artery undergoes repeated subdivisions to form arterioles, and these in turn branch to form the capillary bed, which is a very dense network of narrow vessels with very thin walls. These capillaries permit the diffusion of fluid and gases to and from tissue cells. The capillaries merge to form venules, which in turn join to form veins. As this fusion continues, there is a steady decrease in the number of vessels and increase in the diameter of the vessels, so that ultimately only two, the superior and inferior vena cava, remain to pass the blood into the right atrium, completing the circuit.

### Systole and Diastole

The pumping action of the heart is based on a rhythmic sequence of relaxation called diastole followed by a period of contraction called systole. During diastole the ventricles fill with blood, and during systole they expel it into the large arteries (from the right ventricle into the pulmonary artery and from the left ventricle into the aorta). Before entering the ventricle, the blood passes from the vena cava into the right atrium and from the pulmonary vein into the left atrium. The contraction (or systole) of each atrium precedes that of its ventricle. In this manner the atria work as booster pumps to help fill the ventricle. Figure 9.4 schematizes the blood pressures for the pulmonary and systemic circulation.

## Blood Flow and Blood Pressure

### Blood Flow

Blood flow means the quantity of blood that passes a given point in the circulation in a given period of time. Blood flow is usually measured in milliliters per minute. The overall blood flow in the circulation of an adult person is about 4900 ml/min. This is called the *cardiac output,* because it is

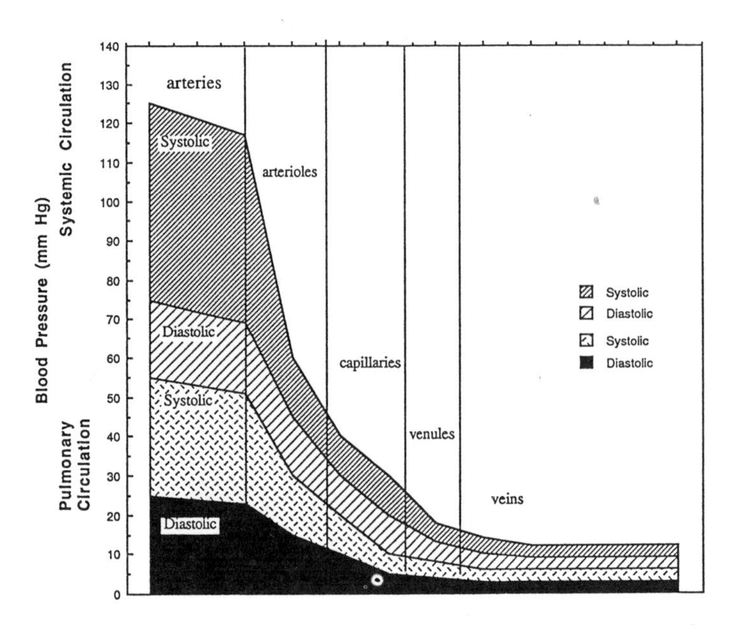


**Fig. 9.4.** Changes in blood pressure throughout the pulmonary and systemic circulation.

the amount of blood pumped by the left ventricle of the heart into the aorta per minute. The cardiac output is influenced by the amount of blood ejected by the left ventricle during each beat, and the number of beats per minute. The amount of blood pumped by a ventricle during each contraction (systole) is called the *stroke volume.* In a resting adult the stroke volume averages about 70 ml. Therefore, if the average heart rate is 72 beats per minute, the average cardiac output is about 5000 ml/min (72 ml/beat × 70 beats/min = 5000 ml/min).

## Blood Pressure

Blood flow through a vessel is influenced by two factors: (1) the pressure difference, that is, blood flow from an area of higher pressure to an area of lower pressure; and (2) the resistance to flow that the blood vessels exert. Blood pressure is usually measured in millimeters of mercury (mm Hg). Blood pressure is defined as the pressure exerted by blood on the wall of any blood vessel. Thus a pressure of 100 mm Hg means that the force exerted by the blood would be sufficient to push a column of mercury up to a level of 100 mm Hg. In the same manner as there are two major types of circulation there are also two major types of blood pressure: pulmonary blood pressure and systemic blood pressure. The pulmonary blood pressure averages approximately 22 mm Hg for systolic and about 8 mm Hg for diastolic pressure. The average systemic blood pressure is about 120 mm Hg for systolic and 80 mm Hg for diastolic pressure. Thus the blood in the pulmonary circulation is at lower pressure than the blood in the systemic circulation.

### Factors That Affect Arterial Blood Pressure

Although blood pressure implies the pressure exerted by blood on the flow of any wall, in clinical use the blood pressure refers to the pressure in the arteries. Blood pressure above 150 mm Hg for systolic and 100 mm Hg for diastolic is indicative of hypertension; conversely, a pressure below 90 mm Hg for systolic and 50 mm Hg for diastolic is indicative of hypotension. Several factors influence blood pressure.

**Blood Volume.** Blood pressure is directly proportional to the volume of blood in the cardiovascular system, so that the higher the blood volume the greater the blood pressure. Thus any environmental condition that leads to an increase in blood volume, such as high salt intake resulting in greater water retention, increases blood pressure.

**Peripheral Resistance.** The amount of resistance that the blood vessels give the blood is referred to as peripheral resistance. As previously indicated, the viscosity of the blood affects the flow of blood. In addition, blood flow is affected by the diameter of the blood vessels, so that the bigger the diameter of the vessels the smaller the resistance. Consequently, the lesser the resistance it offers to the blood. Thus, any condition that increases the viscosity of blood, such as dehydration or an unusually high number of red blood cells, increases blood pressure. The reverse occurs when the viscosity is lowered as in anemia or hemorrhage. To maintain normal blood pressure when there are changes in blood viscosity the diameter of the arterioles is adjusted accordingly; that is, they increase in diameter when there is high viscosity and narrow when there is low viscosity, which together control peripheral resistance.

**Cardiac Output.** Since the blood flow depends upon the amount of blood and the rate with which it is pumped by the heart, the cardiac output affects blood pressure. Thus, if the cardiac output is increased by an increase in stroke volume or increased heart rate per minute, then the blood pressure is increased. Conversely, a decrease in cardiac output decreases blood pressure.

## BLOOD

The blood consists of formed elements, such as red blood cells, white blood cells, platelets, and plasma, which is a liquid containing dissolved substances (fig. 9.5).

### Blood Volume

The average blood volume of an adult male is 5000 ml and 4000 ml in the female. Of this, about 55% consists of plasma and the remainder (45%) of formed elements.

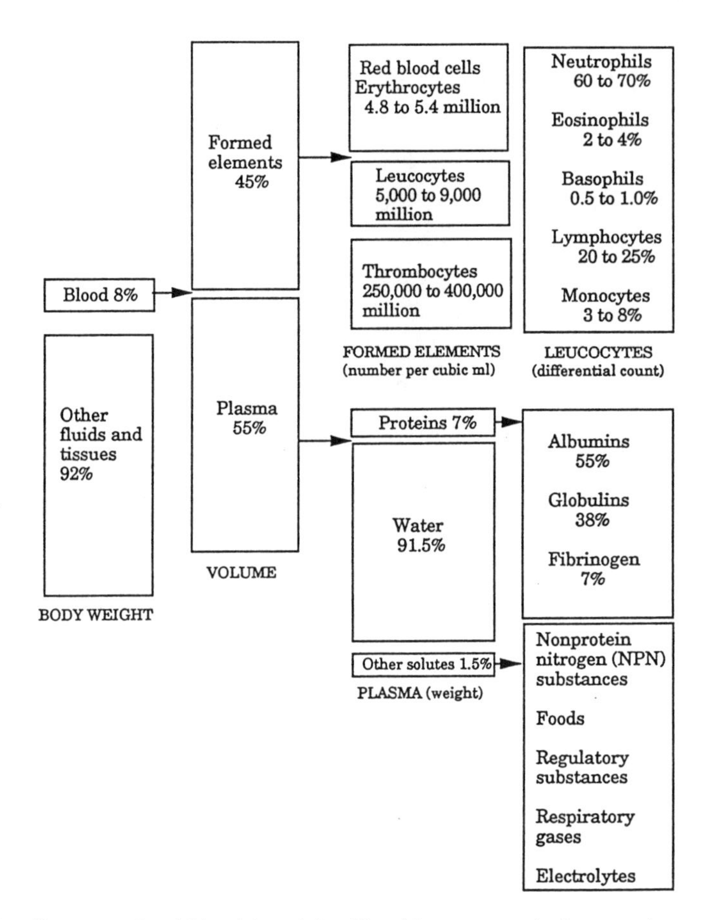


**Fig. 9.5.** Components of blood in adults. Blood is composed of two major portions: formed elements and plasma liquid.

**Effects of Body Composition.** Blood volume is strongly related to body size and body composition. In general, the blood volume varies in direct proportion to body weight, and normally averages about 79 ml/kg. However, the greater the fatness the less the blood volume per unit of weight; because fat tissue has little vascular volume, lean people have a greater proportion of blood to total body weight than fat people.

## Hematocrit

The proportion or percentage of the blood made up of red blood cells is called the hematocrit. Thus, a hematocrit of 40 means that 40 percent of the blood is cells and the rest is plasma. The usual procedure for determining the hematocrit is to centrifuge the blood in "hematocrit tubes" until the cells become packed tightly in the bottom of the tubes. Since it is impossible to remove all the plasma from the packed red blood cells—about 3 to 8% of the plasma remains entrapped among the red blood cells—the true percentage or hematocrit averages about 95% of the measured hematocrit. Therefore, the hematocrit needs to be multiplied by 0.95 to obtain the average hematocrit. Hematocrit values range from an average of 43 to 45% for adult males and 39 to 41% for females.

### Hematocrit and Viscosity

Because resistance to flow rises linearly with viscosity, any pathological increase in hematocrit puts a greater load on the heart and can cause deficient circulation through certain organs. Excessive increase in hematocrit is usually nonadaptive because it increases the blood viscosity. This is due to the fact that the greater the hematocrit the more friction there is between successive layers of blood, and it is friction that determines viscosity.

Conversely, low hematocrits are not adaptive in that they decrease the oxygen-carrying capacity of the blood, as occurs in anemia. Hematocrits below 39% for males and 35% for females are considered indicators of anemia.

## FUNCTIONS OF BLOOD

Blood performs a number of critical functions, among which transport, homeostasis, and protection against hemorrhage and foreign agents are the most important.

### Transport

Blood transports the respiratory gases oxygen and carbon dioxide, both in the chemically bound form and in physical solution—$O_2$ from the lungs to the respiring tissues and $CO_2$ from the tissues to the lungs. Blood transports nutrients from the digestive organs to the cells. The metabolites, or waste materials, produced in the cells are transferred to the excretory organs (such as the kidneys and sweat glands) or to the parts where they can become utilized. Blood transports hormones and enzymes, taking them up at the sites of production or storage and distributing them throughout the intravascular space to the target organs.

### Homeostasis

Blood circulates throughout the body. Its composition and physical properties are continually monitored by certain organs and, if necessary, adjusted to ensure equilibrium of the internal milieu. This includes the regulation of blood pH through buffers and amino acids, and regulation of body temperature because blood contains a large volume of water. This ability to maintain homeostasis is indeed a basic requirement for the normal function of all cells.

### Prevention of Hemorrhage

When bleeding occurs blood has the capacity to coagulate and close small injured vessels. In this manner, the blood prevents loss of valuable fluid.

### Protection against Foreign Agents

Through its capacity to produce antibodies and phagocytic cells the blood protects the organism against toxins and foreign microbes.

## FORMED ELEMENTS OF BLOOD

The formed cell elements of blood include: (1) red blood cells (erythrocytes), (2) white blood cells (leucocytes), and (3) platelets (thrombocytes). (See fig. 9.5.)

### Red Blood Cells

The major function of red blood cells is to transport hemoglobin, which in turn carries oxygen from the lungs to the tissues, and carbon dioxide from the tissues to the lungs. The respiratory functions of hemoglobin have already been discussed, and therefore here we will be concerned mostly with the process of red blood cell production.

#### Shape and Size of Red Blood Cells

Normal red blood cells are biconcave disks having a diameter of about 8 microns and a thickness of about 2 microns. All the red blood cells of an adult man have been estimated to be about 3800 $m^2$. The shape of red blood cells can change as the cells pass through capillaries. The shape plasticity of red blood cells decreases during old age and under pathological conditions such as sickle-cell anemia and spherocytes.

#### Life Span, Number, and Production

The life span of a red blood cell is about 120 days. Under normal circumstances an adult male has 5.4 million red blood cells per cubic millimeter ($mm^3$) of blood, and an adult female has about 4.8 million. To sustain normal quantities of erythrocytes the organism produces new mature cells at a rate of about 2 million per second. In the early embryonic life primitive red blood cells are produced in the yolk sac. During the middle trimester (the fourth through the sixth month) of gestation, red blood cells are produced in the liver, spleen, and lymph nodes. Then during the last months of fetal life and after birth red blood cells are produced almost exclusively by the bone marrow. As shown in figure 9.6 until adolescence red blood cells are produced by the bone marrow of all bones, but after the age of about 20 years the red blood cells are produced mainly by the marrow of the membranous (or spongy) bones such as the vertebrae, sternum, and the ribs.

### Erythropoiesis

The process whereby red blood cells are produced is called erythropoiesis. The red blood cells (or erythrocytes) are highly specialized for oxygen transport. It is estimated that each erythrocyte contains about 280 million molecules of hemoglobin. Once the erythrocyte is formed, it leaves the marrow and enters the bloodstream. Aged erythrocytes are destroyed by the reticuloendothelial cells in the liver and the spleen. The hemoglobin molecules are split apart, the iron is reused, and the rest of the molecule is converted into other substances for reuse and excretion.

Under normal conditions the rate of erythropoiesis keeps pace with the rate of red cell destruction, but any condition that causes the quantity of oxygen transported to the tissues to decrease causes an increase in erythropoiesis. This includes the various types of anemia (nutritional anemia, pernicious anemia, hemorrhagic anemia, hemolytic anemias, aplastic anemia, sickle-cell anemia), which generally reduce the amount of oxygen delivered to the tissues. Similarly, upon exposure to high altitude where the oxygen availability in the air is greatly reduced, the rate of red cell production is increased. Conversely, individuals whose metabolic needs and oxygen requirements are high, such as athletes, maintain a high rate of erythropoiesis. An athlete can have a red blood count as high as 6 million per cubic millimeter whereas the nonathletic person might have one as low as 3 million. It is evident, then, that this mechanism is triggered not by the concentration of red blood cells in the blood but by the functional ability of the cells to transport oxygen to the tissues. The production of red blood cells is not a direct response of the bone marrow, but is mediated through the action of a hormone called erythropoietin. If certain kidney cells become hypoxic they release an enzyme called renal erythropoietic factor that converts a plasma protein into the hormone erythropoietin. This hormone circulates through the blood to the bone marrow, where it stimulates the production of more red blood cells.

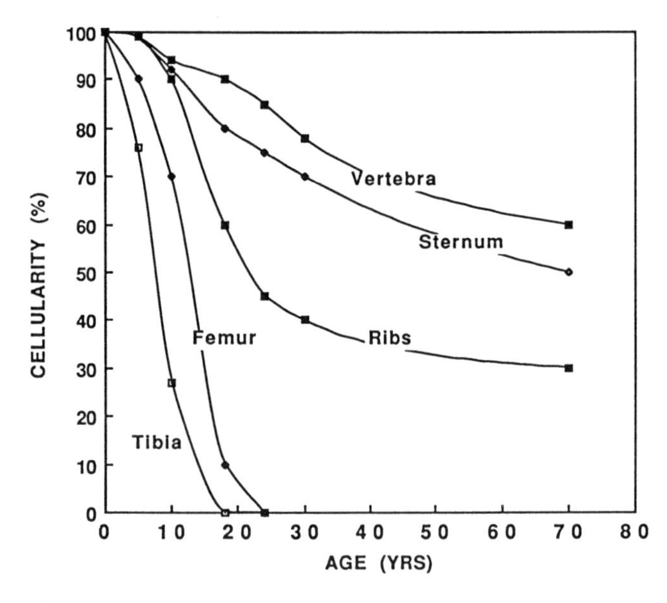


**Fig. 9.6.** Relative rates of red blood cell production in the different bones at different ages. Note that until the age of 20 years the production of red blood cells takes place in both long and flat bones. Thereafter the production of red blood cells continues only in flat bones.

## Nutritional Factors Needed for the Formation of Red Blood Cells

The formation of red blood cells is mediated by the availability of vitamin $B_{12}$, iron, and certain amino acids. Vitamin $B_{12}$ is an essential nutrient for all cells of the body, and growth of tissue cells is depressed in the case of vitamin $B_{12}$ deficiency. Since tissues that produce red blood cells are among the most rapidly proliferating of all tissues, lack of vitamin $B_{12}$ inhibits erythropoiesis. Erythropoiesis can also be inhibited even though vitamin $B_{12}$ is available, due to the lack of a substance known as intrinsic factor. This intrinsic factor, produced by the mucosal cells of the stomach, facilitates absorption of vitamin $B_{12}$ and this in turn helps in the formation of red blood cells.

Iron is important for formation of hemoglobin, the major protein component of red blood cells, and therefore its availability affects the formation of red blood cells. The total quantity of iron in the body averages about 4 grams, 65% of which is present in the form of hemoglobin, 4% in the form of myoglobin, and 15 to 25% stored, mainly in the form of ferritin, and the rest, between 0.1 and 5%, is present in the form of various heme compounds that control intracellular oxidation. When iron is absorbed in the small intestine it immediately combines with a beta globulin to form the compound transferrin, in which form it is transported in the blood plasma. Excess iron in blood is deposited in all cells of the body but especially in the liver cells, where more than 60% of the excess is stored. There it combines with the protein called apoferritin to form ferritin. This iron stored in ferritin is called storage iron. When the quantity of iron in the plasma falls very low, iron is absorbed from ferritin and then transported to the portions of the body where it is needed. When old red blood cells are destroyed the hemoglobin released from the cells is ingested by the reticuloendothelial cells. The free iron is liberated and it can then either be stored in the ferritin pool or reused for formation of hemoglobin.

## White Blood Cells

### Types

The white blood cells or leukocytes, unlike red blood cells, do not contain hemoglobin. Leukocytes are classified into two major groups. The first group is the granular leukocytes. They develop from red bone marrow. Granular leukocytes are in turn subdivided into three kinds: neutrophils (or polymorphs), eosinophils, and basophils.

The second main group of leukocytes is the agranular leukocytes. They are formed by the lymphoid and myeloid tissue. The agranular leucocytes are also subdivided into two kinds: lymphocytes and monocytes.

### Function

The general function of white blood cells is to combat all the microbes, bacteria or antigens that enter the organism. They do so either through phagocytic action or actual production of antibodies that destroy the antigen. *Neutrophils* and *monocytes* destroy foreign bacteria by ingesting them and disposing of the dead matter. Neutrophils are usually the most active leukocytes that respond to tissue destruction caused by bacteria. In addition to carrying on phagocytosis, the neutrophils are able to release the enzyme called lysozyme, which has the capacity to destroy certain bacteria.

Eosinophils are believed to combat the irritants that cause allergies. Eosinophils leave the capillaries, enter the tissue fluid, and produce antihistamines that destroy antigen-antibody complexes.

Basophils are also believed to participate in allergic reactions. Basophils leave the capillaries, enter the tissues, and become the mast cells of the tissues, releasing heparin, histamine, and serotinen.

Lymphocytes are involved in the production of antibodies. Antibodies are proteins produced by the lymphocytes in response to the presence of an antigen. An antigen is any chemical substance that, when introduced into the body, causes the body to produce specific antibodies, which can react with the antigen. Most antigens are proteins, and most are not synthesized by the body. Many of the proteins that make up the cell structures and enzymes of bacteria are also antigens. The toxins released by bacteria are also antigens. When antigens enter the body, they react chemically with substances in the lymphocytes and induce some lymphocytes, called B cells to become plasma cells. The plasma cells then produce antibodies, which are globulin type proteins that attach to antigens in a similar way as enzymes attach to substrates.

These antibodies are usually called immunoglobulins or Ig. Five different types of immunoglobulins are known to exist in humans. These are designated as IgG, IgA, IgM, IgD, and IgE. Each has a distinct chemical structure and a specific biological function. IgG antibodies enhance phagocytosis, neutralize toxins, and protect the fetus and newborn. IgA antibodies provide localized protection on mucosal surfaces of the respiratory and digestive systems. IgM antibodies provide protection by agglutination and lysis. IgD antibodies are involved in the production of other antibodies, and IgE antibodies are involved in allergic reactions.

The mechanisms whereby antibodies inactivate the bacterium or completely dislodge it are several.

### Neutralization of Antigens by Antibodies

When antigens enter the body the antibodies immediately cover the antigens so that the antigens cannot come in contact with other chemicals in the body. In this way bacterial poisons can be sealed up and rendered neutral or harmless. Thereafter the combined antibody and antigen is engulfed by phagocytes of the reticular endothelial system, whereupon both the antigen and antibody are simultaneously digested and thereby destroyed. This process is called the antigen-antibody response. In the same manner that antibodies can attach to antigen molecules, they can also attach one bacterium to the next, thereby binding several bacteria in a clump. This process is called agglutination. The agglutination process prevents the bacteria from invading the tissues and also exposes the bacteria to phagocytosis.

### Life Span and Number

Leukocytes are far less numerous than red blood cells, averaging between 5000 to 9000 cells per cubic millimeter of blood. Since the organism is constantly exposed to foreign bacteria and the life span of antibodies is very short, leukocytes are continuously produced. The granular leukocytes are produced in red bone marrow, and the agranular leukocytes are produced in both myeloid and lymphoid tissue. Excess production of leukocytes (or leukocytosis), greater than 10,000 cells/$mm^3$, usually indicates a pathological condition. Usually an increase in the concentration of monocytes indicates a chronic infection, while increase in eosinophils and basophils indicate allergic reactions. High lymphocyte counts indicate antigen-antibody reactions.

### Platelets

Platelets or thrombocytes are produced in red bone marrow. They are disc-shaped cells without a nucleus.

#### Function

The major function of platelets is prevention of fluid loss by initiating a chain reaction that results in blood clotting.

#### Life Span and Number

Platelets have a short life span, between 5 and 9 days. There are about 200,000 to 400,000 platelets per cubic millimeter of blood.

### Plasma

The plasma is the noncellular fluid portion of the blood. The average amount of plasma is about 3000 ml. It is straw-colored. It is about 90 to 91% water. About 7% of its weight is due to protein present in solute form. These solute proteins, referred to as plasma proteins, include albumins, globulins, and fibrinogen. Albumins, which comprise 55% of plasma proteins, are largely responsible for blood viscosity. Albumin, along with electrolytes, helps regulate blood volume by preventing water in the blood from diffusing into the interstitial fluid. Globulins, which comprise 38% of plasma proteins, are antibody proteins released by plasma cells. The most well-known globulins are the gamma globulins whose function is to combat hepatitis and measles viruses and the tetanus bacterium. Fibrinogen, which comprises 1% of plasma proteins, participates in blood clotting along with the platelets.

## OVERVIEW

The cardiovascular system is a continuous vascular circuit that consists of the heart and a system of blood vessels through which blood is transported to the lungs where it can obtain oxygen and deliver it to the tissue cells for their metabolic activities. The major function of the heart is to transport the blood from the lungs, where it picks up oxygen, and circulate it throughout the body. As a result of metabolic functions, the amount of oxygen present in the arterial blood decreases while the amount of carbon dioxide increases. Therefore, in order to restore the amount of oxygen in the venous blood the heart moves the blood back into the lungs where it picks up oxygen and releases carbon dioxide and so the cycle continues. The process whereby the organism obtains oxygen from the air in the lungs is facilitated by the fact that the red blood cells contain hemoglobin (a protein whose function is to bind to oxygen and carbon dioxide). It is evident then that any factor that interferes with the cardiovascular system will have profound influences on the organism's ability to obtain and utilize oxygen. For example, a weak heart will not be able to transport blood to and from the lungs easily and as such would diminish its capacity to obtain oxygen. Conversely, blood that is deficient in hemoglobin may also impair the organs' ability to obtain oxygen. On the other hand, an excessive amount of hemoglobin may overload the heart by increasing the blood's viscosity and hence decrease oxygen uptake. Concomitant with its function of oxygen transport, blood also plays an important role in providing white blood cells whose function is to combat all the microbes, bacteria, and antigens that enter the organism through either phagocytosis or neutralization of antigens. The blood also contains platelets whose function is to initiate blood clotting and thus prevent bleeding.

All the constituents of blood and their continuing existence are subject to a great number of environmental and inherent biological factors. For example, although the life span of the red blood cell is 120 days, depending on the environmental factors the actual length of life can change drastically, as occurs under conditions of anemia or polycythemia. Thus the organism's ability to maintain its normal metabolic functions depends upon a synchronized function of all components of the cardiovascular system.

## References


1. Tortora, G. J. 1989. Principles of human anatomy. New York, NY: Harper and Row, Publishers.
2. Guyton, A. C. 1986. Textbook of medical physiology. Philadelphia, PA: W. B. Saunders Co.
3. Dunn, C. D. R. 1971. The differentiation of haemopoietic stem cells. Baltimore, MD: Williams and Wilkins.
4. Weatherall, D. J., ed. 1982. Advances in red blood cell biology. New York, NY: Raven Press.
5. Levitzky, M. G. 1986. Pulmonary physiology. New York, NY: McGraw-Hill.
6. Katch, F. I. 1991. Exercise physiology: energy, nutrition, and human performance. Philadelphia, PA: Lea and Febiger.

# CHAPTER 10 Biological Effects and Tolerance to High Altitude

During their conquest of the Incas of Peru, the Spaniards were the first to notice the effects of high-altitude environments on the normal functioning of lowland natives (1). In 1590 the chronicler José de Acosta in his *Historia Natural y Moral de las Indias* gave the first clear description of mountain sickness experienced by sea-level man sojourning at a high altitude (1). Three centuries later, Jourdanet (2) and Bert (3) began their scientific observations of the effects of high altitudes and low barometric pressures. Since then, study of the mechanisms whereby man adapts to the pervasive effects of high-altitude hypoxia has concerned both biological and social scientists.

## ENVIRONMENTAL FACTORS

A high-altitude environment presents several stresses to humans. The most important include (1) hypoxia, (2) high solar radiation, (3) cold, (4) low humidity, (5) high winds, (6) limited nutritional base, and (7) rough terrain. From the physiological point of view hypoxia is the most important, since the other stresses are present in an equal or to a greater degree in other geographical zones. High-altitude hypoxia is a pervasive and ever-present stress that is not easily modified by cultural or behavioral responses. Furthermore, all organ systems and physiological functions are affected by hypoxia. For these reasons, the study of human adaptation to high-altitude hypoxia provides an excellent opportunity to learn about the flexibility and nature of the homeostatic processes that enable an organism to function and survive under extreme environmental stress.

### Nature of Hypoxic Stress and Types of Hypoxia

When tissues receive a deficient supply of oxygen, a physiological situation called hypoxia ("less oxygen" than required) develops. Hypoxia can be produced by any condition, physiological or pathological, that interferes with oxygen acquisition at the cardiopulmonary level and utilization by the cells, hereafter referred to as organic hypoxia. Hypoxia can also be produced by atmospheric conditions, for example, contamination of the air with carbon monoxide or other gases that displace the oxygen, or by low pressure of oxygen in the atmosphere as occurs at high altitudes; the latter will be referred to as high-altitude hypoxia.

#### Organic Hypoxia

Four types of organic hypoxia may be recognized:

1. **Hypoxemic hypoxia.** This term implies that the low oxygen availability

is due to an inadequate uptake by the blood of oxygen. It may be the result of pneumonia, in which the accumulation of fluids in the lungs impairs the transfer of oxygen from the alveolar air into the arterial blood. Other causes are respiratory paralysis, drowning, etc.

2. **Anemic hypoxia.** A reduction in the hemoglobin concentration of the blood causes anemic hypoxia. An inadequate hemoglobin concentration decreases the oxygen capacity of the blood and hence the amount of oxygen available to the cells is decreased.

3. **Stagnant hypoxia.** Stagnant hypoxia refers to the slow flow of blood through the capillaries. This is associated with many disorders of the cardiovascular system, such as cardiac failure, shock, arterial spasm, and emboli. In such cases the tissues extract practically all of the oxygen contained in the blood. But after this quantity of oxygen has been taken, there is no more available and the slow movement of the blood fails to bring in a new supply fast enough to satisfy tissue needs.

4. **Histotoxic hypoxia.** Histotoxic hypoxia occurs due to interference with the utilization of the available oxygen. Many poisons block the essential metabolic processes and therefore oxygen is not used.

## High-Altitude Hypoxia

High-altitude hypoxia is produced mainly by a decrease in the partial pressure of oxygen in the atmosphere proportional to an increase in altitude. As indicated in the previous chapter oxygen reaches the organic cells through the respiratory, cardiovascular, and hematological systems. All of these processes act synergically and enable gas molecules to pass through the tissues in adequate quantities. When tissues receive a deficient supply of oxygen, a physiological situation called hypoxia develops.

As previously indicated, the atmosphere contains 20.95% oxygen, and this quantity is constant up to an altitude of 33,333 m. However, because air is compressible, it contains a greater number of gaseous molecules at low altitudes than at high altitudes; the barometric pressure, which depends on the molecular concentration of air, thus also decreases with an increase in altitude. This is the fundamental problem of high-altitude hypoxia: the oxygen in the air at high altitudes is at a lower pressure than it is at low altitudes. As illustrated in figure 10.1 as the *barometric pressure* decreases with altitude so does the partial pressure of oxygen. Hence, the partial pressure of oxygen at sea level is 159 mm Hg while at 3500 m (11,484 feet) the partial pressure of oxygen is reduced to 103 mm Hg; that is, at an altitude of 3500 m oxygen has about 35% less pressure than at sea level. At 4500 m (14,765 feet) the partial pressure of oxygen ($PO_2 = 91$ mm Hg) is decreased by as much as 40% as compared to sea-level pressure. Because of this decreased partial pressure

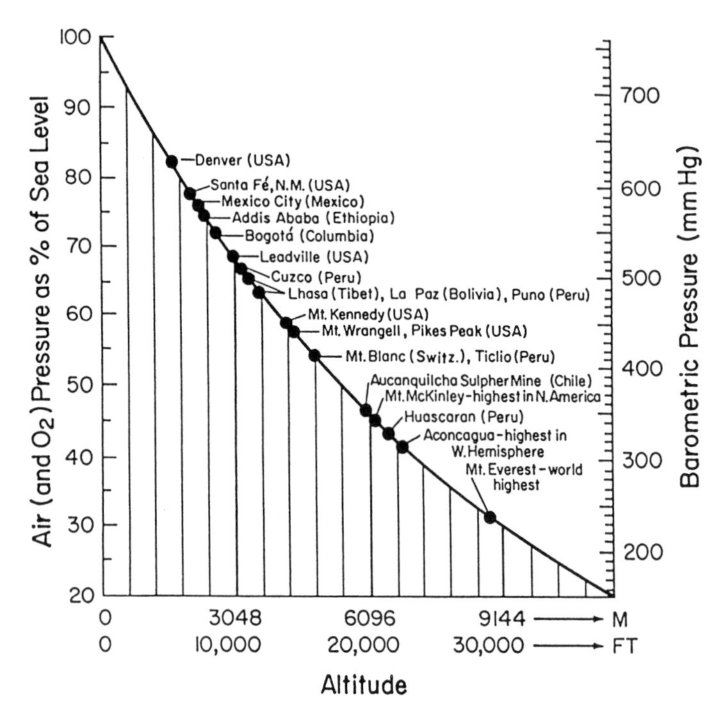


**Fig. 10.1.** Barometric pressure and oxygen pressure at high altitudes. With an increase in altitude there is a percentage decrease in the air and oxygen pressure.

of oxygen in ambient air, the partial pressure of oxygen in air reaching the trachea and alveoli is also reduced. This, in turn, reduces the amount of oxygen available to the tissues.

The drop in partial pressure of oxygen ($PO_2$) at high altitude causes a reduction in oxygen saturation of arterial blood because the proportion of oxyhemoglobin ($HbO_2$) formation in the blood depends on the partial pressure of oxygen in the alveoli. When this pressure is high, as occurs at sea level, a greater proportion of hemoglobin is saturated with oxygen (greater formation of oxyhemoglobin, hence greater arterial oxygen saturation). On the other hand, when the partial pressure of oxygen is low, as occurs at high altitude, a lower proportion of hemoglobin is combined with oxygen (less formation of oxyhemoglobin and, hence, less arterial oxygen saturation). Thus, as shown in table 10.1, if the partial pressure of oxygen in ambient air is 159 mm Hg and in the alveoli it is 104 mm Hg, as it is at sea level, the hemoglobin in arterial blood is 98% saturated with oxygen. On the other hand, if the partial pressure of oxygen in ambient air is 110 mm Hg and in the alveoli is 67 mm Hg, as it is at an altitude of 3048 m (10,000 feet), the hemoglobin in arterial blood is only 90% saturated with oxygen. This means that at an altitude of 3000 m there is a 10% decrease in oxygen for each unit of blood that leaves the lungs. Between 4000 and 5000 m this decrease might be as high as 30%.

## Methods of Studying Hypoxic Stress

The effects of high-altitude hypoxia are considered either acute or chronic. *Acute hypoxia* results from reduced oxygen availability that lasts for a few minutes, hours, or perhaps several weeks. *Chronic hypoxia* is a continuation of this condition for months, years, or a lifetime. The distinction between acute and chronic hypoxia results partly from the different mechanisms with which the organism responds to reduced oxygen availability.

The most common methods of studying the effects of and adaptation to high-altitude hypoxia include field or laboratory studies in a natural high-altitude environment and in low-pressure (decompression) chambers. A third method, sometimes used on laboratory animals, is exposure to gas mixtures low in oxygen concentration but at sea-level barometric pressures. In general, results derived from high-altitude environments are similar to those derived from altitude chambers.

During rest the physiological effects of hypoxia become evident at altitudes above 3000 m (about 10,000 feet), but during physical activity its effects are manifested above 2000 m (6600 feet). In this chapter the initial effects of high-altitude hypoxia, as well as factors affecting tolerance to high altitudes, will be discussed. This will set the stage for a discussion in the next chapter of the process of acclimatization.

## High-Altitude Areas of the World

As shown in figure 10.2 areas located above 3000 m include (1) the Rocky Mountains of the United States and Canada, (2) the Sierra Madre of Mexico,

**TABLE 10.1. Effects of altitude and barometric pressure on partial pressure of oxygen in the air, inspired (tracheal) air, and alveoli and arterial oxygen saturation**

| Altitude | | Barometric Pressure | Partial Pressure of Oxygen $(PO^2)$ | | | Arterial Oxygen Saturation |
|---|---|---|---|---|---|---|
| meters | feet | (mm Hg) | Air (mm Hg) | Tracheal Air (mm Hg) | Alveoli (mm Hg) | (%) |
| 0 | 0 | 760 | 159 | 149 | 104 | 98 |
| 3,048 | 10,000 | 523 | 110 | 100 | 67 | 90 |
| 6,096 | 20,000 | 349 | 73 | 63 | 40 | 70 |
| 9,144 | 30,000 | 226 | 47 | 37 | 21 | 20 |
| 12,192 | 40,000 | 141 | 29 | 19 | 8 | 5 |
| 15,240 | 50,000 | 87 | 18 | 8 | 1 | 1 |

*Source:* Based on data from E. G. Folk. 1974. Textbook of environmental physiology. Philadelphia: Lea and Febiger.

*Note:* The partial pressure of oxygen ($PO_2$) in the trachea is reduced because as inspired air enters the airways, it is warmed to almost body temperature (37°C) and saturated with water vapor, which exerts a vapor pressure of 47 mm Hg, regardless of barometric pressure. Therefore the $PO_2$ of tracheal air at sea level = $(760 - 47) \times 0.2095 = 149$ mm Hg, and 3048 m = $(523 - 47) \times 0.2095 = 100$ mm Hg.

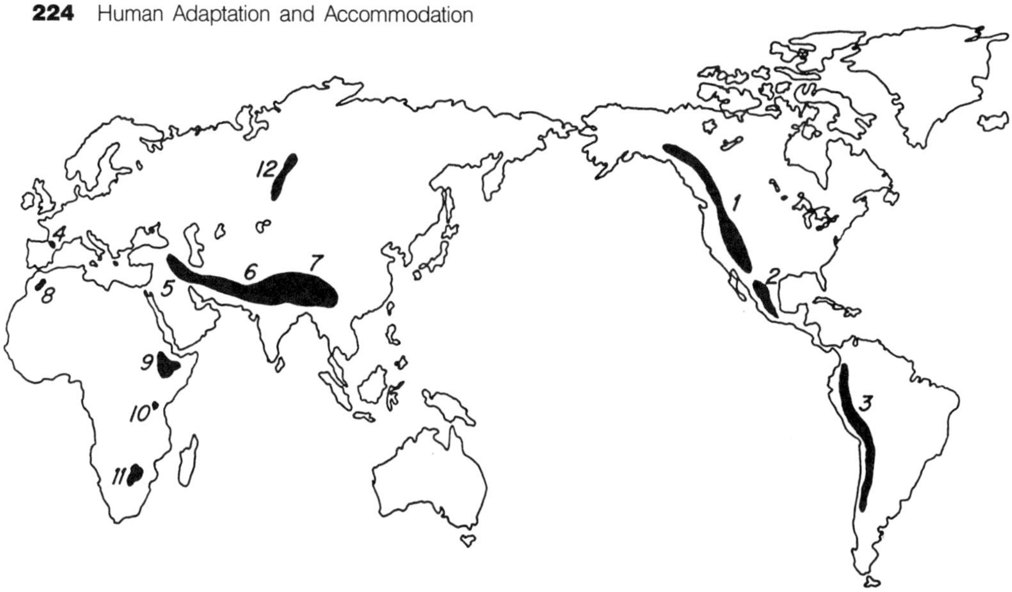

**Fig. 10.2.** Areas of the world exceeding 3000 m. (1) Rocky Mountains; (2) Sierra Madre; (3) Andes; (4) Pyrenees; (5) Mountain ranges of eastern Turkey, Persia, Afghanistan, and Pakistan; (6) Himalayas; (7) Tibetan Plateau and southern China; (8) Atlas Mountains; (9) high plains of Ethiopia; (10) Kilimanjaro; (11) the Basutoland; and (12) Tien Shan Mountains. (Modified from D. Health, and D. R. Williams. 1977. Man at high altitude. The patho-physiology of acclimatization and adaptation. Edinburgh: Churchill Livingstone; and I. G. Pawson and J. Corneille. 1978. The high altitude areas of the world and their cultures. In P. T. Baker, ed., The biology of high altitude peoples. New York: Cambridge University Press.)

(3) the Andes of South America, (4) the Pyrenees between France and Spain, (5) the mountain ranges of eastern Turkey, Persia, Afghanistan, and Pakistan, (6) the Himalayas, (7) the Tibetan Plateau and southern China, (8) the Atlas Mountains of Morocco, (9) the high plains of Ethiopia, (10) Kilimanjaro of east Africa, (11) Basutoland of South Africa, and (12) the Tien Shan Mountains of Russia. These areas are inhabited by more than 25 million people.

## PHYSIOLOGICAL CHANGES ASSOCIATED WITH ASCENT TO HIGH ALTITUDE

An unacclimatized person who ascends rapidly to high altitude may suffer deterioration of the nervous system function. Similar dysfunction can occur upon loss of cabin pressure at high altitude. The changes occur mainly due to hypoxia and may include sleepiness, laziness, a false sense of well-being, impaired judgment, blunted pain perception, increasing errors on simple tasks, decreased visual acuity, clumsiness, and tremors.

### Circulation and Heart Functions

With increasing hypoxia the resting heart rate increases from an average of 70 beats/min at sea level to as many as 105 beats/min at 4500 m. The mechanisms for increased heart rate are not well defined. Some investigators suggest that increased heart rate is related to stimulation of the carotid and aortic chemoreceptors; others attribute it to the influence of arterial hypocapnia and vagal afferent impulses from pulmonary stretch receptors. Concomitant with the increased pulse rate, systolic blood pressure either remains unchanged or decreases (4).

It must be noted, however, that during exercise the heart rate does not increase to the high rates attained at sea level (170 to 210 beats/min). Most studies report a reduction in maximal exercise heart rate as great as 40 to 50 beats/min to a rate of 130 to 150 beats/min. Hence the cardiac output (quantity of blood pumped by the left ventricle of the heart into the aorta per minute) is reduced at high altitude. Furthermore, electrocardiographic studies indicate that among sea-level natives, the strength of the myocardium at high altitude is decreased (5, 6). In some cases, enlargement of the right ventricle has been reported.

### Retinal Circulation

After only 2 hours of exposure to 5330 m the retinal arteries and veins increase in diameter by about one fifth (7). Figure 10.3 shows that compared to sea-level controls diameters increase 130% after 5 days, and after 5 to 7 weeks of exposure the increase still amounts to 100%. That the increased retinal blood flow is caused by hypoxia and not by alkalosis, is inferred from the fact that hypoxia without hypocapnia results in increased dilation of retinal vessels and increased retinal blood flow, whereas hypocapnia results in increased dilation of retinal vessels and increased retinal blood flow, and hypocapnia without hypoxia results in reduced retinal blood flow (8).

### Light Sensitivity and Visual Acuity

High-altitude hypoxia decreases light sensitivity. The higher the altitude, the more is sensitivity to light decreased; above 4500 m (14,765 feet) an intensity about 2.5 times greater than at sea level is required for a light to be seen (9). With hypoxia, the response to light stimulus is not only slow, but also less consistent than at sea level (9). These impairments appear to be caused by the effect of hypoxia on the visual system rather than by any psychological response of the subject. These decreases reach their maximum after about 1 hour of exposure, and in most cases visual performance is recovered by the forty-eighth hour of exposure.

### Memory and Learning

It has been reported that performance of unfamiliar tasks or difficult calculations is often considerably impaired during initial exposure to high altitude. As tested by paired-word association or pattern and position recall,

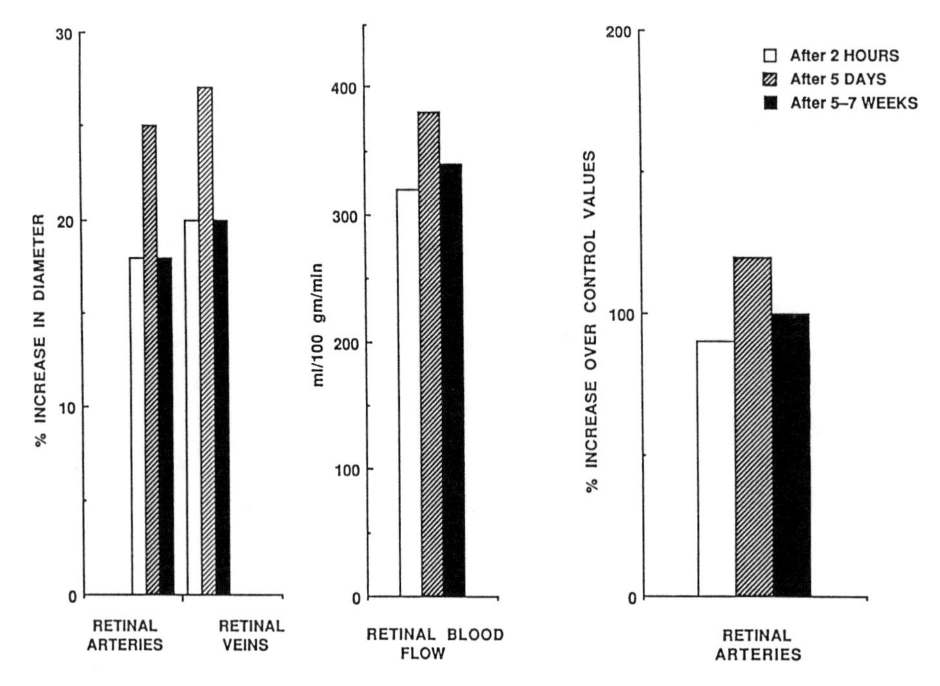


**Fig. 10.3.** Changes in diameter of retinal arteries and retinal veins, and retinal blood flow after 2 hours', 5 days', and 5 to 7 weeks' exposure to high altitude. Exposure to high altitude, especially during first 5 days, results in drastic dilation of retinal arteries and veins and concomitant increase in retinal blood flow. (Modified from R. Frayser, C. H. Houston, G. W. Gray, A. C. Bryan, and I. D. Rennie. 1971. Arch. Intern. Med. 127:708–11.)

immediate and recall memory has been observed to be diminished with an increase in altitude, especially at altitudes over 3660 m (9). However, other studies indicate that the ability to learn a new procedure is affected more severely than the performance of a previously learned procedure (10).

## Hearing

Of the various sensory modalities, hearing is the least sensitive to hypoxia. Indeed, in progressive hypoxia the sense of hearing is the last to disappear, and only above an altitude of 6000 m (19,800 feet) have decreases in hearing acuity been reported (11).

## Motor Function

Several investigators have reported signs of muscular weakness and incoordination on exposure to altitudes above 4500 m. At present, however, it is not known whether these deficiencies represent diminished functional capacity of the muscle itself or absence of muscle stimulation (4).

A clear and well-defined effect of hypoxia is impairment of neuromuscular control. Neuromuscular control impairment increases with increasing altitude; at 6000 m handwriting may become undecipherable (4). The physiological mechanisms responsible for this impairment are not known. It has been suggested that the increase in reflex muscle irritability observed above an altitude of 4000 m may result from the increased neuromuscular excitability associated with hyperventilation-induced alkalosis (4).

### Utilization of Sugars

**Taste.** Exposure to high altitude also influences food preference. In general, at high altitudes preference for sugar increases, and the desire for fat decreases. For unknown reasons sugar tastes less sweet at high altitude, and about two times the normal amount is taken in drinks (12). A high carbohydrate diet, along with increased sugar intake, results in reduced severity and duration of symptoms of the acute effects of high altitude, such as mountain sickness (12). Reflecting the need for more sugar at high altitude, blood glucose levels are lower than at sea level; but when adequate carbohydrate intakes are maintained, the glucose levels are normal (13). Altitude also seems to affect a person's ability to taste differences among four different compounds. It has been demonstrated that gustatory sensitivity to sweet (tested by sucrose), salt (sodium chloride), sour (citric acid or tartaric acid), and bitter (caffeine or quinine) were diminished during exposure to an altitude of 3450 m (14, 15). However, others indicate that rapid exposure to an altitude of 7620 m has no effect on taste perception (16).

**Dependence on Glucose.** Recent studies indicate that exposure and acclimatization to high altitude (4300 m) result in increased dependence on blood glucose (67). This increased dependency is apparent under both resting and exercise conditions and increases with length of exposure to high altitude. The increased dependence on glucose is related to the fact that glucose is an $O_2$-efficient fuel.

### Anorexia and Weight Loss

High-altitude exposure, especially at elevations greater than 3500 m, results in anorexia and, as a consequence of reduced food consumption, weight loss. These responses are most pronounced during the first few days of exposure (12).

In the Mount Everest expeditions the average daily food intake, between 5182 m (17,000 feet) and 6401 m (21,000 feet), was 2000 calories, and above 6401 m it averaged only 1500 calories. One of the major consequences of high-altitude anorexia is an imbalance between energy intake and energy output. The resulting energy deficits are affected by tissue catabolism, such as that of body fat and body protein (12). Because of energy imbalance and body tissue catabolism body weight is reduced. In men the weight loss at Everest (6401 m) for a period of 4 weeks averaged 1 lb/week (0.45 kg) and 2.8 lb/week (1.26 kg) among poorly acclimated men (12). At 4300 m the average weight loss for men for the first 7 days ranged from 3.49% to 5% of prealtitude weight (17, 18). In women, however, the average weight losses were less pronounced, and the loss over 7 days at 4300 m averaged only 1.76% of prealtitude exposure weight (19). Reduced body weight is not only caused by low caloric intake but also by loss of body water (20, 21). Body fluid loss increases at high altitudes because of increased urine output and increased water loss from the lungs associated with increased respiration (20, 21), together with the low humidity present at high altitude. Above 6706 m, despite a decrease in urine output, dehydration becomes progressive because of a blunting of the thirst sensation and the consequent decreased fluid intake (12).

## Adrenal Activity

**Adrenal Cortex.** Reflecting the general stress of the organism, exposure to high altitude results in increased adrenal activity. As shown by measurements of 17-hydroxycorticosteroids (17-OCHS) of the urine and blood plasma during the first week of exposure to altitudes above 2000 m, adrenocortical activity increases rapidly in the first week, and during the second week it returns to sea-level values (4, 22).

**Adrenal Medulla.** On exposure to altitudes above 3000 m adrenal medulla activity is increased as shown by the increased excretion of noradrenaline and catecholamine (4, 12, 23). After about the second week, adrenal medulla activity returns to sea level values.

**Aldosterone.** With controlled dietary sodium and potassium intake, exposure to altitudes above 3500 m results in the reduction of aldosterone secretion (4, 24, 25). The factors responsible for this reduction are not yet well defined. It has been postulated that the mechanism for suppressing aldosterone secretion may involve the right atrium of the heart. According to this hypothesis the increased blood volume that occurs with acclimatization to high altitude stimulates the stretch receptors of the right atrium, that in turn depress aldosterone secretion (25). In support of this hypothesis is the fact that experimental animals exposed to high altitude show reduced width of the adrenal gland's zona glomerulosa, which is the source of aldosterone (26).

**Renin.** As shown by experiments with animals and humans, exposure to high altitude results in an initial increase in plasma renin (27, 28).

## Thyroid Function

Experiments with animals indicate that exposure to high altitude for more than 3 days results in reduced thyroid function (29–32). Iodine retention and thyroid concentrating power are also decreased. The mechanism for this reduced thyroid function has not been clearly identified. It has been postulated that the mechanism is related to deficient secretion of the thyroid-stimulating hormone by the pituitary (29); others suggest that altitude reduces the thyroxine requirements and hence results in a concomitant reduction in hormone synthesis (32, 33).

## Testosterone Secretion

Studies of sea-level subjects transported to 4250 m indicate that during the first 3 days of exposure the 24-hour urinary testosterone secretion was reduced by more than 50% when compared to sea-level values (34, 35). By the end of the first week of exposure to high altitude, testosterone secretion returns to normal. Since it is known that exposure of sea-level subjects to high altitude results in decreased plasma luteinizing hormone (LH), Heath and Williams (4) suggest that decreased testosterone secretion at high altitude is caused by low levels of LH.

## Sexual Function

**Spermatogenesis.** Experiments with animals and humans indicate that exposure to high altitude results in decreased spermatozoid production and increased abnormal forms (35, 36). Studies of sea-level subjects at sea level and during 8 and 13 days' exposure to an altitude of 4330 m indicate that (fig. 10.4) by the eighth day of exposure to high altitude the sperm count declined from a mean of 216.2 million/ml to 150.2 million/ml and to 98.2 million/ml by the thirteenth day. Along with this decrease there was also an increase in abnormal forms from 0% at sea level to 39.3% by the eighth day, and a decrease in the motile forms from 70% at sea level to 50% by the eighth day at high altitude. Subsequent studies (35, 36) on the semen of sea-level subjects transported from sea level to 4270 m for 4 weeks indicated that the sperm count (expressed as total number of sperm) from the first day to the twenty-eighth day of exposure dropped gradually from a control level of $7.42 \times 10.8$. This gradual decline continued even 15 days after the descent to sea level. Furthermore, the motile cells decreased from control values of 85.8% to 53.4% at the end of the experimental period. The frequency of abnormal forms increased from a control level of 15.5% to 31.6%. Thus, it appears that high-altitude hypoxia has profound effects on spermatogenesis.

**Histological changes in the testes.** The effects of hypoxia on spermatogenesis appear related to alterations in the tissues of the testes. Experiments on rats (37)indicate that after the third day of exposure to a simulated altitude of 7580 m the germinal epithelium suffered considerable damage, which progressed through the end of the experiment.

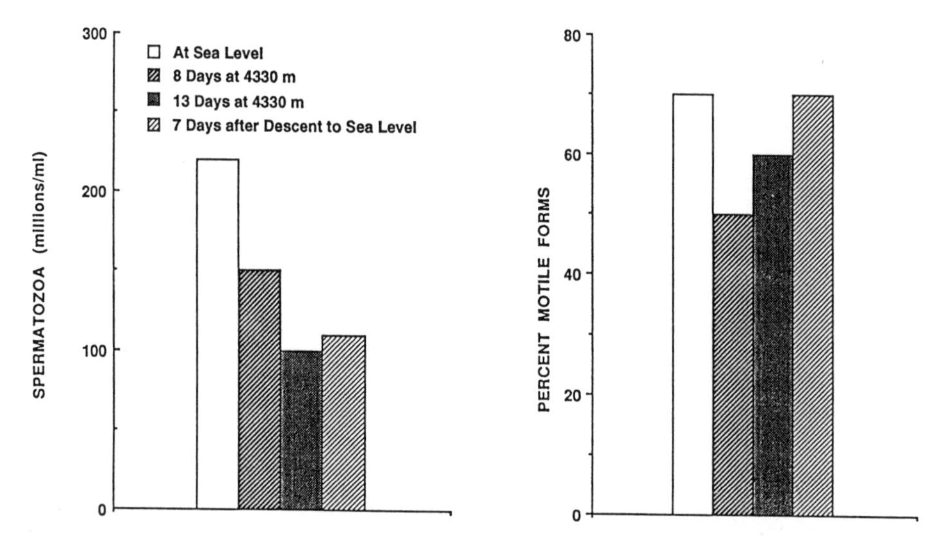


**Fig. 10.4.** Changes in spermatozoa count and motile forms after 8 days and 13 days of residence at high altitude and subsequent descent to sea level. Exposure to high altitude is associated with decreased spermatozoid count and reduced motility of spermatozoa, which return to normal with descent to sea level. (Modified from J. Donayre. 1966. Population growth and fertility at high altitude. In Life at high altitudes. Pub. 140. Washington, DC: Pan American Health Organization.)

Tissue alterations appear to continue for longer periods; even after 6 months exposure to an altitude of 4510 m the germinal epithelium of cats showed various degrees of destruction (38). Similarly, guinea pigs after 2 weeks exposure to an altitude of 4330 m exhibited profound alterations in the epithelium of the seminiferous tubules with a marked decrease of all cellular types (34).

**Estrus Cycle.** As shown in studies with rats, high-altitude hypoxia along with cold decreased the incidence of estrus. Rats exposed to cold in addition to hypoxia exhibited fewer instances of estrus than those rats exposed to hypoxia alone (39). In other words, cold superimposes an anestrous effect on that produced by hypoxia, but these changes are reversible after return to sea level and temperate conditions.

**Menstruation.** Studies on females residing at high altitude indicated an increased incidence of menstrual disturbances such as amenorrhea, dysmenorrhea (36), and increased menstrual flow (36). However, the extent to which these variations in menstrual cycle are the result of hypoxia is not well defined because the majority of the subjects questioned were taking contraceptive pills.

## PATHOLOGICAL RESPONSES TO HIGH-ALTITUDE HYPOXIA

High-altitude hypoxia elicits direct and indirect responses, some of which are evidently adaptive in reducing the severe oxygen deficiency in the cells to permit the organism to function normally. However, the adaptive significance of other responses is less obvious, and some of them can cause from mild to severe malfunction, eventually becoming deleterious and even fatal to the organism. Three clinical entities specific to a high-altitude environment have been identified. These include acute mountain sickness, pulmonary edema, and Monge's disease (fig. 10.5). In this section the symptoms, cause, and treatment of these diseases will be briefly summarized. Extensive discussions of these topics are given in other works (40–43).

## Acute Mountain Sickness

### Symptoms

Acute mountain sickness (AMS), or soroche, occurs during the first few days of exposure to high-altitude hypoxia. There is great individual variability in the altitude threshold at which it occurs. The usual symptoms include anorexia, nausea and vomiting, marked dyspnea, physical and mental fatigue, interrupted sleep, and headaches intensified by activity. There may also occur some digestive disorders, which result in weight loss and dehydration. In some cases the individual may feel an increased sensitivity to cold, dizziness, weakness, palpitation, transient leg, back, and chest pains, nasal congestion, and rhinorrhea. In very rare cases at altitudes above 4500 m a diminution of visual acuteness, painful menstruation, and bleeding of the gums may occur (41).

In general the onset of acute mountain sickness may be sudden when ascent is very rapid, but more often it develops gradually over a period of hours, with maximal severity reached after 24 to 48 hours. The symptoms then decrease even more slowly and tend to disappear within 6 to 8 days.

There also appears to be a certain individual predisposition for the symptoms. Some persons may experience acute sickness as low as 1500 m (4922 feet), whereas occasionally subjects are unaffected at 4500 m (14,765 feet). At 5000 m all nonacclimatized subjects are affected. At any altitude the onset of mountain sickness is increased by physical activity and alcohol consumption. The physiological limits of human tolerance appear to be reached at 8545 m (28,036 feet).

The symptoms of acute mountain sickness probably occur because of poor adjustment to the normal physiological reflex activities that are triggered by hypoxia. The onset and severity of the symptoms are governed by the body's inability to respond rapidly to internal homeostatic disruption. The severity of acute mountain sickness is less in females than in males (42).

### Cause

Although the cause of acute mountain sickness has been attributed specifically to hypoxia, the mechanisms that induce its symptoms are not well

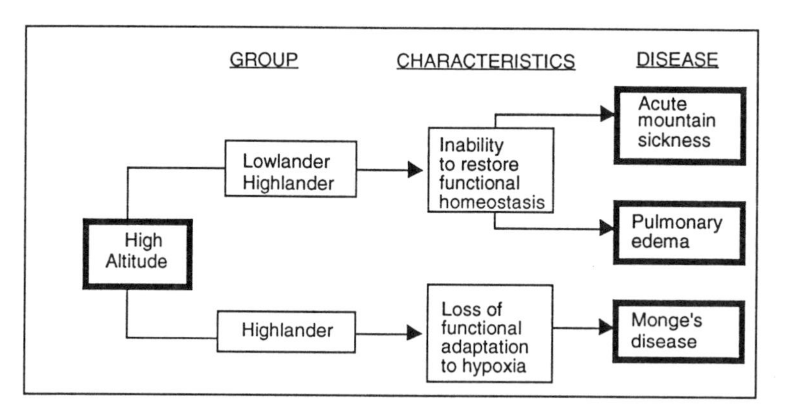


**Fig. 10.5.** Schematization of group incidence and general characteristics of diseases associated with high-altitude environment. Acute mountain sickness and pulmonary edema occur mostly among sea-level natives or highlanders returning from sea level. Both diseases are characterized by an inability to restore functional homeostasis when exposed to stress of high altitude. In contrast, Monge's disease occurs mostly among highlanders who for some reason lose their functional adaptation.

defined (41, 44). It has been postulated that the onset of acute mountain sickness is related to hypoxic effects on the brain, which is highly sensitive to changes in oxygen level. Exposure to high altitude results in a shift of water from the extracellular compartment into the cells, together with a slight increase in total body water (42). The result is an intracellular edema that, in the case of the brain, may be responsible for some of the neurological symptoms of acute mountain sickness (45). However, other studies have found little relationship between body dehydration and the incidence and severity of acute mountain sickness (45). Recent investigations report that subjects who developed acute mountain sickness had multiple coagulation abnormalities (46) and suggest that problems of coagulation may be a contributing factor in the etiology of acute mountain sickness. Others attribute the onset of acute mountain sickness to hypocapnia and respiratory alkalosis resulting from hyperventilation (47).

According to Singh et al. (41), on ascending to high altitude there is an initial hypersecretion of adrenal corticosteroids and vasopressin. Excessive secretion of vasopressin results in the initial retention of urine (oliguria), which coupled with the effects of hypoxia, brings about reduced peripheral blood flow and increased pulmonary blood volume and leads to pulmonary congestion. Both oliguria and the resulting increased sodium retention are associated with increased cerebral edema and elevated cerebrospinal fluid pressure. All these factors cause headache, retinal hemorrhages, and neurological disorders that characterize mountain sickness (41).

### Treatment

Because the cause of acute mountain sickness is not known, aside from oxygen administration, its treatment and prevention have been empiric. The symptoms and severity are usually diminished by avoiding sudden ascent, elevation, alcohol, and by restraint in eating, especially fats. Carbonic anhydrase inhibitors, such as acetazolamide (*Dimox*), have been used with some success (48, 49). The effectiveness of acetazolamide varies between individuals. In some it may ameliorate symptoms or decrease the frequency of occurrence of AMS if used prophylactically, while in others it has minimal positive results. The effective dosage recommended is 500 mg to 1000 mg daily given in partial doses taken 24 to 48 hours prior to ascent and for 48 hours, or longer to control symptoms, after beginning ascent. Acetazolamide appears to have little effect when given after symptoms appear.

The prophylactic function of acetazolamide is related to the fact that it lowers plasma bicarbonate ion concentration and lowers the alkaline blood pH normally brought about by hypoxia-induced hyperventilation. Acetazolamide also increases diuresis resulting in the excretion of excess bicarbonate ions, which may facilitate restoring the blood pH. The side effects include paraesthesia in the limbs and diuresis (50), but acetazolamide produced no demonstrable cerebral changes in dogs (50) and no cognitive impairment was found in men (61).

Diuretics such as furosemide (*Lasix*) have been used to counteract the initial urine retention and to treat acute mountain sickness (41). However, some studies cast doubt on the efficacy of furosemide (42–46, 51) because subjects suffering from acute mountain sickness are usually already dehydrated from the efforts of climbing, low humidity, and vomiting. It has been found that high intakes of potassium along with low intakes of sodium decrease the symptoms of acute mountain sickness (47). A high carbohydrate and low fat diet also reduces the symptoms of acute mountain sickness (42, 52). Experimental studies have found that a high carbohydrate diet increases ventilation and improves oxygenation during acute exposure to simulated high altitude (52) and increases endurance in heavy physical work at high altitude (53). In any event an important way to decrease the symptoms of acute mountain sickness is to ascend to higher altitudes gradually and if possible spend many days at around 2000 or 3000 m before ascending to higher altitudes.

## Pulmonary Edema

### Symptoms

Since the early 1930s it has been recognized that some high-altitude natives and mountain climbers, who after a visit to sea level return to the mountains, develop acute pulmonary edema that declines after descent to sea level (54).

The main pathological features of pulmonary edema include widespread edema of the alveoli, extensive plugging of alveolar capillaries with sludged red blood cells, and uneven pulmonary vascular constriction, giving a patchy distribution of edema. At sea level pulmonary edema usually involves a disturbance in the Starling forces, such as increased pulmonary capillary pressure or increased pulmonary capillary permeability, that normally permits leakage of plasma proteins and thus facilitates the transudation of fluid from pulmonary capillaries.

The risk of pulmonary edema appears greater among sea-level subjects well acclimatized to high altitude who return to high altitude after a stay of 1 or 2 weeks at low altitudes (55). When pulmonary edema occurs among highlanders, the condition is usually associated with pronounced right ventricular hypertrophy (56–58), which is indicative of pulmonary hypertension. The main period of risk is between 12 and 72 hours after ascent. Males in the late teens and early twenties are known to be more at risk than females (42–43, 53). Rapid exposure to high altitude when associated with increased physical activity is also a precipitating situation for pulmonary edema. Most reported cases occur between 3500 and 4500 m.

### Cause

It has been suggested that at high altitude the likely cause of pulmonary edema is increased capillary pressure (59). It has been suggested that the pathogenesis of pulmonary edema is related to the unequal pulmonary vasoconstriction induced by high-altitude hypoxia, which in turn results in reduced blood flow and sludging of platelets and erythrocytes in some areas and increased blood flow to nonvasoconstricted areas (59). These two factors cause an increased filtration pressure in open capillaries and a patchy distribution of edema. According to others, the likely factor for the pathogenesis of pulmonary edema is increased plasma volume that occurs when highland subjects descend to sea level for a short period and thereafter return to high altitude (55). These investigators point out that highland subjects on return to sea level during the first 2 weeks experience a decreased red cell volume, and this is accompanied by an excessive rise in plasma volume. Therefore, if a highlander returns from sea level to high altitude during this period, he may become more susceptible to the development of pulmonary edema than the newcomer who will have a lower plasma volume. In fact, most subjects who developed pulmonary edema at high altitude were at sea level from 5 to 21 days, and only in few cases did pulmonary edema occur in newcomers (55).

### Treatment

Improvement is usually rapid with bed rest, oxygen administration, treatment with corticosteroids, and transport to lower altitudes. Use of *Nifedipine,* a calcium antagonist, has been shown to lessen the pulmonary vasoconstriction associated with pulmonary edema (50). The most important preventive measure includes avoidance of excessive physical activity for a few days following ascent to high altitude for the first time or after returning to high altitude from sea level (42–43, 55, 60).

## Chronic Mountain Sickness or Monge's Disease

### Characteristics

In 1928 Monge described a complex clinical syndrome as the sickness of the Andes, and it is currently termed *chronic mountain sickness,* or *Monge's disease,* in his honor. Monge's disease is a complex pathophysiological condition that occurs when individuals normally acclimatized to high altitude lose their ability to adapt to the altitude at which they have been living for long periods without symptoms. It is prevalent only among Andean populations, and no cases have been reported among Himalayan populations (43). It occurs at altitudes exceeding 3000 m, and in most cases the disease seems to occur in young and middle-aged males (40, 42–43).

### Clinical Features

Patients with Monge's disease appear cyanotic with nearly black lips and dark purple ear lobes and facial skin (40, 42–43). The conjunctival vessels of the eyes are extremely suffused, congested, and dark red-purple in color, reflecting excessive polycythemia and diminished arterial oxygen saturation (42, 43). Hemorrhages beneath the fingernails are also characteristic of subjects with Monge's disease (43).

Patients with Monge's disease exhibit an exaggeration of the normal hematological and cardiovascular characteristics of the high-altitude native. In general hemoglobin exceeds 22 g/100 ml, the hematocrit level is greater than 70%, and the mean arterial pressure is about 50 mm Hg compared to 25 mm Hg of the high-altitude native (40, 42–43). Similarly, compared to healthy highland natives the right ventricle of the heart and overall size of the ventricles is excessively enlarged (56). The arterial blood oxygen saturation is usually lower than that predicted for the altitude, and the arterial partial pressure of carbon dioxide is high, indicating poor ventilation (40, 61).

### Cause

The clinical features of Monge's disease are usually associated with other clinical traits such as neuromuscular disorders affecting the thoracic cage, pulmonary tuberculosis, pneumoconiosis, kyphoscoliosis, and pulmonary emphysema (40). These characteristics by themselves are capable, even at sea level, of producing chronic hypoxia (43). For this reason, it has been concluded that Monge's disease is a collective term for a heterogenous group of conditions and does not indicate a distinct pathological entity (43).

#### Treatment

An effective treatment for patients showing the signs and symptoms of Monge's disease is transport to lower altitudes. On removal of the patients to sea level the cyanosis, fatigue, right ventricular hypertrophy, hematocrit, and hemoglobin decrease rapidly; in 2 months normal values are restored (57).

## INDIVIDUAL FACTORS AND TOLERANCE TO HIGH ALTITUDE

### Age

From studies on animals and humans it is known that tolerance to acute high-altitude hypoxia is affected by age. It has been found that on exposure to 10,360 m in a decompression chamber for 6 hours the survival of young rats (birth to 110 days) and old rats (older than 2 years) was poorer than those of adult rats (110 to 330 days) (62). Studies on humans showed that the decrease in light and auditory sensitivity associated with human aging is enhanced when sea-level subjects are exposed to a simulated altitude of 4880 m (9). Similarly, the resistance of healthy subjects (pilots) to acute oxygen deficiency in a decompression chamber simulating 7500 m increases between the ages of 18 and 40 years, and after 40 years of age tolerance decreases slowly (63). Acclimatization studies indicate that among six 58- to 71-year-old subjects, five (83%) subjects exhibited a decrease in hemoglobin concentration during the first few days at high altitude (64); this was in contrast to their responses 27 years earlier when all showed an increased hemoglobin concentration beginning with arrival at high altitude. This response appears to result from the fact that the mobilization of plasma water into interstitial or extracellular spaces and the associated increased hemoconcentration that occurs in young subjects exposed to high altitude does not occur in older subjects under similar conditions. Furthermore, it has been found that plasma aldosterone levels and probably secretion rates are reduced among older, but not among younger, subjects when exposed to high altitude (65).

### Physical Fitness

Some of the major adaptive effects of physical training include increased vascularization, increased size of the striated and cardiac muscles, and increased maximum aerobic capacity. Although the influence of training has not been documented, one would expect that degree of physical fitness would affect degree of tolerance and rate of acclimatization to high altitude stress. Individuals with high fitness would be able to tolerate the stress of hypoxia better. At all ages among the Nepalese Sherpas from Namche Bazar (3440 m) the most active natives such as porters, herders, and farmers attain a greater aerobic capacity than their less active counterparts (storekeepers, traders, and artisans) residing at the same altitude (66).

### Sex

Experimental studies indicate that female rats have a better survival rate than male rats when exposed to a simulated altitude of 10,360 m for 6 hours (62). Circumstantial evidence from humans also indicates that the incidence and severity of acute mountain sickness and high-altitude pulmonary edema is greater in males than in females (4, 12).

## OVERVIEW

The most important environmental stress at high altitude is the low oxygen availability that results from the decrease in barometric pressure. Therefore, in this chapter the physiological effects and responses that the organism makes to the stress of hypoxia associated with high altitude have been emphasized.

The physiological effects of hypoxia are multifaceted and complex. In response to low oxygen availability the organism increases its rate of pulmonary ventilation, as well as increasing the oxygen-carrying capacity of the blood. These adjustments are not completely adequate, as shown by the fact that sea-level natives exposed to high altitude do experience disturbances of the sensory nervous system, adrenal activity, and reproductive function and reduced work capacity. Along with physiological disturbances, high-altitude hypoxia also affects appetite, producing anorexia which in turn results in weight loss. The weight loss is also caused by dehydration or loss of body water, which occurs because of inhibition of fluid intake and excessive fluid loss through the increased altitude-induced ventilation and low humidity of the high altitudes. Eventually, the sea-level organism is able to function at high altitude. This is achieved through the long-term process of acclimatization discussed in the next chapter.

Three clinical entities specific to altitude have been identified: acute mountain sickness, pulmonary edema, and Monge's disease. Acute mountain sickness has been clearly associated with hypoxic stress and is characterized by headache, malaise, dizziness, shortness of breath, sleep difficulties, and stomach upset. As schematized in figure 10.5 acute mountain sickness occurs in lowlanders as well as highlanders who are unable to restore functional homeostasis when exposed to hypoxia. Several factors precipitate the onset of acute mountain sickness, among which rapid ascent, dehydration, and excessive activity are the most important. Usually symptoms of acute mountain sickness disappear without treatment. In contrast, pulmonary edema is a severely debilitating disease and can be fatal if the patient is not treated rapidly. The disease is associated with excessive pulmonary hypertension, impaired pulmonary oxygen exchange, and histological alterations of the pulmonary vessels. As a result the lowlander or highlander with pulmonary edema suffers a greater loss of functional homeostasis than with acute mountain sickness. Furthermore, there are multiple causes of the inability to restore functional homeostasis with pulmonary edema. For this reason treatment includes oxygen administration, corticosteroids, and transport to lower altitude. Preventive measures include initial avoidance of excessive physical activity when ascending to higher altitude either for the first time or when returning from sea level. Monge's disease or chronic mountain sickness is a severely debilitating disease that occurs mostly among highland natives with prolonged residence at high altitude and is associated with a loss of functional adaptation. It is characterized by an excessive hypoxemia and polycythemia, but the etiology is not well defined. Treatment includes transfer to a lower altitude.

Tolerance to high-altitude hypoxic stress is influenced by age, physical fitness, and sex. Thus very young and very old subjects are more impaired than adults, less active individuals experience a greater decline in aerobic capacity than more active individuals, and tolerance to hypoxic stress appears greater in females than in males.

## References

1. Kellogg, R. H. 1968. Altitude acclimatization, a historical introduction emphasizing the regulation of breathing. Physiologist 11:37-57.
2. Jourdanet, D. 1861. Les altitudes de l'Amérique tropicale comparées au niveau des mers au point de vue de la constitution médicale. Paris: Baillière.
3. Bert, P. 1878. La pression barométrique: recherches de physiologie expérimentale. Paris: Masson.
4. Heath, D., and D. R. Williams. 1977. Man at high altitude. The pathophysiology of acclimatization and adaptation. Edinburgh: Churchill Livingstone.
5. Jackson, F., and H. Davies. 1960. The electro-cardiogram of the mountaineer at high altitude. Br. Heart J. 22:671–85.
6. Peñaloza, D., and M. Echevarria. 1957. Electrocardiographic observations on ten subjects at sea level and during one year of residence at high altitudes. Am. Heart J. 54:811–22.
7. Frayser, R., C. S. Houston, G. W. Gray, A. C. Bryan, and I. D. Rennie. 1971. The response of the retinal circulation to altitude. Arch. Intern. Med. 127:708–11.

8. Hickam, J. B., and R. Frazer. 1966. Studies of the retinal circulation in man. Circulation 33:302–8.
9. McFarland, R. A. 1972. Psychophysiological implications of life at altitude and including the role of oxygen in the process of aging. In M. K. Yousef, S. M. Horvath and R. W. Bullard, eds., Physiological adaptations: Desert and mountain. New York, NY: Academic Press.
10. Cahoon, R. L. 1972. Simple decision making at high altitude. Ergonomics 15: 157–63.
11. Curry, E. T., and F. Boys. 1956. Effects of oxygen on hearing sensitivity of simulated altitudes. Eye Ear Nose Throat Mon. 35:239–45.
12. Ward, M. 1975. Mountain medicine. A clinical study of cold and high altitude. New York, NY: Van Nostrand Reinhold Co.
13. Consolazio, C. F., L. O. Matoush, H. L. Johnson, H. J. Krzywicki, T. A. Daws, and G. J. Isaac. 1969. Effects of high-carbohydrate diets on performance and clinical symptomatology after rapid ascent to high altitude. Fed. Proc. 28:937–43.
14. Grandjean, E. 1955. The effect of altitude on various nervous functions. Proc. R. Soc. (Lond.) 143:12–13.
15. Maga, J. A., and K. Lorenz. 1972. Effect of altitude on taste thresholds. Percept. Mot. Skills 34:667–70.
16. Finkelstein, B., and R. G. Pippett. 1958. Effect of altitude upon primary taste perception. J. Aviat. Med. 29:386–91.
17. Surks, M. I., K. S. K. Chinn, and L. O. Matoush. 1966. Alterations in body composition in man after acute exposure to high altitude. J. Appl. Physiol. 21:1741–46.
18. Krzywicki, H. J., C. F. Consolazio, L. O. Matoush, H. L. Johnson, and R. A. Barnhart. 1969. Body composition changes during exposure to altitude. Fed. Proc. 28:1190–94.
19. Hannon, J. P., J. L. Shields, and C. W. Harris. 1969. Anthropometric changes associated with high altitude acclimatization of women. Am. J. Phys. Anthropol. 31:77–83.
20. Hannon, J. P., G. J. Klain, D. M. Sudman, and F. J. Sullivan. 1976. Nutritional aspects of high altitude exposure. Am. J. Clin. Nutr. 29:604–13.
21. Krzywicki, H. J., C. F. Consolazio, H. L. Johnson, W. C. Nielsen, and R. A. Barnhart. 1971. Water metabolism in humans during acute high-altitude exposure (4,300 m). J. Appl. Physiol. 30:806–9.
22. Halhuber, M. J., and F. Gabl. 1964. 17-OHCS excretion and blood eosinophils at an altitude of 2000 M. In W. H. Weihe, ed., The physiological effects of high altitude. Oxford: Pergamon Press.
23. Pace, N., R. L. Groswold, and B. W. Grunbaum. 1964. Increase in urinary norepinephrine excretion during 14 days sojourn at 3800 meters elevation. Fed. Proc. 23:521–25.
24. Ayres, P. J., R. C. Hunter, E. S. Williams, and J. Rundo. 1961. Aldosterone excretion and potassium retention in subjects living at high altitude. Nature 191:78–80.
25. Williams, E. S. 1966. Electrolyte regulation during the adaptation of humans to life at high altitudes. Proc. R. Soc. Lond. (Biol.). 165:266–80.
26. Hartroft, P. M., M. B. Bischoff, and T. J. Bucci. 1969. Effects of chronic exposure to high altitude on the juxtaglomerular complex and adrenal cortex of dogs, rabbits and rats. Fed. Proc. 28:1234–37.
27. Gould, A. B., and S. A. Goodman. 1970. Testosterone metabolism in men exposed to high altitude. Acta Endocrinol. (Panama). 2:55–58.

28. Frayser, R., I. D. Rennie, G. W. Gray, and C. S. Houston. 1975. Hormonal and electrolyte response to exposure to 17,500 ft. J. Appl. Physiol. 38:636–42.
29. Surks, M. J. 1966. Effects of hypoxia and high altitude on thyroid iodine metabolism in the rat. Endocrinology 78:307–15.
30. Tryon, C. A., W. R. Kodric, and H. M. Cunningham. 1968. Measurement of relative thyroid activity on free-ranging rodents along an altitudinal transect. Nature 218:278–80.
31. Martin, L. G., G. E. Westenberger, and R. W. Bullard. 1971. Thyroidal changes in the rat during acclimatization to simulated high altitude. Am. J. Physiol. 221: 1057–63.
32. Galton, V. A. 1972. Some effects of altitude on thyroid function. Endocrinology 91:1393–94.
33. Kotchen, T. A., E. H. Mougey, R. P. Hogan, A. E. Boyd III, L. L. Pennington, and J. W. Mason. 1973. Thyroid responses to simulated altitude. J. Appl. Physiol. 34:165–68.
34. Guerra-Garcia, R. 1971. Testosterone metabolism in men exposed to high altitude. Acta Endocrinol. (Panama). 2:55–62.
35. Donayre, J. R., R. Guerra-Garcia, F. Moncloa, and L. A. Sobrevilla. 1968. Endocrinological studies at high altitude. IV. Changes in the semen of men. J. Reprod. Fertil. 16:55–58.
36. Donayre, J. 1966. Population growth and fertility at high altitude. In Life at high altitudes. Pub. 140. Washington, DC: Pan American Health Organization.
37. Atland, P. D. 1949. Effects of discontinuous exposure to 25,000 feet simulated altitude on growth and reproduction of the albino rat. J. Exp. Zool. 110:1–18.
38. Monge, M., and C. Mori-Charez. 1942. Fisiologia de la reproduccion en la altura. An. Fac. Med. Lima (Peru). 25:34–42.
39. Donayre, J. 1969. The oestrus cycle of rats at high altitude. J. Reprod. Fertil. 18:29–32.
40. Monge, M. C., and C. Monge. 1966. High-altitude diseases: mechanism and management. Springfield, IL: Charles C. Thomas, Publisher.
41. Singh, I., P. K. Khanna, M. C. Srivastava, M. Lal, S. B. Roy, and C. S. V. Subramanyan. 1969. Acute mountain sickness. N. Engl. J. Med. 280:175–84.
42. Ward, M. 1975. Mountain medicine, a clinical study of cold and high altitude. New York, NY: Van Nostrand Reinhold Co.
43. Heath, D., and R. R. Williams. 1977. Man at high altitude. The pathophysiology of acclimatization and adaptation. Edinburgh: Churchill Livingstone.
44. Hansen, J. E., and W. O. Evans. 1970. A hypothesis regarding the pathophysiology of acute mountain sickness. Arch. Environ. Health. 21:666–69.
45. Aoki, V. S., and S. M. Robinson. 1971. Body hydration and the incidence and severity of acute mountain sickness. J. Appl. Physiol. 31:363–67.
46. Maher, J. T., P. H. Levine, and A. Cymerman. 1976. Human coagulation abnormalities during exposure to hypobaric hypoxia. J. Appl. Physiol. 41:702–7.
47. Waterlow, J. C., and H. W. Bunje. 1966. Observations on mountain sickness in the Colombian Andes. Lancet 2:655–61.
48. Carson, R. P., W. D. Evans, J. L. Shields, and J. P. Hannon. 1969. Symptomatology, pathophysiology and treatment of acute mountain sickness. Fed. Proc. 28: 1085–91.
49. Gray, G. W., A. C. Bryan, R. Frayser, C. S. Houston, and I. D. Rennie. 1970. Control of acute mountain sickness. Aerospace Med. 42:81–84.

50. Beeckman, D., and E. R. Buskirk. 1988. A review of drug use at high terrestrial altitude and in cold climates. Human Biology 60:663–77.
51. Wilson, R. 1973. Acute high-altitude illness in mountaineers and problems of rescue. Ann. Intern. Med. 78:421–27.
52. Hansen, J. E., L. H. Hartley, and R. P. Hogan. 1972. Arterial oxygen increase by high carbohydrate diet at altitude. J. Appl. Physiol. 33:441–45.
53. Consolazio, C. F., L. O. Matoush, H. L. Johnson, H. J. Krzywicki, T. A. Daws, and G. J. Isaac. 1969. Effects of high carbohydrate diets on performance and clinical symptomatology after rapid ascent to high altitude. Fed. Proc. 28:937–43.
54. Hurtado, A. 1937. Aspectos fisicos y patológicos de la vida en las alturas. Lima, Peru: Imprenta Rimac.
55. Hultgren, H. N., W. B. Sprikard, J. Hellriegel, and C. S. Houston. 1961. High altitude pulmonary edema. Medicine 40:289–313.
56. Peñaloza, D., and F. Sime. 1971. Chronic cor pulmonale due to loss of altitude acclimatization (chronic mountain sickness). Am. J. Med. 50:728–43.
57. Peñaloza, D., F. Sime, and L. Ruiz. 1971. Cor pulmonale in chronic mountain sickness: present concept of Monge's disease. In R. Porter and J. Knight, eds., High altitude physiology: cardiac and respiratory aspects. Ciba Foundation Symposium. Edinburgh: Churchill Livingstone.
58. Marticorena, E., F. A. Tapia, J. Dyer, J. Severino, N. Banchero, R. Gamboa, H. Kruger, and D. Peñaloza. 1964. Pulmonary edema by ascending to high altitudes. Dis. Chest 45:275–279.
59. Fred, H. L., A. M. Schmidt, T. Bates, and H. H. Hecht. 1962. Acute pulmonary edema of altitude. Clinical and physiologic observations. Circulation 25:929–37.
60. Menon, N.D. 1965. High-altitude pulmonary edema. N. Engl. J. Med. 273:66–73.
61. Lozano, R., and M. C. Monge. 1965. Renal function in high-altitude natives and in natives with chronic mountain sickness. J. Appl. Physiol. 20:1026–27.
62. Atland, P. D., and B. Highman. 1964. Effects of age and exercise on altitude tolerance in rats. In W. H. Weihe, ed., The physiological effects of high altitude. Oxford: Pergamon Press.
63. Klein, K. E. 1964. Discussion of paper by Atland and Highman (1964). In W. H. Weihe, ed., The physiological effects of high altitude. New York, NY: Macmillan Co.
64. Dill, D. B., J. W. Terman, and F. G. Hall. 1963. Hemoglobin at high altitude as related to age. Clin. Chem. 9:710–16.
65. Jung, R. C., D. B. Dill, R. Horton, and S. M. Horvath. 1971. Effects of age on plasma aldosterone levels and hemoconcentration at altitude. J. Appl. Physiol. 31:593–97.
66. Weitz, C. A. 1973. The effects of aging and habitual activity on exercise performance among a high altitude Himalayan population. University Park, PA: Ph.D. Thesis. Pennsylvania State University.

CHAPTER 11

# Adult Acclimatization, Developmental Acclimatization, and the High-Altitude Native

After the initial effects of and responses to high altitude, usually characterized by the onset and disappearance of the symptoms of acute mountain sickness, gradual adaptive responses develop, some of which require months or many years for complete development. The various systemic and cell responses that together permit the organism to function normally with low oxygen availability develop at different rates. Some increase progressively for many months; others reach an early peak and then subside; and others require exposure during the organism's period of growth and development. For this purpose, the physiological information derived from sea level natives exposed to high altitudes during adulthood will be presented first. In the next section the functional characteristics of the high-altitude native and their developmental components will be discussed.

## ADULT ACCLIMATIZATION

Many of the observed adjustments begin in the first few hours at high altitude, even though the results are manifested days or months later. For this reason, in this chapter the emphasis is not on tracing the chronological manifestation of a given response but on evaluating the relative interaction of various systemic and cell changes in the adaptive process. The various adaptive mechanisms that enable sea-level natives to acclimatize to high altitude involve changes in pulmonary function and oxygen transport.

### Pulmonary Ventilation

On exposure to high-altitude hypoxia, sea-level natives show, both at rest and during exercise, a progressive increase in pulmonary ventilation that may reach as much as 100% of sea-level values (1–3) (see fig. 11.1). Since the increase in pulmonary ventilation permits the newcomer to maintain an increase in $PO_2$, at the alveolar level and an increase in arterial oxygen saturation (4), it would appear that a hyperventilatory response is critical to the newcomer's acclimatization. As a result of hyperventilation, the partial pressure of $CO_2$ in the alveoli decreases. As shown in figure 11.2 there is a linear decrease in alveolar $PCO_2$ as the altitude increases. Such hyperventilation is adaptive because it increases the partial pressure of oxygen at the alveolar and arterial levels and consequently increases the diffusion gradient between blood and tissues. In fact, as indicated by the studies on Mount Everest (5–6) hyperventilation is the most important mechanism whereby the body protects itself against the severe hypoxia of high altitudes.

A consequence of hyperventilation and concomitant decrease in alveolar $PCO_2$ is that it changes the pH of the blood from a normal (pH = 7.4) to an

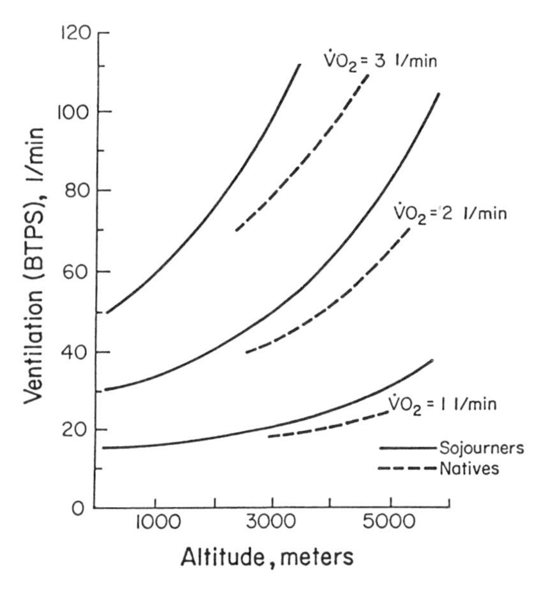


**Fig. 11.1.** Pulmonary ventilation in relation to altitude in lowlander sojourners to high altitude and highland natives measured at rest and at two levels of exercise. With exercise ($VO_2$ = oxygen consumption of 2 and 3 l/min) ventilation increases much more at high altitude than at sea level; this increase is much greater in sojourners than in highland natives. (Modified from C. Lenfant and K. Sullivan. 1971. N. Engl. J. Med. 284:1298–1309.)

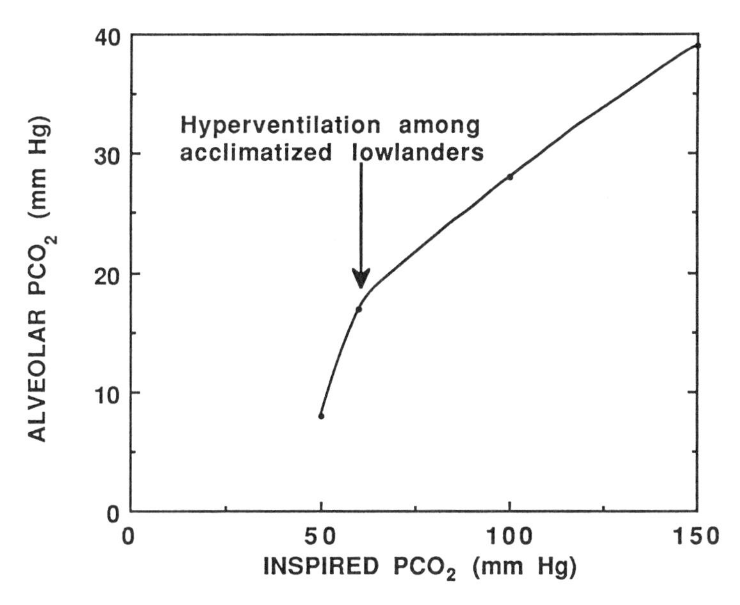


**Fig. 11.2.** Relationship between inspired $PO_2$ and alveolar $PCO_2$ at extreme altitudes. Note that as the altitude increases the pressure of $CO_2$ decreases. For example, at the $PO_2$ of 50 mm Hg (8040 m) the $PCO_2$ is 8 mm Hg, but at the $PO_2$ of 100 mm Hg (sea level) the $PCO_2$ is about 30 mm Hg. (Modified from R. B. Santolaya, S. Labin, R. T. Alfaro, and R. B. Schoene. 1989. Respiration physiology 77:253–62.)

alkaline state (pH>7.4), resulting in respiratory alkalosis. As schematized in figure 11.3, since the pH is a function of hydrogen concentration, the sea-level native increases the rate of bicarbonate excretion, which shifts the pH of the blood and cerebrospinal fluid back to normal during acclimatization to high altitude in an attempt to correct this problem. However, this respiratory alkalosis is never completely compensated among sea-level natives acclimatized to high altitude (7) and as shown below appears beneficial.

### Oxygen-Hemoglobin Dissociation Curve

One basis for oxygen transport from lung to tissues is that oxygen combines strongly with hemoglobin when the partial pressure of oxygen is high, as in the pulmonary capillaries; when the partial pressure of oxygen is low, as in the tissue capillaries, oxygen is released freely. In other words, the hemoglobin affinity for oxygen is inversely related to the partial pressure of oxygen. Physiologists express this affinity in terms of the partial pressure of oxygen in plasma ($PO_2$) associated with 50% oxygen saturation of blood at 37°C and a pH of 7.4, known as the *oxygen-hemoglobin dissociation curve.*

Several studies have shown that at high altitude hemoglobin has a decreased affinity for oxygen when the pressure is low as it occurs in the venous blood, resulting in a *right* shift of the oxygen dissociation curve (2, 3, 8–12). The decrease in hemoglobin affinity for oxygen was thought to be related to an increase in intraerythrocytic 2,3-diphosophoglycerate (DPG), which in turn seems related to changes in blood pH (2, 3, 8–13). Lenfant, Torrance, and Reynafarje (13) found that in subjects made acidotic with acetazolamide when exposed for 4 days to an altitude of 4400 m there were no changes in plasma pH, no increase in 2,3-DPG, and no shift in the oxygen dissociation curve (13). In contrast, in the control subjects (who were not acidotic) the oxygen-hemoglobin dissociation curve shifted rapidly to the right when these subjects were exposed to high altitudes. This shift seemed to be caused by an increase in 2,3-DPG, which in turn was associated with an increase in plasma pH above sea-level values. However, as shown by studies conducted on Mount Everest (7) due to the stronger effect of alkalinity (pH>7.4) resulting from the low concentration of alveolar $CO_2$ the oxygen-dissociation curve is shifted to the *left*. This left shift protects arterial oxygen saturation at extreme altitude. Furthermore, studies using rodents and camelids (vicuna, alpaca, and llama) that are native to the highlands indicate a higher hemoglobin-oxygen affinity than those at sea level (14, 15). Similarly, experimental studies on rats indicate that a decreased hemoglobin-oxygen affinity is not adaptive in extreme high altitudes (16).

### Lung Volume and Pulmonary Diffusion Capacity

On initial exposure to high-altitude hypoxia the vital capacities and residual lung volumes of sea-level natives are reduced, but after 1 month at high

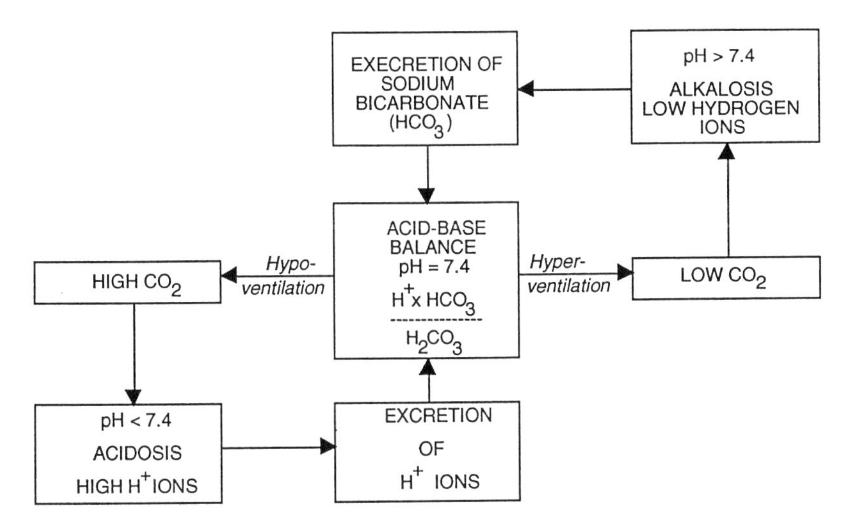


**Fig. 11.3.** Schematization of acid-base balance maintenance with acclimatization to high-altitude hypoxia. During acclimatization to high altitude excretion of sodium bicarbonate is increased to counteract alkalosis and maintain acid-base balance.

altitudes, such subjects attain values comparable to those they had at sea level (17–21).

As shown by several studies, acclimatization to high altitudes does not change the oxygen pulmonary diffusing capacity of sea-level subjects (22–24).

## Transport of Oxygen in the Blood

### Erythropoietic Activity

In response to delivery of insufficient oxygen, the bone marrow is stimulated by an erythropoietic factor to increase production of red blood cells (25–27). Within 2 hours of exposure to high altitude (4540 m) the average iron turnover rate increases on the average from 0.37 to 0.54 mg/day/kg body weight (28). Furthermore, the iron turnover rate increases to a maximum of 0.91 mg/day/kg in 1 to 2 weeks; during this period erythropoietic activity is about three times that at sea level. The degree of erythropoietic stimulation has been found to be proportional to severity of the hypoxia; it is high during initial hours of exposure to high altitude and falls in a few days to a value intermediate between the initial peak response and that at sea level (28).

This secondary decrease in erythropoietic activity is related to the fact that after the first day several acclimatization processes operate and thus decrease the hypoxic stimuli. Furthermore, erythropoietic activity is not linearly related to altitude exposure because above 6000 m the rate of erythrocytic and hemoglobin formation is decreased (29).

## Red Blood Cell and Hemoglobin

As a result of the increased erythropoietic function the mean red cell count during acclimatization changes from about 5 million to 7 million/cu mm at 5000 m. Along with the increased red blood cell count, hemoglobin concentration also increases. The highest increase in hemoglobin takes place from 7 to 14 days after exposure, and during this time erythropoietic activity is about three times that observed at sea level. This process may continue for as long as 6 months, but the hemoglobin level stabilizes between 18 and 20 g/100 ml (30-31) (see table 11.1).

## Blood Volume

The pronounced polycythemia with exposure to high altitude results in an increased red cell volume. As shown in figure 11.4 the red cell volume, after a residence at 4540 m for 1 to 3 weeks, increases from 40 ml/kg at sea level to 50 ml/kg at high altitude (32). Therefore, in spite of red cell volume increases, total blood volume increases minimally. Consequently, the hematocrit increases from an average of 43% at sea level to about 60% after 2 weeks' exposure to high altitude. This increased viscosity results in greater strain on the heart; at a high altitude the heart must compensate for increased blood viscosity by doing more work. This explains the increased heart rate observed during initial exposure to high altitude.

## Increased Oxygen-Carrying Capacity of the Blood

The effect of increased hemoglobin concentration is translated in an increased oxygen-carrying capacity of the blood at high altitude. For example, as shown in table 11.2, when at 4545 m the hemoglobin increases to about 17.0 g/dl and 81% of the hemoglobin is saturated with oxygen the total quantity of oxygen that the arterial blood carries is about 18.5 ml of oxygen/100 ml of blood ($0.81 \times 1.34 \times 17.0$ g Hb = 18.5). In contrast, at sea level the quantity of oxygen that the arterial blood carries is about 19.7 ml of oxygen/100 ml of

**TABLE 11.1. Hematological responses to high-altitude hypoxia**

| Altitude | *N* | Mean ± SD Hb (g/dl) | Mean ± SD Hct (%) | Mean ± SD MCHC (g/dl) |
|---|---|---|---|---|
| Sea level | 18 | 14.5 ± 0.7 | 43.8 ± 2.3 | 33.2 ± 1.0 |
| 5400 m | 20 | 17.8 ± 1.0 | 50.8 ± 2.6 | 35.1 ± 1.4 |
| 6300 m | 19 | 18.8 ± 1.5 | 53.4 ± 4.0 | 35.2 ± 1.1 |

*Source:* Adapted from R. M. Winslow, M. Samaja, and J. B. West. 1984. Red cell function at extreme altitude on Mount Everest. J. Appl. Physiol.: Respirat. Environ. Exercise Physiol. 56:109–16.

*Note:* Hb = hemoglobin; Hct = hematocrit; MCHC = mean corpuscular hemoglobin concentration.

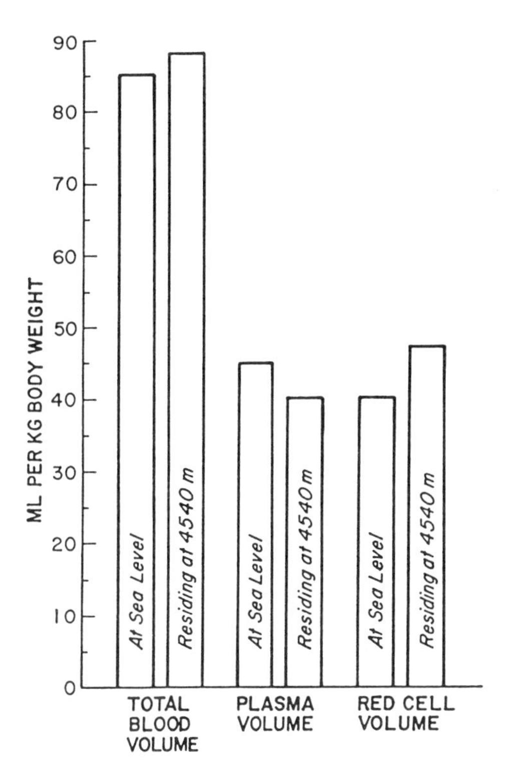


**Fig. 11.4.** Comparison of total blood volume, plasma volume, and red cell volume among sea-level subjects and after residence for 1 to 3 weeks at high altitude. Exposure to high altitude results in increased red cell volume without increase in plasma volume. (Based on data from C. F. Merino. 1950. Blood. 5:1–8.)

blood (0.98 × 1.34 × 15.0 g Hb = 19.7). Thus, despite the increase in hemoglobin concentration sea-level natives acclimatized to high altitude are not successful in maintaining normal levels of blood oxygen availability.

## Work Capacity and Maximal Oxygen Intake

Oxygen consumption, or oxygen uptake, is a basic biological function that reflects the total metabolic work being done by the organism. With increased metabolic requirements during maximal exercise oxygen consumption may increase from an average resting value between 200 and 350 ml/min to as much as 5 l/min. The rate of oxygen consumption is a product of the cardiac

**TABLE 11.2. Changes in oxygen-carrying capacity of the blood with altitude**

| | Sea Level | Andean 4545 m | Andean 5950 m |
|---|---|---|---|
| %Hb Saturation (%) | 98.0 | 81.0 | 65.3 |
| Hb (gr/dl) | 15.0 | 17.0 | 20.7 |
| Arterial (ml $O_2$/100 of blood) | 19.7 | 18.5 | 18.1 |

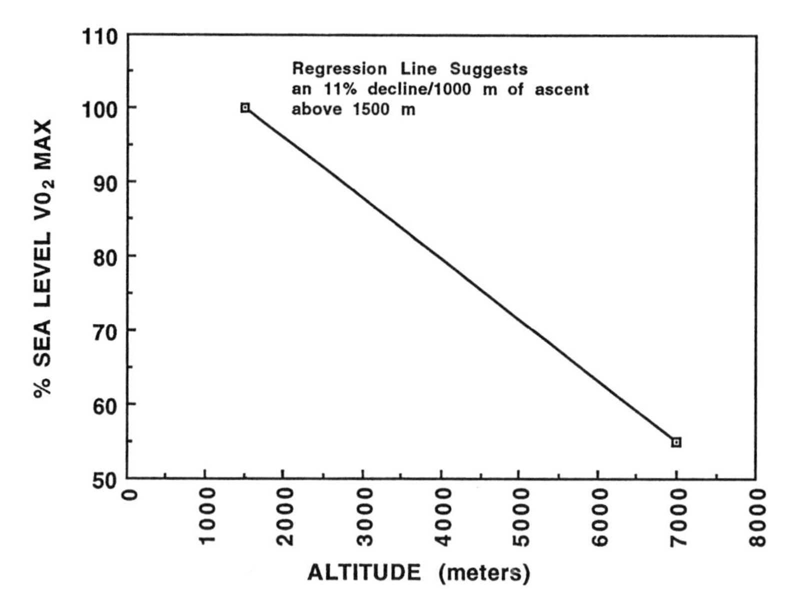


**Fig. 11.5.** Mean percentage reduction in maximal oxygen utilization ($VO_2$ max) from sea-level values for groups abruptly exposed to a high altitude of 4000 m at Nuñoa, Peru. The aerobic capacity of sea-level subjects on exposure to high altitude declines 11% for every 1000 m above 1500 m. (From E. R. Buskirk. 1976. Work performance of newcomers to the Peruvian highlands. In P. T. Baker and M. A. Little, eds., Man in the Andes: A multi-disciplinary study of high-altitude Quechua natives. Stroudsburg, PA: Dowden, Hutchinson, and Ross.)

output and the difference in oxygen content between the venous and arterial blood. The rate of oxygen consumption can also be expressed as a product of heart rate, stroke volume, and the difference in oxygen content between the venous and arterial blood. Therefore, the rate of oxygen consumption during exercise can be affected by changes in heart rate, stroke volume, and rate of oxygen extraction determined by the difference in oxygen content between the venous and arterial blood. Thus, during severe exercise the metabolic requirement for oxygen increases drastically, and all the processes involved in the transport, delivery, and utilization of oxygen are required to work at their maximum. For this reason, the effects of high-altitude hypoxia are most evident during periods of hard work. Thus measurements of an individual's work capacity indicate the degree of success of the various adaptive responses he or she has made.

It is generally agreed that the maximum oxygen intake per unit of body weight (or aerobic capacity) during maximal work is a measure of the individual's work capacity because it reflects the capacity of the working muscles to use oxygen and the ability of the cardiovascular system to transport and deliver oxygen to the tissues. The rate of oxygen consumption increases linearly with the magnitude of work. As an exercising subject approaches the point of exhaustion or fatigue, his or her oxygen consumption will reach a maximum and remain at that level even with further increase in work. This peak value is referred to as the individual's *maximal oxygen intake*. Maximum oxygen intake may be obtained directly or indirectly. By the direct approach maximum oxygen intake, along with respiratory volume, is obtained while the subject is performing a maximal or exhaustive work task for short time periods (9 to 15 minutes), either on a stationary bicycle ergometer or a treadmill. By the indirect approach maximum oxygen intake is derived from measurements of oxygen intake and pulse rate (or respiratory frequency) while performing submaximal work for longer periods (20 to 45 minutes). The maximum oxygen intake is then applied to previously established regression equations. In general the indirect method is less accurate and gives lower values for maximum oxygen intake than those obtained by the direct method.

Table 11.3 summarizes the available data on maximum oxygen consumption during exercise. As shown in figure 11.5 the maximum aerobic capacity of well-trained sea-level subjects, when expressed as percent of the sea-level values, declined by 11% for every 1000 m (3000 feet) ascended beyond 1500 m (5000 feet). Similarly, studies among sea-level newcomers to high altitude

**TABLE 11.3. Comparison of maximum aerobic power ($VO_2$ max: ml/kg/min) attained by samples tested at high and low altitudes (LA)**

| | | | | $VO_2$ max (ml/kg/min) | | |
|---|---|---|---|---|---|---|
| Sample | *N* | Tested at Altitude (m) | Time at Altitude | Attained at Altitude | Attained or Expected at Sea Level | D (%) |
| | | | Sedentary or Untrained | | | |
| Sedentary— | | | | | | |
| U.S. | 12 | 4000 | 4 weeks | 38.1 | 50.4 | 24.4 |
| Urban—Bolivia | 28 | 3600 | Life | 46.0 | 40-50 | 0.0 |
| Rural—Chile | 37 | 3650 | Life | 46.4 | 40-50 | 0.0 |
| Rural—Peru | 8 | 4000 | Life | 51.8 | 40-50 | 0.0 |
| Rural—Peru | 5 | 4000 | 10 years | 49.2 | 40-50 | 0.0 |
| Urban—Peru | 20 | 3400 | 2-15 years | 46.3 | 40-50 | 0.0 |
| Urban—Peru | 8 | 4350 | 2 weeks | 49.0 | 53.6 | 8.6 |
| Urban—Peru | 5 | 4540 | 23 weeks LA | 50.0 | 50.2 | 0.4 |
| | | | Active | | | |
| Urban—Bolivia | 29 | 3700 | Life | 37.7 | 50-60 | 0.0 |
| Urban—Peru | 28 | 4500 | Life | 51.2 | 50-60 | 0.0 |
| Urban—Peru | 20 | 3750 | Life | 50.9 | 50-60 | 0.0 |
| | | | Trained or Athletes | | | |
| Athletes—U.S. | 6 | 4000 | 7 weeks | 49.0 | 63.0 | 22.2 |
| Athletes—U.S. | 6 | 3090 | 2 weeks | 59.3 | 72.0 | 17.6 |
| Athletes—U.S. | 5 | 3100 | 4 weeks LA | 45.5 | 61.7 | 26.3 |
| Athletes—Peru | 10 | 3700 | 6 months | 55.0 | 70.0 | 21.4 |

*Source:* Adapted from A. R. Frisancho. 1983. Perspectives on functional adaptation of the high-altitude nature. In R. J. Sutton, C. S. Houston, and N. L. Jones, eds. Hypoxia, exercise, and altitude. New York: Alan R. Liss.

demonstrated a reduction in aerobic capacity of 13% to 22% (45). It must also be noted that the observed decreases in aerobic capacity at high altitude are influenced very little by physical conditioning. In fact, the decrease in aerobic capacity at high altitude has been observed to be proportional to the extent of training; athletes show the greatest decrease, sedentary individuals the least, and active subjects an intermediate decrease (45–47). Thus, sea-level natives, as judged by their inability to recuperate their work capacity, attain only a partial functional adaptation to high altitude (see fig. 11.15)

## THE HIGH-ALTITUDE NATIVE AND DEVELOPMENTAL ACCLIMATIZATION

The high-altitude native has adapted to the altitude environment through a modification and synchronized interdependence of the respiratory, circulatory, and cardiovascular systems oriented at improving oxygen delivery and oxygen utilization. At present, the evidence suggests that this level of functional adaptation is acquired during growth and development and is not unique to the high-altitude native. For this reason, the present section will summarize physiological information derived from high-altitude natives living at high altitudes. Furthermore, the discussion will also show that changes occurring during growth and development are of major importance in explaining the functional adaptation attained by the high-altitude native. The various adaptive mechanisms that enable high-altitude natives to acclimatize to high altitude involve changes directed toward increasing the availability of oxygen and the pressure of oxygen at the tissue level. This is accomplished through modifications in (1) pulmonary ventilation, (2) lung volume and pulmonary diffusing capacity, (3) transport of oxygen in the blood, (4) diffusion of oxygen from blood to tissues, and (5) utilization of oxygen at the tissue level. (See fig. 11.14.)

### Pulmonary Ventilation

A distinguishing feature of the high-altitude native is that at a given altitude, as shown in figure 11.1, the pulmonary ventilation during exercise of the sea-level native (sojourning on a short-term or long-term basis at high altitudes) is invariably less than that of the high-altitude native (45–56). Studies on ventilatory sensitivity indicate that high-altitude natives have a hypoxic chemoreceptor insensitivity (28, 30–32, 57–58). This insensitivity can be acquired by sea-level natives if they are exposed to high altitudes throughout growth and development (29, 59). However, other studies indicate that it can also be acquired by adults through prolonged residence at high altitudes (60–61).

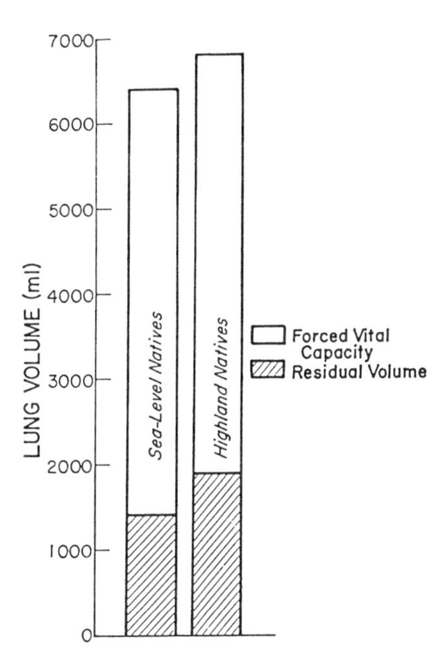


**Fig. 11.6.** Comparison of lung volumes of sea-level and high-altitude natives. Lung volume, especially residual lung volume, is increased among highland natives. (From A. Hurtado. 1964. Animals in high altitudes: resident man. In D. B. Dill, E. F. Adolph, and C. G. Wilber, eds., Handbook of physiology, vol 4. Adaptation to the environment. Washington, DC: American Physiological Society.)

## Lung Volume

**Andeans.** As shown in figure 11.6 highland natives have larger lung volumes and residual volumes than sea-level natives when adjustments are made for differences in body size (48–49, 62).

**Himalayans (Tibetans).** Studies of Tibetans who were born and raised above 3600 m and Hans who were born at sea level and acclimatized to high altitude during adulthood (116–117) are presented in figure 11.7. This figure shows that the high-altitude native Tibetans had both larger forced vital capacity and residual lung volumes than the sea-level Hans.

**Ethiopians.** Investigations in Ethiopia demonstrated that highlanders living above 3000 m have a significantly greater forced vital capacity than their lower altitude counterparts (fig. 11.8). Furthermore, evaluation of the respiratory function of migrant groups demonstrated that the enlarged forced vital capacity of the highland Ethiopian can be acquired by lowland natives who migrate to high altitudes (63).

Thus, high-altitude natives irrespective of the geographic region have larger lung volumes than low-altitude populations (see also chap. 12).

### Developmental Component

Current evidence suggests that the enlarged lung volume of the high-altitude native results from adaptations acquired during the developmental period (59,

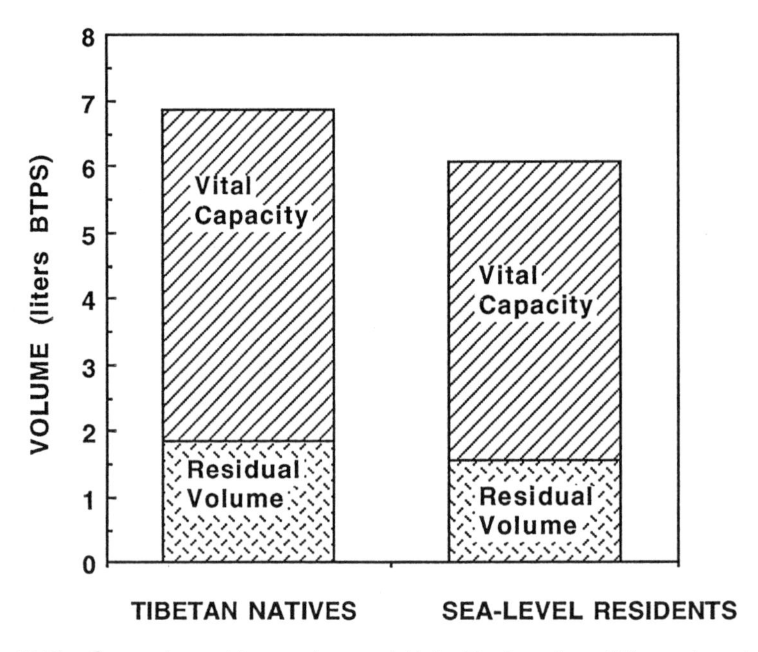


**Fig. 11.7.** Comparison of lung volumes of high-altitude natives (Tibetans) and sea-level natives (Hans) residing at high altitudes in the Himalayas. (Adapted from Droma et al. 1991. Increased vital and lung capacities in Tibetan compared to Hans residents of Lhasa (3,658m). Am. J. Phys. Anthrop. 86:341–51.)

62, 64–67). As shown in figure 11.9 the sea-level natives who were acclimatized to high altitudes during growth, when adjustments were made for variations in body size, attained the same residual lung volume as the high-altitude natives. In contrast, sea-level natives acclimatized as adults had significantly lower residual lung volumes than high-altitude natives (136).

The hypothesis that the enlarged lung volume of high-altitude natives results from developmental adaptation is supported by experimental studies on animals. Various studies (68–72) have demonstrated that young rats, after prolonged exposure to high-altitude hypoxia (3450 m), exhibited an accelerated proliferation of alveolar units and accelerated growth in alveolar surface area and lung volume. In contrast, adult rats, after prolonged exposure to high-altitude hypoxia, did not show changes in quantity of alveoli and lung volume (70–71). Figure 11.10 shows that rats acclimatized to a high altitude from birth to the age of 131 days develop a greater lung volume than sea-level controls. In contrast, adult rats who were the mothers of the young rats studied, were acclimatized to a high altitude for the same number of days and had lung volumes similar to the sea-level controls. These findings suggest that in experimental animals and humans, the attainment of an enlarged lung volume at high altitude is probably mediated by developmental factors.

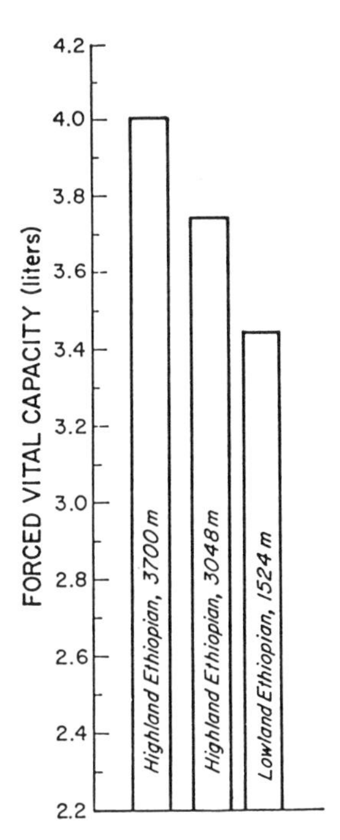


**Fig. 11.8.** Comparison of forced vital capacity of lowland and highland Ethiopian natives. The higher the altitude, the greater the lung volume. (Based on data from G. A. Harrison, C. F. Kuchemann, M. A. S. Moore, A. J. Bouyce, T. Baju, A. E. Mourant, J. J. Godber, B. G. Glasgow, A. C. Kopec, D. Tills, and E. J. Clegg. 1969. Philos. Trans. R. Soc. Lond. 256:147–82.)

## Pulmonary Diffusion Capacity

All studies indicate that the pulmonary diffusing capacity of highland natives is systematically greater than that attained by lowland natives at sea-level (24, 73–76). This difference is probably caused by at least three factors. First, since pulmonary diffusing capacity is related in part to alveolar area, the enhanced pulmonary diffusing capacity of the highland native is probably caused by a greater alveolar area and increased capillary volume. Second, increased oxygen pulmonary diffusing capacity of the highland native may also be related to his high pulmonary pressure. It has been suggested that the high pulmonary pressure that occurs at high altitude is associated with a more uniform perfusion of the lung that may distend patent capillaries and perhaps perfuse capillaries unopened at sea level of sea-level natives becoming acclimatized to a high altitude (77). Third, another probable contributing factor is increased pulmonary ventilation that, along with the increased alveolar and capillary volumes, improves the ratio of ventilation to perfusion.

## Oxygen-Hemoglobin Dissociation Curve

Earlier studies have shown that among high-altitude natives the hemoglobin in the venous blood has a decreased affinity for oxygen which results in a

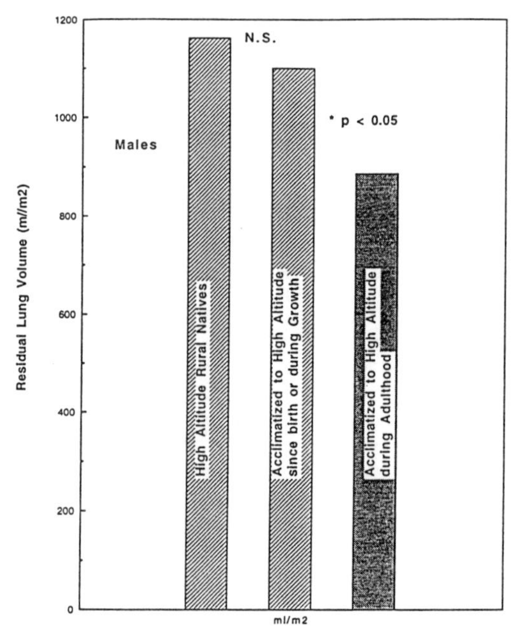


**Fig. 11.9.** Comparison of residual lung volume (ml/m² of body surface area) of high-altitude natives, Peruvian sea-level subjects acclimatized to high altitude during growth, and European sea-level subjects acclimatized to high altitude during adulthood. Acclimatization to high altitude during growth results in lung volumes similar to those of high-altitude natives. (From data of Frisancho et al. 1993. Am. J. Phys. Anthropol. 16).

*right* shift of the oxygen dissociation curve (2, 3, 8–12). Recent studies of Winslow et al. (7) among high-altitude natives from Peru (living at 4540 m) demonstrate that the in vivo oxygen affinity of hemoglobin is not lower than that of sea-level controls. Hence, the oxygen-hemoglobin curve is not shifted to the right. According to this research, many physiological and biochemical effectors of oxygen affinity balance each other resulting in normal oxygen affinity. For example, alkalosis stimulates 2,3-DPG accumulation, which reduces hemoglobin oxygen affinity, but at the same time increases the oxygen affinity due to the Bohr effect (see chap. 8).

## Transport of Oxygen in the Blood

### Red Blood Cell and Hemoglobin

At altitudes above 3500 m both sea-level and high-altitude natives have red blood cell counts ranging from 5 to 8 million/cu mm compared to the 4.5 million at sea level. Along with an increase in red blood cells, the hemoglobin is augmented; among Andean high-altitude natives living above 3500 m the average concentrations range from 16 to 20 g/100 ml (26–27, 78–81). It must be noted, however, that among high-altitude natives there are marked differences in hemoglobin concentration depending on the region of residence.

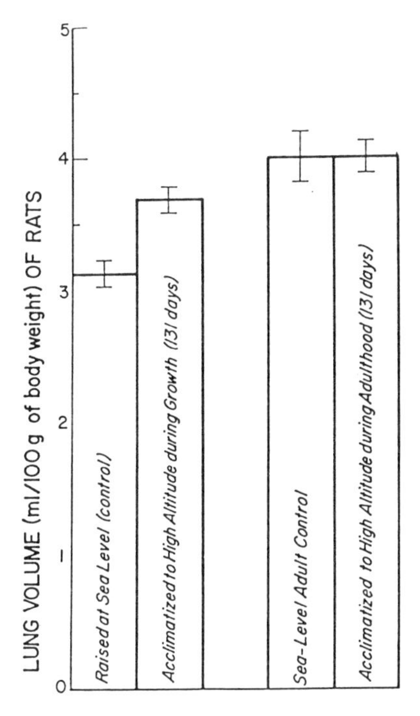


**Fig. 11.10.** Comparison of lung volume of rats raised at sea level and acclimatized to high altitude during growth and during adulthood. Like humans, acclimatization to high altitude during growth results in enlargements of lung volumes. (Based on data from P. H. Burri, and E. R. Weibel. 1971. Morphometric evaluation of changes in lung structure due to high altitudes. In R. Porter and J. Knight, eds., High altitude physiology. Cardiac and respiratory aspects. Edinburgh: Churchill Livingstone.)

Thus, in Peru the mean hemoglobin concentration for natives of the central highlands residing at 4000 m is about 20 g/dl (78, 82), whereas for natives of the southern highlands residing at the same altitude the hemoglobin concentration does not exceed 18 g/dl (80). These differences can be explained in part by the fact that in the central highlands high-altitude hypoxia is compounded by additional hypoxia from mining activities. Furthermore, populations in the central highlands are less heterogenous than those in the southern highlands.

As documented by the studies of Beall and colleagues (83–84) the hemoglobin concentration among Himalayan natives is between 1 and 2 g/dl lower than that found among Andean populations living at comparable or lower altitudes. For example, the mean hemoglobin concentration for Tibetans living between 4850 and 5450 m is about 18.2 g/dl (83–84) while for Bolivians living at about 3700 m is about 18.8 g/dl (85). However, as shown by Beall et al. (86), urban highland Himalayans have hemoglobin concentrations which are equal to those attained by urban highland Andeans. Therefore, the source of the observed differences in hemoglobin concentration may be more a function of sampling than of different hemopoietic responses to hypoxia.

### Increased Oxygen-Carrying Capacity of the Blood

The effect of increased hemoglobin concentration is translated in an increased oxygen-carrying capacity of the blood at high altitude. As shown in table 11.2 this increase may result in an excessive compensation. For example, at 4545 m the total quantity of oxygen that the arterial blood carries is about 21.8 ml of oxygen/100 ml of blood ($0.84 \times 1.34 \times 20.1$ g Hb $= 21.8$) which exceeds that attained at sea level. On the other hand, at extreme high altitudes high-altitude natives who work at 5950 m (87) are not successful in maintaining normal levels of blood oxygen availability despite the increase in hemoglobin concentration.

### Diffusion of Oxygen from Blood to Tissues

**Capillaries.** For oxygen to be used it must reach the cell mitochondria through a process of physical diffusion. The rate of such diffusion depends on the partial pressure of the oxygen. Because oxygen is consumed as it goes through successive layers, the partial pressure of oxygen declines with the increase in the distance that the oxygen has to travel. Since the partial pressure of oxygen in the ambient air is already low at high altitudes, the organism must respond by shortening the distance the oxygen has to travel. This is accomplished by the opening up of existing capillaries and the formation of new capillaries. Through microscopic studies of experimental animals it has been found that the number of open muscle capillaries at high altitude is increased more than 40% when compared to sea-level values (41–44). This finding has recently been confirmed by studies on the Himalayans (134) that demonstrated that the number of capillaries per square millimiter of muscle cross section in both Himalayan Sherpas and elite high-altitude climbers was significantly higher than that of sedentary lowlanders. A very important effect of an increased capillary bed is that it increases the blood perfusion; thus, oxygen is more readily diffused into tissue in spite of a lowered oxygen tension of the blood before it reaches the capillaries (88).

**Myoglobin.** The rate of diffusion of oxygen from the capillaries to the cell mitochondria is also aided by increased amounts of tissue myoglobin. Myoglobin is a protein found in high concentrations in muscles that carry out sustained or periodic work. It has a molecular weight of about 17,500 and consists of 152 amino acid residues and one iron-containing heme group. Studies on humans and animals at high altitudes indicate a high concentration of myoglobin in the sartorius muscle (82, 89–91). Proteins of low molecular weight such as myoglobin facilitate diffusion of oxygen from capillaries to the mitochondria by random movement interspersed with larger advances of translational and rotational movements (92–93). It is quite reasonable to assume that at high altitude the increased concentration of myoglobin is an adaptive response to facilitate diffusion of oxygen from the capillaries to the cell mitochondria.

### Utilization of Oxygen

The last step in the process of adaptation to hypoxia involves variations in the rate of oxygen utilization and generation of energy at the cellular level. Based on studies with guinea pigs, it is postulated that at high altitude the carbohydrate metabolism, during its glycolytic (anaerobic) breakdown, goes through the pentose-phosphate pathway rather than the Embden-Meyerhof pathway (94–97). The advantage in using the pentose-phosphate pathway appears to be the fact that no additional adenosinetriphosphate (ATP) is required to generate glyceraldehyde-3-phosphate as is necessary in the Embden-Meyerhof pathway. According to Reynafarje (97) at high altitude, by relying on the pentose-phosphate pathway, the organism saves energy (ATP) or produces more chemical energy with the same oxygen consumption. This hypothesis is supported by the finding that the activity of oxidative enzymes in the sartorius muscles is greater at high altitudes than at sea level (94). For example, in the homogenate of whole cells the reduced diphosphopyridine nucleotide (DPNH)-oxidase system and triphosphopyridine nucleotide (TPNH)-cytochrome-c reductase and transhydrogenase are significantly more active in the highland native than in the sea-level native (97). Recent studies of Andean Quechuas indicate that for a given work rate highland natives at 4,200 m accumulated substantially less lactate than lowlanders at sea level (135). This lower accumulation of lactate is said to be an expression of a metabolic organization that maximizes the amount of ATP obtained per mole of carbon substrate catabolized (135). Thus, among high-altitude natives the chemical and morphological characteristics related to energy utilization and production appear qualitatively and quantitatively different from those of lowland natives. It is not known whether such characteristics may be acquired by sea-level natives residing for long periods at high altitudes.

## Cardiovascular Traits

### Pulmonary Circulation

Andean adults maintain an increased arterial pulmonary pressure (39, 98–100). As shown in figure 11.11, in both resting and exercise conditions the systolic and diastolic arterial pulmonary pressure of Peruvian highland natives living at 4330 m is nearly twice that of sea-level natives tested at sea level. Histological studies have demonstrated that after the first months of postnatal development, children born at high altitudes show a thickening of the muscular layer and muscularization of the pulmonary arteries and arterioles that

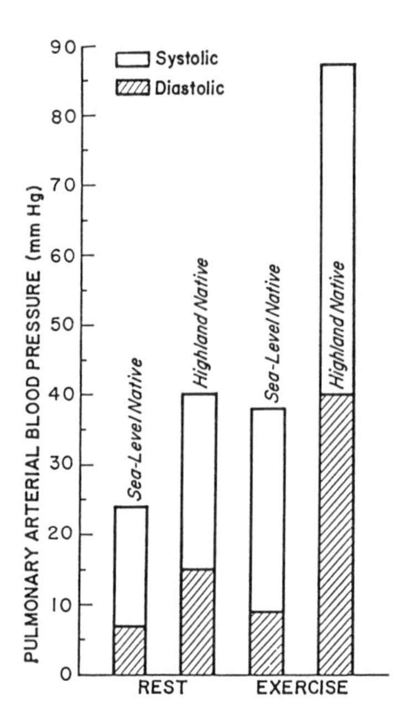


**Fig. 11.11.** Comparison of pulmonary blood arterial pressure among sea-level and high-altitude natives. Both at rest and during exercise high-altitude natives exhibit higher pulmonary pressure than sea-level natives. (Based on data from D. Peñaloza, F. Sime, N. Banchero, and R. Gamboa. 1962. Med. Thorac. 19:449–60.)

resembles the development of the fetal pulmonary vascular tree (101, 102). These characteristics, along with the viscosity accompanying polycythemia, contribute to the development of greater pulmonary vascular resistance or pulmonary hypertension in the high-altitude resident and native.

Based on studies of steers the hypothesis has been put forward that pulmonary hypertension at high altitudes would favor a more effective perfusion of all the pulmonary areas and, therefore, increase the effective blood gas interfacial area of the alveoli (77). In this manner perfusion of the entire lung coupled with an increased vascularization would enhance the diffusing capacity of the lung and should decrease the difference between arterial and alveolar blood. These changes would permit more effective arterial blood oxygenation (77). However, this has not been proven in the acclimatized human, nor can one assume that there will be a consequent decrease in the arterial-alveolar gradient. Therefore, the application of this hypothesis to human adaptation remains to be proven.

## Heart

As a result of increased pulmonary resistance or hypertension, the workload of the right ventricle of the heart is increased at high altitude. As demonstrated by anatomical and electrocardiographic studies this increased workload leads to an enlargement or hypertrophy of the right ventricle (98–99,

103–106). Figure 11.12 shows that after the age of 3 months the weight of the right ventricle is greater than that of the left ventricle (ratio is less than 1), and this characteristic is maintained through childhood to adulthood. On the other hand, at sea level from infancy to adulthood the left ventricle weighs more than the right ventricle (ratio greater than 1).

Another factor that may contribute to the enlargement of the right ventricle is the high incidence of patent ductus arteriosus. Of 5000 school children of both sexes born at high altitudes, 0.72% had a patent ductus arteriosus compared to 0.04% of sea-level children (107–108). This increased incidence of patent ductus arteriosus is probably a consequence of fetal and newborn hypoxia. It should be noted that during the prenatal stage when the lungs are not yet expanded, blood pressure in the pulmonary arteries exceeds that in the aorta so that blood flows from the pulmonary artery to the aorta through the ductus arteriosus. After birth, with the interruption of umbilical circulation and expansion of the lungs, the blood pressure falls sharply in the pulmonary arteries and rises in the systemic circulation. As a result the flow of blood is from the aorta to the pulmonary artery (rather than the prebirth flow from the pulmonary artery to the aorta). Thus, if the ductus arteriosus remains open it acts as a shunt from the aorta to the pulmonary artery where the pressure is lower, and hence the work of the right ventricle of the heart is increased. For these reasons it is quite possible that the incidence of patent ductus arteriosus may be another source of the right ventricular hypertrophy of high-altitude populations. Lowland natives with patent ductus arteriosus commonly suffer from right ventricular hypertrophy and pulmonary stenosis. These findings demonstrate the influence of developmental factors in the acquisition of the cardiovascular characteristics of highland dwellers. This does not mean, however, that preponderance of the right ventricle cannot be acquired during adulthood because adult rats in a hypobaric chamber to 5500 m developed right ventricular enlargement over a period of only 5 weeks along with muscularization of the small pulmonary arterial vessels (109).

## Cardiac Output and Stroke Volume

The stroke volume (SV) is defined as the volume of blood pumped from the heart with each beat. The cardiac output (Q) is the total volume of blood pumped by the heart each minute. It is the product of stroke volume and heart rate.

On initial exposure to high-altitude hypoxia the resting pulse rate of the sea-level native, because of a generalized increase in sympathetic activity, increases rapidly from an average of 70 beats/min to as much as 105 beats/min. This increase is associated with an abrupt augmentation of the resting

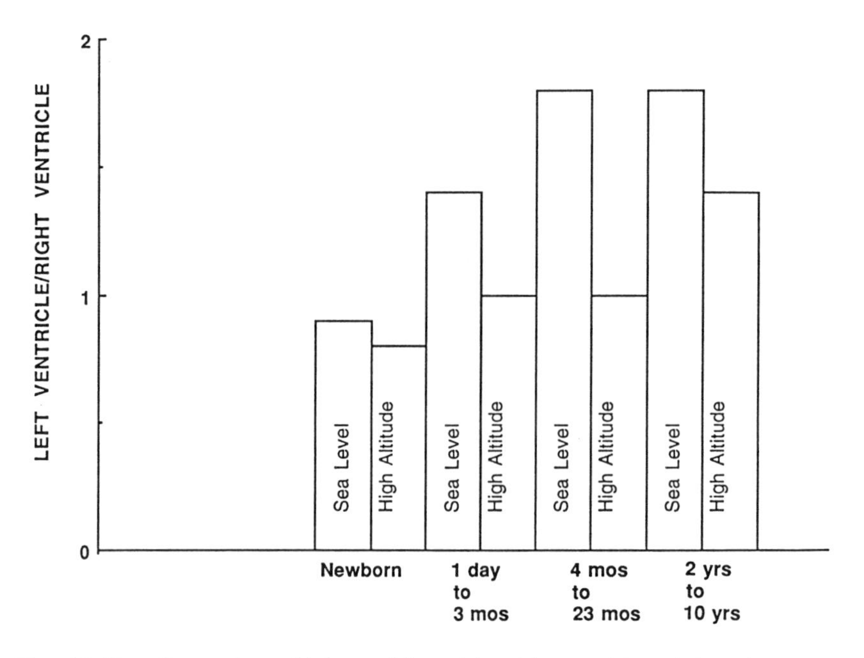


**Fig. 11.12.** Comparison of left ventricle weight, right ventricle weight ratio among sea-level and high-altitude natives at birth, 1 day to 3 months, 4 to 23 months, and 2 to 10 years. Among high-altitude subjects the right ventricle weighs less than left ventricle, a characteristic that becomes apparent after the age of 3 months. (Modified from J. Arias-Stella, and S. Recavarren. 1962. Am. J. Pathol. 41:55–59.)

cardiac output (33–34). With acclimatization the cardiac output and stroke volume decline; in about 1 week they are equal to or below those attained at sea level (4, 35–38). This decline in cardiac output and stroke volume seems associated with a decrease in heart rate, which usually remains above sea-level values. The cardiac output of high-altitude natives during rest and exercise was found to be equal to that attained at sea level (39). Therefore oxygen requirements of the body at high altitude appear to be met by greater oxygen extraction rather than by greater blood flow.

## Systemic Blood Pressure

Various studies indicate that the systemic blood pressure levels in adult high-altitude natives are lower than those observed at sea level (110). Epidemiological studies report that among high-altitude natives the frequency of essential hypertension and ischemic heart disease are significantly lower than among their counterparts at sea level (111). Furthermore, prolonged residence at high altitude results in lowering of systolic and diastolic pressure after a residence of 2 to 15 years declines by as much as 14 mm Hg and diastolic pressure by as much as 7 mm Hg. The decrease in systemic blood pressure indicates that chronic hypoxia has a relaxing effect on smooth muscle—the effect on the arterial media is vasodilation (40). Furthermore, because exposure to high-altitude hypoxia results in increased vascularization (41–44), it is possible that the prevalence of low blood pressure at high altitude may be related to a reduction in peripheral vascular resistance to blood flow. In other words, lower blood pressure may also be considered a by-product of tissue adaptation to high-altitude hypoxia.

## Work Capacity and Maximal Oxygen Intake

As shown by measurements of oxygen consumption during maximal exercise the aerobic capacity of high-altitude natives is comparable to that attained by lowland natives tested at sea level (1, 45–47, 51–56, 112–114). Thus, high-altitude natives, as judged by their ability to attain an aerobic capacity similar to that at sea level, have acquired a full (complete) functional adaptation to high altitude.

### Developmental Component

As demonstrated by studies of sea-level migrants residing at high altitude, the attainment of a normal aerobic capacity at high altitude is the result of adaptations acquired during the period of growth (53–65). As shown in figure 11.13, sea-level subjects, when acclimatized to high altitude during childhood and adolescence, attained an aerobic capacity (maximum oxygen intake) that was equal to that of the highland natives. In contrast sea-level natives (Peruvian and white United States subjects), when acclimatized to high altitudes as adults, attained significantly lower aerobic capacities than the highland natives. Evaluations of the data indicated that in the sea-level subjects who were acclimatized young the volume of air ventilated per unit of oxygen consumed (ventilation equivalent), maximum pulse rate, and the volume of oxygen consumption per pulse rate were comparable to those of the highland natives. On the other hand, the lowland subjects (Peruvian and United States) attained significantly higher ventilation ratios and lower heart rates than the highland natives (53).

Studies of Tibetans indicate that the aerobic capacity (maximum oxygen intake) of high-altitude native Tibetans was higher than that of the Hans who were born at sea level and acclimatized to high altitude after the age of 8 years (132). However, in this study the proportion of smokers (45%) among the sea-level Hans was much higher than that in the Tibetans (38%). Therefore, the lower aerobic capacity of the Hans probably reflects the negative effects of smoking rather than differences in the process of acclimatization to high altitude. In other words, if the comparison would have included either the same proportion of smokers (or excluded the smokers) the sea-level

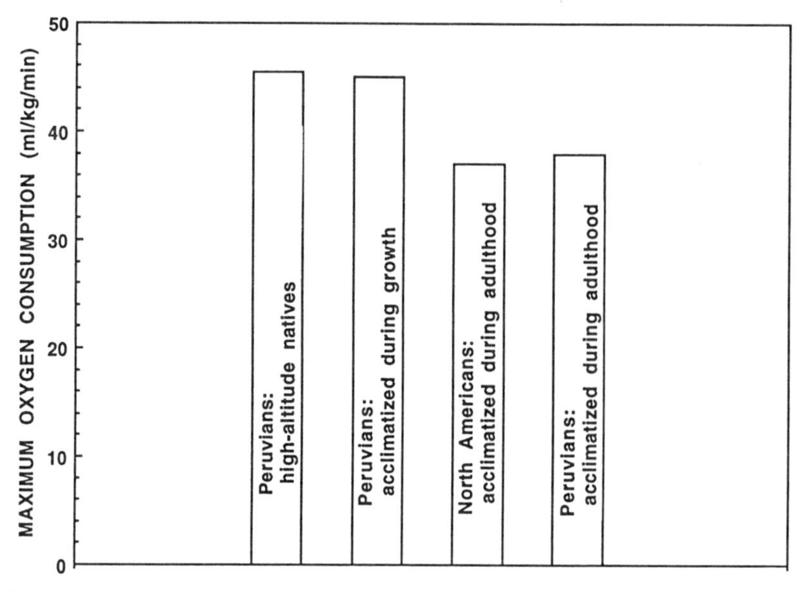


**Fig. 11.13.** Comparison of aerobic capacity among subjects acclimatized to high altitude during growth and adulthood and subjects native to high altitude. Acclimatization to high altitude during developmental period is associated with attainment of aerobic capacity similar to high-altitude native. (Based on data from A. R. Frisancho, C. Martinez, T. Velasquez, J. Sanchez, and H. Montoye. 1973. J. Appl. Physiol. 34:176–80.)

Hans would have attained a similar aerobic capacity as the high-altitude Tibetan natives, which would suggest a developmental role on the attainment of full functional adaptation to high altitude.

The extent to which developmental factors influence the attainment of aerobic capacity at high altitude is also illustrated in figure 11.14. These data show that among sea-level natives acclimatized to high altitude in the developmental period, the attainment of aerobic capacity is directly related to age at migration and length of residency. In contrast, when subjects were acclimatized to high altitude as adults, age at migration and length of residency did not influence the attainment of aerobic capacity. From these investigations it appears that the attainment of normal aerobic capacity at high altitude is influenced by adaptations acquired during the developmental period (53, 65).

## ACCLIMATIZATION TO HIGH ALTITUDE AND ATHLETIC PERFORMANCE

### Athletic Performance at High Altitude

The 1968 Olympics held in Mexico City called attention to the possibility of altitude effects on athletic performance. The fact that the moderate altitude

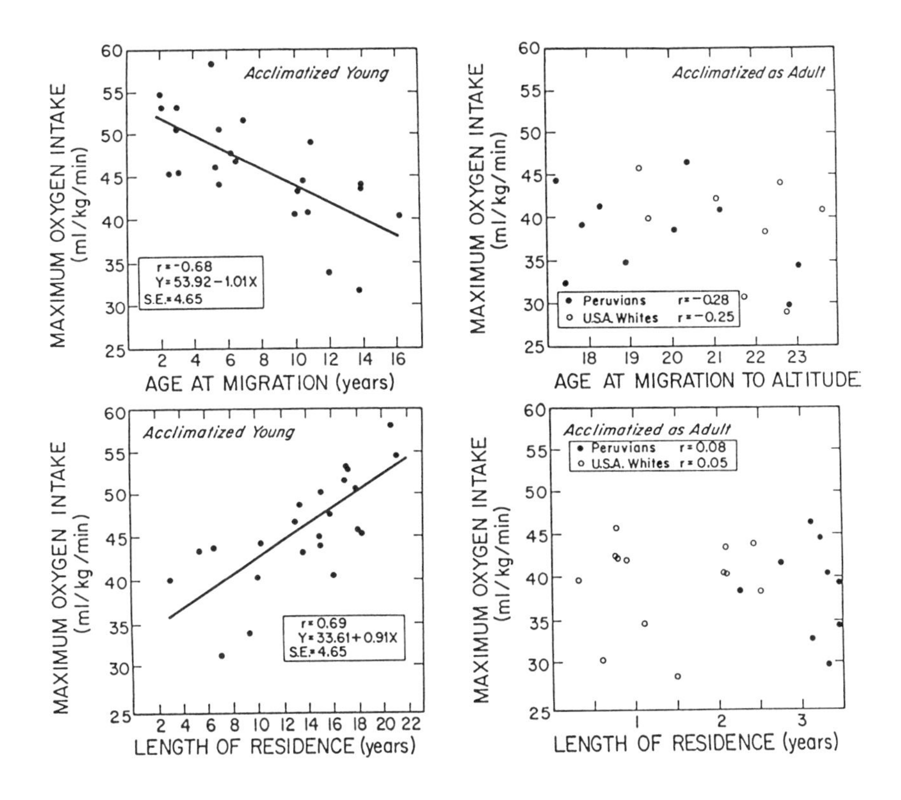


**Fig. 11.14.** Relationship between age at migration to high altitude and attainment of aerobic capacity among Peruvian and United States sea-level natives. Among subjects acclimatized to high altitudes during developmental period, age at migration is significantly correlated with aerobic capacity; this is not the case if subjects are acclimatized as adults. (Modified from A. R. Frisancho, C. Martinez, T. Velasquez, J. Sanchez, and H. Montoye. 1973. J. Appl. Physiol. 34:176–80.)

of 2380 m of Mexico City might affect athletic performance also raised the possibility that the effects of high altitude might not be the same for all events, and not all athletes would be equally affected. For some the altitude effects would be an advantage. This concern is exemplified by a letter in the *Times* signed by twenty-six British Olympic medalists in April, 1966 (40). They pointed out the possibility that natives or long-term residents at high altitude would have an advantage over those ascending from sea level. Exercise physiologists were called on to make predictions about the advantages and disadvantages of athletic performances in Mexico City. These predictions were based both on empirical information about altitude work physiology and statistics of athletic performance during the 1955 Pan American Games held in Mexico City and those held at sea level (128–31).

Tables 11.4 and 11.5 give the performance and running speed equivalents for various altitudes to the current sea-level records for men and women (133).

**TABLE 11.4. Performance equivalent to actual sea-level world record for men at selected altitudes**

| | Sea Level | Munich | Calgary | Albuquerque | Colorado Springs | Mexico City | La Paz | |
|---|---|---|---|---|---|---|---|---|
| Altitude (m): | 0 | 520 | 1,015 | 1,507 | 1,823 | 2,240 | 3,658 | 4,000 |
| Ps. torr: | 760 | 711 | 670 | 611 | 610 | 578 | 483 | 462 |
| Air density: (kg/m$^3$) | 1.204 | 1.112 | 1.061 | 1.001 | 0.066 | 0.017 | 0.766 | 0.733 |
| 60 m | 6.41 | 6.39 | 6.37 | 6.35 | 6.34 | 6.32 | 6.28 | 6.27 |
| 100 m | 9.92 | 9.88 | 9.83 | 9.79 | 9.77 | 9.74 | 9.66 | 9.64 |
| 200 m | 19.75 | 19.64 | 19.52 | 19.43 | 19.37 | 19.30 | 19.12 | 19.10 |
| 400 m | 43.29 | 43.14 | 42.97 | 42.84 | 42.76 | 42.70 | 42.96 | 43.16 |
| 800 m | 1:41.73 | 1:42.11 | 1:42.40 | 1:42.73 | 1:43.07 | 1:43.71 | 1:48.67 | 1:50.78 |
| 1,500 m | 3:29.46 | 3:31.24 | 3:32.76 | 3:34.33 | 3:35.71 | 3:38.10 | 3:54.10 | 4:00.64 |
| 1,609 m | 3:46.32 | 3:50.05 | 3:50.05 | 3:51.82 | 3:53.37 | 3:56.04 | 4:13.82 | 4:21.06 |
| 5,000 m | 12:58.39 | 13:09.07 | 13:18.61 | 13:28.14 | 13:36.16 | 13:49.58 | 15:14.93 | 15:49.28 |
| 10,000 m | 27:08.23 | 27:32.52 | 27:54.36 | 28:15.97 | 28:34.00 | 29:01.89 | 32:10.83 | 33:25.38 |
| Marathon | 2:06:50.00 | 2:08:57.58 | 2:10:52.93 | 2:12:46.21 | 2:14:19.67 | 2:16:53.15 | 2:32:37.87 | 2:38:52.13 |

*Source:* Adapted from F. Peronnet, G. Thibault, and D. L. Cousineau. 1991. A theoretical analysis of the effect of altitude on running performance. J. Appl. Physiol.: Respirat. Environ. Exercise Physiol. 70(1):399–404.

**TABLE 11.5. Performance equivalent to actual sea-level world record for women at selected altitudes**

| | Sea Level | Munich | Calgary | Albuquerque | Colorado Springs | Mexico City | La Paz | |
|---|---|---|---|---|---|---|---|---|
| Altitude (m): | 0 | 520 | 1,015 | 1,507 | 1,823 | 2,240 | 3,658 | 4,000 |
| Ps. torr: | 760 | 711 | 670 | 611 | 610 | 578 | 483 | 462 |
| Air density: ($kg/m^3$) | 1.204 | 1.112 | 1.061 | 1.001 | 0.066 | 0.017 | 0.766 | 0.733 |
| 60 m | 7.00 | 6.97 | 6.94 | 6.92 | 6.90 | 6.88 | 6.82 | 6.81 |
| 100 m | 10.49 | 10.44 | 10.38 | 10.33 | 10.30 | 10.27 | 10.16 | 10.14 |
| 200 m | 21.34 | 21.20 | 21.06 | 20.95 | 20.87 | 20.78 | 20.56 | 20.53 |
| 400 m | 47.60 | 47.42 | 47.24 | 47.08 | 47.00 | 46.94 | 47.29 | 47.53 |
| 800 m | 1:53.28 | 1:53.75 | 1:54.11 | 1:54.53 | 1:54.94 | 1:55.72 | 2:01.64 | 2:04.17 |
| 1,500 m | 3:52.47 | 3:54.50 | 3:56.26 | 3:58.06 | 3:59.64 | 4:02.37 | 4:20.67 | 4:28.15 |
| 1,609 m | 4:15.80 | 4:18.17 | 4:20.22 | 4:22.33 | 4:24.16 | 4:27.33 | 4:48.43 | 4:57.05 |
| 5,000 m | 14:37.33 | 14:49.65 | 15:00.67 | 15:11.67 | 15:20.92 | 15:36.40 | 17:14.89 | 17:54.58 |
| 10,000 m | 30:13.74 | 30:40.99 | 31:05.49 | 31:29.74 | 31:49.95 | 32:23.46 | 35:53.04 | 37:16.63 |
| Marathon | 2:21:06.00 | 2:23:28.42 | 2:25:37.22 | 2:27:43.72 | 2:29:28.12 | 2:32:19.58 | 2:49:55.60 | 2:56:54.31 |

*Source:* Adapted from F. Peronnet, G. Thibault, and D. L. Cousineau. 1991. A theoretical analysis of the effect of altitude on running performance. J. Appl. Physiol.: Respirat. Environ. Exercise Physiol. 70(1):399–404.

These data show that the improvement or deterioration in performance with increasing altitude occurs in three categories.

**(1) Short sprinting distances.** For distances of 60 to 200 meters the contribution of aerobic metabolism to the total energy expenditure is less than 10% (133). Therefore, the negative effects of altitude on aerobic capacity have no effect on the energy available to the runner. On the other hand, the reduction in air density with increasing altitude has a positive effect. For this reason, the performances on short sprinting distances steadily improve with increase in altitude. The gain in average running speed at a moderate altitude of Mexico City (2240 m) for the 60, 100, and 200 m was about 1.4, 1.9, and 2.4% for men and 1.7, 2.2, and 2.7% for women. The greater improvement in running speed of women is due to the fact that the aerodynamic cost is less than for men due to their lower body mass.

**(2) 400 m.** The 400 m is run at a comparatively high speed (90% of the speed sustained over 200 m). Therefore, a reduction in air density is an advantage. For this reason, up to an altitude of 2400–2500 m the reduction in air density leads to an improvement in performance. Beyond this altitude the reduction of maximal aerobic capacity is large enough to counterbalance the reduction of air resistance, and the 400 m mark slowly deteriorates.

**(3) Middle and long distances.** For middle and long distance runs (>800 m) the aerobic metabolism to the total energy expenditure is quite high (75–99%). Consequently, the reduction in maximal aerobic capacity with increasing altitude has a strong effect on the amount of energy available to the runner. Accordingly, an elevation in altitude leads to a deterioration in performance, which increases 1) as BP decreases and 2) as the running distance increases. At the altitude of Mexico City, the reduction in the average speed sustained is 2% for 800 m, 4% for 1500 m, 6% for 5000 m, and 7% for the marathon, for both men and women.

## Training at High Altitude

Because the effects of altitude exposure produce cardiovascular changes similar to that of physical conditioning, several investigations have been conducted to determine the extent to which exposure to altitude improves work capacity at sea level. Earlier investigations reported that altitude training improved running performance and increased maximal aerobic capacity on return to sea level (118–120). Recent investigations showed that well-trained athletes, on return to sea level after training at about 3000 m, have an improved work capacity (121–122). On the other hand, other studies have found little or no effect (123–126), and any differences in work performance have been attributed to a significant training effect during the course of the experiments (125). In fact a recent study indicates that there is no major effect of hard endurance training at 2300 m over equivalent sea level training on aerobic power or on 2-mile run times at sea level in already well-trained middle-distance runners (127).

Thus, the hypothesis that training at high altitude has a potential effect on sea level performance appears inconclusive. This is not surprising when one considers that at high altitudes, especially above 4000 m, it is quite difficult for a sea-level subject to maintain and tolerate intensive training equivalent to that of sea level. Therefore any altitude potentiating effects are counterbalanced by the lower intensity of physical conditioning. An answer to this important question could be obtained by studying sea-level natives well acclimatized to high altitude, so intensive training could be maintained at both high altitude and sea level.

### Natural Acclimatization

Various studies pointed out the possibility that natives or long-term residents at high altitude would have an advantage over those ascending from sea level. The prediction that high-altitude natives would be at an advantage in long-distance events has been confirmed by the results of the Olympics. The first five places in the 10,000 m race were taken by athletes who were either born or acclimatized for several years at around 1500 m to 2500 m. First was Naftali Temu from Kenya (1500 m to 2000 m), second Mamowolde from Ethiopia (2000 to 2500 m), third Mohamed Gammoudi from Tunisia (1500 to 2000 m), fourth Juan Martinez from Mexico (2380 m), and fifth Nikolay Sviridov from Leminakan (1500 m) of the Soviet Union. However, the actual times were slower than those in previous Olympics. The first place was won with 29 minutes, 27.4 seconds, which is almost 2 minutes slower than Ron Clarke's world record, proving that moderate altitude does affect athletic performance even in those with previous acclimatization.

## OVERVIEW

Acclimatization to high-altitude hypoxia is a complex phenomenon that develops through the modification and synchronized interdependence of the respiratory, circulatory, and cardiovascular systems to improve oxygen delivery and oxygen utilization. As judged by evaluations of oxygen intake during maximal exercise, sea-level natives who have resided for as long as 2 years at high altitudes have acquired only a partial acclimatization to high altitude. In contrast, the high-altitude native or those raised at high altitude during growth and development have acquired a full acclimatization to high altitude.

The various adaptive mechanisms that enable both partially and fully acclimatized high-altitude residents to cope with the hypoxic stress of high altitudes are represented schematically in figure 11.15.

In both partial and full acclimatization, adaptation to the low availability of oxygen results in the operation of a variety of coordinated mechanisms oriented toward increasing the oxygen supply and oxygen delivery to the tissue cells (fig. 11.15). However, the physiological mechanisms used by those who attain complete functional acclimatization are different from those who are partially acclimatized.

The partially acclimatized rely on exaggerated pulmonary ventilation and an increased oxygen-carrying capacity of the blood resulting from a high polycythemic response which sometimes may become excessive. But these responses are not sufficient, so that the rate at which oxygen diffuses across the pulmonary membranes and into the capillary blood is low, and consequently the organism's ability to meet the oxygen requirements during maximal exercise is impaired. Hence, the sea-level native cannot attain the same aerobic work capacity as that attained at sea level.

The fully acclimatized rely on a moderate increase in pulmonary ventilation and moderate polycythemic response, yet are able to get an aerobic work capacity comparable to that attained at sea level. The low dependence on hyperventilation and moderate polycythemic response would suggest that in full acclimatization the diffusion of oxygen from the blood to the tissues is increased. However, at present the physiological mechanisms whereby this occurs are not known.

The available information suggests that full acclimatization to high altitude is acquired through growth and development in a hypoxic environment. The hypothesis that developmental adaptations account for the greater part of the differences in physiological performance and morphology between fully and partially acclimatized residents living at high altitudes is supported by the evidence. It is also based on the well-grounded assumption that the respective contributions of genetic and environmental factors vary with the developmental stage of the organism. In general, the earlier the developmental stage, the greater the influence of the environment. Furthermore, studies on chemoreceptor sensitivity indicate that the hyposensitivity that characterizes the high-altitude native is acquired during growth and development (29, 59). From these studies emerges the conclusion that differences between highland and sea-level natives in physiological performance and morphology of the organs of the oxygen transport system are caused in part by adaptations acquired during the developmental period. This does not mean, however, that all physiological characteristics exhibited by the high-altitude native are due to developmental adaptive responses. For example, it is now generally accepted that the attainment of low systemic blood pressure at high altitude

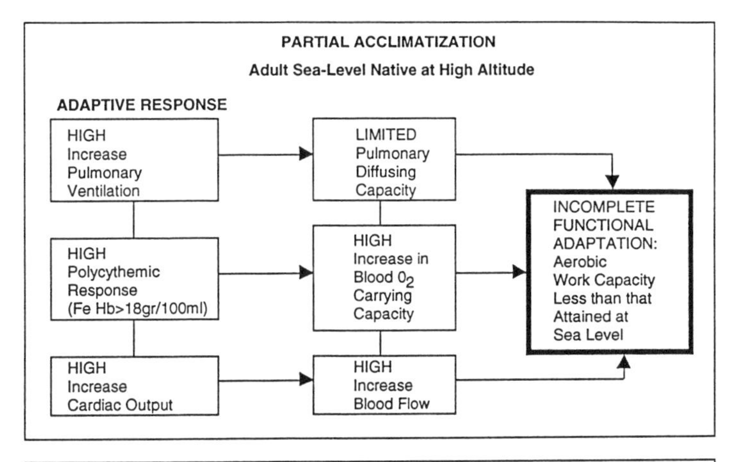


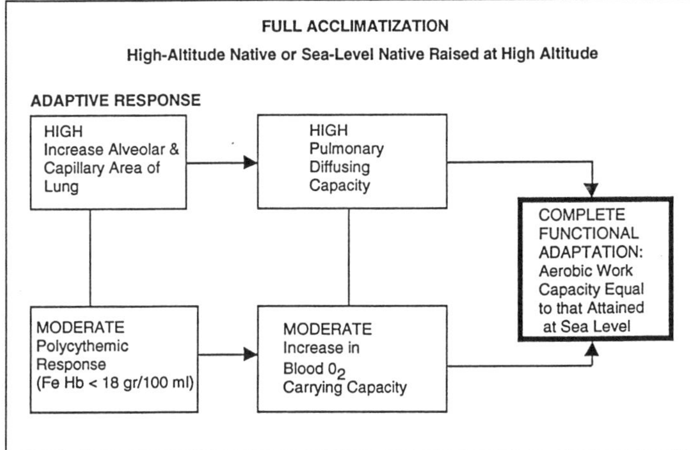


**Fig. 11.15.** Schematization of the functional adaptation associated with partial and full acclimatization to high altitude.

does not depend on developmental factors because it can be acquired by long-term residency at high altitude (115). Furthermore, we do not know the developmental modifications that occur within each component of the oxygen transport system (such as ventilation, pulmonary diffusion, and oxygen delivery to the tissues) that enable a sea-level native to attain a complete functional adaptation to high altitude. Similarly, it is not known whether other sea-level populations who do not have indigenous admixture such as the Peruvian populations can acquire a full functional adaptation. Future research must be addressed to answering these questions, for they have importance to all areas concerned with the organism's coping with low oxygen availability.

During growth and development environmental factors are constantly conditioning and modifying the expression of inherited potentials. The environmental influences felt by the organism depend on the type of stress imposed and especially on the age at which the individual is subjected to the stress. Hence, the respective contributions of genetic and environmental factors vary with the developmental stage of the organism; in general, the earlier the age, the greater the environmental influence. Thus, it would be surprising if developmental processes did not influence the functional performance and morphology of the high-altitude native. However, the principle of developmental sensitivity and plasticity does not necessarily imply greater adaptive responses in all biological parameters, which may depend on the developmental stage of the organism, the type of organism, and the particular functional process that is affected. Homeostatic interdependence does not imply uniform correlations for all physiological and morphological variables. The finding that acclimatization to high altitude in humans and experimental animals results in an enhancement of respiratory volumes and pulmonary diffusing capacity without the direct participation of chest dimensions, with which one would expect them to be associated, is a good example of the organism's functional plasticity and independent response of morphological traits. In other words, successful acclimatization to high altitudes depends primarily on the adjustments and synchronized interdependence of environmentally modifiable functions that relate to oxygen transport such as the respiratory and circulatory functions and less on morphological traits such as skeletal dimensions that are affected more by ethnicity and energy availability than by high-altitude hypoxia.

The reduction in air density of altitude, caused by a decrease in barometric pressure, is an advantage for the runner, because it decreases aerodynamic resistance, thus the energy cost of running at a given speed. This advantage is more evident in short sprinting distances (less than 200 m), where the aerodynamic resistance is greater. On the other hand, because high-altitude hypoxia reduces the maximal aerobic capacity, for runs over distances of 400 m an elevation in altitude leads to a deterioration in performance. The optimal altitude estimated to improve the record is comparatively low (1130 m vs. 2400–2500 m).

Athletic performance at sea level does not appear enhanced by acclimatization to high altitude. On the other hand, the advantages of acclimatization to moderate altitude were evident in the middle and long distance runs of the 1968 Olympics held in Mexico in which highland athletes excelled.

## References

1. Grover, R. F., J. T. Reeves, E. B. Grover, and J. S. Leathers. 1967. Muscular exercise in young men native to 3,100 m altitude. J. Appl. Physiol. 22:555–64.
2. Torrance, J. D., C. Lenfant, J. Cruz, and E. Marticorena. 1970–1971. Oxygen transport mechanisms in residents at high altitude. Respir. Physiol. 11:1–15.
3. Lenfant, C., J. D. Torrance, E. English, C. A. Finch, C. Reynafarje, J. Ramos,

and J. Faura. 1968. Effect of altitude on oxygen binding by hemoglobin and on organic phosphate levels. J. Clin. Invest. 47:2652–56.
4. Lenfant, C., and K. Sullivan. 1971. Adaptation to high altitude. N. Engl. J. Med. 284:1298–1309.
5. West, J. B. 1984. Human physiology at extreme altitudes on Mount Everest. Science 223:784–88.
6. Schoene, R. B., S. Lahiri, P. H. Hackett, R. M. Peters, Jr., J. S. Milledge, C. J. Pizzo, F. H. Sarnquist, S. J. Boyer, D. J. Graber, K. H. Maret, and J. B. West. 1984. Relationship of hypoxic ventilatory response to exercise performance on Mount Everest. J. Appl. Physiol. 56:1478–83.
7. Winslow, R. M., M. Samaja, and J. B. West. 1984. Red cell function at extreme altitude on Mount Everest. J. Appl. Physiol. 56:109–16.
8. Aste-Salazar, H., and A. Hurtado. 1944. The affinity of hemoglobin for oxygen at sea level and at high altitudes. Am. J. Physiol. 142:733–43.
9. Brewer, G. J., J. W. Eaton, J. V. Weil, and R. F. Grover. 1970. Studies of red cell glycolysis and interactions with carbon monoxide, smoking and altitude. In G. Brewer, ed., Red cell metabolism and function. New York, NY: Plenum Press.
10. Rorth, M., S. Nygaard, and H. Parving. 1972. Effects of exposure to simulated high altitude on human red cell phosphates and oxygen affinity of hemoglobin. Influences of exercise. Scand. J. Clin. Lab. Invest. 29:329–33.
11. Moore, L.G. 1973. Red blood cell adaptation to high altitude: mechanism of the 2,3 diphosphoglycerate response. Ph.D. Thesis. Ann Arbor, MI: University of Michigan.
12. Moore, L. G., G. Brewer, and F. Oelshlegel. 1972. Red cell metabolic changes in acute and chronic exposure to high altitude. In G. Brewer, ed., Hemoglobin and red cell structure and function. New York, NY: Plenum Press.
13. Lenfant, C., J. D. Torrance, and C. Reynafarje. 1971. Shift of the $O_2$-Hb dissociation curve at altitude: mechanisms and effect. J. Appl. Physiol. 30:625–31.
14. Banchero, N., and R. F. Grover. 1972. Effects of different levels of simulated altitude on $O_2$ transport in llama and sheep. Am. J. Physiol. 222:1239–45.
15. Monge, C. M., and C. Monge. 1968. Adaptation to high altitude. In E. S. E. Hafez, ed., Adaptation of domestic animals. Philadelphia, PA: Lea and Febiger.
16. Eaton, J. W., T. D. Skelton, and E. Berger. 1974. Survival at extreme altitude: protective effect of increased hemoglobin-oxygen affinity. Science 183:743–44.
17. Tenney, S. M., H. Rahn, R. C. Stroud, and J. C. Mithorfer. 1952. Adaptation to high altitude: changes in lung volumes during the first seven days at Mt. Evans, Colorado. J. Appl. Physiol. 5:607–13.
18. Rahn, H., and D. Hammond. 1952. The vital capacity at reduced barometric pressure. J. Appl. Physiol. 4:715–22.
19. Ulvedal, F., T. E. Morgan, Jr., R. G. Cutler, and B. E. Welch. 1963. Ventilatory capacity during prolonged exposure to simulated altitude without hypoxia. J. Appl. Physiol. 18:904–8.
20. Consolazio, C. F., H. G. Johnson, L. O. Matoush, R. A. Nelson, and G. J. Isaac. 1967. Respiratory function in normal young adults at 3475 and 4300 meters. Rep. 300. Denver, CO: U.S. Army Medical Research and Nutrition Laboratory, Fitzsimons General Hospital.
21. Shields, J. L., J. P. Hannon, C. W. Harris, and W. S. Platner. 1968. Effects of altitude acclimatization on pulmonary function in women. J. Appl. Physiol. 25:606–9.

22. Kreuzer, F., and P. Van Lookeren Campagne. 1965. Resting pulmonary diffusing capacity for CO and $O_2$ at high altitude. J. Appl. Physiol. 20:519–24.
23. DeGraff, A. C., Jr., R. F. Grover, J. W. Hammond, Jr., J. M. Miller, and R. L. Johnson, Jr. 1965. Pulmonary diffusing capacity in persons native to high altitude. Clin. Res. 13:74.
24. DeGraff, A. C., Jr., R. F. Grover, R. L. Johnson, J. W. Hammond, and J. M. Miller. 1970. Diffusing capacity of the lung in Caucasians native to 3,100 m. J. Appl. Physiol. 29:71–76.
25. Abbrecht, P. H., and J. K. Littell. 1972. Plasma erythropoietin in men and mice during acclimatization to different altitudes. J. Appl. Physiol. 32:54–58.
26. Reynafarje, C. 1957. The influence of high altitude on erythropoietic activity. Brookhaven Symp. Biol. 10:132–46.
27. Reynafarje, C. 1959. Bone-marrow studies in the newborn infant at high altitudes. J. Pediatr. 54:152–67.
28. Severinghaus, J. W., C. R. Bainton, and A. Carcelen. 1966. Respiratory insensitivity to hypoxia in chronically hypoxic man. Respir. Physiol. 1:308–34.
29. Forster, H. V., J. A. Dempsey, M. L. Birnbaum, W. G. Reddan, J. Thoden, R. F. Grover, and J. Rankin. 1971. Effect of chronic exposure to hypoxia on ventilatory response to $CO_2$ and hypoxia. J. Appl. Physiol. 31:586–92.
30. Chiodi, H. 1957. Respiratory adaptation to chronic high altitude hypoxia. J. Appl. Physiol. 10:81–87.
31. Lefrançois, R., H. Gautier, and P. Pasquis. 1968. Ventilatory oxygen drive in acute and chronic hypoxia. Respir. Physiol. 4:217–28.
32. Velásquez, T., C. Martinez, W. Pezzia, and N. Gallardo. 1968. Ventilatory effects of oxygen in high altitude natives. Respir. Physiol. 5:211–20.
33. Stenberg, J., B. Ekblom, and R. Messin. 1966. Hemodynamic response to work at simulated altitude, 4,000 m. J. Appl. Physiol. 21:1589–94.
34. Vogel, J. A., J. E. Hansen, and C. W. Harris. 1967. Cardiovascular responses in man during exhaustive work at sea level and high altitudes. J. Appl. Physiol. 23:531–39.
35. Klausen, K. 1966. Cardiac output in man in rest and work during and after acclimatization to 3,800 m. J. Appl. Physiol. 21:609–16.
36. Saltin, B., and G. Grimby. 1968. Physiological analysis of middle-aged and old former athletes: comparison with still active athletes of the same ages. Circulation 38:1104–15.
37. Hartley, L. H., J. Alexander, M. Modelski, and R. F. Grover. 1967. Subnormal cardiac output at rest and during exercise in residents at 3,100 m altitude. J. Appl. Physiol. 23:839–48.
38. Hoon, R. S., V. Balasubramanian, O. P. Mathew, S. C. Tiwari, S. S. Sharma, and K. S. Chadha. 1977. Effect of high-altitude exposure for 10 days on stroke volume and cardiac output. J. Appl. Physiol. 42:722–27.
39. Banchero, N., F. Sime, D. Peñaloza, J. Cruz, R. Gamboa, and E. Marticorena. 1966. Pulmonary pressure, cardiac output, and arterial oxygen saturation during exercise at high altitude and at sea level. Circulation 33:249–62.
40. Heath, D., and D. R. Williams. 1977. Man at high altitude. The pathophysiology of acclimatization and adaptation. Edinburgh: Churchill Livingstone.
41. Becker, E. L., R. G. Cooper, and G. D. Hathaway. 1955. Capillary vascularization in puppies born at simulated altitude of 20,000 feet. J. Appl. Physiol. 8:166–68.
42. Valdivia, E., M. Watson, and C. M. Dass. 1960. Histologic alterations in muscles of guinea pigs during chronic hypoxia. Arch. Pathol. 69:199–208.

43. Cassin, S., R. D. Gilbert, and E. M. Johnson. 1966. Capillary development during exposure to chronic hypoxia. SAM-TR-66-16. Brooks Air Force Base, TX: U.S. Air Force School of Aerospace Medicine.
44. Tenney, S. M., and L. C. Ou. 1970. Physiological evidence for increased tissue capillarity in rats acclimatized to high altitude. Respir. Physiol. 8:137–50.
45. Buskirk, E. R. 1976. Work performance of newcomers to the Peruvian highlands. In P. T. Baker and M. A. Little, eds., Man in the Andes: a multidisciplinary study of high-altitude Quechua natives. Stroudsburg, PA: Dowden, Hutchinson, and Ross.
46. Kollias, J., E. R. Buskirk, R. F. Akers, E. K. Prokop, P. T. Baker, and E. Piconreategui. 1968. Work capacity of long-time residents and newcomers to altitude. J. Appl. Physiol. 24:792–99.
47. Mazess, R. B. 1969. Exercise performance of Indian and white high altitude residents. Hum. Biol. 41:494–518.
48. Hurtado, A. 1964. Animals in high altitudes: resident man. In D. B. Dill, E. F. Adolph, and C. G. Wilber, eds., Handbook of physiology, vol. 4. Adaptation to the environment. Washington, DC: American Physiological Society.
49. Hurtado, A., T. Velasquez, C. Reynafarje, R. Lozano, R. Chavez, H. Aste-Salazar, B. Reynafarje, C. Sanchez, and J. Muñoz. 1956. Mechanisms of natural acclimatization: studies on the native resident of Morococha, Peru, at an altitude of 14,000 feet. Rep. 56-1. Randolph Field, TX: U.S. Air Force School of Aviation Medicine.
50. Lahiri, S., and J. S. Milledge. 1965. Sherpa physiology. Nature 207:610–12.
51. Lahiri, S., J. S. Milledge, H. P. Chattopadhyay, A. K. Bhattacharyya, and A. K. Sinha. 1967. Respiration and heart rate of Sherpa highlanders during exercise. J. Appl. Physiol. 23:545–54.
52. Mazess, R. B. 1969. Exercise performance at high altitude (4000 meters) in Peru. Fed. Proc. 28:1301–6.
53. Frisancho, A. R., C. Martinez, T. Velásquez, J. Sanchez, and H. Montoye. 1973. Influence of developmental adaptation on aerobic capacity at high altitude. J. Appl. Physiol. 34:176–80.
54. Velásquez, T. 1966. Acquired acclimatization to sea level. In Life at high altitudes. Pub. 140. Washington, DC: Pan American Health Organization.
55. Velásquez, T. 1970. Aspects of physical activity in high altitude natives. Am. J. Phys. Anthropol. 32:251–58.
56. Velásquez, T., and B. Reynafarje. 1966. Metabolic and physiological aspects of exercise at high altitude. Part 2. Response of natives to different levels of workload breathing air and various oxygen mixtures. Fed. Proc. 25:1400–1402.
57. Lahiri, S., F. F. Kao, T. Velásquez, C. Martinez, and W. Pezzia. 1969. Irreversible blunted respiratory sensitivity to hypoxia in high altitude natives. Respir. Physiol. 6:360–74.
58. Lahiri, S., F. F. Kao, T. Velásquez, C. Martinez, and W. Pezzia. 1970. Respiration of man during exercise at high altitude: highlander vs. lowlander. Respir. Physiol. 8:361–75.
59. Lahiri, S., R. G. Delaney, J. S. Brody, M. Simpser, T. Velásquez, E. K. Motoyama, and C. Polgar. 1976. Relative role of environmental and genetic factors in respiratory adaptation to high altitude. Nature 261:133–35.
60. Byrne-Quinn, E., I. E. Sodal, and J. V. Weil. 1972. Hypoxic and hypercapnic ventilatory drives in children native to high altitude. J. Appl. Physiol. 32:44–46.

61. Weil, J. V., E. Byrne-Quinn, I. E. Sodal, G. F. Filley, and R. F. Grover. 1971. Acquired attenuation of chemoreceptor function in chronically hypoxic man at high altitude. J. Clin. Invest. 50:186–95.
62. Frisancho, A. R., T. Velásquez, and J. Sanchez. 1973. Influences of developmental adaptation on lung function at high altitude. Hum. Biol. 45:583–94.
63. Harrison, G. A., C. F. Kuchemann, M. A. S. Moore, A. J. Boyce, T. Baju, A. E. Mourant, M. J. Godber, B. G. Glasgow, A. C. Kopec, D. Tills, and E. J. Clegg. 1969. The effects of altitudinal variation in Ethiopian populations. Philos. Trans. R. Soc. Lond. 256:147–82.
64. Frisancho, A. R. 1976. Growth and functional development at high altitude. In P. T. Baker and M. A. Little, eds., Man in the Andes: a multidisciplinary study of high-altitude Quechua natives. Stroudsburg, Pa.: Dowden, Hutchinson and Ross.
65. Frisancho, A. R. 1975. Functional adaptation to high altitude hypoxia. Science 187:313–19.
66. Frisancho, A. R. 1978. Human growth and development among high-altitude populations. In P. T. Baker, ed., The biology of high-altitude peoples. New York, NY: Cambridge University Press.
67. Brody, J. S., S. Lahiri, M. Simpser, E. K. Motoyama, and T. Velásquez. 1977. Lung elasticity and airway dynamics in Peruvian natives to high altitude. J. Appl. Physiol. 42:245–51.
68. Burri, P. H., and E. R. Weibel. 1971. Morphometric estimation of pulmonary diffusion capacity, Part 2. Effect of $PO_2$ on the growing lung. Adaptation of the growing rat lung to hypoxia and hyperhypoxia. Respir. Physiol. 11:247–64.
69. Bartlett, D. 1972. Postnatal development of the mammalian lung. In R. Goss, ed., Regulation of organ and tissue growth. New York, NY: Academic Press.
70. Burri, P. H., and E. R. Weibel. 1971. Morphometric evaluation of changes in lung structure due to high altitudes. In R. Porter and J. Knight, eds., High altitude physiology. Cardiac and respiratory aspects. Edinburgh: Churchill Livingstone.
71. Cunningham, E. L., J. S. Brody, and B. P. Jain. 1974. Lung growth induced by hypoxia. J. Appl. Physiol. 37:362–66.
72. Bartlett, D., Jr., and J. E. Remmers. 1971. Effects of high altitude exposure on the lungs of young rats. Respir. Physiol. 13:116–25.
73. Velásquez, T. 1956. Maximal diffusing capacity of the lungs at high altitudes. Rep. 56-108. Randolph Field, TX: U.S. Air Force School of Aviation Medicine.
74. Velásquez, T., and E. Florentini. 1966. Maxima capacidad de difución del pulmon en nativos de la altura. Arch. Inst. Biol. Andina 1:179–87.
75. Remmers, J. E., and J. C. Mithoefer. 1969. The carbon monoxide diffusing capacity in permanent residents at high altitudes. Respir. Physiol. 6:233–44.
76. Guleria, J. S., J. N. Pande, P. K. Sethi, and S. B. Roy. 1971. Pulmonary diffusing capacity at high altitude. J. Appl. Physiol. 31:536–43.
77. Grover, R. F., J. T. Reeves, D. H. Will, and S. G. Blount. 1963. Pulmonary vasoconstriction in steers at high altitudes. J. Appl. Physiol. 18:567–70.
78. Merino, C. F. 1950. Blood formation and blood destruction in polycythemia of high altitude. Blood 5:1–31.
79. Bharadwaj, N., A. P. Singh, and M. S. Malhotra. 1973. Body composition of the high altitude natives of Ladakh. A comparison with sea level residents. Hum. Biol. 45:423–34.

80. Garruto, R. M. 1976. Hematology. In P. T. Baker and M. A. Little, eds., Man in the Andes: a multidisciplinary study of high-altitude Quechua natives. Stroudsburg, PA: Dowden, Hutchinson and Ross.
81. Quillici, J. C., and H. Vergnes. 1978. The hematological characteristics of high-altitude peoples. In P. T. Baker, ed., The biology of high-altitude peoples. New York, NY: Cambridge University Press.
82. Anthony, A., E. Ackerman, and G. K. Strother. 1959. Effects of altitude acclimatization on rat myoglobin: changes in myoglobin of skeletal and cardiac muscle. Am. J. Physiol. 196:512–16.
83. Beall, C. M., and A. B. Reichsman. 1984. Hemoglobin levels in a Himalayan high altitude population. Am. J. Phys. Anthropol. 63:301–6.
84. Beall, C. M., and M. C. Goldstein. 1990. Hemoglobin concentration, percent oxygen saturation and arterial oxygen content of Tibetan nomads at 4,850 to 5,450 m. In J. R. Sutton, G. Coates, and J. E. Remmers, eds., Hypoxia: the adaptations. Philadelphia, PA: B. C. Decker.
85. Tufts, D. A., J. D. Haas, J. L. Beard, H. Spielvogel. 1985. Distribution of hemoglobin and functional consequences of anemia in adult males at high altitude. Am. J. Clin. Nutr. 42:1–11.
86. Beall, C. M., G. M. Brittenham, F. Macuaga, and M. Barragan. 1990. Variation in hemoglobin concentration among samples of high-altitude natives in the Andes and the Himalayas. Am. J. Hum. Biol. 2:639–51.
87. Santolaya, R. B., S. Lahiri, R. T. Alfaro, and R. B. Schoene. 1989. Respiratory adaptation in the highest inhabitants and highest Sherpa mountaineers. Resp. Physiol. 77:253–62.
88. Rahn, H. 1966. Introduction to the study of man at high altitudes: conductance of $O_2$ from the environment to the tissues. In Life at high altitudes. Pub. 140. Washington, DC: Pan American Health Organization.
89. Hurtado, A., A. Rotts, C. Merino, and J. Pons. 1937. Studies of myoglobin at high altitude. Am. J. Med. Sci. 194:708–13.
90. Tappan, D. V., and B. Reynafarje. 1957. Tissue pigment manifestations of adaptation to high altitude. Am. J. Physiol. 190:99–103.
91. Reynafarje, B. 1962. Myoglobin content and enzymatic activity of muscle and altitude adaptation. J. Appl. Physiol. 17:301–5.
92. Wittenberg, J. B. 1965. Myoglobin-facilitated diffusions of oxygen. J. Gen. Physiol. 49 (1):57–74.
93. Scholander, P. F. 1960. Oxygen transport through hemoglobin solutions. Science 131:585–90.
94. Reynafarje, B. 1966. Physiological patterns: enzymatic changes. In Life at high altitudes. Pub. 140. Washington, DC: Pan American Health Organization.
95. Reynafarje, B., L. Loyola, R. Cheesman, E. Marticorena, and S. Jimenez. 1969. Fructose metabolism in sea-level and high-altitude natives. Am. J. Physiol. 216:1542–47.
96. Reynafarje, B., and T. Velásquez. 1966. Metabolic and physiological aspects of exercise at high altitude. Part 1. Kinetics of blood lactate, oxygen consumption and oxygen debt during exercise and recovery breathing air. Fed. Proc. 25: 1397–99.
97. Reynafarje, B. 1971. Mecanismos moleculares en la adaptación a la hypoxia de las grandes alturas. Thesis Doctoral en Medicina. Lima, Peru: Universidad Nacional Mayor de San Marcos.

98. Peñaloza, D., F. Sime, N. Banchero, and R. Gamboa. 1962. Pulmonary hypertension in healthy men born and living at high altitude. Med. Thorac. 19: 449–60.
99. Sime, F., N. Banchero, D. Peñaloza, R. Gamboa, J. Cruz, and E. Marticorena. 1963. Pulmonary hypertension in children born and living at high altitudes. Am. J. Cardiol. 11:143–49.
100. Peñaloza, D., F. Sime, N. Banchero, R. Gamboa, J. Cruz, and E. Marticorena. 1963. Pulmonary hypertension in healthy men born and living at high altitudes. Am. J. Cardiol. 11:150–57.
101. Saldana, M., and J. Arias-Stella. 1963. Studies on the structure of the pulmonary trunk. Part 2. The evolution of the elastic configuration of the pulmonary trunk in people native to high altitudes. Circulation 27:1094–1100.
102. Arias-Stella, J., and S. Recavarren. 1962. Right ventricular hypertrophy in native children living at high altitude. Am. J. Pathol. 41:55–59.
103. Rotta, A., A. Canepa, A. Hurtado, T. Velásquez, and R. Chavez. 1956. Pulmonary circulation at sea level and at high altitudes. J. Appl. Physiol. 9:328–32.
104. Peñaloza, D., R. Gamboa, J. Dyer, M. Echevarria, and E. Marticorena. 1960. The influence of high altitudes on the electrical activity of the heart. Part 1. Electrocardiographic and vectorcardiographic observations in the newborn, infants and children. Am. Heart J. 59:111–28.
105. Peñaloza, D., R. Gamboa, J. Dyer, M. Echevarria, and E. Marticorena. 1961. The influence of high altitudes on the electrical activity of the heart. Electrocardiographic and vectorcardiograhic observations in adolescence and adulthood. Am. Heart J. 61:101–7.
106. Peñaloza, D., J. Arias-Stella, F. Sime, S. Recavarren, and E. Marticorena. 1964. The heart and pulmonary circulation in children at high altitude. Physiological anatomical and clinical observations. Pediatrics 34:568–82.
107. Marticorena, E., D. Peñaloza, J. Severino, and K. Hellriegel. 1962. Incidencia de la persistencia del conducto arterioso en las grandes alturas. Memorias del IV Congreso Mundial de Cardiologia, Mexico. I-A:155–59.
108. Alzamora-Castro, V., G. Battilana, R. Abugatas, and S. Sialer. 1960. Patent ductus arteriosus and high altitude. Am. J. Cardiol. 5:761–65.
109. Heath, D., C. Edwards, M. Winson, and P. Smith. 1973. Effects on the right ventricle, pulmonary vasculature, and carotid bodies of the rate of exposure to, and recovery from, simulated high altitude. Thorax 28:24–28.
110. Monge, M. C., and C. Monge. 1966. High-altitude diseases: mechanism and management. Springfield, IL: Charles C. Thomas, Publisher.
111. Ruiz, L. 1973. Epidemiologia de la hipertencion arterial y de la cardiopatia isquemica en las grandes alturas. Thesis Doctoral en Medicina. Lima, Peru: Universidad Peruana Cayetano Heredia.
112. Elsner, R. W., A. Bolstad, and C. Forno. 1964. Maximum oxygen consumption of Peruvian Indians native to high altitude. In W. H. Weihe, ed., The physiological effects of high altitude. Oxford: Pergamon Press.
113. Baker, P. T. 1976. Work performance of highland natives. In P. T. Baker and M. A. Little, eds., Man in the Andes: a multidisciplinary study of high-altitude Quechua natives. Stroudsburg, PA: Dowden, Hutchinson and Ross.
114. Way, A. B. 1976. Exercise capacity of high altitude Peruvian Quechua Indians migrant to low altitude. Hum. Biol. 48:175–91.
115. Marticorena, E., L. Ruiz, J. Severino, J. Galvez, and D. Peñaloza. 1969. Systemic

blood pressure in white men born at sea level: changes after long residence at high altitudes. Am. J. Cardiol. 23:364–68.
116. Moore, L. G., J. G. Zhuang, R. G. McCullough, A. Cymerman, T. S. Droma, S. F. Sun, Y. Ping, and R. E. McCullough. 1991. Increased lung volumes in Tibetan high altitude residents. Am. J. Phys. Anthropol. 12 (abstract): 134.
117. Droma, T. S., R. G. McCullough, R. E. McCullough, J. G. Zhuang, A. Cymerman, S. F. Sun, J. R. Sutton, G. Rapmund, and L. G. Moore. 1991. Increased vital and lung capacities in Tibetan compared to Hans residents of Lhasa (3,658 m). Am. J. Phys. Anthropol. 86:341–51.
118. Balke, B. 1964. Work capacity and its limiting factors at high altitude. New York, NY: Pergamon Press.
119. Balke, B., F. J. Nagle, and J. T. Daniels. 1965. Altitude and maximum performance in work and sports activity. JAMA 194:646–49.
120. Faulkner, J. A., J. T. Daniels, and B. Balke. 1967. Effects of training at moderate altitude on physical performance capacity. J. Appl. Physiol. 23:85–89.
121. Daniels, J. T., and N. Oldridge. 1970. The effects of alternate exposure to altitude and sea level on world-class middle-distance runners. Med. Sci. Sports 2:107–12.
122. Dill, D. B., and W. C. Adams. 1971. Maximal oxygen uptake at sea level and at 3090 m altitude in high school champion runners. J. Appl. Physiol. 30:854–59.
123. Reeves, J. T., R. F. Grover, and J. E. Cohn. 1967. Regulation of ventilation during exercise at 10,200 ft. in athletes born at low altitude. J. Appl. Physiol. 22: 546–54.
124. Saltin, B. 1967. Aerobic and anaerobic work capacity at an altitude of 2,250 meters. In R. F. Goddard, ed., The effects of altitude on physical performance. Chicago, IL: Athletic Institute.
125. Faulkner, J. A., J. Kollias, C. B. Favour, E. R. Buskirk, and B. Balke. 1968. Maximum aerobic capacity and running performance at altitude. J. Appl. Physiol. 24:685–91.
126. Buskirk, E. R., J. Kollias, R. F. Akers, E. K. Prokop, and E. P. Reategui. 1967. Maximal performance at altitude and on return from altitude in conditioned runners. J. Appl. Physiol. 23:259–66.
127. Adams, W. C., E. M. Bernauer, D. B. Dill, and J. B. Bomar, Jr. 1975. Effects of equivalent sea level and altitude training on $VO_2$ max and running performance. J. Appl. Physiol. 39:262–66.
128. Jokl, E., and P. Jokl. 1968. The effect of altitude on athletic performance. Exercise and altitude. Baltimore, MD: University Park Press.
129. Faulkner, J. A. 1967. Training for maximum performance at altitude. In R. F. Goddard, ed., The effects of altitude and athletic performance. Chicago, IL: Athletic Institute.
130. Faulkner, J. A. 1971. Maximum exercise at medium altitude. In R. J. Shephard, ed., Frontier of fitness. Springfield, IL: Charles C. Thomas, Publisher.
131. Craig, A. B. 1969. Olympic 1968: a post mortem. Med. Sci. Sports 1:177–83.
132. Sun, S. F., T. S. Droma, J. G. Zhuang, J. X. Tao, S. Y. Huang, R. G. McCullough, R. E. McCullough, C. S. Reeves, J. T. Reeves, and L. G. Moore. 1990. Greater maximal $O_2$ uptakes and vital capacities in Tibetan than Han residents of Lhasa. Respiration Physiology 79:151–62.
133. Peronnet, F., G. Thibault, and D. L. Cousineau. 1991. A theoretical analysis of the effect of altitude on running performance. J. Appl. Physiol.: Respirat. Environ. Exercise Physiol. 70(1):399–404.

134. Kayser, B., H. Hoppeler, H. Claassen, and P. Cerretelli. 1991. Muscle structure and performance capacity of Himalayan Sherpas. J. Appl. Physiol.: Respirat. Environ. Exercise Physiol. 70(5):1938–42.
135. Hochachka, P. W., C. Stanley, G. O. Matheson, D. C. McKenzie, P. S. Allen, and W. S. Parkhouse. 1991. Metabolic and work efficiencies during exercise in Andean natives. J. Appl. Physiol.: Respirat. Environ. Exercise Physiol. 70(4):1720–30.
136. Frisancho, A. R., H. G. Frisancho, R. Albalak, M. Villena, and E. Vargas. 1993. Lung volumes of natives and Bolivians of non-native ancestry living at high altitude. Am. J. Phys. Anthropol. 16 (abstract):89.

# CHAPTER 12 Prenatal and Postnatal Growth and Development at High Altitude

From conception an individual's growth and development depend on the interaction of genetic and internal and external environmental conditions. It is a basic principle that the environment provides external factors that make development possible and permit the expression of genetic potentials. Research with animals and humans has demonstrated that high-altitude hypoxia directly or indirectly affects growth and development. In this chapter the results of studies of prenatal and postnatal human growth conducted in the South American Andes, Ethiopian highlands, Himalayas, India, and Tien Shan Mountains of the Soviet Union will be summarized along with the results of experimental studies on animals. The objective is to derive an overview of the mechanisms whereby a high-altitude environment influences human growth and development. Because studies of European children raised at high altitude have proved to be useful for elucidating the role of ethnicity and environmental factors on growth at high altitude, where applicable the data will be presented separately for high-altitude natives and Europeans residing at high altitude.

## PRENATAL GROWTH AND DEVELOPMENT

### Placenta

Even at sea level the fetus develops in a hypoxic environment (1) and, therefore, at high altitude is subject to an even greater hypoxic stress than at sea level. For example, at sea level the intrauterine umbilical arterial oxygen tension is about 20 mm Hg, which corresponds to an atmospheric oxygen tension of 61 mm Hg found at an altitude of 7500 m (2). For this reason, and as shown by experimental studies (3–5) and observations on humans residing at high altitudes (6–13), the prenatal structures and especially the placenta are modified by a high-altitude environment. This modification results in (1) increased surface area available for diffusion of oxygen and transfer between maternal and fetal blood and (2) decreased resistance of the placental barrier to the transfer of oxygen.

Human studies indicate that the frequency of "irregular shape," rather than the usual round or oval placentas, is three times greater at high altitude than at sea level (6–10). Furthermore, the average weight of the placenta is between 10% and 15% greater than the average weight at sea level (6–11). Therefore, as illustrated in figure 12.1, the placental weight/birth weight ratio at high altitude is greater (0.16 to 0.21) than at sea level (0.12 to 0.15). The placentas at high altitude are also thinner, and they contain greater amounts of cord hemoglobin than those at sea level (6, 10–15). In view of the fact that placental surface area is highly correlated with placental capillary area and

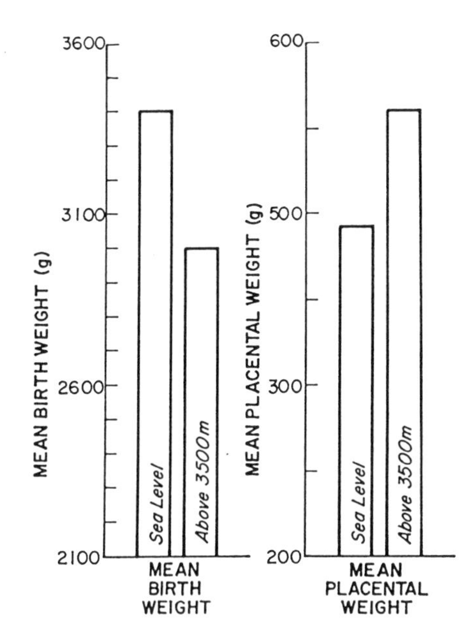


**Fig. 12.1.** Comparison of birth weight and placental weight of sea-level and high-altitude natives. Residence at high altitude is associated with lower birth weight but increased weight of placenta. (Based on data from H. Kruger and J. Arias-Stella. 1970. Am. J. Obstet. Gynecol. 106:586–91.)

with the capacity of the placenta to nourish and oxygenate the fetus (16), the observed morphological characteristics of the placentas of high-altitude populations reflect a compromise between increasing the placental area and diminishing the placental barrier for oxygen transfer (10). Through these modifications, despite a reduction of approximately one half in the oxygen pressure gradient between maternal and fetal bloods, the rate at which oxygen reaches the fetal blood per kilogram of tissue supplied approaches sea-level values (3).

## Birth Weight

Studies in the U.S. and South America indicate that babies born at high altitude on the average weigh less than those born at low altitudes (fig. 12.1). The mean birth weight for infants born to indigenous mothers at 3600 m in La Paz, Bolivia was 3165 grams, while for infants born at low altitude was 3427 grams (17). In Colorado, the birth weight for infants born at 3100 m averaged 3126 grams and 3301 grams for infants born at 1600 m (19). The frequency of low birth weight infants is also increased at high altitude. For example, in Leadville, Colorado, at 3050 m, the frequency of birth weights less than 2500 g ranges between 20% and 31% compared to an average of

10% for sea-level populations (20). Comparing these findings with those found in Peru at equivalent altitudes, it becomes evident that the frequency of birth weight depression in Peru is less severe than in the United States. For example, the frequency of low birth weight infants at 3400 m in the city of Cusco is 10%, which is less than half the frequency reported for comparable altitudes in the United States. This difference is probably related to differences in sampling and in degree of acclimatization between United States populations and native high-altitude populations. Similarly, the frequency of low birth weight infants at 3400 m in the city of Cuzco, Peru, is 10%. A recent study conducted in Tibet (21) indicates that 15 women living at about 3,650 m had infants who were born with a mean weight of 3,307 grams, which is similar to that found at low altitudes. Comparative studies conducted in Nepal indicate that the birth weights of 37 infants were equally reduced at both high and low (99). On the other hand, Wiley (100) reports that in a native sample of 168 women and their newborns from Ladak (situated at 3,500 m) the mean birth weight averaged only 2,764 grams, which is much lower than those reported at sea level. Thus, it would appear that in the Himalayas, as in the Andes, birth weight is reduced.

## Factors

At present the extent to which the reduction in birth weight is related to high-altitude hypoxia per se or to environmental factors associated with high altitude is not well defined. Several factors have been suggested.

**Acclimatization Status**. As shown by studies conducted in the South American Andes the reduction in birth weight may be related to variations in degree of acclimatization of high-altitude populations. Populations that have lived longest at high altitude and therefore have the greatest degree of adaptation can have newborns with weights that approach the weight of sea-level newborns. For example, evaluation of a large number of births occurring in the city of Puno situated at 3850 m in southern Peru (9) indicated that the frequency of low birth weight infants (less than 2500 g) among the low-altitude native mestizo women is 23%, whereas among highland native women the percent of low birth weight approaches sea-level values. Similarly, the mean birth weight for newborns of highland native women of the city of Puno was greater than those of the mestizo women who presumably are low-altitude natives (22–23). Recent studies in La Paz, Bolivia (24), also indicate that mestizo women or descendants of Europeans have newborns with lower birth weights than newborns of native women.

**Nutritional Factors**. Studies of newborn body composition suggest that infants born at high altitude differ from their low-altitude counterparts in the amount of body fat and muscle. According to studies conducted in La Paz, Bolivia, the reduction in birth weight is more a function of reduction in body muscle mass than of body fat. Anthropometric studies indicate that while high-altitude infants were lighter and shorter than low-altitude infants the high-altitude infants were fatter but had equal amounts of body muscle (18). These results have been interpreted to suggest that inadequate delivery of energy was not the cause of the high/low altitude difference in newborn weight and length (18, 25).

**Maternal Respiration and Circulation**. According to the studies of Moore and colleagues (21), among Himalayans the availability of oxygen to the fetus is attained throughout a redistribution of blood flow from the lower extremities to the uterine circulation. On the other hand, among women from Colorado and Peru adaptation to hypoxia involves an increase of arterial oxygen content attained through an increase in pulmonary ventilation. From the energetic point of view redistribution of blood flow is more efficient than pulmonary ventilation. Hence, from these studies it would appear that Himalayans are better adapted than the Colorado and Andean women.

**Fetal Hypoxia**. Since fetal hemoglobin (HbF) has a greater affinity for oxygen than adult hemoglobin (HbA) and its presence is related to hypoxia (26), the higher is the proportion of hemoglobin F the greater is the fetal hypoxia. Ballew and Haas (18) found that the proportion of hemoglobin F in cord blood among newborns in La Paz (3600 m), Bolivia, is greater than in low-altitude newborns. Similarly, experimental studies in rats indicate an increased oxygen affinity of fetal blood is associated with significant reduction in birth weight (27–28). These findings suggest that hypoxia does play an important role in the birth weight reduction at high altitude.

## Optimal Birth Weight and Mortality

In all populations there is a negative relationship between birth weight and mortality. At high altitude, however, according to studies in Peru (29) the mean birth weight associated with the lowest mortality (or optimal birth weight) for Peruvian high-altitude (3860m) populations (3462 grams) is significantly less than that for sea-level populations (3632 grams). This finding suggests that the low birth weight that characterizes the Andean population is not necessarily negative for infant survival. Thus, it would appear that Himalayans and Andeans use different pathways to overcome hypoxic stress. However, these responses are not completely successful because in both regions the birth weight is reduced.

## POSTNATAL GROWTH AND DEVELOPMENT

### Height and Weight

#### High-Altitude Natives

In the previous section we saw that prenatal growth and development is delayed at high altitude. This pattern of slow growth and development continues during the postnatal period and is evident in experimental animals as well as in humans. In general experimental studies on animals exposed to hypoxic conditions, whether it be in altitude chambers with low oxygen mixtures or at high altitudes under favorable nutritional conditions, show that weight increases at a slower rate than at sea level (30–37). This retarded growth has been attributed to anorexia (lack of appetite) that experimental animals suffer from (38) and to a deficiency in the intestinal absorption of nutrients (39–40). However, these factors alone cannot explain the growth pattern in body size, body morphology, and the organ systems concerned with the transport of oxygen. Microscopic studies reveal that the retarded growth associated with high-altitude hypoxia is caused by a lesser number of cells, whereas the retarded growth associated with malnutrition is caused by a decreased amount of cytoplasm (34). Other investigations indicate that high-altitude hypoxia interferes with cellular and protein multiplication of the brain (35–36). From all these studies it is evident that high-altitude hypoxia plays an important role in delayed growth at high altitude.

Studies of Andean populations, such as those from the Peruvian, Chilean, and Bolivian highlands, indicate that postnatal growth is delayed compared to that of low-altitude populations with comparable nutritional and socioeconomic status (22, 41, 46–50). Results of an extensive study based on semi-longitudinal and cross-sectional samples of the Quechua population of the district of Nuñoa situated at a mean altitude of 4250 m are presented in figures 12.2 to 12.4. From these data it is evident that growth in stature in both sexes is delayed, and until the age of 14 years there is no clear sex differentiation. In men growth in height continues until about the age of 22 years and in women until 20 years. The rapid growth associated with adolescence occurs after the age of 16 years in males and 14 years in females. In other words, the growth of highland Quechuas is characterized by (1) late sexual dimorphism, (2) slow growth and prolonged periods of growth, and (3) late adolescent height spurt.

The pattern of slow growth in stature observed among Andean populations can also be found among populations from the Himalayas and the Tien Shan Mountains (51–52). As shown in figure 12.5, between the ages of 12 and 18 Russian highland children are shorter than their low-altitude counterparts (52). However, it must be pointed out that not all populations that live at

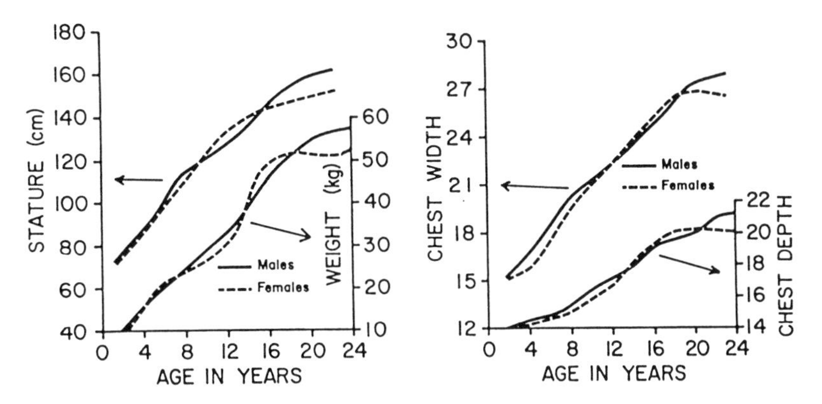


**Fig. 12.2.** Development of stature, weight, chest width, and chest depth of Nuñoa Quechua Indians. Pattern of growth in body size is characterized by late sexual dimorphism. (Modified from A. R. Frisancho and P. T. Baker. 1970. Am. J. Phys. Anthropol. 32:279–92.)

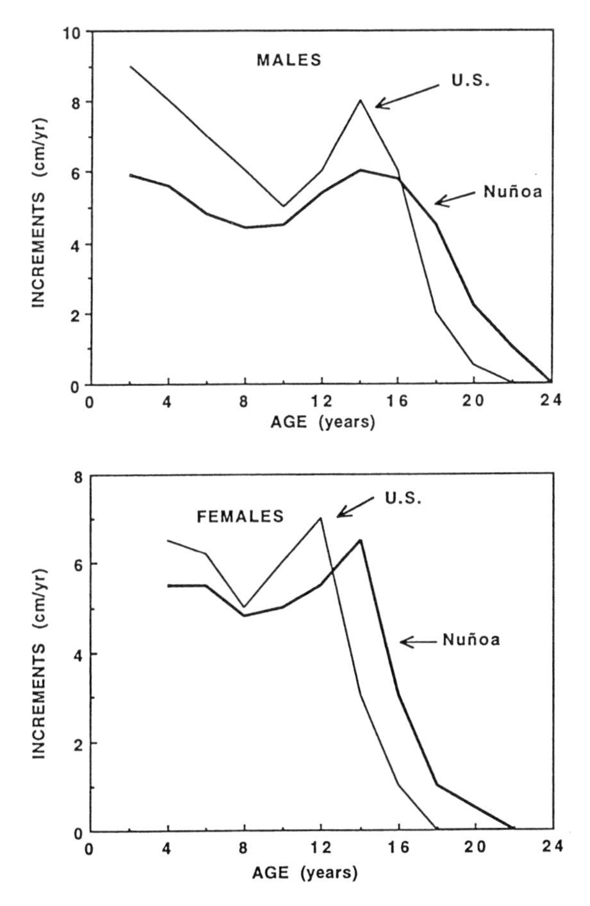


**Fig. 12.3.** Stature growth rate of Nuñoa Indians. Compared to United States Fels Standards, Nuñoa males and females have late and poorly defined spurt. (Modified from A. R. Frisancho and P. T. Baker. 1970. Am. J. Phys. Anthropol. 32:279–92.)

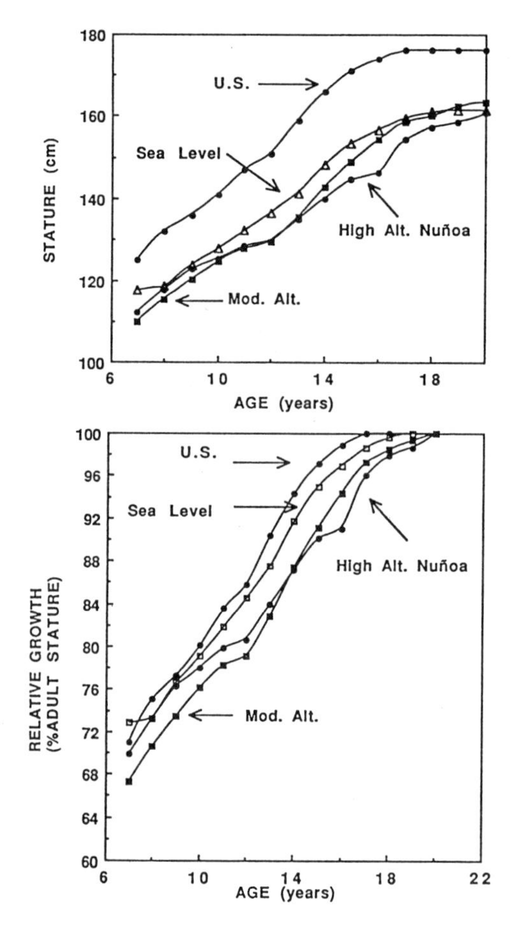


**Fig. 12.4.** Comparison of absolute and relative development in stature of Nuñoa Quechua boys and Peruvians from sea level and a moderate altitude of 2300 m. (Modified from A. R. Frisancho and P. T. Baker. 1970. Am. J. Phys. Anthropol. 32:279–92.)

high altitudes necessarily have slow growth. It has been demonstrated in studies of Africans that the populations of the mountains of Ethiopia, because they live in better nutritional and socioeconomic conditions, have a faster growth rate than their low-altitude counterparts (53). These findings agree with investigations conducted with Peruvian Quechua populations situated in the district of Junin at 4300 m and those situated in the district of Lamas at 1000 m in Peru's eastern lowlands (54). In these studies we found that growth in stature and weight, because of the better nutritional reserves of the highlanders, was faster than those at low altitudes.

## Europeans

Schutte et al. (55) studied the growth in height and weight of European children who lived in Acoma (3200 m) for 9 to 36 months. This study indicates

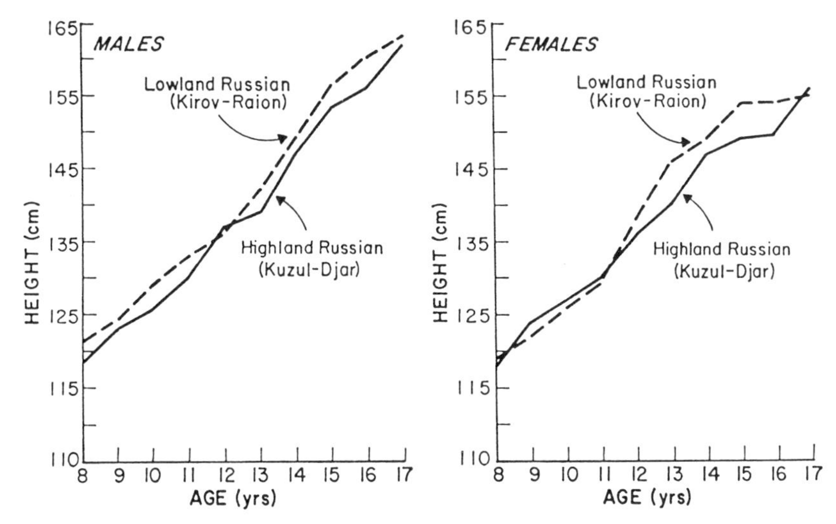


**Fig. 12.5.** Comparison of growth in stature among lowland and highland Russian children from Tien Shan Mountains. During adolescence highland children are shorter than their lowland counterparts. (Modified from N. N. Miklashevskaia, V. S. Solovyeva, and E. Z. Godina. 1972. Vopros Anthropologii. 40:71–91.)

that among healthy, well-nourished children the pattern of growth in height and weight is delayed at high altitude when compared to pre-exposure values. Similarly, as shown by Stinson (56) who studied French children raised in La Paz (ranging in altitude residence from 3200 to 3600 m), Bolivia, there is an inverse relationship between length of residence and growth, so that the longer the residence the smaller are the children (see table 12.1). Furthermore, the height and weight of Bolivian French school children is lower than the corresponding values for U.S. children from Berkeley. However, Stinson (56) and Greksa et al. (57) indicate that the height of French school boys (but not girls) is very similar to that of American school boys of European ancestry. Based on this information both authors concluded that the effect of high altitude on growth is negligible. Since the American school children from Guatemala are also delayed when compared to U.S. standards their similarity

**TABLE 12.1. Comparison of height, weight, and chest depth of European children raised in La Paz (altitude 3700 m), Bolivia**

| Percentage of Life at High Altitude | Number | Adjusted mean[a] Height (cm) | Weight (kg) | Chest Depth (cm) |
|---|---|---|---|---|
| 0–74 | 17 | 144.0 | 37.2 | 15.4 |
| 75–99 | 39 | 141.4 | 34.1 | 15.2 |
| 100 | 110 | 140.3 | 34.2 | 15.5 |
| F-ratio for alt effect | | 3.33* | 2.51 | 1.58 |

*Source:* Adapted from S. Stinson. 1982. The effect of high altitude on the growth of children of high socioeconomic status in Bolivia. Am. J. Phys. Anthropol. 59:61–71.

[a]Height and weight adjusted for age and sex; chest depth adjusted for age, sex, and height.

*Significant at $p<0.05$.

in growth to the French high-altitude children cannot be interpreted as indicative of no effects of high altitude. As shown by the studies of Fellmann et al. (58–59) French children raised in La Paz (3600 m) are significantly shorter than their counterparts from low altitude (see table 12.2).

## Chest Size

### High-Altitude Natives

A distinguishing feature of the highland native is an enlarged thorax (47). As shown by studies done in the central and southern highlands of Peru and Chile the increased chest size of the highland native is acquired through rapid and accelerated growth, especially after the end of childhood (42–46, 47–48, 50, 61) (fig. 12.6). Studies of an Ethiopian population (53, 60) indicate that the trend toward greater chest size is also present among highland children and adults. Figure 12.7 depicts the relationship, as derived from regression analysis, of age to chest circumference (Log 10) of Ethiopian children (53) and shows that at a given age the highland boys and girls have a larger chest size than their low-altitude counterparts. Figure 12.8 shows that the highland adult Ethiopians have a greater maximum chest circumference than their low-altitude counterparts. Furthermore, evaluations of the anthropometric measurements of migrant groups demonstrated that the enlarged chest size of the highland Ethiopians could be attained by lowland natives who migrated to high altitude (60). A recent study by Malik and Singh (62) conducted among the Bod natives of the mountains of Ladakh situated at 3514 m in India demonstrated that the growth in chest circumference of the highlanders is significantly faster than the low-altitude Indian norm. Adult Sherpas also exhibit an enlarged thorax (63). Thus it would appear that, with the exception of the Russians from the Tien Shan Mountains (52), high-altitude natives are characterized by a rapid growth in chest dimension resulting in a relatively large adult chest size.

### Europeans

That prolonged residence at high altitude can result in enlarged chest size is also inferred from the fact that United States adult women who resided at

**TABLE 12.2. Comparison of height and weight of French children raised in La Paz (altitude 3600 m), Bolivia**

| | | Age (yr) | Weight (kg) | Height (cm) |
|---|---|---|---|---|
| Sample | *N* | Mean ± SD | Mean ± SD | Mean ± SD |
| High altitude | 30 boys/21 girls | 11.0 ± 1.2 | 34.9 ± 7.3 | 141.9 ± 5.3 |
| Low altitude | 25 boys/15 girls | 11.4 ± 0.7 | 35.6 ± 5.6 | 145.4 ± 5.4* |

*Source:* Adapted from N. Fellmann, M. Bedu, H. Spielvogel, G. Falgairette, E. Van Praagh, and J. Coudert. 1986. Oxygen debt in submaximal and supramaximal exercise in children at high and low altitude. J. Appl. Physiol. 60: 209–15.

*Significant at $p<0.05$.

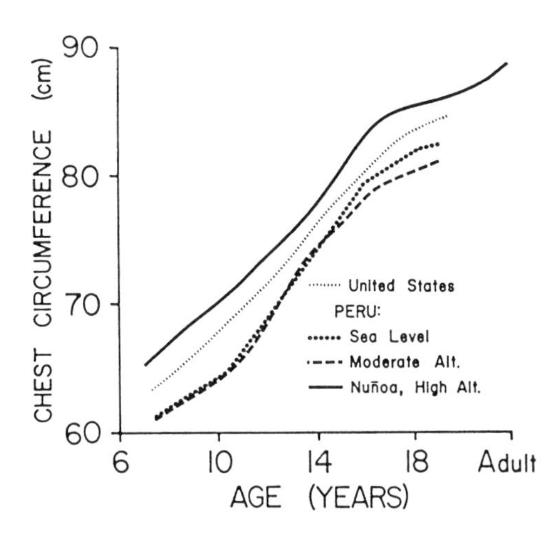


**Fig. 12.6.** Comparison of chest circumference among Peruvian children and adults from sea level, moderate altitude, and high altitude. Highland Quechuas from Nuñoa exhibit accelerated growth in chest size. (From A. R. Frisancho and P. T. Baker. 1970. Am. J. Phys. Anthropol. 32:279–92.)

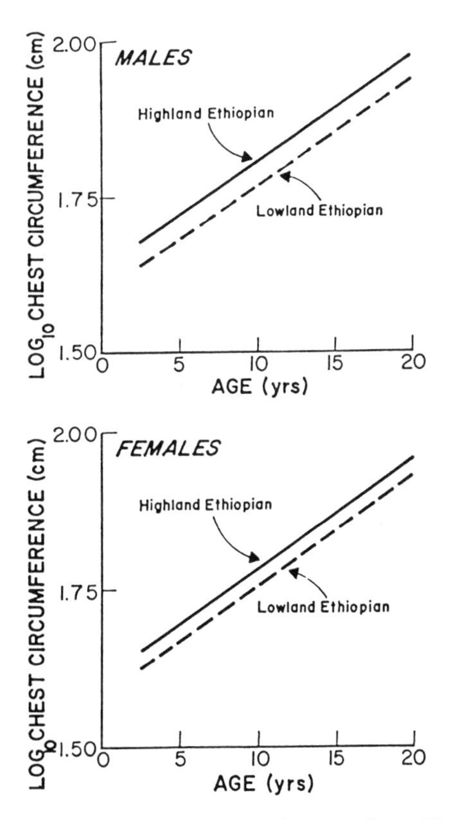


**Fig. 12.7.** Relationship of age to chest circumference (Log 10) among highland and lowland Ethiopian children. At a given age highlanders show greater chest size for their stature than do lowlanders. (Based on data from E. J. Clegg, I. G. Pawson, E. H. Ashton, and R. M. Flinn. 1972. Philos. Trans. R. Soc. Lond. [Biol.] 264:403–37.)

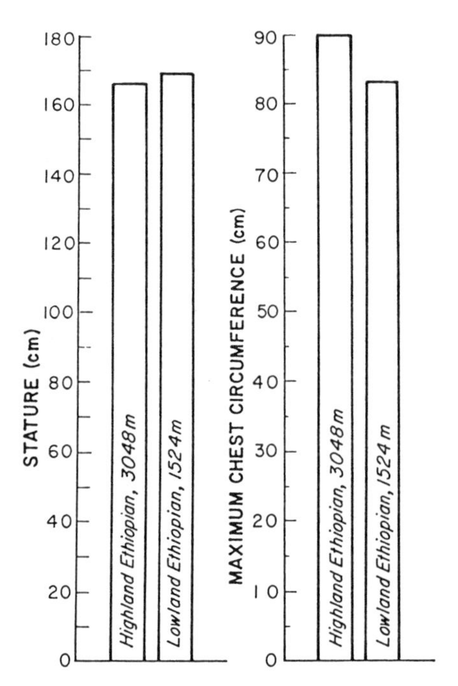


**Fig. 12.8.** Comparison of stature and maximum chest circumference among highland and lowland Ethiopians. Highland adults, despite their similar statures, have greater chest size than lowland counterparts. (Based on data from G. A. Harrison, G. F. Kuchemann, M. A. S. Moore, A. J. Boyce, T. Baju, A. E. Mourant, M. J. Godber, B. G. Glasgow, A. C. Kopec, D. Tills, and E. J. Clegg. 1969. Philos. Trans. R. Soc. Lond. [Biol.] 256B:147–82.)

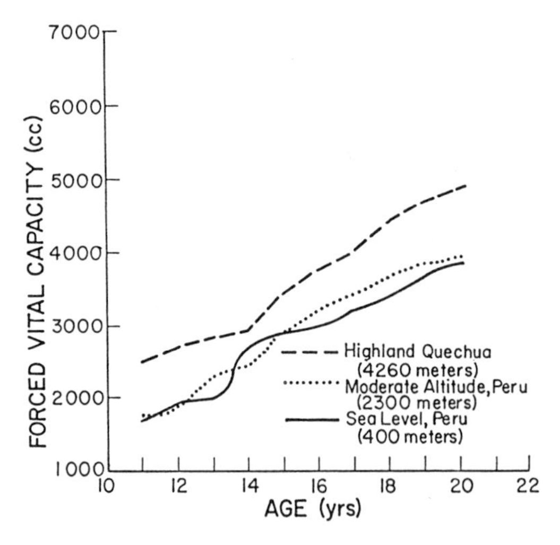


**Fig. 12.9.** Comparison of forced vital capacity among Peruvian boys from sea level, moderate altitude, and high altitude. Highland boys from Nuñoa exhibit accelerated development of lung volume. (Based on data from A. R. Frisancho and P. T. Baker. 1970. Am. J. Phys. Anthropol. 32:279–92.)

Pikes Peak for 2 1/2 months exhibited an increase in maximum chest circumference despite a significant reduction in subscapular skinfold thickness (64). Furthermore, data of European migrants residing in La Paz (3600 m), Bolivia, indicated that lifelong exposure to chronic hypoxia resulted in an average expansion of chest depth of 1.2 to 1.8 cm (65–66).

## Lung Volume

### High-Altitude Natives

Concomitant with increased chest size, the lung volume of high-altitude natives is greater than that for persons at low altitudes. As illustrated in figure 12.9, greater forced vital capacity (FVC) of highland natives is acquired through a rapid and accelerated growth (42–45, 47, 61, 67–68). In view of the fact that during childhood at sea level growth in lung volume is associated with an enhanced quantity of alveolar units and alveolar surface area (69), the rapid growth in lung volume at high altitude is probably also associated with an enhanced quantity of alveolar units and alveolar surface area. Since there is a direct relationship between alveolar surface area and rate of oxygen diffusion from the lungs to the capillary bed, the rapid growth in lung volume at high altitude may reflect an adaptation to high-altitude hypoxia. As discussed in chapter 11, evidence suggests that the low-altitude native through growth and development at high altitude can acquire an enlarged lung volume similar to the high-altitude native (44, 50, 53, 70).

### Europeans

Comparison of European highland children residing at high altitude (La Paz, Bolivia) with lowlanders indicated that exposure to chronic hypoxia resulted in an average increase in forced vital capacity of 150–300 ml (65, 66). Furthermore, data of European migrants indicated that lifelong exposure to chronic hypoxia resulted in an average expansion of 288 to 384 ml in forced vital capacity (65, 66).

In summary, the above studies suggest that children of European ancestry growing under conditions of chronic hypoxia exhibit an enhancement of chest size and lung volumes. However, as indicated by Greksa and Beall (66) there is little evidence to suggest that European children exhibit an accelerated development of chest size and lung volumes during childhood and adolescence. On the other hand, the enhancement of chest dimensions and lung volumes of high-altitude natives appears to be established by early childhood, and therefore the differences between high altitude and low altitude natives are merely maintained during childhood and adolescence. The fact that the

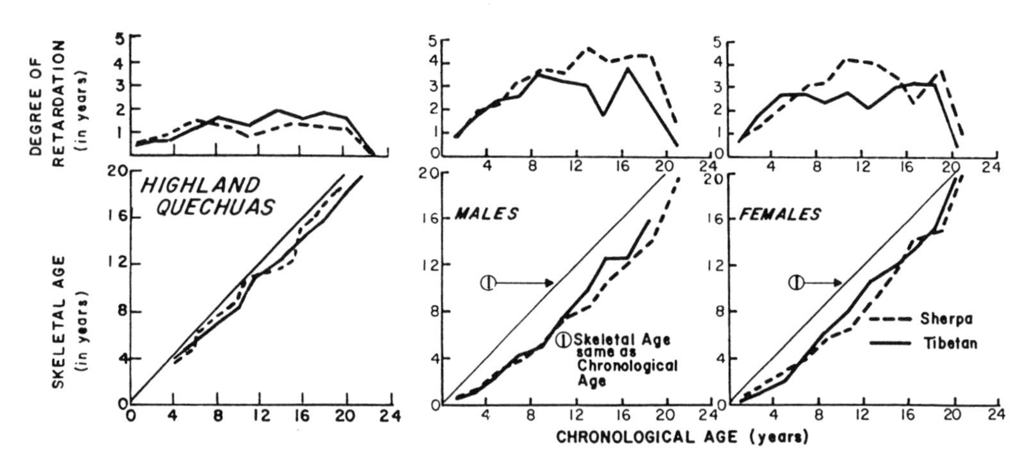


**Fig. 12.10.** Relationship between skeletal age and chronological age among highland Quechuas, highland Sherpas, and lowland Sherpas (Tibetan). In both highland samples skeletal maturation, especially during childhood, is delayed. During adolescence this difference decreases. (Based on data from A. R. Frisancho. 1969. Growth, physique, and pulmonary function at high altitude: a field study of a Peruvian Quechua population. Ph.D. Thesis. University Park, PA: Pennsylvania State University; and I. G. Pawson. 1977. Am. J. Phys. Anthropol. 47:473–82.)

magnitude of the hypoxia-induced expansion of chest size and lung volumes among European children is much smaller than the range of variation found in high-altitude populations suggests that the distinctive thorax of high-altitude Andean natives is mediated by genetic factors.

## Heart

Another characteristic of the highland native is that the heart is bigger and heavier than that of the low-altitude native. Anatomical and electrocardiographic studies conducted mostly in Peru indicate that the larger heart size of the highland native is a result of the lack of involution of the right ventricle (71–73). These studies indicate that at birth the characteristics of the heart in the highlander and the sea-level native are the same; in both environments the newborns show a similar preponderance of the right ventricle, but after the second month regression of the right ventricle that is natural in the sea-level native does not occur in the highlander (71–73). That is, during the period of growth and development at high altitude, fetal characteristics are retained, so in the adult state the preponderance of the left ventricle that occurs in the sea-level native does not occur in the highlander; on the contrary, it is the right ventricle and not the left that is bigger.

The preponderance of the right ventricle of the highlander has been explained both as a compensatory mechanism to counteract pulmonary hypertension present at high altitude (chap. 11) and as part of a systemic adaptive response to increase oxygen transport and delivery.

## Skeletal Maturation

Skeletal maturation, assessed through estimates of skeletal age with reference to the Greulich and Pyle atlas (74) of Andean, Himalayan, and Ethiopian high-altitude populations prior to the age of 16 years is quite retarded (43–45, 51, 53). However, in both the Himalayan Sherpas and Peruvian Quechuas after the age of 16 years the retardation decreases drastically, so by about the age of 20 years in males and 18 years in females the high-altitude native approaches western standards (fig. 12.10). Compared to United States norms, before the age of 16 years the highland Sherpas and Quechuas exhibit an average delay in skeletal age of about 20%, but between the ages of 16 and 20 the difference amounts to only 10%. This indicates that the age at which complete epiphyseal union occurs (reflected by the measurement of skeletal age during adolescence) at high altitude is retarded by 10%. This delay coincides with the increase in age of attainment of adult stature. As pointed out earlier, growth in height continues up to age 22 in males and 20 in females, which compared to United States norms equals a delay of about 10%. For this reason, the origin of the short adult stature of high-altitude populations is explained mostly by the marked growth delay that occurs during childhood, a retardation which is not recuperated by the lengthened period of growth.

## Sexual Maturation

### High-Altitude Natives

Concomitant with retardation in adolescent skeletal maturation, sexual maturation among high-altitude Andean populations is delayed. Investigations of a large sample of Peruvian girls aged 11 to 17 years indicate that the age of menarche at high altitude averaged 13.58 years compared to 12.58 years at sea level (75) (fig. 12.11). Studies in Bolivia indicate that the mean age at menarche for high-altitude native girls equalled 13.4 years but for low altitude samples averaged 12.3 years (76). Furthermore, it has been found that in both males and females the age at which the secondary sexual characteristics develop is markedly delayed at high altitude (75, 77–78). Similarly, measurements of luteinizing hormones indicate that adult values in girls are attained 1 year later at high altitude than at sea level (78). Thus it appears that adolescent maturation by any criteria is delayed among Andean high-altitude populations. On the other hand, among Ethiopian high-altitude populations secondary sexual maturation does not show any altitude-related differences (53). Nevertheless, studies in the Himalayas report that the age at menarche among the highland Sherpas averaged 18.1 years, which is one of the latest ever recorded for a human population (51). The extent to which these data correspond to actual sexual maturation remains to be seen.

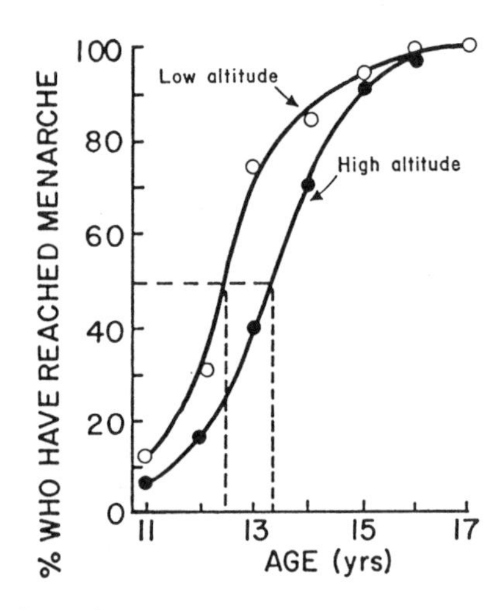


**Fig. 12.11.** Comparison of age at menarche among Peruvian mestizos from sea level and central highlands. The mean age at menarche is later for the highland girls than for those of sea level. (Modified from J. Donayre. 1966. Population growth and fertility at high altitude. In Life at high altitudes. Pub. 140. Washington, DC: Pan American Health Organization.)

## Europeans

Studies in Bolivia indicate that the mean age at menarche for high-altitude Europeans equalled 13.1 years, and 12.3 years for low altitude samples (76) (see table 12.3). Thus this study suggests that growth and development at high altitudes result in a delay in median age at menarche of nearly 0.8 years. On the other hand, Fellmann et al. (59) indicated that salivary testosterone concentration of high-altitude boys in La Paz, Bolivia, was higher than of low-altitude controls and accordingly concluded that at high altitude boys mature at an earlier age than at low altitude. However, it should be noted that high altitude hypoxia is associated with increased testosterone concentrations, which stimulates hemoglobin and 2,3-DPG production (79–80). Therefore, increased testosterone concentration at high altitude cannot be considered as indicative of earlier maturation.

**TABLE 12.3. Comparison of age of menarche of Bolivian girls of European and Aymara ancestry, raised in La Paz (altitude 3600 m), Bolivia**

| | | Age of Menarche (years) | |
|---|---|---|---|
| Sample | *N* | Median | (95% confidence interval) |
| European ancestry | | | |
| Santa Cruz sedentes | 172 | 12.3 | (11.9, 12.7) |
| La Paz sedentes | 199 | 13.1 | (12.8, 13.5) |
| La Paz migrants | 84 | 13.8 | (13.4, 14.3) |
| Aymara ancestry | | | |
| La Paz sedentes | 375 | 13.4 | (13.2, 13.7) |

*Source:* Adapted from L. P. Greksa. 1990. Age of menarche in Bolivian girls of European and Aymara ancestry. Ann. Hum. Biol. 17:49–53.

## DETERMINANTS OF GROWTH

Obviously the pattern of growth and development at high altitude reflects the interaction of genetic and environmental factors. For this reason, to determine the causes of the highland pattern of growth and development it is necessary to consider the role of each of these factors.

### Genetic Factors

One way of determining the role of genetic factors in human growth is to assess the parent-offspring correlation. In general a high correlation coefficient implies that the phenotypic expression is under a high genetic control, whereas a low correlation implies a low genetic control. As shown in table 12.4 the parent-offspring correlations in the Quechua highlanders from Nuñoa range from 0.29 to 0.52 (44). These values are similar to those found among Western populations with good nutritional status. Thus it appears that the genetic contribution to phenotypic variations in growth and stature among high-altitude Quechuas is comparable to that of sea-level populations. With respect to this conclusion, it must be pointed out that comparative studies of Andean Aymara populations from Chile indicate that the most important factor explaining differences in anthropometric measurements is the altitude difference, whereas geographic and genetic differences contribute the least (81). Another approach for determining the role of genetic factors on growth is to compare populations of similar genetic composition living at sea level and at high altitude. Studies of highland and lowland Peruvian populations of the same Quechua stock (23, 48, 67) suggest that most of the differences in growth are traceable to altitude differences and not to genetic differences.

In summary, at the present stage of knowledge it appears that differences in growth between highland and lowland populations are not caused by differences in genetic composition.

**TABLE 12.4. Parent-offspring correlation in height among high-altitude Quechuas from the district of Nuñoa, situated at a mean altitude of 4250 m in southern Peru**

| | Father | | | | Mother | | | |
|---|---|---|---|---|---|---|---|---|
| Age group | Son | | Daughter | | Son | | Daughter | |
| (years) | *N* | r | *N* | r | *N* | r | *N* | r |
| 5–10.9 | 115 | 0.31* | 98 | 0.29* | 96 | 0.36* | 86 | 0.32* |
| 11–16.9 | 80 | 0.40* | 60 | 0.42* | 60 | 0.48* | 50 | 0.52* |

*Source:* Based on data from A. R. Frisancho. 1976. Growth and functional development at high altitude. In P. T. Baker and M. A. Little, eds., Man in the Andes: a multidisciplinary study of high altitude Quechua natives. Stroudsburg, PA: Dowden, Hutchinson and Ross.

*Weighted mean r is derived from Z-transformed age-specific values of R; probability $<0.01$.

## Nutrition

In general, because of the joint effects of hypoxia and cold, agriculture at high altitude is limited, and the nutritional base is less than at low altitudes. Given this, one would expect undernutrition to contribute to delayed growth at high altitude (82). A way of determining the role of nutritional reserves in human growth is to evaluate the amount of nutritional reserves a population has (83). Through anthropometric measurements the caloric and protein reserves are easily determined, and in general a high amount of subcutaneous fat and body muscle implies high calorie and protein reserves. As shown in

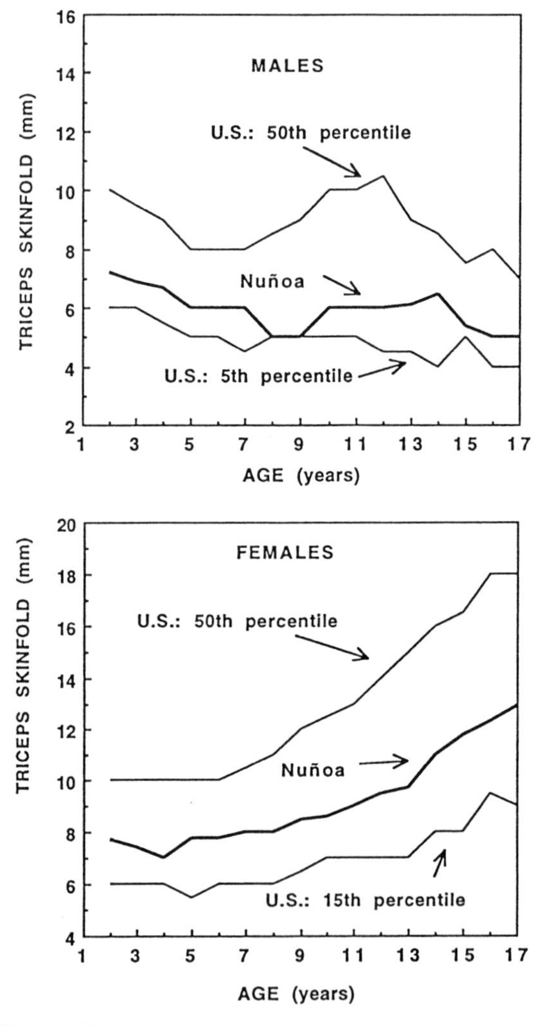


**Fig. 12.12.** Triceps skinfold in the fiftieth percentile of highland Quechua children from Nuñoa and fiftieth and fifteenth percentiles of United States children. The highland children's subcutaneous fat is within lower normal range of United States standards. (Based on data from A. R. Frisancho and P. T. Baker. 1970. Am. J. Phys. Anthropol. 32:279–92, and A. R. Frisancho. 1974. Am. J. Clin. Nutr. 27:1052–58.)

figure 12.12, the mean triceps skinfold thickness of the highland Nuñoa children (43–45) is between the fiftieth and fifteenth percentile of the United States Ten State Nutritional Survey (83). These findings indicate that in this highland population the children have adequate calorie reserves. Both metabolic balance studies (84) and dietary surveys (85–86) demonstrate that the dietary intake of Nuñoa samples meet the United States recommended dietary allowances (fig. 12.13). Furthermore, as shown in figure 12.14, in Nuñoa fatter children are not taller than their leaner counterparts (43) suggesting that greater calorie reserve is not associated with increased dimensional growth. This finding is contrary to those of sea-level populations in which increased fatness is associated with advanced maturity and growth (87). A recent investigation conducted by Leonard, (88) who re-studied the high-altitude children from Nuñoa demonstrates that during the last twenty years there has been a major deterioration of nutritional resources. As a consequence the growth in anthropometric dimensions has been considerably delayed when compared to those observed before. Thus, at present, differences in nutritional status have become important determinants of the pattern of growth at high altitude (89). Furthermore, studies of school children from Puno, Peru

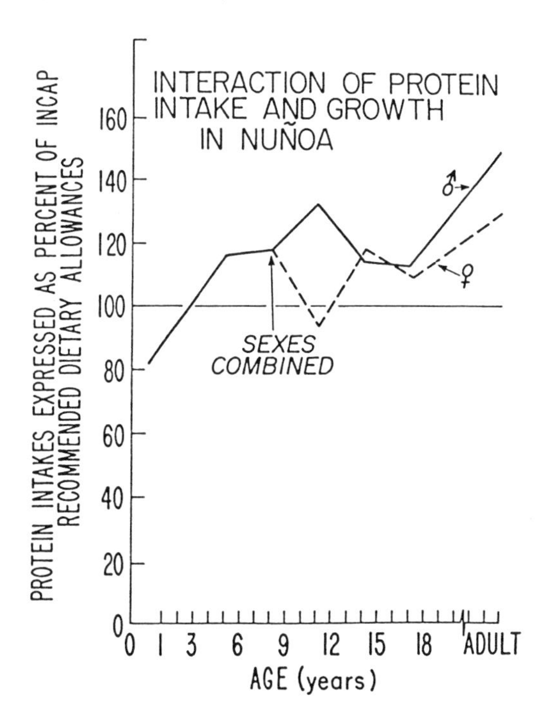


**Fig. 12.13.** Protein intake of Nuñoa highland Quechua children and adults expressed as percent of INCAP (Institute of Nutrition of central America and Panama) recommended dietary allowances. (Calculated from individual values given by M. Gursky. 1969. A dietary survey of three Peruvian highland communities. M.A. thesis. University Park, PA: Pennsylvania State University.)

(3800 m), found a secular trend in growth in height (90) suggesting that the slow growth of high-altitude children in addition to hypoxia may also be due to poor socioeconomic conditions.

Thus it would appear that among those high-altitude Quechuas from Nuñoa nutritional limitation along with high-altitude hypoxia appears to account for their pattern of growth. Furthermore, it is possible that a high-altitude environment may also indirectly affect energy balance. First, as indicated in chapter 5, the natives of Nuñoa, in spite of efficient technological adaptations

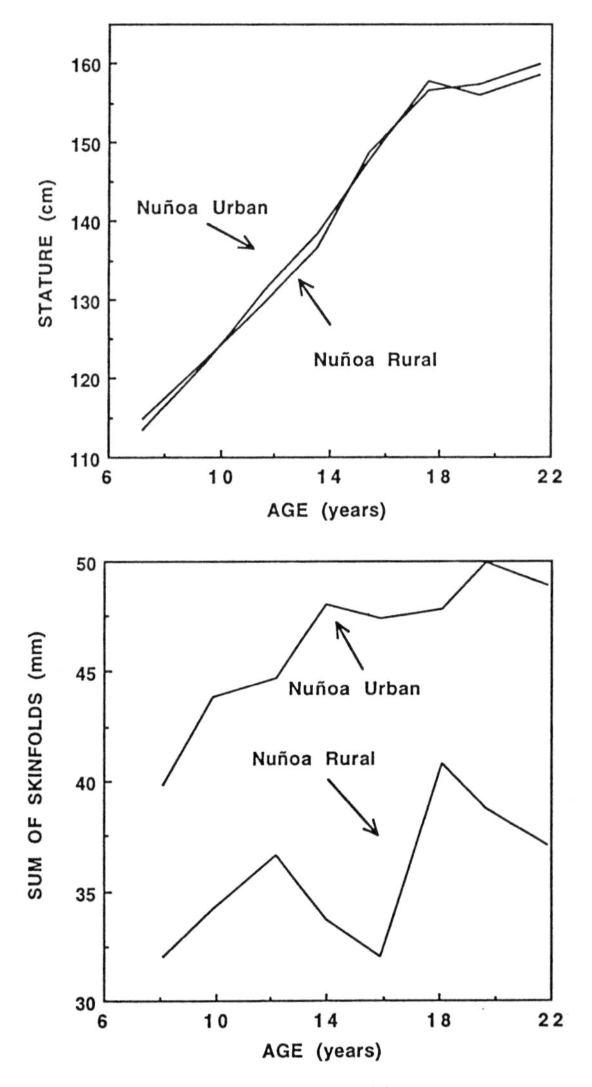


**Fig. 12.14.** Comparison of stature and sum of (7) skinfolds among highland Quechua males from urban and rural areas of district of Nuñoa. Note that although there are marked differences in skinfolds, urban and rural exhibit similar statures. (Based on data from A. R. Frisancho and P. T. Baker. 1970. Am. J. Phys. Anthropol. 32:279–92.)

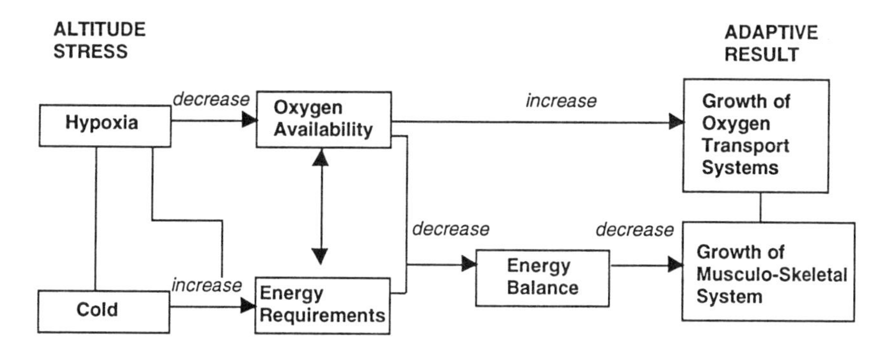


**Fig. 12.15.** Schematization of two-directional patterns of growth at high altitude. At high altitude, because of combined effects of hypoxia, cold, and increased energy requirements, growth of musculoskeletal system is delayed, whereas in response to low oxygen availability, growth of oxygen-transport system organs is accelerated.

to cold, are frequently exposed to severe cold stress because of the requirements of a subsistence economy. Laboratory studies indicate that high-altitude Quechuas have a higher basal metabolic rate (produce more heat) than sea-level residents (91–92) and maintain body temperatures at higher levels when exposed to cold stress than sea-level inhabitants (93–95). Furthermore, when exposed to cold both children and adults are able to maintain hand and foot temperatures at comfortable levels, in marked contrast to sea-level residents (93–95). Such cold exposure is bound to increase the individual's food energy requirements. Most of the metabolic studies conducted on high-altitude Quechuas from Nuñoa indicate that to maintain body weight without modifying body composition an active adult man residing at an altitude of 4000 m must increase his caloric intake by about 14.8% over the requirements of a man with the same physical characteristics who resides at sea level (84). Second, investigations of glucose metabolism demonstrate that among high-altitude adult male and pregnant female natives the rate of blood glucose utilization and glucose tolerance is greater than at sea level (96–98).

## OVERVIEW

At high altitude the pattern of growth and development in body size and the organ systems concerned with oxygen transport differs from the low-altitude pattern. This high-altitude pattern follows two directions of response in which accelerated and slow patterns occur simultaneously. As schematized in figure 12.15, high-altitude hypoxia accelerates the growth of oxygen transport systems organs such as the placenta, lungs, heart, and thorax. On the other hand, joint effects of hypoxia and cold increase energy requirements. This in turn affects the energy balance and results in prenatal and postnatal growth retardation of the musculoskeletal system, which affects both birth weight and stature. Because of this two-directional response, human growth in high-altitude populations must be viewed as the result of interaction and adaptation of the organism to competing stresses of hypoxia, cold, and energy requirements that characterize the high-altitude environment. Recent comparative studies conducted among sea-level subjects acclimatized to high altitude during growth and during adulthood indicate that the Andean chest shape is not acquired by either developmental or adult acclimatization (101). Furthermore, based on studies conducted among high-altitude and low-altitude natives Eckhardt et al. (102) demonstrated that chest dimensions of Andean highlanders have a substantial heritable component (102). These findings together suggest that the attainment of the Andean chest shape is influenced more by genetic than by acclimatization factors.

## References

1. Barcroft, J. 1936. Fetal circulation and respiration. Physiol. Rev. 16:103–8.
2. Dawes, G. S. 1965. Oxygen supply and consumption in late fetal life, and the onset of breathing at birth. In W. O. Fenn and H. Rahn, eds., Handbook of physiology. Section 3. Respiration, vol. 2. Washington, DC: American Physiological Society.
3. Barron, D. H., J. Metcalf, G. Mechia, A. Hilligers, H. Prystovsky, and W. Huckabee. 1964. Adaptations of pregnant ewes and their fetuses to high altitude. In W. H. Weihe, ed., The physiological effects of high altitude. New York, NY: Pergamon Press.
4. Eastman, N. J. 1930. Fetal blood studies. Part 1. The oxygen relationships of umbilical cord blood at birth. Bull. Johns Hopkins Hosp. 47:221–24.
5. Metcalfe, J., G. Meschia, A. Hellegers, H. Prystowski, W. Huckabee, and D. H. Barron. 1962. Observations on the placental exchange of the respiratory gases in pregnant ewes at high altitude. Q. J. Exp. Physiol. 47:74–92.
6. Rendon, H. 1964. Aspectos microscopico de la placenta a 2,300 mts. de altitud. Thesis de Bachillerato. Arequipa, Peru: Universidad Nacional de San Agustin, Facultad de Medicina.
7. Chabes, A., J. Perada, L. Hyams, N. Barrientos, J. Perez, L. Campos, A. Monroe, and A. Mayorga. 1968. Comparative morphometry of the human placenta at high altitude and at sea level. Part 1. The shape of the placenta. Obstet. Gynecol. 31:178–85.
8. McClung, J. 1969. Effects of high altitude on human birth. Cambridge, MA: Harvard University Press.
9. Passano, S. 1969. Observaciones humanas de la placenta en Puno 3812 m. de altura. Ginecologia y Obstetricia 15:45–57.
10. Frisancho, A. R. 1970. Developmental responses to high altitude hypoxia. Am. J. Phys. Anthropol. 32:401–8.
11. Kruger, H., and J. Arias-Stella. 1970. The placenta and the newborn infant at high altitudes. Am. J. Obstet. Gynecol. 106:586–91.
12. Kadar, N., and M. Saldana. 1971. La placenta en la altura. Part 1. Caracteristics

macroscopicas y morfometria. In R. Guerra-Garcia, ed., Estudio sobre la gestacion y el recien nacido en la altura. Lima, Peru: Universidad Peruana Cayetano Heredia.

13. Saldana, M., K. Kadar, and S. Recavaren. 1971. La placenta en la altura. Part 2. Estudio ultraestructural cuantitative de placentas de Cerro de Pasco (4330 m), Puno (3850 m) y Lima (150 m). In R. Guerra-Garcia, ed., Estudio sobre la gestacion y el recien nacido en la altura. Lima, Peru: Universidad Peruana Cayetano Heredia.
14. Sobrevilla, L. A. 1971. Analysis matematico de la relacion ponderal placenta: recien nacido en la altura. In R. Guerra-Garcia, ed., Estudio sobre la gestacion y el recien nacido en la altura. Lima, Peru: Universidad Peruana Cayetano Heredia.
15. Guerra-Garcia, R., R. Lozano, and M. Cateriano. 1971. Bioquimica de la sangre materna y del cordon umbilical en Cerro de Pasco (4200 m). In R. Guerra-Garcia, ed., Estudio sobre la gestacion y el recien nacido en la altura. Lima, Peru: Universidad Peruana Cayetano Heredia.
16. Aherne, W., and M. S. Dunnill. 1966. Morphometry of the human placenta. Br. Med. Bull. 22:5–8.
17. Haas, J. D. 1980. Maternal adaptation and fetal growth at high altitude in Bolivia. In L. S. Green and F. S. Johnston, eds., Social and biological predictors of nutritional status, physical growth and neurological development. New York, NY: Academic Press. 257–90.
18. Ballew, C., and J. D. Haas. 1986. Hematologic evidence of fetal hypoxia among newborn infants at high altitude in Bolivia. Am. J. Obstet. Gynecol. 155:166–69.
19. Leibson, C., M. Brown, S. Thibodeau, D. Stevenson, H. Vreman, R. Cohen, G. Clemons, W. Callen, and L. G. Moore. 1989. Neonatal hyperbilirubinemia at high altitude. Am. J. Dis. Child. 143:983–87.
20. Lichty, J. A., R. Y. Ting, P. D. Bruns, and E. Dyar. 1957. Studies of babies born at high altitude. Part 1. Relation of altitude to birth weight. Am. J. Dis. Child. 93:666–69.
21. Moore, L. G. 1990. Maternal $O_2$ transport and fetal growth in Colorado, Peru, and Tibet high-altitude residents. Am. J. Hum. Biol. 2:627–37.
22. Haas, J. D. 1976. Infant growth and development. In P. T. Baker and M. A. Little, eds., Man in the Andes: a multidisciplinary study of high-altitude Quechua natives. Stroudsburg, PA: Dowden, Hutchinson and Ross.
23. Haas, J. D., P. T. Baker, and E. E. Hunt, Jr. 1977. The effects of high altitude on body size and composition of the newborn infant in Southern Peru. Hum. Biol. 49:611–28.
24. Haas, J. D., D. A. Small, and J. L. Beard. 1978. Prenatal growth at high altitude: ethnic variation related to differential maternal adaptability. Am. J. Phys. Anthropol. 48 (abstract):401.
25. Haas, J. D., P. T. Baker, and E. E. Hunt, Jr. 1977. The effects of high altitude on body size and composition of the newborn infant in Southern Peru. Hum. Biol. 49:611–28.
26. Allen, D. W., and J. Jandl. 1960. Factors influencing relative rates of synthesis of adult and fetal hemoglobin in vitro. J. Clin. Invest. 39:1107.
27. Hebbel, R. P., E. M. Berger, and J. W. Eaton. 1980. Effect of increased maternal hemoglobin oxygen affinity on fetal growth in rats. Blood 55:969–74.
28. Bauer, C., W. Jelkmann, and W. Moll. 1981. High oxygen affinity of maternal blood reduces fetal weight in rats. Resp. Phys. 43:169–78.
29. Beall, C. M. 1981. Optimal birthweights in Peruvian populations at high and low altitudes. Am. J. Phys. Anthropol. 56:209–16.

30. Timiras, P. S., A. A. Krum, and N. Pace. 1957. Body and organ weights of rats during acclimatization to an altitude of 12,470 feet. Am. J. Physiol. 191:598–604.
31. Timiras, P. S. 1964. Comparison of growth and development of the rat at high altitude and at sea level. In W. H. Weihe, ed., The physiological effects of high altitude. New York, NY: Pergamon Press.
32. Metcalfe, J., G. Meschia, A. Hellegers, H. Prystowsky, W. Huckabee, and D. H. Barron. 1962. Observations on the growth rates and organ weights of fetal sheep at altitude and sea level. Q. J. Exp. Physiol. 47:305–13.
33. Delaquerriere-Richardson, L., E. S. Forbes, and E. Valdivia. 1965. Effect of simulated high altitude on the growth rate of albino guinea pigs. J. Appl. Physiol. 20:1022–25.
34. Naeye, R. L. 1966. Organ and cellular development in mice growing at simulated high altitude. Lab. Invest. 15:700–705.
35. Cheek, D. J., A. Graystone, and R. A. Rowe. 1969. Hypoxia and malnutrition in newborn rats: effects on RNA, DNA, and protein tissues. Am. J. Physiol. 217: 642–45.
36. Petropoulos, E. A., K. B. Dabal, and P. S. Timiras. 1972. Biological effects of high altitude on myelinogenesis in brain of the developing rat. Am. J. Physiol. 223: 951–57.
37. Clegg, E. J. 1978. Fertility and early growth. In P. T. Baker, ed., The biology of high-altitude peoples. New York, NY: Cambridge University Press.
38. Schnakenberg, D. D., L. F. Krabill, and P. C. Weiser. 1971. The anorexic effect of high altitude on weight gain, nitrogen retention and body composition of rats. J. Nutr. 101:787-96.
39. Van Liere, E. J., W. V. Crabtree, D. W. Nothup, and J. C. Stickney. 1948. Effect of anoxic anoxia on propulsive activity of the small intestine. Proc. Soc. Exp. Biol. Med. 67:331–32.
40. Chinn, K. S. K., and J. P. Hannon. 1969. Efficiency of food utilization at high altitude. Fed. Proc. 28:944–47.
41. Bouloux, C. J. 1968. Contribution a l'étude biologique des phénomènes pubertaires en très haute altitude (La Paz). Centre d'hématypologie du centre national de la recherche scientifique. Toulouse, France: Centre Régional de Transfusion Sanguine et d'Hématologie.
42. Frisancho, A. R. 1969. Human growth and pulmonary function of a high-altitude Peruvian Quechua population. Hum. Biol. 41:365–79.
43. Frisancho, A. R., and P. T. Baker. 1970. Altitude and growth: a study of the patterns of physical growth of a high altitude Peruvian Quechua population. Am. J. Phys. Anthropol. 32:279–92.
44. Frisancho, A. R. 1976. Growth and functional development at high altitude. In P. T. Baker and M. A. Little, eds., Man in the Andes: a multidisciplinary study of high altitude Quechua natives. Stroudsburg, PA: Dowden, Hutchinson and Ross.
45. Frisancho, A. R. 1978. Human growth and development among high altitude populations. In P. T. Baker, ed., The biology of high-altitude peoples. New York, NY: Cambridge University Press.
46. Hoff, C. 1974. Altitudinal variations in the physical growth and development of Peruvian Quechua. Homo. 24:87–99.
47. Hurtado, A. 1932. Respiratory adaptation in the Indian natives of the Peruvian Andes. Studies at high altitude. Am. J. Phys. Anthropol. 17:137–65.
48. Beall, C. M., P. T. Baker, T. S. Baker, and J. D. Haas. 1977. The effects of high

altitude on adolescent growth in southern Peruvian Amerindians. Hum. Biol. 49:109–24.

49. Stinson, S. 1978. Child growth, mortality and the adaptive value of children in rural Bolivia. Ph.D. Thesis. Ann Arbor, MI: University of Michigan.
50. Mueller, W. H., V. N. Schull, W. J. Schull, P. Soto, and F. Rothhammer. 1978. A multinational Andean genetic and health program: growth and development in an hypoxic environment. Ann. Hum. Biol. 5:329–52.
51. Pawson, I. G. 1977. Growth characteristics of populations of Tibetan origin in Nepal. Am. J. Phys. Anthropol. 47:473–82.
52. Miklashevskaia, N. N., V. S. Solovyeva, and E. Z. Godina. 1972. Growth and development in high-altitude regions of Southern Kirghizia, U.S.S.R. Vopros Anthropologii 40:71–91.
53. Clegg, E. J., I. G. Pawson, E. H. Ashton, and R. M. Flinn. 1972. The growth of children at different altitudes in Ethiopia. Philos. Trans. R. Soc. Lond. (Biol.). 264:403–37.
54. Frisancho, A. R., G. A. Borkan, and J. E. Klayman. 1975. Patterns of growth of lowland and highland Peruvians of similar genetic composition. Hum. Biol. 47:233–43.
55. Schutte, J. E., R. E. Lilljeqvist, and R. L. Johnson, Jr. 1983. Growth of lowland native children of European ancestry during sojourn at high altitude (3,200 m). Am. J. Phys. Anthropol. 61:221–26.
56. Stinson, S. 1982. The effect of high altitude on the growth of children of high socioeconomic status in Bolivia. Am. J. Phys. Anthropol. 59:61–71.
57. Greksa, L. P., H. Spielvogel, and L. Paredes-Fernandez. 1985. Maximal exercise capacity in adolescent European and Amerindian high-altitude natives. Am. J. Phys. Anthropol. 67:209–16.
58. Fellmann, N., M. Bedu, H. Spielvogel, G. Falgairette, E. Van Praagh, and J. Coudert. 1986. Oxygen debt in submaximal and supramaximal exercise in children at high and low altitude. J. Appl. Physiol. 60:209–15.
59. Fellmann, N., M. Bedu, H. Spielvogel, G. Falgairette, E. Van Praagh, J-F. Jarrige, and J. Coudert. 1988. Anaerobic metabolism during pubertal development at high altitude. J. Appl. Physiol. 64:1382–86.
60. Harrison, G. A., G. F. Kuchemann, M. A. S. Moore, A. J. Boyce, T. Baju, A. E. Mourant, M. J. Godber, B. G. Glasgow, A. C. Kopec, D. Tills, and E. J. Clegg. 1969. The effects of altitudinal variation in Ethiopian populations. Philos. Trans. R. Soc. Lond. (Biol.). 256B:147–82.
61. Mueller, W. H., F. Yen, F. Rothhammer, and W. J. Schull. 1978. A multinational Andean genetic and health program. VI. Physiological measurements of lung function in a hypoxic environment. Hum. Biol. 5:489–541.
62. Malik, S. L., and I. P. Singh. 1978. Growth trends among male Bods of Ladakh—a high altitude population. Am. J. Phys. Anthropol. 48:171–76.
63. Sloan, A. W., and M. Masali. 1978. Anthropometry of Sherpa men. Ann. Hum. Biol. 5:453–58.
64. Hannon, J. P., J. L. Shields, and C. W. Harris. 1969. Anthropometric changes associated with high altitude acclimatization in females. Am. J. Phys. Anthropol. 31:77–84.
65. Greksa, L. P., H. Spielvogel, M. Paz-Zamora, and E. Caceres. 1988. Effect of altitude on the lung function of high altitude residents of European ancestry. Am. J. Phys. Anthropol. 75:77–85.

66. Greksa, L. P., and C. M. Beall. 1989. Development of chest size and lung function at high altitude. In M. A. Little and J. D. Haas, eds., Human population biology. New York, NY: Oxford University Press. 222–38.
67. Boyce, A. J., J. S. J. Haight, D. B. Rimmer, and G. A. Harrison. 1974. Respiratory function in Peruvian Quechua Indians. Ann. Hum. Biol. 1:137–48.
68. Anderson, H. R., J. A. Anderson, H. M. King, and J. E. Cotes. 1978. Variations in the lung size of children in Papua, New Guinea: genetic and environmental factors. Ann. Hum. Biol. 5:209–18.
69. Dunnill, M. S. 1962. Postnatal growth of the lung. Thorax 17:329–33.
70. Frisancho, A. R. 1977. Developmental adaptation to high altitude hypoxia. Int. J. Biometeorol. 21:135–46.
71. Arias-Stella, J., and S. Recavarren. 1962. Right ventricular hypertrophy in native children living at high altitude. Am. J. Pathol. 41:55–62.
72. Peñaloza, D., J. Arias-Stella, F. Sime, S. Recavarren, and E. Marticorena. 1964. The heart and pulmonary circulation in children at high altitude. Physiological anatomical and clinical observations. Pediatrics 34:568–82.
73. Peñaloza, D., R. Gamboa, E. Marticorena, M. Echevarria, J. Dyer, and E. Gutierrez. 1961. The influence of high altitudes on the electrical activity of the heart: electrocardiographic and vectocardiographic observations in adolescence and adulthood. Am. Heart J. 61:101–7.
74. Greulich, W. W., and S. I. Pyle. 1959. Radiographic atlas of skeletal development of the hand and wrist, 2nd ed. Stanford, CA: Stanford University Press.
75. Donayre, J. 1966. Population growth and fertility at high altitude. In Life at high altitudes. Pub. 140. Washington, DC: Pan American Health Organization.
76. Greksa, L. P. 1990. Age of menarche in Bolivian girls of European and Aymara ancestry. Ann. Hum. Biol. 17:49–53.
77. Peñaloza, J. B. 1971. Crecimiento y desarrollo sexual del adolescente Andino. Ph.D. Thesis. Lima, Peru: Universidad Nacional de San Marcos.
78. Llerena, L. A. 1973. Determinación de hormona luteinizante por radioinmunoensayo. Variaciones fisiologicas y por efecto de la altura. Ph.D. Thesis. Lima, Peru: Universidad Peruana Cayetano Heredia, Instituto de Investigaciones de Altura.
79. Gurney, C. W., E. L. Simmons, E. O. Gaston, and J. C. Cox. 1980. The erythropoietic effect, in mice, of androgens combined with hypoxia. Exp. Hemat. 8:192–99.
80. Bille-Brahe, N-E., H. Kehlet, S. Madsbad, and M. Rorth. 1976. Effects of androgen or oxygen affinity in vivo and 2,3-diphosphoglycerate content of red cells in peripheral arterial insufficiency. Scan. J. Clin. Lab. Inv. 36:801–4.
81. Rothammer, F., and R. Spielman. 1972. Anthropometric variation in Aymara: genetic, geographic and topographic contributions. Am. J. Hum. Genet. 24:371–80.
82. Malina, R. M. 1974. Growth of children at different altitudes in Central and South America. Am. J. Phys. Anthropol. 40:144.
83. Frisancho, A. R. 1974. Triceps skinfolds and upper arm muscle size norms for the assessment of nutritional status. Am. J. Clin. Nutr. 27:1052–58.
84. Picon-Reategui, E. 1976. Nutrition. In P. T. Baker and M. A. Little, eds., Man in the Andes: a multidisciplinary study of high altitude Quechua natives. Stroudsburg, PA: Dowden, Hutchinson and Ross.
85. Mazess, R. B., and P. T. Baker. 1964. Diet of Quechua Indians living at high altitude: Nuñoa, Peru. Am. J. Clin. Nutr. 15:341–51.
86. Gursky, M. 1969. A dietary survey of three Peruvian highland communities. M.A. Thesis. University Park, PA: Pennsylvania State University.

87. Garn, S. M., and D. J. A. Haskell. 1960. Fat thickness and developmental status in childhood and adolescence. Am. J. Dis. Child. 99:746–51.
88. Leonard, W. R., T. L. Leatherman, J. W. Carey, and R. B. Thomas. 1990. Contributions of nutrition versus hypoxia to growth in rural Andean populations. Am. J. Hum. Biol. 2:613–26.
89. Carey, J. W. 1990. Social system effects on local level morbidity and adaptation in the rural Peruvian Andes. Med. Anthropol. Quart. 4:266–95.
90. Gonzales, G. 1982. Secular change in growth of native children and adolescents at high altitude I. Puno, Peru (3800 meters). Am. J. Phys. Anthropol. 58:191–95.
91. Picon-Reategui, E. 1961. Basal metabolic rate body composition at high altitude. J. Appl. Physiol. 16:431–34.
92. Mazess, R. B., E. Picon-Reategui, R. B. Thomas, and M. A. Little. 1969. Oxygen intake and body temperature of basal and sleeping Andean natives at high altitude. Aerospace Med. 40:6–9.
93. Baker, T. P., E. R. Buskirk, J. Kollias, and R. B. Mazess. 1967. Temperature regulation at high altitude: Quechua Indians and U.S. whites during total body cold exposure. Hum. Biol. 39:155–69.
94. Little, M. A. 1976. Physiological responses to cold. In P. T. Baker and M. A. Little, eds., Man in the Andes: a multidisciplinary study of high altitude Quechua natives. Stroudsburg, PA: Dowden, Hutchinson and Ross.
95. Little, M. A., and J. M. Hanna. 1978. The responses of high-altitude populations to cold and other stresses. In P. T. Baker, ed., The biology of high-altitude peoples. New York, NY: Cambridge University Press.
96. Picon-Reategui, E. 1962. Studies on the metabolism of carbohydrates at sea level and at high altitudes. Metabolism 11:1148–54.
97. Picon-Reategui, E. 1963. Intravenous glucose tolerance test at sea level and at high altitudes. J. Clin. Endocrinol. 23:1256–61.
98. Calderon, L., A. Llerena, L. Munive, and F. Kruger. 1966. Intravenous glucose tolerance test in pregnancy in women living in chronic hypoxia. Diabetes 15: 130–32.
99. Smith, C. F. 1993. Comparative birthweight data for Sherpa women living at high and low altitudes in Nepal. Am. J. Phys. Anthropol. 16 (abstract):183.
100. Wiley, A. S. 1993. Neonatal anthropometric characteristics in a high altitude population of the western Himalaya. Am. J. Hum. Biol. 5 (abstract):141.
101. Frisancho, A. R., H. G. Frisancho, R. Albalak, M. Villena, and E. Vargas. 1993. Lung volumes of natives and Bolivians of non-native ancestry living at high altitude. Am. J. Phys. Anthropol. 16 (abstract):89.
102. Eckard, R. B., and T. W. Melton. 1992. Population studies on human adaptation and evolution in the Peruvian Andes. In Occasional Papers in Anthropology. Museum of Anthropology. University Park, The Pennsylvania State University.

PART 5

# Accommodation to Energy Variability

# CHAPTER 13 Accommodation to the Energy Demands of Pregnancy

During pregnancy there are anatomical and physiological changes that affect almost every function of the body. These changes occur both as an adaptation to facilitate the appropriate milieu for the development of the fetus and as a preparation of the mother for the process of labor, birth, and lactation. The physiological changes include a 50% increase in plasma volume and a 5 to 10% decrease in hematocrit and hemoglobin concentrations. As a result of the increased plasma volume the work of the heart, measured by cardiac output, increases during pregnancy by about 50%. Owing to the increased demands for oxygen during the growth of maternal and fetal tissues ventilation increases through an increase in tidal volume (deep breath). Along with these changes, adjustments in kidney function through increased blood flow and glomerular filtration rate allow for the excretion of fetal and maternal metabolic by-products. Similarly, throughout the course of pregnancy the gastrointestinal function is altered. The most noticeable changes include increased appetite and gastrointestinal discomfort. Changes in taste also occur.

All of the above changes result in increased energy demands. It is estimated that an extra 68,000 Kcalories (250 kcal/day) must be ingested to support the total cost of pregnancy. Since the energy needs are not uniform throughout pregnancy the mother makes adjustments in metabolic activity. These adjustments are oriented toward the efficient utilization of nutrients and storage of energy to be used for the development of the fetus throughout the course of pregnancy and lactation. The metabolic adjustments are made through the action of hormones and energy-sparing changes.

## METABOLISM DURING PREGNANCY

### Resting Metabolic Rate

Studies of well-nourished women indicate that the resting metabolic rate (RMR) increases throughout pregnancy by an average of about 13% (1). This reflects the increased oxygen demands of the uterine-placental-fetal unit. As such, there is a direct relationship between RMR and birth weight. The relationship between RMR and birth weight is evident after the first trimester of pregnancy because it is then that rapid fetal growth occurs.

### Cost of the Increased Resting Metabolic Rate

Estimates of the cumulative cost of increased metabolic rate and maintenance are variable. Studies of well-nourished British and Dutch women indicate a cost of 36,000 kcal and 34,350 kcal (2, 3), while studies conducted in Sweden indicate a cost of about 46,000 kcal (4).

### Weight Gain and Its Components

The weight gained during pregnancy is the by-product of physiological processes oriented at promoting maternal and fetal growth. Estimates of body composition of well-nourished women (5) indicate that the average weight gain during pregnancy is about 11.6 kg (25.5 lb). The components of the weight gain are given in table 13.1. Much of the gain can be accounted for by the product of gestation. About half of the weight gain resides in the fetus, placenta, and amniotic fluid; the remainder represents the weight of maternal reproductive tissues such as the uterus, mammary gland, blood, and extracellular fluid that is excreted in the urine during birth. The net maternal fat mass gain of 2.5 kg represents the maternal fat stores. However, since the fat content of adipose tissue is 80%, the net weight of fat gain is only 2.0 kg ($2.5 \times .80 = 2.0$ kg or 4.4 lb).

### Pattern of Maternal Fat Gain

As illustrated in figure 13.1, during the first 32 weeks of gestation the maternal percentage of body fat increases by about 4.3%. Thereafter, the rate of fat gain remains stable or even declines, depending on the mother's energy intake. This pattern of fat gain, as previously indicated, is an opportunistic adaptation that enables the mother to store energy to support the needs of fetal growth and lactation.

### Energy Cost of Pregnancy: Including Cost of the Synthesis of New Tissue

Estimates of the cost of synthesis of fetal and maternal tissues are given in table 13.2. These data show that the energy cost of pregnancy is about 68,000 kcal (3, 6). Maternal fat gain accounts for 35% of the energy cost (about 2.0 kg). The synthesis of fat and protein for the development of the fetus, placenta, and maternal tissues accounts for 17.5% of the energy cost. The cumulative effects of basal metabolic rate and an increased metabolic rate

**TABLE 13.1. Maternal pregnancy gain and its components**

| Component | Weight (kg) (lb) |
|---|---|
| Fetus | 3.300 (7.26) |
| Placenta | 0.66 (1.45) |
| Uterus, amniotic fluid, mammary gland, blood, and extracellular fluid | 5.1 (11.2) |
| Net maternal fat mass gain | 2.5 (5.0) |
| Total | 11.56 (25.0) |

*Source:* Adapted from J. M. A. van Raaij, M. E. M. Peek, S. H. Vermaat-Miedema, C. M. Schonk, and J. G. A. J. Hautvast. 1988. New equations for estimating body fat mass in pregnancy from body density or total body water. Am. J. Clin. Nutr. 48:24–29 (5).

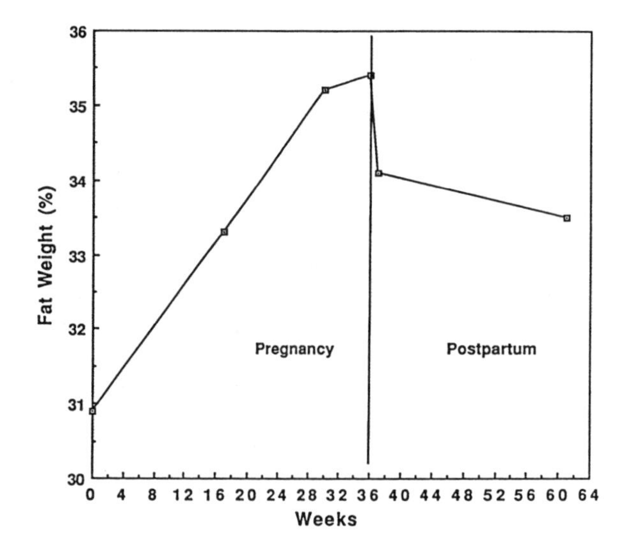


**Fig. 13.1.** Changes in maternal body composition with pregnancy. During the first eight months of pregnancy there is a dramatic increase in energy storage to be used by the fetus during the last stages of pregnancy.

account for 48% of the energy cost. Earlier estimates of the cost of pregnancy were as high as 80,000 kcal (2), but these estimates assumed a net fat gain of about 4.0 kg which is 2.0 kg higher than the present estimates (see table 13.3). In any event, an extra 68,000 Kcalories (250 kcal/day) must be ingested to support the total cost of pregnancy. Since the energy needs are not uniform throughout pregnancy the mother makes metabolic adjustments oriented at both the efficient utilization and storage of energy. These metabolic adjustments will be discussed below.

## Utilization and Derivation of Energy

Glucose provides the energy needs of the developing fetus. By term the daily fetal glucose uptake is about 30 g. To satisfy the demand for increased energy

**TABLE 13.2. Energy cost of pregnancy including cost of increased metabolic rate and tissue synthesis**

| Variable<br>Tissue Deposition | Scotland Energy Cost (kcal) | Netherlands Energy Cost (kcal) | Average Energy Cost (kcal) | Percent Total (%) |
|---|---|---|---|---|
| Maternal fat gain | 25,334 | 21,988 | 23,661 | 34.9 |
| Fetus | 8,126 | 8,222 | 8,174 | 12.1 |
| Placenta | 729 | 741 | 735 | 1.1 |
| Expanded maternal tissues | 2,891 | 2,940 | 2,915 | 4.3 |
| Cumulative cost of increased RMR | 30,114 | 34,416 | 32,265 | 47.6 |
| Total cost of pregnancy | 67,194 | 68,307 | 67,751 | |

*Source:* Adapted from J. V. G. A. Durnin. 1987. Energy requirements of pregnancy: An integration of the longitudinal data from the five-country study. Lancet 2:1131–33 (6).

availability a series of adjustments in carbohydrate, protein, and fat metabolism occur. These changes are oriented toward storing energy during the first half of pregnancy and providing a continuous supply of energy during the second half of pregnancy. The action of hormones such as insulin, *human chorionic somatomammotropin* (HCS), estrogen, progesterone, and maternal cortisol adjust the supply of energy. The quantity and metabolic activity of these hormones differ during the first and second half of pregnancy (7).

### During First Half of Pregnancy

During the first half of pregnancy the mother prepares for the anticipated fetal glucose demand by storing fat. As schematized in figure 13.2 the first half of pregnancy is characterized by an anabolic phase wherein an increase in the secretion of *progesterone* and *estrogen* leads to an increase in sensitivity to the action of insulin (referred to as insulin sensitivity). This results in reduced maternal plasma levels of glucose, amino acids, free fatty acids, and glycerol. Furthermore, carbohydrate and amino acid loads are readily assimilated. During this phase lipogenesis (the conversion of carbohydrates and proteins into fatty acids) is favored while lipolysis (the conversion of fatty acids into glucose) in maternal adipose tissue is stopped. In this way, glycogen stores in the mother's liver and muscle are increased, and protein synthesis is enhanced. The net effect is to stimulate the deposition of fat in maternal tissues and the growth of the breasts, uterus, and essential musculature in the mother. These changes are reflected in maternal fat gain, an energy store from which the mother and conceptus will draw during later pregnancy.

### During Second Half of Pregnancy

Figure 13.3 schematizes the metabolic changes during the last trimester of pregnancy. During this phase the placenta increases the secretion of the hormone *human chorionic somatomammotropin* (HCS), also called human placental lactogen (HPL). HCS is a polypeptide similar in structure and function to human growth hormone. This hormone causes the gravida to become

**TABLE 13.3. Fat gain during pregnancy**

| Population | *N* | Mean ± SE |
|---|---|---|
| Gambia | 52 | 0.6 (0.3) |
| Philippines | 51 | 1.3 (0.3) |
| Thailand | 44 | 1.4 (1.1) |
| Netherlands | 57 | 2.0 (0.3) |
| Scotland | 88 | 2.3 (0.3) |

*Source:* Adapted from J. V. G. A. Durnin. 1987. Energy requirements of pregnancy: An integration of the longitudinal data from the five-country study. Lancet 2:1131–33 (6).

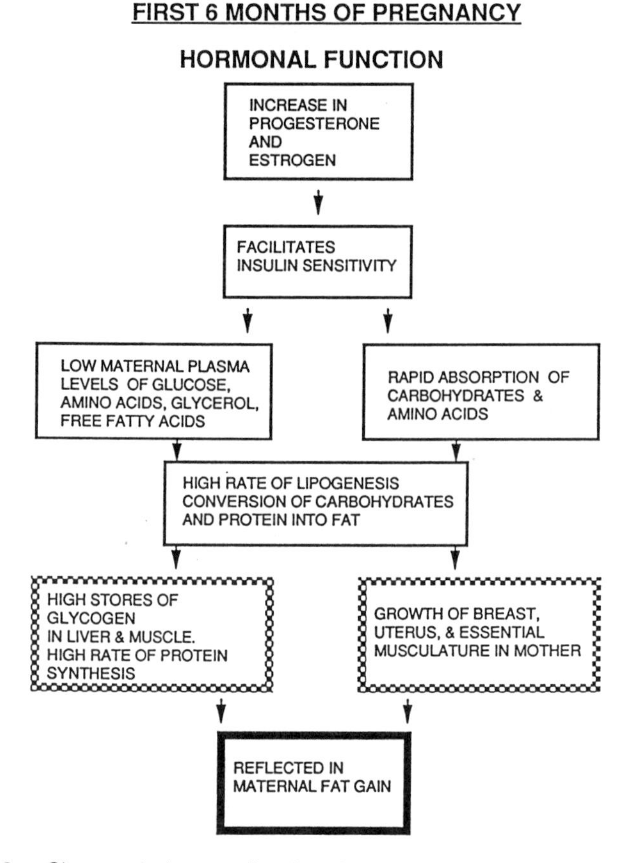


**Fig. 13.2.** Changes in hormonal action during the first six months of pregnancy. The increase in progesterone and estrogen concentrations leads to a rapid uptake of energy to be used for the growth of maternal reproductive organs and energy storage reflected in maternal fat gain.

resistant to the action of insulin. As a consequence of this change the assimilation of dietary carbohydrate, protein, and fat by maternal peripheral tissues slows, so that postprandial plasma levels of glucose and amino acids increase. During this phase the plasma concentration of cortisol is also increased. By antagonizing insulin and stimulating both gluconeogenesis (formation of new glucose) and lipolysis (conversion of fatty acids into glucose) this hormone elevates maternal blood glucose levels. This enhances the rate of glucose diffusion and facilitates amino acid transport across the placenta into the fetus. Since glucose is the major fuel of the fetus, and the amino acids are required for fetal protein synthesis, rapid diffusion of these components from maternal to conceptus causes rapid fetal growth. During the last trimester of pregnancy, the mother may feel more tired than during the first two trimesters as a result of the continuing energy demands of the fetus. Along with the other changes in maternal metabolism, plasma cholesterol and triglyceride levels rise throughout pregnancy. Cholesterol is partly used for estrogen and progesterone synthesis.

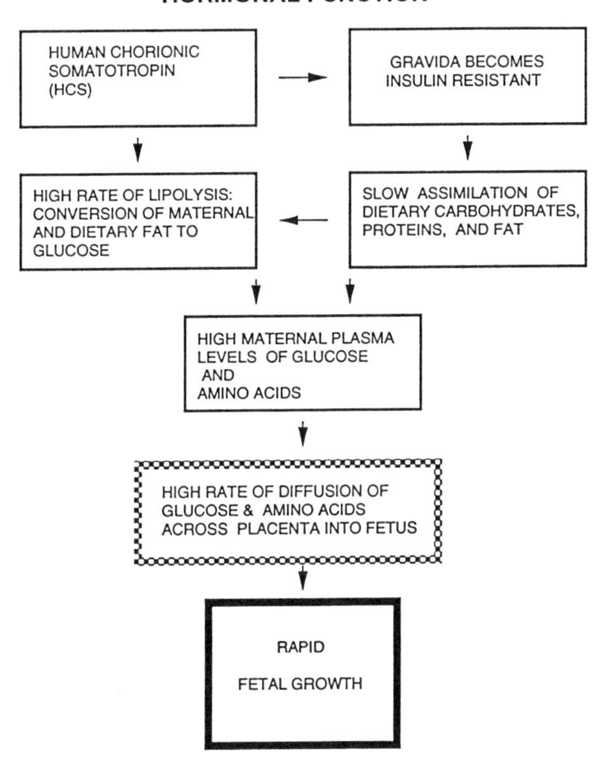


**Fig. 13.3.** Changes in hormonal action during the last three months of pregnancy. During this stage the placenta increases the secretion of the hormone human chorionic somatomammotropin, which in turn decreases the uptake of energy by the peripheral tissues but increases the diffusion of energy in the form of glucose into the fetal tissues. During the last trimester of pregnancy the fetus grows very rapidly as a consequence of these changes in hormonal activity.

## BIRTH WEIGHT

Birth weight is the final common pathway for the expression of pregnancy. Figure 13.4 illustrates that during the first 20 weeks of gestation the fetus grows very slowly, but between 28 and 37 weeks it grows rapidly (at about 200 g per week), so that between 37 and 43 weeks white newborns weigh on the average about 3300 g (7.26 lb) (see table 13.4).

### Classification of Newborns

Newborns are usually classified by gestational age, by birth weight, and by both birth weight and gestational age.

(1) **By Gestational Age.** Based on gestational age at delivery, newborns are classified into three categories (8): (i) preterm or premature, infants born

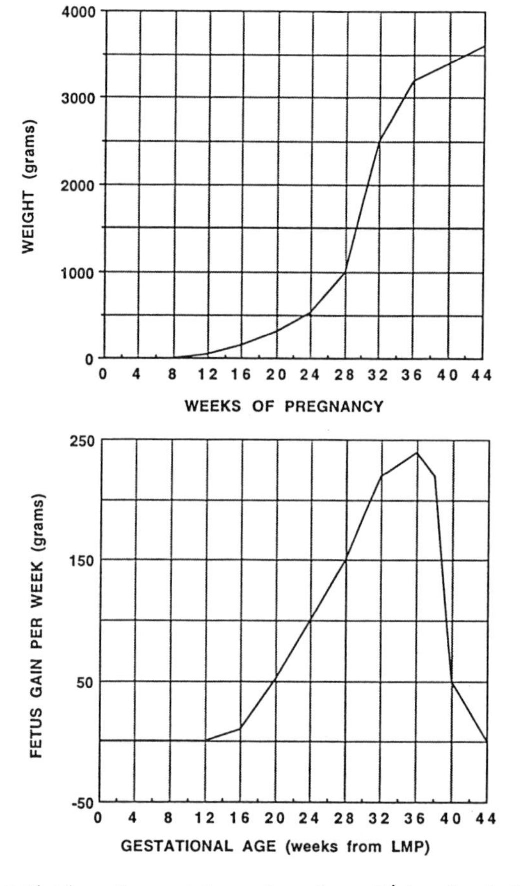


**Fig. 13.4.** Schematization of prenatal growth pattern. During the first twenty weeks of gestation the fetus grows very slowly, but thereafter it grows very rapidly.

with gestational age of less than 37 weeks ( 259 days); (ii) term or full term, infants born with gestational age of more than 37 weeks (>259 to 293 days) but less than 42 completed weeks; and (iii) post-term, infants born with gestational age of more than 42 completed weeks (>294 days).

(2) **By Birth Weight.** Based on birth weight, newborns are classified into four categories: (i) low birth weight, infants born with a weight of less than 2500 g (5.5 lb); (ii) average birth weight, infants whose birth weight ranges from 2501 to 3500 g (5.51 to 7.70 lb); (iii) large birth weight, infants whose birth weight ranges between 3500 g and 4540 g (7.71 to 9.98 lb); and (iv) very large birth weight, infants whose birth weight exceeds 4540 g (>10 lb).

(3) **By Birth Weight and Gestational Age.** Using as reference the percentile distribution of the weight by age, newborns are classified into three categories: (i) small-for-gestational-age (SGA), infants whose birth weight is below (or does not correspond) to the expected gestational age. Using the 10th percentile of the weight-by-age reference as the cutoff point, infants in the SGA category are considered small because of intra-uterine growth retardation; (ii) adequate-for-gestational-age (AGA), infants whose birth weight

corresponds to their gestational age even if their weight is below 2500 g; and (iii) large-for-gestational-age (LGA), infants whose birth weight is above (or does not correspond to) the expected gestational age. Infants in this category are large because of rapid intra-uterine growth.

## Determinants of Birth Weight Variability

The sources of variability in birth weight are numerous. Current research indicates that gestational age, sex of the newborn, birth order, parity, and the biological and anthropometric traits of the mother such as age, prepregnancy height, prepregnancy weight, weight gain, and ethnicity are important sources of variability in birth weight. While socioeconomic factors such as income, education, and access to prenatal care are known to influence birth weight, the effects of these factors are usually expressed through their influence on maternal biological traits.

### Gestational Age

By far, gestational age is the most important component of variability in birth weight. The influence of environmental factors on gestational age becomes apparent in the third trimester of pregnancy (see fig. 13.5).

**TABLE 13.4. Percentiles of birth weight (g) and crown heel length (cm) from 28 weeks of gestation to term, based on infants who survived**

| Gest. | Percentile | | | | | | | | | |
|---|---|---|---|---|---|---|---|---|---|---|
| Age | 10th | 25th | 50th | 75th | 90th | 10th | 25th | 50th | 75th | 90th |
| (weeks) | Birth Weight (g) | | | | | Crown Heel Length (cm) | | | | |
| 28 | 967 | 1085 | 1165 | 1245 | 1363 | 36.2 | 37.3 | 38.0 | 38.7 | 39.8 |
| 29 | 1124 | 1226 | 1295 | 1364 | 1466 | 37.3 | 38.3 | 39.0 | 39.7 | 40.7 |
| 30 | 1232 | 1356 | 1440 | 1524 | 1648 | 38.7 | 39.7 | 40.3 | 40.9 | 41.9 |
| 31 | 1386 | 1515 | 1601 | 1687 | 1816 | 39.4 | 40.6 | 41.4 | 42.2 | 43.4 |
| 32 | 1534 | 1675 | 1769 | 1863 | 2004 | 40.6 | 41.7 | 42.5 | 43.3 | 44.4 |
| 33 | 1703 | 1854 | 1955 | 2056 | 2207 | 41.8 | 43.1 | 43.9 | 44.7 | 46.0 |
| 34 | 1791 | 2012 | 2160 | 2308 | 2529 | 42.8 | 44.3 | 45.3 | 46.3 | 47.8 |
| 35 | 2008 | 2235 | 2387 | 2539 | 2766 | 44.0 | 45.6 | 46.7 | 47.8 | 49.4 |
| 36 | 2121 | 2420 | 2621 | 2822 | 3121 | 45.2 | 46.9 | 48.0 | 49.1 | 50.8 |
| 37 | 2354 | 2667 | 2878 | 3089 | 3402 | 46.1 | 47.8 | 48.9 | 50.0 | 51.7 |
| 38 | 2570 | 2898 | 3119 | 3340 | 3668 | 46.7 | 48.5 | 49.7 | 50.9 | 52.7 |
| 39 | 2676 | 3039 | 3283 | 3527 | 3890 | 47.0 | 48.8 | 50.0 | 51.2 | 53.0 |
| 40 | 2776 | 3143 | 3388 | 3633 | 4000 | 47.1 | 49.1 | 50.4 | 51.7 | 53.7 |
| 41 | 2778 | 3190 | 3466 | 3742 | 4154 | 47.1 | 49.3 | 50.7 | 52.1 | 54.3 |
| 42 | 2778 | 3209 | 3498 | 3787 | 4218 | 47.1 | 49.3 | 50.8 | 52.3 | 54.5 |
| 43 | 2780 | 3210 | 3499 | 3788 | 4218 | 47.2 | 49.4 | 50.9 | 52.4 | 54.6 |

*Source:* Adapted from R. L. Naeye and J. B. Dixon. 1978. Distortions in fetal growth standards. Pediat. Res. 12:987–91.

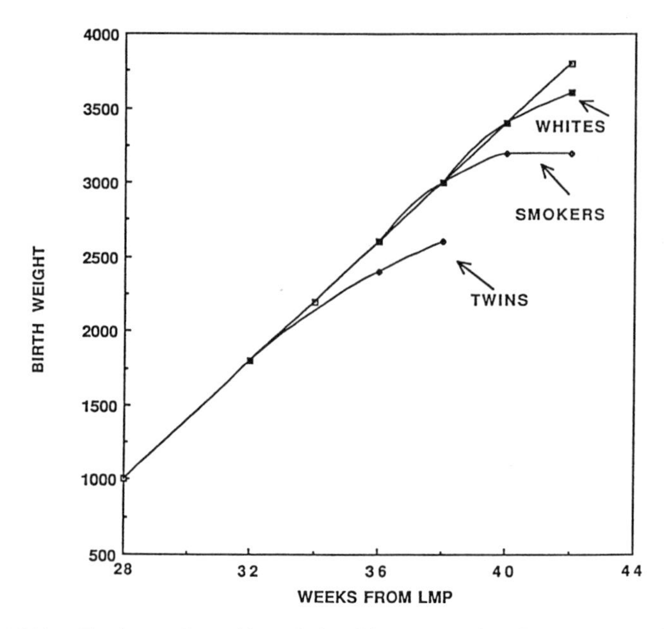


**Fig. 13.5.** Fetal growth and its relationship to gestational age and environmental factors. The effects of environment on gestational age become apparent in the third trimester of pregnancy.

## Sex

Boys and girls grow at similar rates until the 28th week of gestation. Thereafter, the growth rate of boys is faster than that of girls, so that at term, boys weigh an average of about 150 g (0.33 lb) more than girls. Despite the lower birth weight, girls have a lower mortality rate than boys. This difference is related to a great extent to the fact that at a given birth weight girls, for reasons intrinsic to the genome, are on the average more mature than boys.

## Birth Order and Parity

In general, irrespective of sex, second and third borns are about 100–400 g heavier than first-born infants. From the third to the fifth pregnancy birth weight remains stable, but after the sixth pregnancy it tends to decline, especially if the pregnancy is associated with increased maternal age. Among well-nourished women, the effects of parity on the second born probably reflect the increased capacity of maternal circulation and reproductive organs to nourish and contain the growing fetus (2). This capacity, however, appears to level off between the second and fifth pregnancy, and probably starts to decline thereafter. This is because high parities are usually associated with increased maternal age. It should be noted that parity per se does not affect neonatal mortality. In fact, at a given birth weight the risk of mortality is the same across parities.

### Ethnicity and Optimum Birth Weight

Optimum birth weight is defined as the mean birth weight associated with the lowest mean mortality (9–15). For most populations, the optimum birth weight is higher than the average birth weight. As shown in figure 13.6 the lowest mortality rates for whites and blacks are in the ranges of 3001–3500 g. This figure also shows that at birth weights below 2500 g nonwhites had lower mortality rates than whites. On the other hand, at birth weights greater than 4000 g whites had lower mortality rates than nonwhites.

### Ethnicity and Low Birth Weight

According to recent vital statistics in the U.S. the incidence of low birth weight (<2500 g) for black infants is about 12.7% compared to 5.7% for whites. The greater proportion of low birth weight in black infants indicates that a greater number of black infants than white infants are at a higher risk of neonatal mortality (even though the neonatal mortality for low birth weight infants is lower in blacks than whites). This accounts for the increased neonatal mortality of black infants when compared to that of whites.

### Ethnicity and Mean Birth Weight

At birth, black infants weigh between 150 and 200 g less than white infants depending on the sample. The average birth weight for full-term blacks is about 3100 g (6.82 lb) while that of whites is about 3400 g (7.48 lb). It has been known for some time that black mothers represent a high-risk group. However, as shown in table 13.5, even when adjusted for differences in socioeconomic conditions and prenatal care, black infants are born with a lower weight than white infants. It appears that improvement in socioeonomic conditions lowers the percent of low birth weight infants but does not change the mean birth weight of blacks. At present, we do not know whether the lower birth weight reflects a true genetic difference or simply a difference in the mother's response to the growth retarding factors.

### Paternal Size

Since half of the fetal genes are inherited from the father one would expect that variability in birth weight must be related to paternal characteristics. However, most studies indicate that variability in birth weight is related to the size of the mother rather than that of the father. Studies of heritability among Japanese (16) infants indicate that the correlation in birth weight among siblings who have the same mother but a different father was significantly higher (r = 0.58) than the correlation (r = 0.102) of siblings who had

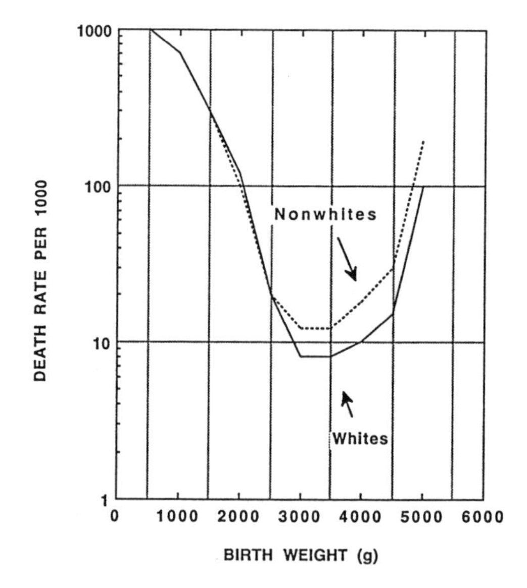


**Fig. 13.6.** Fetal mortality by birth weight among ethnic groups. The lowest mortality (optimum birth weight) for most populations occurs at a birth weight that is higher than the average birth weight. The mortality rate for nonwhites at birth weights below 2500 g is lower than those of whites, but the mortality rate at birth weights above 4000 g is greater for nonwhites than for whites.

the same father but a different mother. Similarly, MacKeown and Record (17) showed that birth weights were correlated with maternal height (r = 0.31) but not with paternal height (r = – 0.03). Based on these findings Hytten and Leitch (2) concluded that birth weight depends more upon the maternal environment than upon the characteristics of the father. However, as shown below, the father's ethnicity does play a role in birth weight variability.

## Ethnicity and Paternal Ethnicity

Recent studies indicate that an infant's birth weight is related to ethnicity and stature of both the mother and father (18–24). As previously indicated, black infants have a lower birth weight than white infants. An important question is whether the low birth weight of black infants is influenced by

**TABLE 13.5. Comparison of birth weight (g) of black and white infants adjusted for differences in maternal age, prenatal care, education, and per capita income**

| | Whites | | Blacks | | |
|---|---|---|---|---|---|
| Sex | *N* | Mean ± SE | *N* | Mean ± SE | F-test |
| Males | 1497 | 3578.9 ± 11.9 | 489 | 3363.0 ± 21.2 | 74.1 |
| Females | 1337 | 3449.2 ± 11.9 | 514 | 3246.2 ± 19.9 | 72.3 |

*Source:* A. R. Frisancho, and S. L. Smith. 1990. Reduction in birth weight associated with smoking among young and older-age women. Am. J. Hum. Biol. 2:85–88 (97).

the father's ethnicity. Since as a group black fathers were probably of lower birth weight than white fathers, their contribution to variability in birth weight can be evaluated by comparing birth weights by parental ethnicity. Based on data analysis of the U.S. Vital Statistics of 1983 (25), table 13.6 shows that having a white father seems to increase birth weight by 116 g while having a black father seems to decrease birth weight by about 101 g. These findings suggest that birth weight also depends, although to a lesser extent, upon the ethnicity of the father.

## Maternal Age

**Younger than 16 Years.** Third World populations are experiencing a massive increase in the migration of rural populations into the urban centers. One consequence of migration is the disruption of cultural norms that govern marriage and reproduction. As a result of this disruption the number of pregnancies among teenagers has drastically increased. One consequence of the increased incidence of teenage pregnancy is the increase in the frequency of low birth-weight infants. Several investigations have shown that teenagers (younger than 16 years) have infants whose birth weight is about 100 to 250 grams less than infants born to adult women (26) of the same pregnancy weight. Infants born to mothers younger than 16 years of age are at a high risk of being low birth weight. As illustrated in figure 13.7, this reduction is the result of a maternal-fetal competition for nutrients. Contrary to previous assumptions, pregnant teenagers younger than 16 years continue to grow (27). Therefore, to complete their own growth they compete for nutrients with their fetus, reducing birth weight by 100 to 250 grams. The reduction in birth weight might also be related to the small size of the birth canal that is associated with early maturation. The pelvis continues to grow through adolescence; therefore, girls who mature and get pregnant earlier have smaller and less-mature pelvises than do average or late maturers (28).

**TABLE 13.6. Comparison of birth weight (g) by parental ethnicity**

| | White Mother White Father | Black Mother White Father | White Mother Black Father | Black Mother Black Father |
|---|---|---|---|---|
| *N* | 24,076 | 18,025 | 5,619 | 15,245 |
| Mean[a] | 3,426 | 3,382 | 3,325 | 3,266 |
| Difference | | | | – 160[b] |
| Effect of white father | | + 116[c] | | |
| Effect of black father | | | – 101[d] | |

*Source:* Adapted from A. Migone, I. Emanuel, B. Mueller, J. Daling, and R. E. Little. 1991. Gestational duration and birth weight in white, black, and mixed race babies (unpublished manuscript) 24.

[a] Adjusted for differences in maternal age, education, marital status, live birth order, previous fetal deaths, and baby's sex.

[b] When compared to the birth weight of infants born to white mother–white father.

[c] When compared to the birth weight of infants born to black mother–black father.

[d] When compared to the birth weight of infants born to white mother–white father.

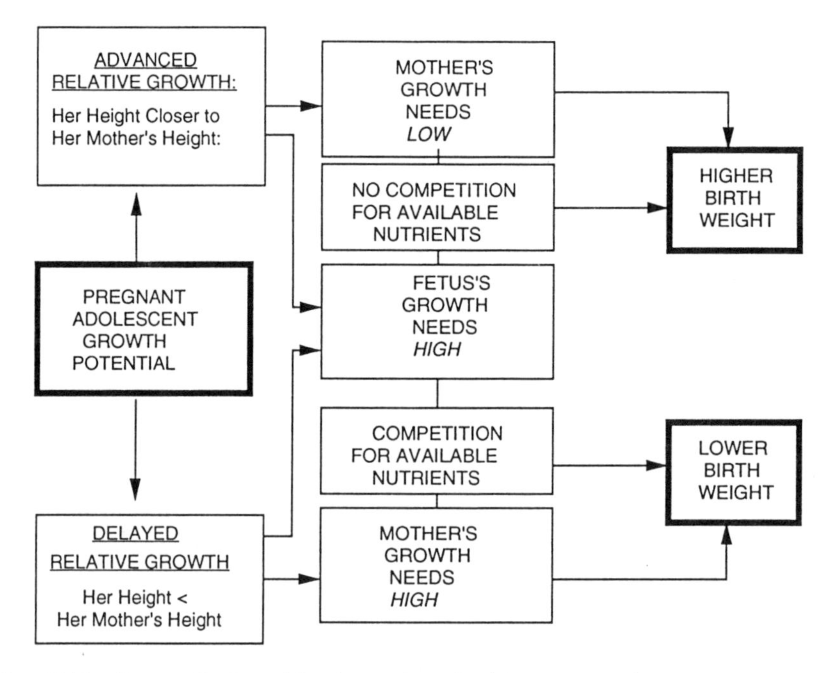


**Fig. 13.7.** Schematization of the interaction of adolescent growth status, physiological maturity, placenta function, and maternal and fetal growth needs. Because pregnant teenagers younger than 16 years continue growing, they compete with their fetuses for nutrients, resulting in a lower birth weight of their infants.

**Optimal Age.** Among well-nourished populations the highest mean birth weight occurs among women whose age is between 20 and 34 years of age. Mothers within this age range are least likely to have a low birth-weight infant.

**Older Age.** Primiparous women older than 40 years have infants whose weight is about 100–200 g less than those born to women aged 20 to 34 years. Infants born to mothers over 40 years are at a high risk of being low birth weight. This reduction has been attributed to a decline of hormonal function associated with aging. However, the fact that multiparous older women have infants whose weight is similar to that of younger women indicates that aging may not influence birth weight.

## Prepregnancy Weight and Weight Gain

Because prepregnancy weight is a measure of maternal fat stores and weight gain an indicator of food intake during pregnancy, one would expect that variability in these components would have an effect on birth weight variability. In fact, the evidence suggests that both prepregnancy weight and weight gain during pregnancy have an independent effect on birth weight. When the factors are combined there is an interactive effect whereby greater weight gain will be most effective in increasing birth weight among women of low prepregnancy weight (see table 13.7). This effect diminishes in normal or overweight women (see fig. 13.8). In general, prepregnancy weight and weight gain explains 5–10% of the variance in birth weight. Studies in well-nourished populations have shown that body mass index (BMI) is no more effective than prepregnancy weight in predicting the size of the newborn (29).

### Maternal Height

Since height is primarily indicative of past nutritional status while prepregnancy weight is a measure of both current and past nutritional status the effects of these factors on birth weight should differ depending on the nutritional history of the sample. The effect of height on birth weight should be greater among chronically undernourished populations than among well-nourished ones. Studies in developed nations indicate that height has very little influence on birth weight. On the other hand, among chronically undernourished populations short stature has been found to be more predictive of low birth weight than prepregnancy weight. Studies investigating the effects of supplementation on birth weight in Guatemala (30–31), Gambia (32), and East Java (33) demonstrate that after controlling for supplement (kcal), parity, and gestational age, pregnancy height had a significant effect on birth weight variability.

### Maternal Fat and Muscle

Since differences in the amounts of subcutaneous fat reflect differences in caloric reserves and since muscle growth requires adequate protein intake, variability in body muscle indicates variability in body protein reserves (34, 35). One of the functions of maternal weight gain during pregnancy is to provide energy stores for the growth of the fetus as well as for future lactation. Measurements of skinfold thickness, arm circumference, and estimates of mid-arm muscle area have been used as indicators of body composition. Figure 13.9 summarizes the results of an extensive study on the relationship between

**TABLE 13.7. Pregnancy weight gain by category of weight and body mass index**

| Category | BMI (kg/cm$^2$) | Weight Gain (kg) | (lb) |
|---|---|---|---|
| Underweight | <19.8 | 12.7-18.2 | 28-40 |
| Average weight | 19.8-26.0 | 11.4-16.0 | 25-35 |
| Overweight | >26.0-29.0 | 6.8-11.4 | 15-25 |

*Source:* Adapted from the Subcommittee on Nutritional Status and Weight Gain During Pregnancy. 1990. Nutritional status and weight gain during pregnancy of the food and nutrition board. Institute of Medicine. Part 1. Washington, DC: National Academy Press.

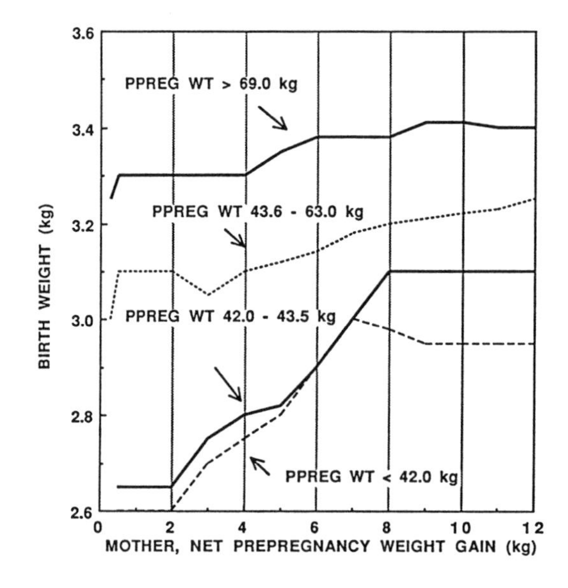


**Fig. 13.8.** Relationship between prepregnancy weight, pregnancy weight gain, and birth weight. An increase in pregnancy weight is most effective in increasing birth weight among women of low prepregnancy weight, but least effective among overweight women.

maternal nutritional status and prenatal growth in a Peruvian urban population (34, 35). The data show a direct relationship between the type of maternal nutritional reserves and prenatal growth. High muscle and high fat (indicative of increased maternal reserves of protein and calories) are associated with increased weight and recumbent length of the newborn. Viegas (36) demonstrated that Asian mothers living in England who increased their triceps skinfolds by less than 0.02 mm/week during the second trimester were nutritionally at risk of having a low birth-weight infant. Similarly, triceps and subscapular skinfolds (37) were significantly different between normal and small-for-gestational-age infants. Arm circumference, when measured at the appropriate time during pregnancy, can accurately predict birth weight and is useful for monitoring high risk, low birth-weight infants (38). Similarly, in cases where weight gain data are not available, arm circumference has been found to be useful as a proxy for prepregnancy weight. In Guatemala (39) maternal arm circumference was found to be significantly related to birth weight. In Bangladesh (40) arm circumference along with pregnancy weight gain was found to be significantly related to neonatal mortality. These findings support the view that variability in skinfold thickness is a good indicator of maternal caloric reserves and as such a good predictor of prenatal growth.

## Maternal Drug Habits

Despite the recent increased awareness of the deleterious effects of drugs on the newborn nearly 20% of the women sampled smoke and consume

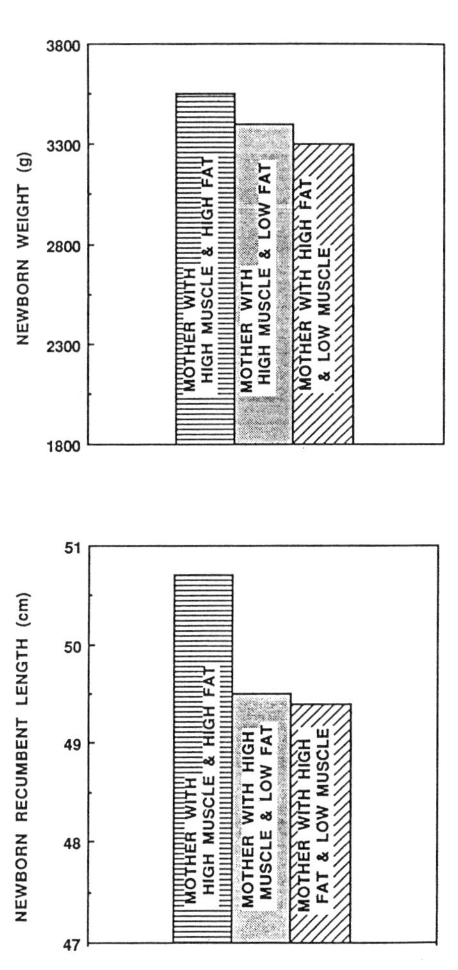


**Fig. 13.9.** Relationship of muscularity and fatness to newborn weight and recumbent length. Muscular mothers have longer and heavier newborns than fat mothers.

alcohol during pregnancy. A similar proportion is known to smoke marijuana and use cocaine. The effects of these drugs on birth weight appear to be independent of socioeconomic factors and maternal biological factors.

**Smoking.** In the industrialized countries, smoking is the most consistent and important determinant of birth weight. The reduction in birth weight associated with smoking ranges from an average of 150 g to 200 g. As illustrated in figure 13.10 (41) the reduction in mean birth weight is inversely related to the number of cigarettes smoked. Similarly, the frequency of low birth-weight infants is proportional to the number of cigarettes smoked (42). A unique feature of smoking is the reduction in birth weight associated with muscle but not fat (43). The mechanisms whereby smoking causes intrauterine growth retardation have been attributed to hypoxia (44–50). It is well known that smoking increases the blood's concentration of carbon monoxide, which in turn binds itself with hemoglobin forming carboxyhemoglobin and,

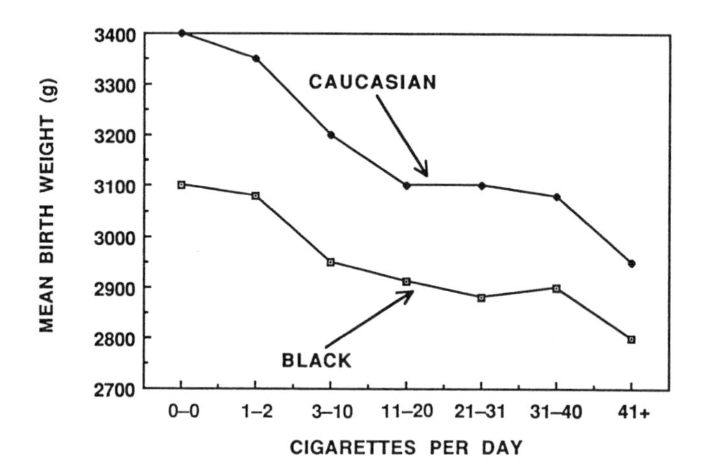


**Fig. 13.10.** Maternal cigarette use and infant birth weight. In both white and black groups the reduction in birth weight is inversely associated with the number of cigarettes smoked.

as such, decreases the oxygen-carrying capacity of fetal blood. In addition it has been postulated that the hypoxia caused by carbon monoxide impairs the synthesis of proteinic tissues (47). Various experimental studies on animals indicate that hypoxia decreases the rate of protein synthesis in vivo in the brain and certain visceral tissues (47). Studies in humans using direct and indirect measures of body composition have demonstrated that infants born to smokers as compared to nonsmokers had a significant reduction of body muscle (48, 49) but not fat (48–50). It should also be noted that studies at high altitude found that the birth weight and muscle area of infants born at high altitude were reduced as compared to those born at low altitude, but the skinfold thicknesses were similar at high and low altitude (51) (see chap. 12). Others postulate that the reduction in birth weight may be related to reduction in uterine blood flow and changes in blood pressure resulting from the absorption of nicotine in tobacco smoke (52–54) and from changes in energy expenditure associated with the nicotine in tobacco smoke (55). It should be noted, however, that the reduction in birth weight associated with smoking is primarily related to reduction in muscle, while reduction in birth weight associated with chronic malnutrition is related with reduction of fat tissue and not of muscle area (56).

**Alcohol.** Infants born to chronic alcoholics exhibit a series of specific morphological and physiological anomalies known as the ''fetal alcohol syndrome.'' This syndrome encompasses two somatic effects—growth retardation and dysmorphic faces accompanied by mental dullness. These anomalies include low birth weight, flat nasal bridge, epicanthal folds, flattened maxilla, cleft lip with or without cleft palate, small teeth, malformed ears, etc. (57). At present, the physiological mechanisms whereby alcohol affects prenatal growth are not well defined. It is quite possible that fetal growth retardation results from the peripheral vasodilation and reduction of blood flow to the internal organs as a result of alcohol consumption. Since alcohol consumption is associated with smoking, the role of alcohol has not been separated from that of smoking and the probable role of individual susceptibility remains to be defined. Furthermore, the timing and duration of the exposure necessary to produce the fetal alcohol syndrome is not clear.

**Marijuana.** Experimental studies on rats indicate that cannabiloids (from marijuana) ingested during pregnancy cause prenatal growth retardation. Epidemiologic studies indicate that infants born to marijuana users had a higher incidence of nervous system anomolies than infants born to nonusers (58). Studies have also found a direct relationship between marijuana use during pregnancy and the incidence of low birth-weight infants (59). Similarly, marijuana users were five times more likely than nonusers to deliver infants with characteristics of fetal alcohol syndrome. However, other studies have found that the effects of marijuana are not easily separated from the effects of alcohol, other habits, and the individual's medical history (60). Evaluations of fetal growth indicate that after exclusion of potentially confounding factors marijuana use during pregnancy is associated with reduction of 79 g in birth weight and 0.5 cm in birth length (61).

**Cocaine.** Clinical studies indicate that cocaine users have a higher incidence of specific complications during pregnancy including miscarriages, placental abruptions, and preterm labor as compared to methadone users and non–drug users (62). Furthermore, cocaine-exposed infants had significantly lower birth weights, shorter gestational periods, and smaller head circumferences than controls (62, 63). A major deterioration of prenatal growth has been reported among infants born to cocaine and crack users. Infants of all drug abusers weighed on average 423 g less than controls (64), amounting to a reduction of nearly 13% in prenatal growth. This reduction is of special significance since prenatal growth retardation is also associated with increased frequency of congenital cardiac anomalies (65).

## OVERVIEW

During pregnancy there are anatomical and physiological changes that affect almost every function of the body. These changes occur both as an adaptation to facilitate the appropriate milieu for the development of the fetus and as a preparation of the mother for the processes of labor, birth, and lactation. Owing to the increased demands for oxygen required for the growth of the maternal and fetal tissues, maternal energy needs increase drastically. It is estimated that an extra 68,000 kcalories (250 kcal/day) must be ingested to support the total cost of pregnancy. Since the energy needs are not uniform throughout pregnancy, the mother makes adjustments in metabolic activity. These are oriented toward both the efficient utilization of nutrients and the storage of energy through the action of hormones and energy-sparing functions to be used for the development of the fetus throughout the course of pregnancy and lactation. For example, during the first trimester of pregnancy the mother stores in her body a large supply of energy in the form of fat gain, while during the second half of pregnancy the mother decreases the consumption of glucose and amino acids so as to facilitate their use by the fetus. All of these enable the fetus to grow rapidly, so that during the last two months of pregnancy it doubles its weight.

Beside gestational age and sex, ethnicity is an important component of variability in birth weight. In the U.S. black infants weigh on average between 100 to 200 g less than white newborns. This difference persists even when adjusted for differences in pregnancy weight gain and socioeconomic characteristics. Recent analyses indicate that the reduction in birth weight is associated with maternal ethnicity and, to a lesser extent, with the ethnicity of the father. The major determinants of variability in birth weight of full-term infants associated with the biological characteristics of the mother include age, prepregnancy height, prepregnancy weight, and weight gain. There is a great deal of information indicating that socioeconomic factors such as income, education, and access to prenatal care influence birth weight, but the effects of these factors are usually expressed through their influence on maternal prepregnancy weight and pregnancy weight gain. Thus, the poorer the socioeconomic status the lower the prepregnancy weight, pregnancy weight gain, and birth weight. Cigarette smoking and drug abuse including cocaine and alcohol consumption are associated with a high incidence of low birth-weight infants.

## References

1. Forsum, E., A. Sadurskis, and J. Wager. 1988. Resting metabolic rate and body composition of healthy Swedish women during pregnancy. Am. J. Clin. Nutr. 47:942–47.
2. Hytten, F. E., and I. Leitch 1971. The physiology of human pregnancy, 2nd ed. Oxford: Blackwell Scientific Publications.
3. van Raaij, J. M. A., C. M. Schonk, S. H. Vermaat-Miedema, M. E. M. Peek, and J. G. A. J. Hautvast. 1989. Body fat mass and basal metabolic rate in Dutch women before, during, and after pregnancy: a reappraisal of energy cost of pregnancy. Am. J. Clin. Nutr. 49:765–72.
4. Forsum, E., A. Sadurskis, and J. Wager. 1985. Energy maintenance cost during pregnancy in healthy Swedish women. Lancet 1:107–8.
5. van Raaij, J. M. A., M. E. M. Peek, S. H. Vermaat-Miedema, C. M. Schonk, and J. G. A. J. Hautvast. 1988. New equations for estimating body fat mass in pregnancy from body density or total body water. Am. J. Clin. Nutr. 48:24–29.

6. Durnin, J. V. G. A. 1987. Energy requirements of pregnancy: an integration of the longitudinal data from the five-country study. Lancet 2:1131–33.
7. Genuth, S. M. 1983. The endocrine system. In R. M. Berne and M. N. Levy, eds., Physiology. St. Louis, MO: C. V. Mosby. 895–1069.
8. World Health Organization. 1977. Recommended definitions, terminology and format for statistical tables related to the perinatal period and use of a new certificate for cause of perinatal deaths (modifications recommended by FIGO as amended October 14, 1976). Acta Obstet. Gynaecol. Scand. 56:247–53.
9. Karn, M., and L. S. Penrose 1951. Birth weight and gestation in relation to maternal age, parity and infant survival. Ann. Eugen. 16:147–64.
10. Fraccaro, M. 1956. A contribution to the study of birth weight based on an Italian sample. Ann. Hum. Genet. 20:282–87.
11. Millis, J. 1959. Distribution of birth weights of Chinese and Indian infants born in Singapore: Birth weight an index of maturity. Ann. Hum. Genet. 23:164–70.
12. Jayant, K. 1964. Birth weight and some other factors in relation to infant survival. A study on an Indian sample. Ann. Hum. Genet. Lond. 27:261–70.
13. Hollingsworth, M. J. 1965. Observations on the birth weights and survival of African babies: single births. Ann. Hum. Genet. Lond. 28:291–300.
14. Promboon, S., M. P. Mi, and K. Chaturachinda. 1983. Birth weight, placental weight and gestation time in relation to natural selection in Thailand. Ann. Hum. Genet. 47:133–41.
15. Terrenato, L., M. F. Gravina, and L. Ulizzi. 1981. Natural selection associated with birth weight. I. Selection intensity and selective deaths from birth to one month of life. Ann. Hum. Genet. 45:55–63.
16. Morton, N. E. 1955. The inheritance of human birthweight. Ann. Hum. Genet. 200:125–34.
17. McKeown, T., and R. G. Record. 1954. Influence of pre-natal environment on correlation between birth weight and parental height. Am. J. Hum. Genet. 6:457–63.
18. Ounsted, M., V. A. Moar, and A. Scott. 1985. Children of deviant birthweight: The influence of genetic and other factors on size at seven years. Acta Paediatr. Scand. 74:707–12.
19. Magnus, P. 1984. Further evidence for a significant effect on fetal genes on variation in birth weight. Clin. Genet. 26:289–96.
20. Little, R. E., and C. F. Sing. 1987. Genetic and environmental influences on human birthweight. Am. J. Hum. Genet. 40:512–26.
21. Ounsted, M., and C. Ounsted. 1968. Rate of intrauterine growth. Nature 220: 599–600.
22. Hackman, E., I. Emanuel, G. van Belle, and J. Daling. 1983. Maternal birthweight and subsequent pregnancy outcomes. JAMA 250:2016–19.
23. Klebanoff, M. A., B. I. Graubard, S. S. Kessell, and H. W. Berendes. 1984. Low birthweight across generations. JAMA 252:2423–517.
24. Little, R. E. 1987. Mother's and father's birthweight as predictors of infant birth weight. Paediat. Perinat. Epid. 1:19–31.
25. Migone, A., I. Emanuel, B. Mueller, J. Daling, and R. E. Little. 1991. Gestational duration and birthweight in white, black, and mixed race babies. Personal communication.
26. Frisancho, A. R., J. Matos, W. R. Leonard, and L. A. Yarochi. 1985. Developmental and nutritional determinants of pregnancy outcome among teenagers. Am. J. Phys. Anthropol. 66:247–61.

27. Hediger, M. L., T. O. Scholl, I. G. Ances, D. H. Belsky, and R. W. Salmon. 1990. Rate and amount of weight gain during adolescent pregnancy: associations with maternal weight-for-height and birth weight. Am. J. Clin. Nutr. 52:793–99.
28. Moerman, M. L. 1982. Growth of the birth canal in adolescent girls. Am. J. Obstet. Gynecol. 148:528–32.
29. Garn, S. M., and S. D. Pesick. 1982. Relationship between various maternal body mass measures and size of the newborn. Am. J. Clin. Nutr. 36:664–68.
30. Lechtig, A., H. Delgado, C. Yarbrough, J. P. Habicht, R. Martorell, and R. E. Klein. 1976. A simple assessment of the risk of low birthweight to select women for nutritional intervention. Am. J. Obstet. Gynecol. 125:25–34.
31. Villar, J., and J. Rivera. 1988. Nutritional supplementation during two consecutive pregnancies and the interim lactation period: effect on birthweight. Pediatrics 81:51–57.
32. Prentice, A. M., T. J. Cole, F. A. Foord, W. H. Lamb, and R. G. Whitehead. 1987. Increased birthweight after prenatal dietary supplementation of rural African women. Am. J. Clin. Nutr. 46:912–25.
33. Kardjati, S., J. A. Kusin, and C. De With. 1988. Energy supplementation in the last trimester of pregnancy in East Java: I. Effect on birthweight. Brit. J. Obstet. and Gynec. 95:783–94.
34. Frisancho, A. R., J. E. Klayman, and J. Matos. 1977. Influence of maternal nutritional status on prenatal growth in a Peruvian urban population. Am. J. Phys. Anthropol. 46:265–74.
35. Frisancho, A. R., J. E. Klayman, and J. Matos. 1977. Newborn body composition and its relationship to linear growth. Am. J. Clin. Nutr. 30:704–11.
36. Viegas, O. A. C., T. J. Cole, and B. A. Wharton. 1987. Impaired fat deposition in pregnancy: an indicator for nutritional intervention. Am. J. Clin. Nutr. 45:23–28.
37. Bissenden, J. G, P. H. Scott, J. King, J. Hallum, H. N. Mansfield, and B. A. Wharton. 1981. Anthropometric and biochemical changes during pregnancy in Asian and European mothers having light for gestational age babies. Brit. J. Obstet. and Gynec. 88:999–1008.
38. Anderson, G. D., I. N. Blidner, S. McClemont, and J. C. Sinclair. 1984. Determinants of size at birth in a Canadian population. Am. J. Obstet. Gynecol. 150:236–44.
39. De Vaquera, M. V., J. W. Townsend, J. J. Arroyo, and A. Lechtig. 1983. The relationship between arm circumference at birth and early mortality. J. Trop. Pediatrics 29:167–74.
40. Krasovec, K. A. 1989. An investigation into the use of maternal arm circumference for nutritional monitoring of pregnant women. Thesis, Doctor of Science. Baltimore, MD: Johns Hopkins University.
41. Nishwander, K. R., and M. Gordon. 1972. The collaborative perinatal study of the National Institute of Neurological Diseases and Stroke: The women and their pregnancies. U.S. Dept. of Health, Education and Welfare, NIH, Philadelphia, PA: W. B. Saunders Co.
42. Frisancho, A. R., and S. L. Smith. 1990. Reduction in birth weight associated with smoking among young and older-age women. Am. J. Hum. Biol. 2:85–88.
43. Frisancho, A. R., and A. A. Compton. 1991. Reduction in newborn muscle and fat associated with maternal smoking. Am. J. Hum. Biol. In press.
44. Frisancho, A. R. 1981. Human adaptation: A functional interpretation. Ann Arbor, MI: University of Michigan Press.

45. Longo, L. D. 1977. The biological effects of carbon monoxide on the pregnant woman, fetus and newborn infant. Am. J. Obstet. Gynecol. 129:69–103.
46. Stillman, R. J., M. J. Rosenberg, and B. P. Sachs. 1986. Smoking and reproduction. Fertility and Sterility 46:545–66.
47. Serra, I., M. Alberghina, M. Viola, and A. M. Giuffrida. 1981. Effect of hypoxia on nucleic acid and protein synthesis in different brain regions. Neurochem. Res. 6:595–605.
48. Harrison, G. G., R. S. Branson, and Y. E. Vaucher. 1983. Association of maternal smoking with body composition of the newborn. Am. J. Clin. Nutr. 38:757–62.
49. Spady, D. W., M. A. Atrens, and W. A. Szymanski. 1986. Effects of mothers' smoking on their infants' body composition as determined by total body potassium. Pediatr. Res. 20:716–19.
50. D'Souza, S. W., P. Black, and B. Richards. 1981. Smoking in pregnancy: associations with skinfold thickness, maternal weight gain, and fetal size at birth. Br. Med. J. 282:1661–63.
51. Haas, J. D. 1976. Prenatal and infant growth and development. In P. T. Baker and M. A. Little, eds., Human population biology: a transdisciplinary science. Stroudsburg, PA: Dowden, Hutchinson, and Ross. 161–79.
52. Resnik, R., G. W. Brink, and M. Wilkes. 1979. Catecholamine-mediated reduction in uterine blood flow after nicotine infusion in the pregnant ewe. J. Clin. Invest. 63:1133–36.
53. Suzuki, K., T. Horiguchi, A. C. Comas-Urrutia, E. Mueller-Heubach, H. Morishima, and K. Adamsons. 1971. Pharmacologic effects of nicotine upon the fetus and mother in the rhesus monkey. Am. J. Obstet. Gynecol. 111:1092–1101.
54. Monheit, A. G., H. van Vunakis, T. C. Key, and R. Resnik. 1983. Maternal and fetal cardiovascular effects of nicotine infusion in pregnant sheep. Am. J. Obstet. Gynecol. 145:290–96.
55. Hofstetter, A., Y. Schutz, E. Jéquier, and J. Wahren. 1986. Increased 24-hour energy expenditure in cigarette smokers. N. Engl. J. Med. 314:79–82.
56. Martorell, R., C. Yarbrough, A. Lechtig, H. Delgado, and R.E. Klein. 1976. Upper arm anthropometric indicators of nutritional status. Am. J. Clin. Nutr. 29:46–53.
57. Jones, K. L., D. W. Smith, C. N. Ulleland, and A. P. Streissguth. 1973. Pattern of malformation in offspring of chronic alcoholic mothers. Lancet 1:1267–71.
58. Fried, P. A. 1980. Marijuana use by pregnant women: neurobehavioral effects in neonates. Drug Alcohol Depend. 6:415–24.
59. Hingson, R., J. J. Alpert, N. Day, E. Dooling, H. Kayne, S. Morelock, E. Oppenheimer, and B. Zuckerman. 1982. Effects of maternal drinking and marijuana use on fetal growth and development. Pediatrics 70:539–46.
60. Linn S., S. C. Schoenbaum, R. R. Monson, R. Rosner, P. C. Stubbefield, and K. J. Ryan. 1983. The association of marijuana use with outcome of pregnancy. Am. J. Pub. Health 73:1161–64.
61. Zuckerman, B., D. A. Frank, R. Hingson, H. Amaro, S. M. Levenson, H. Kayne, S. Parker, R. Vinci, K. Aboagye, L. E. Fried, H. Cabral, R. Timperi, and H. Bauchner. 1989. Effects of maternal marijuana and cocaine use on fetal growth. N. Engl. J. Med. 370:762–68.
62. Chasnoff, I. J., D. E. Lewis, D. R. Griffith, and S. Willey. 1989. Cocaine and pregnancy: clinical and toxicological implications for the neonate. Clin. Chem. 35:1276–78.
63. Chouteau, M., P. B. Namerow, and P. Leppert. 1988. The effect of cocaine abuse on birth weight and gestational age. Obstet. Gynecol. 72:351–54.

64. Kaye, K., L. Elkind, D. Goldberg, and A. Tytun. 1989. Birth outcomes for infants of drug abusing mothers. N.Y. State J. Med. 89:256–61.
65. Little, B. B., L. M. Snell, V. R. Klein, and L. C. Gilstrap, 3d. 1989. Cocaine abuse during pregnancy: maternal and fetal implications. Obstet. Gynecol. 73:157–60.

CHAPTER 14

# Accommodation to the Energy Demands of Lactation

---

**Development of the Breast and Maintenance of Lactation**

Anatomy

Maturation of the Breast

Hormonal Interaction in the Initiation and Maintenance of Lactation

Lactation and Ovulation

**Composition of Immature and Mature Milk**

Major Constituents of Milk

Dietary Pattern and Milk Composition

Maternal Nutritional Status and Milk Production and Milk Composition

**Energy Cost of Lactation**

Sources of Energy for Lactation

Maternal Depletion of Energy Reserves

**Overview**

The production of milk and the process of lactation require the synchronized action of various hormonal control mechanisms and the development and maturation of the mammary glands. The production of milk involves the synthesis of nutrients and antibodies, all of which are essential for the survival of the newborn. As such, the initiation and maintenance of lactation have profound importance for studies of human adaptation as they pertain to human nutrition. In this section the development and maturation of the breast, the physiology of lactation, the composition of milk, and the role of environmental factors that modify its production will be discussed.

## DEVELOPMENT OF THE BREAST AND MAINTENANCE OF LACTATION

### Anatomy

The *breasts* are made of epithelium lined ducts which converge at the *nipples.* These ducts branch all through the breast tissue and terminate in sacklike structures typical of eccrine glands, called *alveoli.* The alveoli, which secrete the milk, look like bunches of grapes with stems terminating in the ducts. The alveoli and the ducts immediately adjacent to them are surrounded by specialized contractile cells called *myoepithelial cells* (1) (see fig. 14.1).

### Maturation of the Breast

#### The Breast and Menstrual Cycle

The breasts begin to develop at puberty and their maturation is stimulated by the estrogens of the monthly sexual cycles (2). These estrogens stimulate growth of the stroma and ductal systems and fat deposition gives mass to the breasts. Furthermore, during pregnancy the breast and its structures grow rapidly and the glandular tissue becomes completely developed for milk production. The maturation of the breasts occurs through the action of placental hormones that stimulate the growth of the ductal and lobule alveolar systems to produce the basic architecture of the adult breast. Normal breast development at puberty also requires *prolactin* and *growth hormone,* both secreted by the anterior pituitary. During each menstrual cycle, breast morphology fluctuates in association with the changing blood concentrations of estrogen and progesterone, but these changes are small compared to the marked breast enlargement which occurs during pregnancy as a result of the stimulatory effects of estrogen and progesterone on both ducts and alveoli.

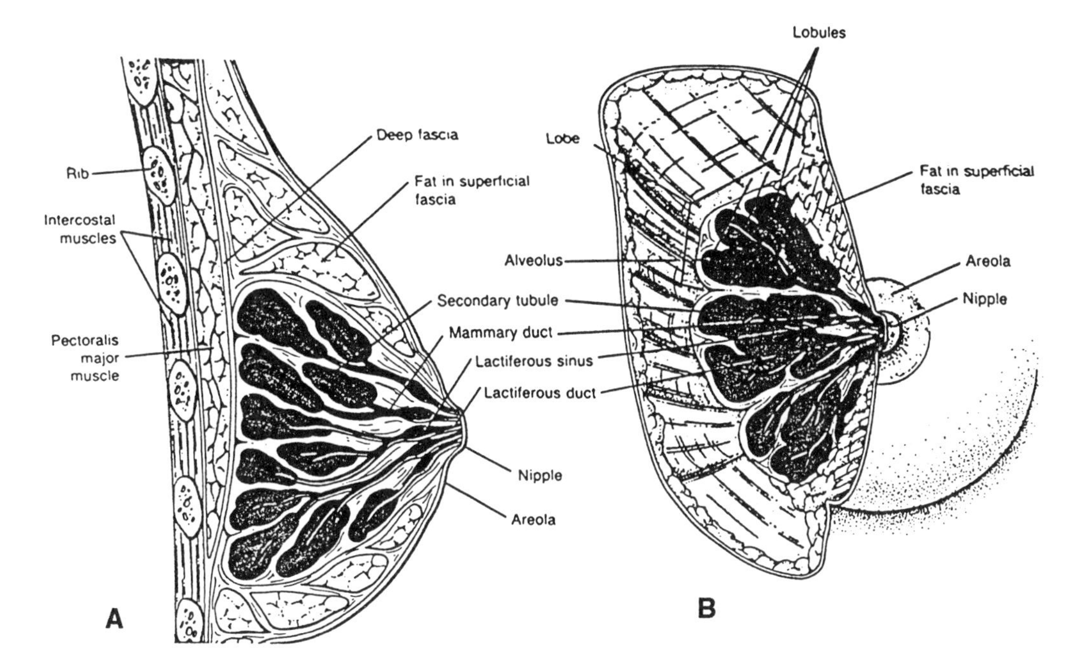


**Fig. 14.1.** Anatomy of mammary glands. **A**, Diagram in sagittal section. **B**, Diagram in anterior view. (Adapted from G. J. Tortora. 1989. Principles of human anatomy, 5th edition. New York: Harper and Row Publishers.

### Growth of the Ductal System

Throughout pregnancy large quantities of *estrogens* are secreted by the placenta which cause the ductal system of the breasts to grow and to branch. Concurrently, the stroma of the breasts increases in quantity, and large quantities of fat are laid down in the stroma. The growth of the ductal system is also assisted by the increased secretion of hormones such as *growth hormone, prolactin,* the *adrenal glucocorticoids,* and *insulin* that play a role in protein metabolism.

### Development of the Lobule-Alveolar System

Once the ductal system has developed, progesterone acting synergistically with all the above hormones (estrogen, growth hormone, prolactin, the adrenal glucocorticoids, and insulin) causes growth of the lobules, budding of alveoli, and development of secretory characteristics in the cells of the alveoli (2). These changes are analogous to the secretory effects of progesterone on the endometrium of the uterus during the latter half of the female sexual cycle.

## Hormonal Interaction in the Initiation and Maintenance of Lactation

The initiation and maintenance of lactation results from the synchronization and interaction of neuroendocrine processes. It involves the sensory nerves in the nipples and the adjacent skin of the breast and chest well, the spinal cord, the hypothalamus, the pituitary gland which secretes prolactin, and other hormones. As indicated previously, estrogen and progesterone are responsible for the development and maturation of the breasts during pregnancy. These hormones also inhibit the secretion of milk during pregnancy (2). However, postpartum the concentration of estrogen and progesterone decreases drastically, while the concentration of prolactin secreted by the mother's pituitary gland rises steadily from the fifth week of pregnancy onwards, so that by birth the concentration is about ten times the normal nonpregnant level (see fig. 14.2). In addition, the placenta secretes large quantities of human *chorionic somatomammotropin,* which also has mild lactogenic properties, thus supporting the prolactin from the mother's pituitary.

During the first days postpartum the mother secretes an immature milk called *colostrum,* which has a different composition than mature milk (see below). Within two to three days after delivery, the sudden loss of both estrogen and progesterone secretion by the placenta allows the lactogenic effect of the prolactin from the mother's pituitary gland to assume its natural milk-promoting role. The breasts can now begin to secrete copious quantities of mature milk instead of colostrum.

### Suckling and the Hypothalamic Control of Prolactin Secretion

**Suckling and Prolactin.** As shown in figure 14.2 within four weeks postpartum the level of prolactin secretion returns to the nonpregnant level. However, each time the mother nurses her baby, nervous signals from the nipples to the hypothalamus cause approximately a tenfold surge in prolactin secretion which lasts about one hour. The prolactin in turn acts on the breasts to provide the milk for the next nursing period. If the prolactin surge is absent, if it is blocked as a result of hypothalamic or pituitary damage, or if nursing does not continue, the breasts lose their ability to produce milk within a few days due to the atrophy of the alveolar cells. However, successful lactation can continue as long as adequate sucking stimulation is maintained (see fig. 14.3). In anthropological populations breast-feeding continues until about the age of 3 or 4 years (3). In fact even in the absence of childbirth, women can lactate with the stimulus of suckling. In some areas, induced lactation in nonpuerperal women is still a well-recognized and accepted method of feeding infants whose mothers could not breast-feed or who died during childbirth (4). However, after 12 months even with continued suckling the volume of milk production declines.

**Hypothalamic Control of Prolactin Secretion.** One important function of the hypothalamus is that it *inhibits* prolactin production. Consequently,

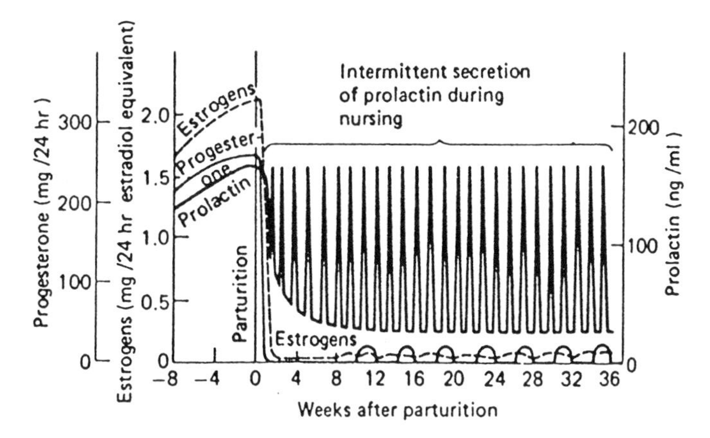


**Fig. 14.2.** Changes in rates of secretion of estrogens, progesterone, and prolactin for 8 weeks prior to parturition and for 36 weeks thereafter. Note especially the decrease of prolactin secretion back to basal levels within a few weeks, but also the intermittent periods of marked prolactin secretion (for about one hour at a time) during and after periods of nursing. (Adapted from A. C. Guyton. 1986. Textbook of medical physiology. 7th ed. Philadelphia: W. B. Saunders Co.)

damage to the hypothalamus or blockage of the hypothalamic-hypophysial portal system increases prolactin secretion and depresses the secretion of other anterior pituitary hormones (because the hypothalamus usually *stimulates* the production of all the other hormones). However, when the baby suckles the breast the hypothalamus releases a *prolactin-releasing factor* (PRF) that intermittently increases prolactin secretion.

## Lactation and Ovulation

In about 50% of lactating mothers, the ovarian cycle does not resume until a few weeks after cessation of lactation. The exact cause of this is not known.

**Role of GnRH.** Some researchers maintain that suckling stimulates the secretion of *prolactin-releasing factor* and simultaneously inhibits the secretion of gonadotrophin releasing hormone (GnRH) (previously referred to as luteinizing hormone, releasing hormone or LHRH) by the hypothalamus. This in turn suppresses the secretion of the pituitary gonadotropic hormones, luteinizing hormone, and follicle-stimulating hormone. As a result ovulation and cycling is inhibited.

**Role of Endorphins.** Other investigators postulate that endogenous opioids such as beta-endorphins rather than prolactin may be the factors responsible for the lactation-induced suppression of ovulation (5). This hypothesis is based on the evidence that beta-endorphins have the dual action of inhibiting LH secretion and stimulating prolactin secretion. Experimental studies in laboratory animals have found an increased secretions of endorphins in response to suckling. Studies in humans suggest that exercise-induced

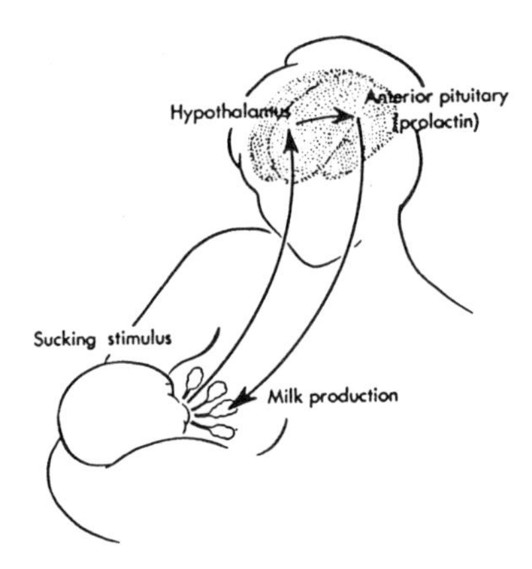


**Fig. 14.3.** Diagrammatic representation of the interaction between the suckling stimulus and milk production. The suckling provided by the baby sends a message to the hypothalamus. The hypothalamus, in turn, stimulates the anterior pituitary to release prolactin, the hormone that promotes milk production by alveolar cells of the mammary glands. (Adapted from B. S. Worthington-Roberts, ed. 1989. Nutrition in pregnancy and lactation. 4th ed. St. Louis, MO: Times Mirror/Mosby College Publishing.)

amenorrhea associated with anorexia nervosa may be related to the increased secretion of endorphins (6–8) rather than to undernutrition per se. After several months of lactation, the pituitary once again begins to secrete sufficient gonadotropic hormones to reinstate the monthly sexual cycle in about half of the mothers. Thus, because milk secretion can be maintained without the high levels of prolactin necessary to prevent ovulation, pregnancy can occur while lactating.

## The Ejection or "Let-Down" Process in Milk Secretion

**Function of oxytocin.** Milk is secreted continuously into the alveoli of the breasts, then it must be ejected or "let down," from the alveoli to the ducts. This process is caused by a combined neurogenic and hormonal reflex involving the posterior pituitary hormone *oxytocin.* The stimulus is suckling on the nipple. When the baby suckles the breast, somatic nerves from the nipples to the spinal cord and then to the hypothalamus send sensory impulses to the alveoli ducts, which in turn cause oxytocin and prolactin to be secreted concurrently. The oxytocin is then carried in the blood to the breasts where it causes the *myoepithelial cells* that surround the outer walls of the alveoli to contract and push the milk from the alveoli into the ducts, where it is easily available to the infant. It is estimated that within 30 seconds to a minute after a baby begins to suckle the breast, milk begins to flow. This process is called milk ejection, or milk let-down. Suckling affects both breasts, so that suckling on one breast causes milk flow not only in that breast but also in the opposite breast. It should be noted that milk ejection can also occur without actual stimulation of the breast. For example, fondling the baby by the mother or hearing the baby cry often gives enough of a signal to the mother's hypothalamus to cause milk ejection.

## COMPOSITION OF IMMATURE AND MATURE MILK

As a result of all these processes, the mammary alveolar cells are capable of extracting water, protein, fat, and the carbohydrate lactose from milk. The raw materials—amino acids, fatty acids, glycerol, glucose, etc.—are extracted from the gravidae blood and converted into the higher molecular-weight substances. These synthetic processes require the synchronized participation of insulin, growth hormone, cortisol, and probably other hormones. As shown in table 14.1 the concentration of lactose in human milk is approximately 50% greater than that in cow's milk. On the other hand, the concentration of protein (casein, lactalbumin) in human milk is about three times less than in cow's milk. Similarly, the concentration of ash, which contains the calcium and other minerals, is only one third as great in human milk as in cow's milk (2).

In the first days after birth the mammary glands secrete a small amount of thick fluid called *colostrum*. The volume varies between 2 and 10 ml/feeding/day in the first three days. Women who have had other pregnancies and have nursed babies previously have a greater volume than primipara women. Colostrum is rich in carotene (which gives it its yellow color) and protein but poor in sugar and fat. The adaptive significance of colostrum is that it facilitates the establishment of ''bifidus flora'' in the infant's digestive system and contains large quantities of antibodies that protect the infant against gastrointestinal tract infections (9).

### Major Constituents of Milk

The major constituents of mature milk and colostrum are summarized in table 14.2, and will be briefly discussed here. A thorough review of milk composition can be found elsewhere (10–12).

**TABLE 14.1. Comparison of human and cow's milk**

| | Human Milk | Cow's Milk |
|---|---|---|
| Water | 88.5 | 87.0 |
| Fat | 3.3 | 3.5 |
| Lactose | 6.8 | 4.8 |
| Casein | 0.9 | 2.7 |
| Lactalbumin and other protein | 0.4 | 0.7 |
| Ash | 0.2 | 0.7 |

*Source:* Adapted from A. C. Guyton. 1989. Textbook of medical physiology. Philadelphia: W. B. Saunders. 996.

**TABLE 14.2. Composition of human colostrum and mature breast milk**

| Constituent (per 100 ml) | Colostrum (1–5 d) | Mature (>30 d) | Constituent (per 100 ml) | Colostrum (1–5 d) | Mature (>30 d) |
|---|---|---|---|---|---|
| Energy (kcal) | 58 | 70 | Cholesterol (mg) | 27 | 16 |
| Total solids (g) | 12.8 | 12.0 | **Vitamins, fat soluble** | | |
| Lactose (g) | 5.3 | 7.3 | Vitamin A (μg)[a] | 89 | 47 |
| **Total nitrogen** (mg) | 360 | 171 | β-carotene | 112 | 23 |
| Protein nitrogen (mg) | 313 | 129 | Vitamin D (μg) | — | 0.04 |
| NPN (mg) | 47 | 42 | Vitamin E (μg)[b] | 1280 | 315 |
| **Total protein** (g) | 2.3 | 0.9 | Vitamin K1 (μg) | 0.23 | 0.21 |
| Casein (mg) | 140 | 187 | **Vitamins, water soluble** | | |
| α-Lactalbumin (mg) | 218 | 161 | Thiamine (μg) | 15 | 16 |
| Lactoferrin (mg) | 330 | 167 | Riboflavin (μg) | 25 | 35 |
| IgA (mg) | 364 | 142 | Niacin (μg) | 75 | 200 |
| **Amino acids** (total) | | | Folic acid (μg) | — | 5.2 |
| Alanine (mg) | — | 52 | Vitamin $B_6$ (μg) | 12 | 28 |
| Arginine (mg) | 126 | 49 | Biotin (μg) | 0.1 | 0.6 |
| Aspartate (mg) | — | 110 | Pantothenicacid (μg) | 183 | 225 |
| Cystine (mg) | — | 25 | Vitamin $B_{12}$ (ng) | 200 | 26 |
| Glutamate (mg) | — | 196 | Ascorbic acid (mg) | 4.4 | 4.0 |
| Glycine (mg) | — | 27 | **Minerals** | | |
| Histidine (mg) | 57 | 31 | Calcium (mg) | 23 | 28 |
| Isoleucine (mg) | 121 | 67 | Magnesium (mg) | 3.4 | 3.0 |
| Leucine (mg) | 221 | 110 | Sodium (mg) | 48 | 15 |
| Lysine (mg) | 163 | 79 | Potassium (mg) | 74 | 58 |
| Methionine (mg) | 33 | 19 | Chlorine (mg) | 91 | 40 |
| Phenylalanine (mg) | 105 | 44 | Phosphorus (mg) | 14 | 15 |
| Proline (mg) | — | 89 | Sulphur (mg) | 22 | 14 |
| Serine (mg) | — | 54 | **Trace elements** | | |
| Threonine (mg) | 148 | 58 | Chromium (ng) | — | 39 |
| Tryptophan (mg) | 52 | 25 | Cobalt (μg) | — | 1 |
| Tyrosine (mg) | — | 38 | Copper (μg) | 46 | 35 |
| Valine (mg) | 169 | 90 | Fluorine (μg) | — | 7 |
| Taurine (free) (mg) | — | 8 | Iodine (μg) | 12 | 7 |
| **Urea** (mg) | 10 | 30 | Iron (μg) | 45 | 40 |
| **Creatine** (mg) | — | 3.3 | Manganese (μg) | — | 0.4,1.5 |
| **Total fat** (g) | 2.9 | 4.2 | Nickel (μg) | — | 2 |
| **Fatty acids** (% total fat) | | | Selenium (μg) | — | 2.0 |
| 12:0 lauric (%) | 1.8 | 5.8 | Zinc (μg) | 540 | 166 |
| 14:0 myristic (%) | 3.8 | 8.6 | | | |
| 16:0 palmitic (%) | 26.2 | 21.0 | | | |
| 18:0 stearic (%) | 8.8 | 8.0 | | | |
| 18:1 oleic (%) | 36.6 | 35.5 | | | |
| 18:2 n-6 linoleic (%) | 6.8 | 7.2 | | | |
| 18:3 n-3 linolenic (%) | — | 1.0 | | | |
| $C_{20}$ and $C_{22}$ polyunsat. | 10.2 | 2.9 | | | |

*Source:* From C. E. Casey and K. M. Hambidge. 1983. Nutritional aspects of human lactation. In M. C. Neville and M. R. Neifert, eds., Lactation: physiology, nutrition and breast-feeding. New York: Plenum Press. 94.

[a] retinol equivalents

[b] total tocopherols

### Protein and Antimicrobial Substances

**Proteins.** The major proteins found in breast milk are casein (curd protein), alfa-lactalbumin, lactoferrin, serum albumin, and secretory immunoglobulins (sIgA). Other proteins, including other immunoglobulins and glycoproteins, are present in low concentrations.

**Amino Acids.** Human milk is relatively low in several amino acids that are known to be detrimental if found in the bloodstream at high levels (i.e., phenylalanine); it is high in other amino acids that the infant cannot synthesize well, such as cystine and taurine.

**Antimicrobial Substances.** In both colostrum and mature milk the concentration of antimicrobial substances such as lactoferrin, lysozyme enzyme, and secretory immunoglobulins (sIgA) is very high. The high concentration of these substances provides a passive immunity to the newborn. As such, breast-fed infants experience fewer infectious diseases of both the digestive and respiratory tracts than do formula-fed infants. Furthermore, human milk with its low casein content forms a flocculent suspension that is easily digested and consequently better tolerated by the neonate.

**Lactoferrin.** Lactoferrin is an iron-binding protein that inhibits the growth of certain iron-dependent bacteria in the gastrointestinal tract and, as such, helps protect against certain gastrointestinal tract infections in breast-fed infants.

**Lysozyme.** Lysozyme is an enzyme found in human milk that attaches to the cell wall and the outer membrane of a variety of microorganisms causing lysis of the cell (dissolving the cell membrane). Human milk also contains several fat-digesting enzymes or lipases (13) such as lipoprotein lipase, and lipolytic milk enzymes. These enzymes are shown to be stable and active in the intestine of infants. They contribute to the hydrolysis of milk triglycerides and partly account for the ease in fat digestion that is commonly demonstrated by breast-fed babies (14).

**sIgA.** sIgA provides passive immunological protection against antigens of microorganisms present in the gastrointestinal tract of the mother. When in close contact with the intestinal epithelium of the infant, sIgA prevents binding of the antigen. Thus, sIgA provides a protective defense against infection by retarding viral and bacterial invasion of the mucosa (12).

### Lipids

**Total Fat.** Mean values for the total fat content of colostrum are 2.9 g/100 ml and 4.2 g/100 ml for mature milk. About 90% of the lipid in human milk is present in the form of triglycerides, but small amounts of phospholipids, cholesterol, diglycerides, monoglycerides, glycolipids, sterol esters, and free fatty acids are also found.

**Fatty Acids.** The fatty acid composition of human milk varies proportionally with the diet of the mother, so that the higher the concentration of polyunsaturated fats (such as corn and cotton-seed oil) in the diet the higher the content of polyunsaturated fats in breast milk (15). Similarly, the lower the calorie intake the greater the reduction in adipose tissue and the more the fatty acid composition of human milk resembles adipose fat.

**Cholesterol.** Human milk contains more cholesterol than cow's milk and much more cholesterol than commercial infant formulas. The cholesterol content of human milk ranges from about 10 to 20 mg/dl, and decreases as lactation progresses. The adaptive significance of the high cholesterol level is that (1) it is needed by the rapidly growing central nervous system for myelin synthesis, and (2) it stimulates in early life the development of enzymes necessary later for cholesterol degradation (16). However, there is no evidence that infant consumption of breast milk rich in cholesterol provides any protection during adulthood (17).

## Carbohydrates

**Lactose.** Lactose (or milk sugar) is the main carbohydrate in human milk. In addition, human milk contains trace amounts of glucose, galactose, glucosamines, and other nitrogen-containing oligosaccharides. Lactose is relatively insoluble and is slowly digested and absorbed in the small intestine. The presence of lactose in the gut of the infant stimulates the growth of microorganisms, which produce organic acids and synthesize many of the B vitamins. The acid milieu that is created helps to check the growth of undesirable bacteria in the infant's gut and improves the absorption of calcium, phosphorus, magnesium, and other metals. Although all newborns are able to digest lactose, lactase sufficiency develops more slowly in some infants (18, 19). The rare infant with congenital lactase deficiency may actually be able to profit from the use of either cow or human milk provided the lactose is first broken down by adding three drops of lactase to each 200 ml bottle of human or cow's milk.

## Vitamins

### *Fat Soluble*

As shown in table 14.2 all the required vitamins are present in human milk. The concentrations of vitamins vary markedly from one person to another. The variability is related to differences in dietary patterns and environmental factors (10).

**Vitamin D.** The mean content of Vitamin D is about 0.04 micrograms/

100 ml. The concentration of Vitamin D can increase markedly depending on the amount of exposure to the sun and the amount of Vitamin D supplementation.

**Vitamin A and Beta-Carotene.** Vitamin A and beta-carotene are present in large quantities in colostrum but decrease in quantity in mature milk. The concentrations vary with maternal dietary patterns. They increase with increased consumption of leafy and yellow vegetables and vitamin supplements.

**Vitamin E.** There is a great deal of variability in the concentration of vitamin E in mature milk. The variability is related mostly to stage of lactation and to dietary patterns. After the first three months of lactation the concentration declines.

#### *Water Soluble*

The water soluble vitamins such as thiamine, riboflavin, etc. (see table 14.2) are present in required quantities in milk. The concentration of these vitamins, with the exception of folacin, decline as pregnancy progresses.

### Minerals and Trace Elements

As it has less protein and fat, human milk also has fewer minerals than cow's milk. Human milk has six times less phosphorus, four times less calcium, three times less total ash, and three times less protein than cow's milk. Other minerals such as manganese, zinc, magnesium, aluminum, iodine, chromium, selenium, and fluorine are found in trace amounts. The mineral content of human milk varies with the nutritional status of the mother and the length of lactation (20). Because iron is necessary for the formation of hemoglobin, breast-fed infants are usually given iron supplements. Some studies suggest that infants who are breast-fed during the first 6 months of life and receive little or no dietary iron other than that found in human milk are not iron deficient at 6 months of age (21). On the other hand, nonsupplemented breast-fed infants were found to be in negative iron balance between 3 and 6 months of age (21).

### Carnitine

Human milk contains about 50 to 100 nmole/ml of carnitine. Carnitine assists in the oxidation of long chain fatty acids by facilitating their transport across the mitochondrial membrane. It also plays an important role in the regulation of thermogenesis in brown adipose tissue, and together with malonyl-CoA in the initiation of ketogenesis.

Summary

In summary, compared to other species human milk has less protein but more essential fatty acids. Furthermore, human milk contains important immunoglobulins that provide protection against gastrointestinal infections in the infant. Because of the large quantity of milk production the metabolic substrates are drained from the mother. For instance, each day the mother loses approximately 50 grams of fat and 100 grams of lactose that enter the milk Similarly, each day about 2 to 3 grams of calcium phosphate may be lost. To supply the needed nutrients, the dietary intake of the mother is increased and the energy reserves that are stored in the adipose tissue during pregnancy are utilized (see section on energy cost of lactation).

## Dietary Pattern and Milk Composition

**Lipids and Fat Intake.** A clear relationship exists between the type of fat consumed and the fatty acid pattern of milk lipids (22). It has been found that in five lactating women, mean milk lipid concentration tended to be higher while on the high-fat diet than while on the low-fat diet (33 vs. 25 gm/l) (23). In another study of three lactating women who were at least 6 months postpartum, milk lipid concentrations were significantly lower during the high-fat diet (26 vs. 41 gm/l) (24). The caloric content of the high-fat diet was 50% fat, 15% protein, and 35% carbohydrate whereas the low-fat diet was 15% fat, 20% protein, and 65% carbohydrate. Other investigators, however, found no relationship between maternal dietary fat intake (either as percentage of calories or total intake) and milk lipid composition (25, 26).

**Lipids and Protein Intake.** During the later months of lactation milk lipid concentration has been found to be related to maternal protein intake. Furthermore, dietary-protein intake (8% vs. 22% of calories from protein) was found to be significantly associated with milk nitrogen, total protein, and nonprotein nitrogen concentrations (27). At present the physiological basis for the relationship between dietary protein and milk lipid concentration remains undefined.

## Maternal Nutritional Status and Milk Production and Milk Composition

In the developing nations a large proportion of children are exclusively breastfed, and in the industrialized countries the proportion of breast-fed children is increasing. Comparisons of the lactation perfomance of well-nourished and undernourished women have provided valuable information for ascertaining the role of maternal nutritional status in milk production and milk composition.

### Well-nourished

Well-nourished women can lactate without additional dietary intake and without depletion of their energy sources because of their abundant energy stores, as shown by their high net pregnancy fat gain (see table 14.3), during the first six months postpartum. For this reason, there is no relationship between maternal fatness and milk concentrations of protein, lactose, or lipid, during the first 6 months of lactation. However, when lactation lasts more than 6 months, the nutrient content of milk is directly related to the amount of maternal fat gain (25, 28). Studies in central Michigan indicate that a decline in maternal adiposity was associated with a breast-feeding pattern of short, frequent feeds (29). Accordingly, it was postulated that fat mobilization during lactation was caused by prolactin-mediated suppression of lipoprotein lipase in the peripheral adipose tissue and augmentation of activity in the mammary gland. Frequent feeds promote elevated levels of prolactin and thus facilitate fat uptake by the mammary gland at the expense of adipose tissue (30). Other investigations conducted in Texas, however, found no relationship between maternal adiposity and breast-feeding pattern (31).

### Undernourished

As soon as undernourished mothers start lactating their already low energy stores are gradually depleted. Studies in Bangladesh (32) indicate that the milk lipid concentration depends on the availability of maternal fat stores right from the beginning of lactation. Similarly, among women from Gambia (33) and Thailand (34) the lipid composition of milk is inversely related to feeding frequency, so that the mean milk lipid concentration for a single

**TABLE 14.3. Milk volume and composition**

| | Month of Lactation | | | |
|---|---|---|---|---|
| | 3 mo ($n=58$) Mean ± SD | 6 mo ($n=45$) Mean ± SD | 9 mo ($n=28$) Mean ± SD | 12 mo ($n=21$) Mean ± SD |
| Volume consumed by infant (g/d) | 811 ± 133 | 780 ± 185 | 674 ± 236 | 514 ± 238 |
| Volume produced (g/d) | 895 ± 200 | 844 ± 237 | 750 ± 252 | 516 ± 232 |
| Protein (g/l) | 12.1 ± 1.5 | 11.4 ± 1.5 | 11.6 ± 1.8 | 12.4 ± 1.5 |
| Lipid (g/l) | 36.2 ± 7.0 | 37.7 ± 9.6 | 38.1 ± 8.0 | 37.2 ± 11.3 |
| Lactose (g/l) | 74.4 ± 1.5 | 74.4 ± 1.9 | 73.5 ± 2.9 | 74.0 ± 2.7 |
| Gross energy (kcal/l) | 697 ± 67 | 707 ± 92 | 709 ± 74 | 706 ± 110 |

*Source:* Adapted from: L. A. Nommsen, C. A. Lovelady, M. J. Heinig, B. Lonnerdal, and K. G. Dewey. 1991. Determinants of energy, protein, and lactose concentrations in human milk during the first 12 mo. of lactation. Am. J. Clin. Nutr. 53:457–65 (25).

feeding declines the more frequent and the shorter the interfeed interval (34, 35). Furthermore, studies of Gambian women indicate that the production of milk declines markedly during the wet season, which is characterized by limited energy availability (36). Studies of undernourished rural women from Zaire indicate that milk production in these rural women averaged only 307 ml/day while milk production in their well-nourished urban counterparts averaged 612 ml/d (37). Thus, it appears that despite the strong drive toward milk production maternal undernutrition does lead to reduction of milk output. However, the concentration of lactoferrin, lysosyme, and sIgA in milk is unaffected by maternal nutritional status (37, 38). Therefore, children breast-fed by undernourished mothers receive fewer antimicrobial agents than children breast-fed by well-nourished mothers because of a much lower milk production in undernourished mothers.

In summary, both the production and composition of milk appears to be influenced by the availability of energy stored in the adipose tissue resulting from maternal pregnancy fat gain. Mothers with high energy reserves are able to produce milk without additional dietary intake and with little reduction of their reserves. In contrast, undernourished mothers produce less milk than well-nourished mothers despite the rapid utilization of their energy reserves. Another indication of the role of maternal nutritional status on milk composition is the finding that among marginally nourished women from Gambia the milk lipid concentrations were highest among primiparous women and leveled off at parity four or higher (28). Similarly, protein concentration declines significantly with increasing parity among Egyptian women (39). As a result of the lower production of milk by undernourished mothers, children of undernourished mothers receive fewer antimicrobial agents than children breast-fed by well-nourished mothers. In terms of dietary pattern there is evidence to suggest that the composition of human milk is directly related to the quantity and type of the maternal diet.

## ENERGY COST OF LACTATION

In general, energy requirements for lactation are proportional to the quantity of milk produced. This includes the energy content of milk and the energy required to produce it. According to estimates of the World Health Organization, under conditions of good nutrition milk production during lactation averages about 800 ml/d, energy content averages about 700 kcal/liter and the efficiency of the conversion of food energy to milk energy is assumed to be 80% (105). Therefore, the average energy cost of lactation over 6 months is about 700 kcal/d ($800 \times .70/0.80 = 700$ kcal/d). As shown in table 14.3 these values coincide with those found in well-nourished U.S. women. However, recent analysis among well-nourished lactating women from the Netherlands (40, 41) indicates that milk production averages only 720 ml/d. Therefore, the energy cost of lactation is about 630 kcal/d ($720 \times .70/.80 = 630$ kcal/d).

## Sources of Energy for Lactation

To meet the increased energy demands that occur during lactation, well-nourished women have several options. They may increase their energy intake, mobilize their fat stores, or lower the nonlactational component of their energy expenditure.

**Energy Intake.** To meet the energy cost of lactation women increase their energy intake by about 415 kcal/d (41).

**Utilize Fat Stores.** The net maternal fat mass gain averages about 2.3 kg (see table 13.3) for well-nourished women. This net fat gain represents a reserve of about 18,400 kcal ($2.3 \times 8{,}000 = 18{,}400$ kcal). During the first six months of lactation this reserve can be utilized at a rate of about 200 kcal/day. However, in practice only 75 kcal/d are derived from fat tissue utilization. Hence, lactation per se among well-nourished women rarely leads to the mobilization of all energy reserves (42) unless dietary intake is decreased. On the other hand, among undernourished women from Guatemala the energy cost of lactation is met to a greater extent by fat loss than by either increased dietary intake or reduced energy expenditure (43).

**Reduction of Diet-Induced Thermogenesis.** During the first 3 months of lactation, for yet unknown reasons, the thermic response to a meal and energy for physical activity is reduced when compared to the thermic response after lactation is stopped. For example,the DIT and physical activity during the first 3 months of lactation cost about 590 kcal/d but after 12 months postpartum they increase to about 765 kcal/d (41). Therefore, during the first 3 months the energy requirements decrease by about 175 kcal/d.

## Maternal Depletion of Energy Reserves

To meet the increased energy demand well-nourished women may either increase their energy intake, mobilize their fat stores, or lower the nonlactational component of their energy expenditure. As indicated in the previous chapter (see table 13.3) the net pregnancy fat gain for well-nourished women averages between 2.0 and 2.3 kg while for undernourished women from Gambia, Philippines, and Thailand the average ranges from 0.6 to 1.4 kg (44). This difference implies that right from the beginning the ability to subsidize milk production in undernourished mothers is less than that of their well-nourished counterparts. In addition, for obvious reasons dietary intakes are not easily increased among undernourished populations. In fact, among undernourished women from Guatemala the energy cost of lactation is met to a greater extent by fat loss than by either increased dietary intake or reduced energy expenditure (43). Under these conditions frequent pregnancy and lactation, especially when lactation occurs concurrently with pregnancy, may lead to the development of the ''maternal depletion syndrome'' (45).

### Parity

As shown in figure 14.4 among the Au lowland population of New Guinea there is a inverse relationship between maternal adiposity and parity (46, 47). Similar relationships were also reported for the populations of New Guinea highlands (45) and Karkar island (50). From these studies it is evident that at a given age the greater the number of pregnancies the lesser the maternal energy reserves. This relationship is evident only among women of low socioeconomic status, suggesting that when the availability of energy sources is limited the maternal depletion syndrome is expressed. Therefore, under conditions that permit the accumulation of adequate energy stores during pregnancy there is no association of parity and maternal depletion (51, 52).

### Concurrent Pregnancy and Lactation

Analysis of birth weights by lactation and supplementation status in four villages of Guatemala shows that the increased demands of pregnancy concomitant with lactation results in the depletion of maternal energy (subcutaneous fat) stores (48, 49). As illustrated in figure 14.5, during pregnancy and lactation the skinfold thickness (thigh) for the women who had a (1) long interval (>6 months) between pregnancy and lactation is higher than in the women who had a (2) short interval (<6 months) between pregnancy and lactation. That is, the women with low energetic stress maintained greater body fatness that their counterparts who had a high energetic stress. This difference occurred even though the samples were provided with a protein and calorie supplement and were corrected for maternal age and parity. The decline in skinfold thickness found among lactating women from Lesotho (Africa) and rural Taiwan (51, 52) probably also reflects the depletion of maternal energy reserves.

In summary, it is evident that the increased energy demands of pregnancy result in the depletion of maternal energy stores in populations living under conditions of chronic undernutrition. Therefore, populations living under these circumstances, as shown by studies of the Turkana pastoralists (53) have developed adaptive strategies that enable the survival of offspring under otherwise difficult conditions.

## OVERVIEW

The maintenance and continuing production of milk is directly related to suckling frequency. Human milk is rich in growth-promoting compounds and

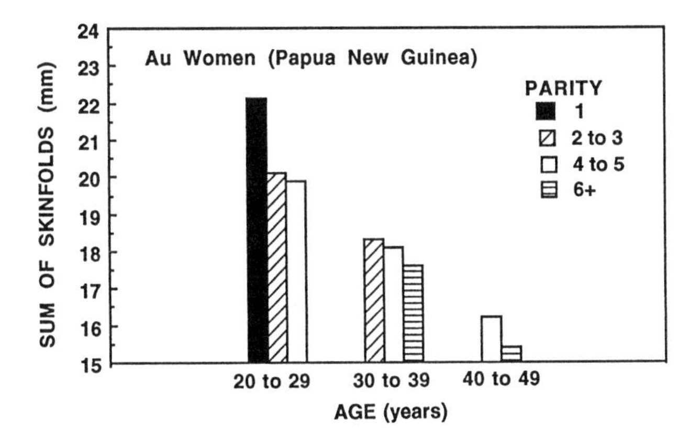


**Fig. 14.4.** Sum of skinfolds by age and parity among 304 Au mothers. At a given age, a high parity is associated with lower skinfold thickness. (Redrawn from data from D. P. Tracer. 1991. Fertility-related changes in maternal body composition among the Au of Papua New Guinea. Am. J. Phys. Anthropol. 85:393–406.)

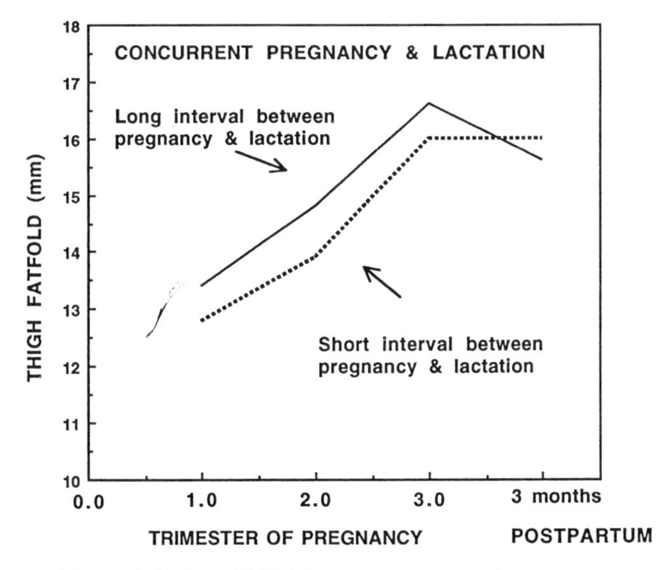


**Fig. 14.5.** Maternal thigh fatfold-thickness measurements across pregnancy and early postpartum for overlap and no-overlap groups. Means are adjusted for maternal parity, age, relative measurement date, and study month. (Redrawn from data from K. Merchant, R. Martorell, and J. Haas. 1990. Maternal and fetal responses to the stresses of lactation concurrent with pregnancy and of short recuperative intervals. Am. J. Clin. Nutr. 52:280–88.)

immunoglobulins. These provide passive immunity and a reduced risk of allergy to breast-fed infants. The composition of human milk is directly related to the quantity and type of the maternal diet and to the maternal nutritional status which is reflected in the amount of energy stores present in the form of adipose tissue. Furthermore, the lipid and protein content of milk is inversely related to length of lactation and feeding frequency. Maternal chronic undernutrition is associated with a reduction of milk production. However, despite the fact that maternal undernutrition does not affect the concentration of lactoferrin and sIgA, children breast-fed by undernourished mothers receive fewer antimicrobial agents than children breast-fed by well-nourished mothers because of a much lower milk production in undernourished women.

The energy cost of lactation averages about 650 kcal/d. Among well-nourished women these energy needs are met by increasing the dietary intake, by moderate utilization of energy stored in the form of maternal fat, and by reduction in the nonlactational component of energy expenditure, such as DIT and physical activity. As inferred from studies of populations living under conditions of chronic undernutrition, the increased energy demands of reproduction result in the depletion of maternal energy stores. When lactation results in excessive depletion of maternal energy stores it is not adaptive in an evolutionary context. Therefore, the maternal depletion syndrome can be viewed as a temporary compromise between what the child needs and what the mother needs rather than as an adaptive response. In the context of sociobiology the depletion of maternal nutritional reserves during pregnancy and lactation can be viewed as responses that enable the survival of the offspring.

## References

1. Tortora, G. J. 1989. Principles of human anatomy, 5th ed. New York, NY: Harper and Row Publishers.
2. Guyton A. C. 1989. Textbook of medical physiology. Philadelphia, PA.: W. B. Saunders.
3. Ford, C. S. A. 1945. Comparative study of human reproduction. New Haven, CT: Yale University Press.
4. Smith, I. D., R. P. Shearman, and A. R. Korda. 1972. Lactation following therapeutic abortion with prostaglandin $F_{2a}$ Nature 240:411–12.
5. McNeilly, A. S. 1988. Suckling and the control of gonadotropin secretion. In E. Knobil and J. Neill et al., eds., The physiology of reproduction. New York, NY: Raven Press.
6. Guazzelli, R., M. Piazzini, L. Papi, M. G. Martinetti, G. Strazzulla, A. Brocchi, and U. Bigozzi. 1989. Plasma beta-endorphin in patients affected by obesity and by anorexia nervosa. Acta Med. Auxol. 21:73–77.
7. Olson, B. R. 1989. Exercise-induced amenorrhea. Am. Fam. Phys. 39:213–21.
8. Ruffin, M. T., R. E. Hunter, and E. A. Arendt. 1990. Exercise and secondary amenorrhoea linked through endogenous opioids. Sports Medicine 10:65–71.
9. Ogra, S. S., and P. L. Ogra. 1978. Immunologic aspects of human colostrum and milk. 1. Distribution characteristics and concentrations of immunoglobulins at different times after the onset of lactation. J. Pediatr. 92:550–55.
10. Casey, C. E., M. C. Neville, and K. M. Hambidge. 1989. Studies in human lactation: secretion of zinc, copper, and manganese in human milk. Am. J. Clin. Nutr. 49:773–85.

11. Gibson, R. A., and G. M. Kneebone. 1981. Fatty acid composition of human colostrum and mature breast milk. Am. J. Clin. Nutr. 34:252–57.
12. Worthington-Roberts, B. S., and S. R. Williams. 1989. Nutrition in pregnancy and lactation, 4th ed. St. Louis, MO: Times Mirror/Mosby College Publishing.
13. Jensen, R. G., R. M. Clark, F. A. deJong, M. Hamosh, T. H. Liao, and N. R. Mehta. 1982. The lipolytic triad: human lingual, breast milk and pancreatic lipases: physiological implications of their characteristics in digestion of dietary fats. J. Pediatr. Gastroenterol. Nutr. 1:243–55.
14. Hernell, O., and L. Blackberg. 1982. Digestion of human milk lipids: physiologic significance of sn-2 monoacylglycerol hydrolysis by bile salt-stimulated lipase. Pediatr. Res. 16:882–85.
15. Finley, D. A., B. Lonnerdal, K. G. Dewey, and L. E. Grivetti. 1985. Breast milk composition: fat content and fatty acid composition in vegetarians and nonvegetarians. Am. J. Clin. Nutr. 41:787–800.
16. Jelliffe, D. B. 1975. Unique properties of human milk. J. Reprod. Med. 14:133–37.
17. Lampkin, B. C., N. A. Shore, and D. Chadwick. 1966. Megaloblastic anemia of infancy secondary to maternal pernicious anemia. N. Engl. J. Med. 274:1168–71.
18. Lifschitz, C. H., E. O. Smith, and C. Garza. 1983. Delayed complete functional lactase sufficiency in breast-fed infants. J. Pediatr. Gastroenterol. Nutr. 2:478.
19. Lifschitz, C. H., E. O. Smith, C. Garza, and B. L. Nichols. 1982. Colonic metabolism of malabsorbed lactose in breast-fed infants. Am. J. Clin. Nutr. 35:849 (abstract).
20. Karra, M. V., S. A. Udipi, A. Kirksey, and J. L. B. Roepke. 1986. Changes in specific nutrients in breast milk during extended lactation. Am. J. Clin. Nutr. 43:495–503.
21. Duncan, B., R. B. Schifman, J. J. Corrigan Jr., and C. Schaefer. 1985. Iron and the exclusively breast-fed infant from birth to six months. J. Pediatr. Gastroenterol. Nutr. 4:421–25.
22. Lonnerdal, B. 1986. Effects of maternal dietary intake on human milk composition. J. Nutr. 116:499–513.
23. Hachey, D. L., G. H. Silber, W. W. Wong, and C. Garza. 1989. Human lactation II: Endogenous fatty acid synthesis by the mammary gland. Pediatr. Res. 25:63–68.
24. Harzer, G., I. Dieterich, and M. Haug. 1984. Effects of the diet on the composition of human milk. Ann. Nutr. Metab. 28:231–39.
25. Nommsen, L. A., C. A. Lovelady, M. J. Heinig, B. Lonnerdal, and K. G. Dewey. 1991. Determinants of energy, protein, lipid, and lactose concentrations in human milk during the first 12 mo. of lactation: the DARLING Study. Am. J. Clin. Nutr. 53:457–65.
26. Motil, K. J., C. M. Montandon, and C. Garza. 1986. Effect of dietary intake on milk production in lactating women. Am. J. Clin. Nutr. 43:677 (abstract).
27. Forsum, E., and B. Lonnerdal. 1980. Effect of protein intake on protein and nitrogen composition of breast milk. Am. J. Clin. Nutr. 33:1809–13.
28. Butte, N. F., C. Garza, J. E. Stuff, E. O. B. Smith, and B. L. Nichols. 1984. Effect of maternal diet and body composition on lactational performance. Am. J. Clin. Nutr. 39:296–306.
29. Quandt, S. A. 1983. Changes in maternal postpartum adiposity and infant feeding patterns. Am. J. Phys. Anthropol. 60:455–61.
30. Bradshaw, M. K., and S. Pfeiffer. 1988. Feeding mode and anthropometric changes in primiparas. Hum. Biol. 60:251–61.
31. Butte, N. F. and C. Garza. 1986. Anthropometry in the appraisal of lactation

performance among well-nourished women. In M. Hamosh and A. S. Goldman, eds., Human lactation 2: maternal and environmental factors. New York, NY: Plenum Press.
32. Brown, K. H., N. A. Akhtar, A. D. Robertson, and M. G. Ahmed. 1986. Lactational capacity of marginally nourished mothers: Relationships between maternal nutritional status and quantity and proximate composition of milk. Pediatrics 78:909–19.
33. Prentice, A. M., R. Whitehead, S. B. Roberts, and A. A. Paul. 1981. Long-term energy balance in child-bearing Gambian women. Am. J. Clin. Nutr. 34:2790–99.
34. Jackson, D. A., S. M. Imong, A. Silprasert, S. Ruckphaopunt, M. W. Woolridge, J. D. Baum, and K. Amatayakul. 1988. Circadian variation in fat concentration of breast-milk in a rural northern Thai population. Br. J. Nutr. 59:349–63.
35. Prentice A., A. M. Prentice, and R. G. Whitehead. 1981. Breast-milk fat concentrations of rural African women. 1. Short-term variations within individuals. Br. J. Nutr. 45:483–94.
36. Prentice, A., P. Alison, A. Prentice, A. Black, T. Cole, and R. Whitehead. 1986. Cross-cultural differences in lactational performance. In M. Hamosh and A. S. Goldman, eds., Human lactation 2: maternal and environmental factors. New York, NY: Plenum Press.
37. Hennart, P. F., D. J. Brasseur, J. B. Delogne-Desnoeck, M. M. Dramaix, and C. E. Robyn. 1991. Lysozyme, lactoferrin, and secretory immunoglobulin A content in breast milk: Influence of duration of lactation, nutrition status, prolactin status, and parity of mother. Am. J. Clin. Nutr. 53:32–39.
38. Miranda, R., N. G. Saravia, R. Ackerman, N. Murphy, S. Berman, and D. N. McMurray. 1983. Effects of maternal nutritional status on immunological substances in human colostrum and milk. Am. J. Clin. Nutr. 37:632–40.
39. Kader, M. M. A., R. Bahgat, and M. T. Aziz. 1972. Lactation patterns in Egyptian women. II. Chemical composition of milk during the first year of lactation. J. Biosoc. Sci. 4:403–9.
40. World Health Organization. 1985. Energy and protein requirements. Geneva: WHO (WHO technical report series 724).
41. van Raaij, J. M. A., C. M. Schonk, S. H. Vermaat-Miedema, M. E. M. Peek, and J. G. A. J. Hautvast. 1991. Energy cost of lactation, and energy balances of well-nourished Dutch lactating women: Reappraisal of the extra energy requirements of lactation. Am. J. Clin. Nutr. 53:612–19.
42. Brewer, M. M., M. R. Bates, and L. P. Vannoy. 1989. Postpartum changes in maternal weight and body fat depots in lactating vs nonlactating women. Am. J. Clin. Nutr. 49:259–65.
43. Schutz, Y., A. Lechtig, and R. B. Bradfield. 1980. Energy expenditures and food intakes of lactating women in Guatemala. Am. J. Clin. Nutr. 33:892–902.
44. Durnin, J. V. G. A. 1987. Energy requirements of pregnancy: an integration of the longitudinal data from the five-country study. Lancet 2:1131–33.
45. Jelliffe, D. B., and I. Maddocks. 1964. Notes on ecologic malnutrition in the New Guinea highlands. Clin. Pediatr. 3:432–38.
46. Tracer, D. P. 1991. The interaction of nutrition and fertility among Au forager-horticulturalists of Papua New Guinea. Ph.D. Dissertation. Ann Arbor, MI: University of Michigan.
47. Tracer, D. P. 1991. Fertility-related changes in maternal body composition among the Au of Papua New Guinea. Am. J. Phys. Anthropol. 85:393–406

48. Merchant, K., R. Martorell, and J. Haas. 1990. Maternal and fetal responses to the stresses of lactation concurrent with pregnancy and of short recuperative intervals. Am. J. Clin. Nutr. 52:280–288.
49. Merchant, K., R. Martorell, and J. D. Haas. 1990. Consequences for maternal nutrition of reproductive stress across consecutive pregnancies. Am. J. Clin. Nutr. 52:616–20.
50. Harrison, G. A., A. J. Boyce, and C. M. Platt. 1975. Body composition changes during lactation in a New Guinea population. Ann. Hum. Biol. 2:395–98.
51. Adair, L. S. 1984. Marginal intake and maternal adaptation: the case of rural Taiwan. In E. Pollitt and P. Amante, eds., Energy intake and activity. New York, NY: Alan R. Liss. 33–55.
52. Miller, J. E., and R. Huss-Ashmore. 1989. Do reproductive patterns affect maternal nutritional status?: An analysis of maternal depletion in Lesotho. Am. J. Hum. Biol. 1:409–19.
53. Little, M. A., P. W. Leslie, and K. L. Campbell. 1992. Energy reserves and parity of nomadic and settled Turkana women. Am. J. Hum. Biol. 4:729–38.

# CHAPTER 15 Accommodation to Acute and Chronic Malnutrition during Growth

# EFFECTS OF PROTEIN-CALORIE MALNUTRITION ON CHILDREN

Malnutrition is one of the most pressing problems confronting the developing nations. A nutritional survey of over 190,000 children in forty-six communities in South America, Africa, and Asia between 1963 and 1972 indicates that between 0.5% and 20% of the population suffer from severe forms of protein-calorie malnutrition (PCM) while about 3% to 74% suffer from moderate malnutrition (1). Furthermore, the inter-American investigation of mortality in childhood in Latin America points out that malnutrition was either directly or indirectly responsible for over 50% of the deaths of children under 5 years of age (2). An understanding of the effects of malnutrition on human growth is of prime importance since an individual's adult biological and behavioral performance is profoundly influenced by environmental factors operating during growth and development. In this chapter the characteristics of protein-calorie malnutrition and their relationship to brain growth and the malnutrition-infection complex and the environment that generates it are discussed.

## Types of Protein-Calorie Malnutrition

The term *protein-calorie malnutrition,* abbreviated PCM, has been widely used to designate the spectrum ranging from pure protein deficiency to deficiencies of both protein and calories. PCM, a disease with multiple etiologies, is found principally in developing nations or populations undergoing cultural transition and/or urbanization. In all forms of PCM there is a reduction of both protein and calorie intake, but the ratio of protein to total calorie intake may vary widely in different conditions. In general two main forms of PCM are recognized: marasmus and kwashiorkor.

### Marasmus

Marasmus refers to exhausted protein and calorie stores in children. It results from a chronic, symmetrical reduction of all nutrients that approaches starvation levels. Its origin is traceable to the prenatal period, but it usually manifests itself during the first year of postnatal life. It causes a drastic reduction in growth and development such that the marasmic child is short and lightweight for its age, as well as retarded in skeletal maturation. The child exhibits extreme muscular wasting and almost no subcutaneous fat (fig. 15.1). These conditions reflect an attempt by the organism to use its own tissues as a source of nutriment in the face of a chronic and generalized decrease in nutrient resources. The marasmic child also has a reduced brain

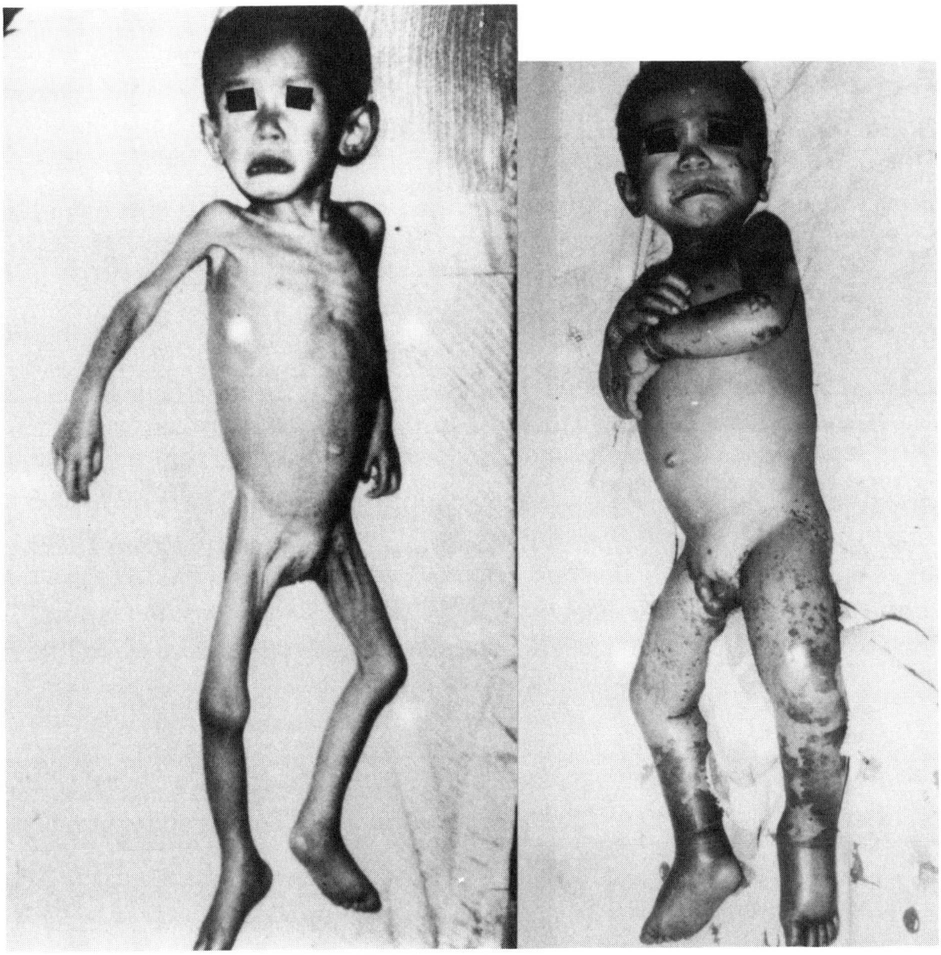

**Fig. 15.1.** Typical appearance of children with, **A**, marasmus and, **B**, kwashiorkor. Note the drastic reduction of subcutaneous fat and muscle wasting of the marasmic child. In contrast, the child with kwashiorkor exhibits edema and changes in skin and hair pigmentation. (From R. M. Suskind. 1975. Pediatr. Clin. North Am. 22:873–83. Photographs courtesy of medical staff, Anemia and Malnutrition Research Center, Chiang Mai, Thailand.)

weight, cortical atrophy, hypotonia, reduced activity, and displays constant hunger behavior.

### Kwashiorkor

Kwashiorkor is caused primarily by an acute protein deficiency occurring in the presence of relatively adequate caloric intake, often from foods poor in protein but rich in carbohydrates such as starch. The name kwashiorkor is used by the Ga people of Ghana to whom the disorder is well known. This disease is usually manifested after the first year, most often between the second and fourth years of postnatal life. The kwashiorkor child displays normal growth and development through the first postnatal year, but after the first year growth in height and weight is drastically reduced to between 60% and 80% of the expected standard. A typical kwashiorkor (fig. 15.1.b) child is the so-called "sugar baby," with a round moon face, pitting edema, variable degrees of dermatosis with depigmentation, and hyperkeratosis. The child's hair is often depigmented, and when malnutrition alternates with periods of relatively adequate dietary intake, depigmented bands often referred to as "flag signs" appear in the hair. In addition, hair implantation is affected so that it falls out spontaneously or can be painlessly removed. Serum albumin and protein levels are also reduced.

The kwashiorkor child is apathetic, lethargic, anorexic (not hungry), withdrawn, and highly irritable. Sometimes the child becomes immobile, lying quietly in a fetal position with open, nonfixating eyes. The kwashiorkor child often maintains a monotonous whimper. Like the marasmic infant, the kwashiorkor child exhibits hypotonia and poorly developed motor skills. In some cases, the poorly developed motor skills are so severe that the child does not respond in any measurable way to standard psychomotor stimuli derived from mental developmental scales.

### Marasmus and Kwashiorkor

The term *marasmus-kwashiorkor* is used to describe conditions in which signs of both syndromes are present. The child with marasmus-kwashiorkor has reduced growth in stature and weight and a drastic reduction in subcutaneous fat and muscle.

## Critical Periods of Brain Vulnerability

At present the dominant thesis is that the effects of malnutrition on brain growth depend on the developmental stage at which the individual is subjected to the stress; if malnutrition occurs during the period of high growth velocity or "growth spurt," the effects are permanent, but if it occurs after this critical period, the effects are reversible. Evidence in support of this thesis has been derived from experimental studies on animals and autopsies of children who died from severe malnutrition.

### Animal Studies

Experimental studies on rats have demonstrated that the growth and structure of the animal brain may be profoundly altered by events occurring during critical periods of development (3–9). These studies have shown that body growth retardation at the time of the brain growth spurt results in a brain-growth deficit associated with a disproportionate reduction of the cerebellum and neuron cells that resist subsequent nutritional rehabilitation. On the other hand, if body growth is retarded after the brain growth spurt, the brain-growth deficit and structural components fully recover with restoration of a good diet. On the basis of these investigations it has been suggested that growth and development of the organs follow a sequential pattern of cell multiplication and growth in cell size that is specific to a given organ system. The development of the brain follows three phases of growth: (1) hyperplasia—in this stage there is an increase in brain weight, protein content, and DNA content; (2) hyperplasia and concomitant hypertrophy—the increase in DNA content falls behind the increase in brain protein content and weight; and (3) hypertrophy—there is no additional increase in DNA content, but the existing cells enlarge resulting in net protein synthesis and brain weight at the same rate. The theory of vulnerable periods maintains that, because of the sequential pattern of growth, malnutrition inflicted during hyperplastic (proliferative stage) growth results in an organ with a permanent deficit in cell number. Biochemically, there may not be complete recovery. On the other hand, malnutrition imposed during the hypertrophic (cell enlargement stage) phase results in a reduction of cell size that can be restored to normal size by rehabilitation.

### Human Studies

Postmortem studies of children who suffered from severe malnutrition support in part the findings from experimental animal studies (6–10). These studies show that the brains of children who suffer from severe malnutrition are characterized by a reduction in weight, size, and cell number. Such a deficit in cell number occurs only among children who died during the first year but not among children who died of kwashiorkor after the first year, presumably because they were well-nourished during the first year of life. Furthermore, in brains of children who died during the first year of life, total cholesterol and total phospholipid content was reduced in proportion to the reduction in cell number. If the malnutrition extended into the second year of life, an actual decrease in lipid quantity occurred.

### Summary

In summary, both experimental studies on animals and postmortem observations of children provide evidence for lasting structural changes in the brain being related to the timing of early malnutrition. It appears that the brain growth spurt is by far the most vulnerable period, and malnutrition at this stage may result in quantitative and qualitative physical distortion.

### Critical Periods and Intellectual Development

As shown in figure 15.2, the brain growth spurt in humans begins during the last trimester of fetal life and terminates by the second postnatal year. Thus, in humans the vulnerable period for the permanent effects of malnutrition is from the last trimester of the prenatal period to the end of the first year. Conversely, the effects of malnutrition on brain growth and structural distortions after the second year are not permanent. An obvious question is whether the quantitative and structural changes of the brain are also found among survivors of severe malnutrition.

Studies during treatment and recovery of Mexican children hospitalized for severe malnutrition revealed that behavioral recovery was less complete in the children who were hospitalized before 6 months of age than in older children (11). In this study the test of neurological development was based on biological quantifiable measurements of brain growth and function that are not influenced by social environmental factors. This finding supports the hypothesis that during the critical period of infancy the individual is more susceptible to permanent effects of malnutrition and thus more resistant to intellectual recovery.

On the other hand, studies of Jamaican (12) and Ugandan (13) school-age children who suffered an episode of acute or severe malnutrition prior, during, and after 2 years of age found no relationship between the scores obtained in psychological or intellectual tests and the age when the subjects were hospitalized. Similarly, a study of school-age Nigerian children who suffered from severe malnutrition (kwashiorkor) when they were 1 to 3 years old found no relationship between scores on psychological tests and the age at which they were hospitalized (14). The failure to find a relationship between the severity of childhood malnutrition and neurological development (15–18) is related to the fact that the measures of cognitive or intellectual development reflect more the individual's social and environmental experience than brain function. This is especially the case where malnutrition coexists with poor socioeconomic conditions.

## Malnutrition and Infections

### Malnutrition and Infant Mortality

Infectious disease, because of its debilitating effects, plays an important role in the development of malnutrition, morbidity, and mortality of young children in developing nations. As shown in table 15.1 infant mortality rates (defined as the number of children $<1$ year old per 1000 live births who died during the first year) in the Third World countries such as Guatemala, Brazil, Ecuador, Honduras, Peru, Haiti and Bolivia are about six to ten times greater

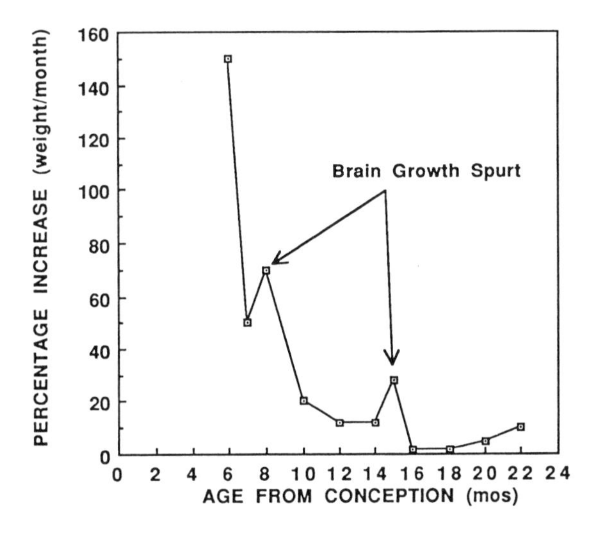


**Fig. 15.2.** Brain growth expressed by percent of increase in weight in relation to conceptual age for males and females combined. There are two "spurts" of brain growth, one during the last trimester of gestation and another at 5 months of postnatal period. (Modified from D. B. Cheek, A. B. Holt, and E. D. Mellits. 1972. Malnutrition and the nervous system. Pub. 251. Washington, DC: Pan American Health Organization.)

than the rate for industrialized countries such as Sweden, Canada, and the United States. The leading causes of death in Third World countries are malnutrition induced infectious and parasitic diseases. Evaluations carried out in poor areas of 10 countries of continental America between 1968 and 1970 by the Inter-American Investigation on Child Mortality (74) showed that of the 33,826 deaths registered in the Latin American and Caribbean countries (eight countries in all), 14,513 (42.9%) were due to infectious diseases. Of the deaths due to infectious diseases, the principal cause of death was diarrhea in 10,037 cases (69.2%) and measles in 2107 cases (14.5%).

## Nutritional Status and Incidence and Duration of Infections

**Measles and Case Fatality Rates.** The case fatality rate for measles in Third World countries ranges from 3.7% to 15% (20–25). An important component

**TABLE 15.1. Infant mortality rates in selected countries**

| Country | IMR[a] | Country | IMR | Country | IMR |
|---|---|---|---|---|---|
| Sweden | 7.0 | Costa Rica | 19.1 | Ecuador | 81.0 |
| Switzerland | 9.1 | Argentina | 41.0 | Honduras | 87.0 |
| France | 9.6 | Mexico | 55.0 | Peru | 87.0 |
| Canada | 10.4 | Colombia | 56.0 | Haiti | 113.0 |
| United States | 11.4 | Guatemala | 65.9 | Bolivia | 130.0 |
| Cuba | 18.5 | | | | |

*Source:* From the 1983 World Population Data Sheet. Washington, DC: Population Reference Bureau.
[a] Infant Mortality Rate = number of children <1 year old per 1000 live births who died during the year. Figures presented are for countries in which the United Nations considers the data to be reliable and are for 1980 or 1981.

of the high fatality rate appears to be related to the poor nutritional status of children in developing countries. In fact, it has been reported that the presence of kwashiorkor notably increases the fatality rates for measles (26), and that measles complications are more frequent in malnourished children (27). High fatality rates for some other diseases, such as chicken pox (28) and diarrhea (27), have also been reported for malnourished children.

**Infections and Diarrhea.** Several epidemiological studies conducted in Guatemala (29), Nigeria (39), Peru (31) and Bangladesh (32-33) have demonstrated that the incidence and duration of diarrheal infections is greater in malnourished than in well-nourished children. Furthermore, the number of diarrheal episodes accompanied by high fever and blood was significantly higher in malnourished children, as was the number of cases of complication by respiratory infections (34). Similarly, the rates of hospitalization and death for the malnourished children were higher than those for the well-nourished ones.

It is evident that nutritional status is directly related to the incidence and duration of infections. This conclusion is supported by the fact that nutritional supplementation studies conducted in Colombia (35) and Mexico (36) found a reduction in the incidence and duration of infections in children who received supplements.

In summary, all infections have an adverse effect on the nutritional status and growth of the child, but they especially affect the malnourished child because they reduce food intake and increase urinary metabolic losses of nitrogen, ascorbic acid, iron, and other important nutrients that are already in short supply (37, 38). Therefore the requirements for protein and other nutrients are increased during both the infection and the recovery phases. If these increased requirements are not met by the insufficient diet, the individual is left in a depleted state, becoming increasingly more vulnerable to the development of clinical and nutritional diseases as a result of additional infectious episodes. Furthermore, these infections may in themselves depress cell-mediated immunity. As a result, new infections may emerge or existing ones may become more severe. In other words, there is a reciprocal and synergistic interaction between infection and malnutrition in which each exacerbates the other. The combination of the two has more profound consequences for the individual than would be predicted from the presence of either one alone (38).

## Breast-feeding and Childhood Malnutrition

In many Third World countries, especially in urban areas, breast-feeding has decreased significantly. This decrease has had detrimental effects on the child's health and nutrition.

**Infant Weight.** As shown in figure 15.3, during the first 2 to 5 months of life the growth curves of breast-fed children from low socioeconomic groups in Third World countries are similar to those derived from studies of healthy children in well-to-do nations. However, after 5 months of life, marked growth retardation is evident. The adequate rates of growth during early infancy in developing countries are related to breast-feeding which provides sufficient levels of energy and nutrients during the first 2 to 5 months of life. Due to the lack of breast-feeding and initiation of weaning, nutrient intakes become deficient and thus growth is delayed after 6 months of age.

**Diarrhea.** As shown in table 15.2 completely weaned children had higher incidence of diarrhea than children with mixed feeding in all age groups. This difference is related to the fact that children who are breast-fed during the first months of life, followed by mixed feeding for several months receive both the benefits of the most appropriate food for their nutritional needs during the first months of life, and the anti-infective properties of maternal milk. As such, the greater the duration of full or partial breast-feeding during infancy the greater the infant survival (39). These data suggest a protective effect of maternal milk even under poor sanitary conditions. This inference is supported by clinical and epidemiologic evidence that show a higher incidence of gastrointestinal and respiratory infections in bottle-fed children as compared with breast-fed children in Canada, the United States, and India (40).

The physiological mechanism whereby breast-feeding protects the infant is related to the fact that maternal milk contains immunoglobulins, macrophages, lymphocytes, neutrophils, components of the complement system, lactoferrin, and lysozyme all of which provide the child with immunity. Furthermore, breast milk promotes the growth of lactobacillus bifidus in the intestine, which generates an acidic environment and in this way inhibits the growth of pathogenic microorganisms. Therefore, the replacement of breast-feeding with artificial feeding, particularly at an early age when the child is immunologically immature and more susceptible to infections, compounds the already nutritionally stressed child.

## Weaning and Child Infections

**Housing Conditions.** Throughout the Third World, where poverty and malnutrition often exist in a symbiotic form, there is an association between the beginning of weaning and the occurrence of infections.

The high incidence of infections during the weaning period is almost exclusively associated with children who live in extreme poverty (34). This association exists because the houses of the very poor (1) lack drainage and latrines which leads to indiscriminate defecation, (2) do not have a piped

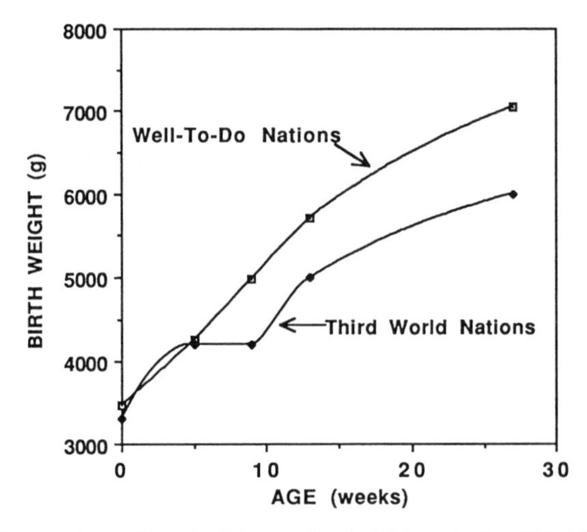


**Fig. 15.3.** Comparison of postnatal growth of children in the Third World and well-to-do nations. The adequate rates of growth during early infancy in developing countries is related to breast-feeding, which provides sufficient levels of energy and nutrients during the first 2 to 5 months of life. Due to the lack of breast-feeding and initiation of weaning, nutrient intakes become deficient and thus growth is delayed after 6 months of age.

water supply, (3) lack refrigeration, causing a significant increase in the number of microorganisms in foods that are not consumed immediately after cooking. In some rural areas, the presence of animals roaming freely about the home is another source of contamination. For example, in a study of rural Guatemalan families it was found that almost all diarrheal outbreaks originated in children younger than 3 years old. This finding suggests that those children who usually stay in or near the home are especially susceptible to the microorganisms that cause diarrhea. Older children and adults are probably either less susceptible or are immune to most of the agents that cause illness in the child, yet they are carriers of these microorganisms (table 15.2) (41).

**Foods Consumed.** The foods used and the consumption time after food preparation are important factors affecting the transmission of gastrointestinal infections during weaning. Rowland et al. (42) found that the gruel used in infant weaning in Gambia (flour and water) was contaminated with E. coli. The highest E. coli counts were found in infant foods 8 hours after

**TABLE 15.2. Incidence of acute diarrheal disease (cases per 1000 children per year)**

| Age (mo) | Wholly Breast-fed | Mixed Feeding | Weaned |
|---|---|---|---|
| 6– 8 | 183.7 | 207.3 | 236.2 |
| 9–11 | 127.5 | 172.4 | 220.1 |
| 12–14 | 90.3 | 122.5 | 207.9 |

*Source:* Adapted from N. S. Scrimshaw, C. E. Taylor, and J. E. Gordon. 1968. Interactions of nutrition and infection. Geneva: World Health Organization.

preparation. E. coli was especially prevalent in foods that contained milk products and in cereal gruels. Similarly, in Bangladesh Black et al. (43) found that 40% of 470 samples of foods prepared for children during weaning were contaminated with E. coli and that the levels of contamination were higher on hot days. More than two-thirds of various samples of milk examined 1 hour after boiling contained E. coli.

**Enteropathogenic E. Coli and Diarrhea.** Studies of children being treated for diarrhea in rural clinics in Bangladesh (44) found that 87% of the children had an enteropathogenic organism. At present, the evidence suggests that E. coli is transmitted primarily by contaminated foods (34), which often have high concentrations of microorganisms.

## Infections and Growth Retardation

From studies conducted in Third World countries there is conclusive evidence indicating a negative relationship between infection and growth (45).

Numerous clinical reports indicate that the appearance of kwashiorkor is frequently immediately preceded by episodes of diarrhea, measles, and other infections in children with histories of nutritional deficits. The same has been observed in children with clinical signs of specific deficiencies such as keratomalacia, beriberi, and scurvy (45). Several studies have found a negative association between the percent of time during which children were ill with diarrhea and increases in weight and height (41, 45). Other investigations have also found negative relationships between growth and diarrhea (46–52), between growth and measles (48, 52), between growth and whooping cough (52–53), between growth and fever (48), and between growth and asymptomatic viral infections (54). Similarly, a study carried out in children from Gambia (55) found that the amount of time the children were ill with diarrhea during a given period was negatively related to increases in both height and weight during the same period. Malarial infections were also negatively related to increases in weight but not height. None of the various other infections studied (including respiratory infections) was found to be related to growth. In summary, clinical and epidemiologic evidence indicates that some infectious diseases, including diarrheal diseases, cause growth retardation.

On the other hand, most studies carried out in industrialized countries have not shown an association between infection and growth. The different results generally obtained in developed countries as opposed to Third World countries reflect important ecologic differences. In Third World countries, children suffer from a higher incidence of infections and probably more severe infections than children in industrialized countries. Furthermore, due to the high frequency of reinfections and inadequate diets during the convalescent period, children with inadequate nutrition do not recover their growth as well as children with adequate nutrition. There is evidence suggesting that with nutritional supplementation Colombian children can offset the negative effects of diarrheal diseases on physical growth (47).

It is quite evident that childhood infections have a direct effect on growth retardation. The physiological mechanisms whereby infections affect infant growth are probably related to reductions in food intake. In general, children suffering from gastrointestinal diseases such as diarrhea decrease their food intake. Several studies have found an inverse correlation between infectious diseases and caloric consumption during weaning (23, 56–57). These studies indicate that the reduction in food intake was a function of both anorexia and the practice many mothers have of reducing the child's food intake during infections (especially diarrhea). Furthermore, the anorexia induced by infection is probably related to decreased nutrient absorption. Acute diarrhea leads to a loss of intestinal microflora resulting in the decreased absorption of carbohydrates (glucose, xylose, and lactose), fats, nitrogen, and some vitamins and minerals (58). Because infections are usually associated with fever, this pathology is compounded by an increased metabolic rate. For every increase of 1°C in body temperature, energy requirements increase from 5% to 8.2% (57). As a result, during the acute phase of various infections, catabolic processes such as gluconeogenesis, glycogenolysis, and the secretion of insulin and glucagon are increased.

## Nutrition and Immunological Response

Several studies have shown conclusively that severe malnutrition affects an individual's immunologic response (28, 59–60). The effects of malnutrition on immunological response include reduction in cell-mediated immunity, decreased humoral immunity, and, to a lesser degree, decreased complement system activity (59, 61) and bactericidal activity (62).

### Cell-Mediated Immunity

Current evidence suggests that severe malnutrition inhibits the cell-mediated response independently of infection (61, 63–66).

**Capacity of Thymus.** The cell-mediated immune response is affected by the capacity of the thymus glands, lymph nodes, and tonsils to produce lymphocytes. The thymus of malnourished children is atrophied, and the lymph nodes, tonsils, and spleen are smaller than in well-nourished children (66–69); hence, their capacity for the production of lymphocytes is impaired. This is manifested by the loss of lymphocyte killer function. The resulting immuno-function deficiency is manifested by an increased susceptibility to infections and decreased reactivity to delayed skin test antigens (64, 65). In fact, the degree of reduced cellular immune response in malnourished children is comparable to that observed in patients with primary cellular immunodeficiency. This depressed cellular immunity contributes to the marked susceptibility of malnourished children to many infections such as measles, herpes simplex, tuberculosis, and gram-negative infections (70).

**Rate of Change.** Other studies have found that severe malnutrition decreases the transformation rate of active lymphocytes to lymphoblasts (71–72), decreases the proportion of T lymphocytes in the peripheral blood (73–74) and, in most cases, increases the incidence of delayed type sensitivity tests (34, 71–72, 74–77).

## Humoral Immunity (Antibody Production)

The humoral response is less altered in moderate malnutrition than in severe malnutrition, although a decrease in the antibody response to tetanus toxin and to Salmonella tiphi has been found (78). There are also decreased levels of secretory IgA in saliva and tears, the respiratory tract, and the mucosa of the jejunum (79). Only one of the studies of humoral response found reduced serum IgA levels in children with moderate malnutrition (77).

**Production.** Malnutrition per se decreases production of immunoglobulins (IgA) in nasopharyngeal salivary secretions, such as those found in the mucosa, tears, and saliva (63, 80). The low production of immunoglobulins facilitates colonization of the mucosal surface by microorganisms and in this way predisposes the malnourished child to an increased incidence of infection.

**Capacity.** Malnutrition does not affect the capacity to produce antibodies (63, 80). In fact, once infected, the malnourished child responds with an elevated production of serum immunoglobulins (IgA, IgM, IgG, IgD).

## Complement System

Some other components of the immunologic system, such as the complement system, phagocyte function (73), and the levels of secretory IgA in tears were found to be depressed in children with moderate malnutrition (81).

In summary, severe and moderate malnutrition have been found to be associated with decreased cellular immune responses, reductions in secretory IgA levels, and defects in some components of the complement system. The fact that malnourished populations are more vulnerable to the effects of infectious diseases suggests that malnutrition results in a defective immune response. In other words, the malnourished child is more susceptible to infections. This is schematized in figure 15.4.

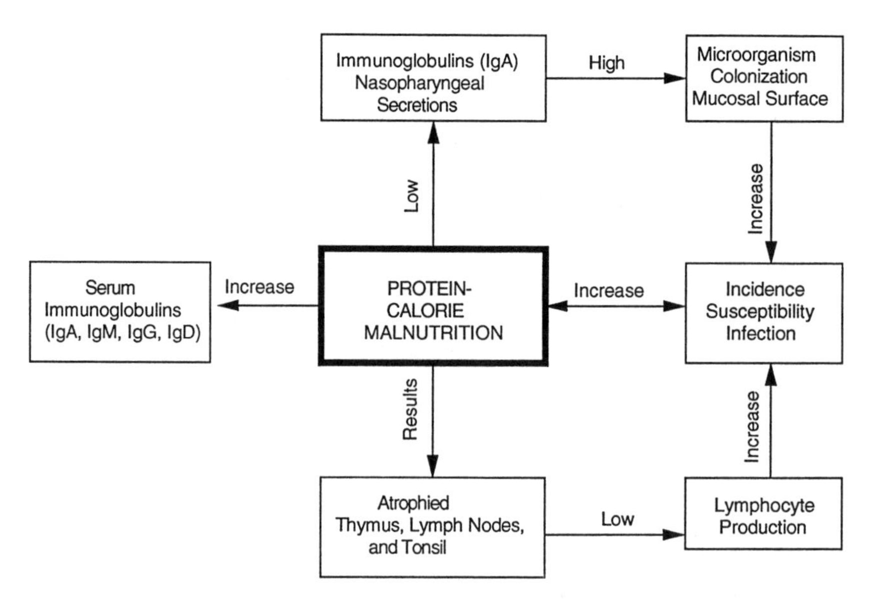


**Fig. 15.4.** Schematization of synergistic interaction of protein-calorie malnutrition and infection. Although malnutrition does not impair one's capacity to produce serum antibodies, it does decrease production of antibodies in nasopharyngeal secretions, which, in turn, becomes a focus of infection. The resulting infection exacerbates malnutrition effects.

## GROWTH WITH CHRONIC UNDERNUTRITION

### Birth Weight

Among populations living in poor socioeconomic conditions in which daily dietary intakes do not exceed 2000 calories and are derived mainly from carbohydrates, which are of poor protein quality, prenatal growth is retarded. In the rural villages of Mexico the average birth weight was 2999 g, and the percentage of prematurity (birth weight less than 2500 g) equaled 12.3 (82). Among four rural Guatemalan villages about 40% of all newborns weighed less than 2500 grams, 11% of these were born prior to 37 weeks gestational age (true prematures), and 20% were born at term (after 37 weeks of gestation) (37). These values are significantly different from those found in well-nourished populations in which the birth weight averages between 3300 and 3500 g, and the percentage of births below 2500 g is approximately less than 8% (83, 84).

### Height

From a recent summary of world-wide variation in growth (85) it can be inferred that the populations living with chronic low dietary intakes have a pattern of growth characterized by (1) slow growth during childhood and adolescence, (2) late adolescence growth spurt, and (3) prolonged period of growth (fig. 15.5).

On the average the slow growth rate in height becomes evident after birth and continues through the early 20s. Most populations attain adult height by about 22 to 24 years for males and 20 to 22 years for females. When compared to the United States standards these values amount to an increase of 10% in the duration of the period of growth. Despite this increase in the period of growth the adult height of undernourished populations is reduced by 10%. This difference is mostly accounted for by the uniform slow growth rate that characterizes undernourished populations (85–87, 102–5).

## Skeletal Maturation

Along with retarded growth in height undernourished populations, when compared to the Gruelich-Pyle, Tanner-Whitehouse, or Garn-Rohman-Silverman standards, exhibit a drastic retardation in skeletal maturity (87). As inferred from studies conducted among Central American rural populations, the delay in skeletal maturation between the ages of 0 to 10 years averages about 20%, whereas the delay during adolescence does not exceed 10% (87). This 10% delay (which reflects epiphyseal closure) coincides with the increase in the age of attainment of adult stature that, as previously indicated, is prolonged by about 10% in most undernourished populations. Since the age at epiphyseal closure and termination of growth is proportionally less retarded than the delay experienced during childhood, environmentally related differences in adult stature are mostly caused by environmental influences that retard growth and maturation during childhood.

## Cortical Bone Thickness

Radiogrammetric measurements of the total bone width of the second metacarpal and width of the medullary cavity have demonstrated that malnutrition reduces the absolute and relative amount of cortical bone. Severely malnourished (kwashiorkor) Guatemalan children, when compared to rural nonmalnourished Guatemalan children, were found to be deficient in cortical bone (90, 91). Further studies have shown that the loss of cortical bone occurs mainly at the endosteal (inner) surface of the bone, whereas the periosteal (outer) surface of the bone continues to increase (92). For this reason during malnutrition the medullary cavity is enlarged, and the total width of the bone is either maintained or continues to grow.

## Sexual Development

Menarche is the most widely used indicator of sexual maturity in females. In general, age at menarche is derived either through the status quo or through

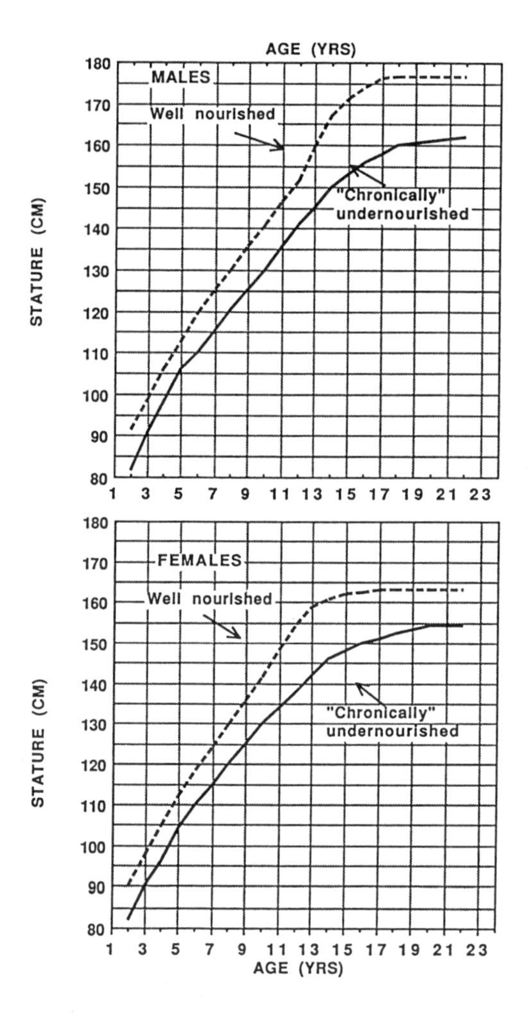


**Fig. 15.5.** Comparison of patterns of growth of chronically undernourished and well-nourished populations. With chronic undernutrition growth rate for all ages is slow and continues into the early twenties. (Adapted from A. R. Frisancho. 1990. Anthropometric standards for the evaluation of growth and nutritional status. Ann Arbor, MI: University of Michigan Press.)

retrospective or prospective methods. Age at menarche, by whatever method of assessment, is delayed among populations in poor socioeconomic conditions (85, 101). However, within a population the effects of socioeconomic indicators on age at menarche are less than on childhood maturity markers. For example, based on the status quo method, age at menarche for well-off Bantus varies between 13 and 15 years, and for poor or lower-class Bantus it ranges from 14.8 to 15.4 years (85, 93). In the same manner, the median age at menarche in United States high income whites averaged 13 years and 13.5 years for the low income groups, and United States undernourished white girls averaged 14.4 years compared to 12.4 in the well-nourished controls (94). For the well-off, middle-class, and lower-income Chinese (Hong Kong) the age of menarche averaged 12.5, 12.8, and 13.3 years respectively (95). In India and Mexico similar differences have been indicated (96, 97). In other words, delay among the poor or undernourished does not appear to exceed 15% of the well-off average age. However, this does not mean that age of menarche is not delayed in other populations. Among New Guinea natives and Himalayan Sherpas the age of menarche, based on the less reliable retrospective method, has been reported to be as late as 18 and 19 years respectively (88, 98).

## OVERVIEW

At present there is considerable evidence indicating that the effects of protein-calorie malnutrition on brain growth and function might have long-lasting consequences depending on the critical period at which the nutritional insult occurs. Both experimental studies on animals and autopsies of children who died of malnutrition indicate that if malnutrition occurs during the period of brain growth spurt, the effects on neuronal cells and biochemical composition might be permanent; but if it occurs after the growth spurt or critical period, the effects can be reversed. The extent to which these findings are applicable to living populations or survivors of malnutrition is not known.

Using behavior or psychological measurements, investigators have unsuccessfully attempted to test the critical period hypothesis. Results of studies involving children who were hospitalized for severe acute malnutrition at different ages during infancy and childhood are inconclusive. They neither support nor reject the hypothesis that systematic differences in intellectual level in school-age children are related to severe malnutrition that occurred at different vulnerable age periods in human development. This inconclusiveness is not surprising in view of the fact that severely malnourished children are exposed to decreased learning opportunities because of both the inherent inactivity and lethargy associated with malnutrition and the poor socioeconomic conditions with which malnutrition coexists. In addition, the tests employed to measure cognitive processes are based on traits that are particularly sensitive to social-environmental influences. Therefore, the possible effects of undernutrition on biological-neurological development are confounded by the influence of external environmental factors.

Laboratory and epidemiological studies suggest that there is a synergistic interaction between malnutrition and infection, whereby malnutrition predisposes the organism either through low production of immunoglobulins on the nasopharyngeal secretions or inhibition of the cell-mediated immunoresponse. The infection in turn exacerbates the effects of malnutrition, and therefore new infections may emerge or existing ones become more severe. Thus a vicious cycle between malnutrition, infections, and immunodeficiency is established (see fig. 15.4). This synergistic interaction calls attention to the

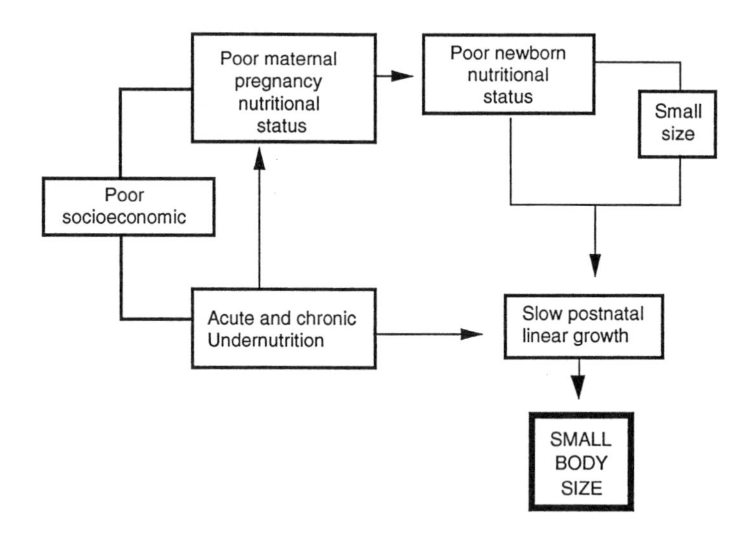


**Fig. 15.6.** Schematization of effects and interaction of prenatal and postnatal undernutrition on human growth and adult body size. Reduction in adult body size among populations living under poor socioeconomic conditions results from compound effects of prenatal and postnatal undernutrition, manifested in acute and chronic forms.

fact that nutritional programs oriented at improving and understanding the etiology of malnutrition must also include programs for control of infection.

There is conclusive evidence indicating that maternal chronic undernutrition retards prenatal growth; thus variations in maternal calorie and protein reserves are the most important factors affecting newborn body size and composition. These findings and those derived from the Dutch famine (99, 100) suggest that in conditions of drastic dietary restriction maternal calorie and protein reserves are not sufficient to maintain adequate prenatal growth. The relationship between maternal nutritional status and prenatal growth is so well defined that any changes in maternal calorie intake or change in calorie and protein reserves are reflected in dramatic changes in birth weight. Indeed, it can be safely concluded that about one-half of the incidence of prematurity (10%) observed among developing nations (12% to 20%) is either directly or indirectly related to malnutrition. In view of the negative concomitants of prematurity, improvements of maternal nutritional status must be a prime priority of public health programs.

During the postnatal period populations living in chronic undernutrition or poor socioeconomic conditions exhibit slow growth in stature and weight, decreased bone cortex, and late age of menarche. Although the period of growth is extended, undernourished populations do not attain the adult stature of well-nourished populations. Experimental studies indicate that an animal's postnatal growth is to a large extent determined by the size and nutritional status attained at birth. Those born small who did not have a tall parent and continue to experience undernutrition are likely to grow slowly and remain small for their age in spite of the fact that they may continue to grow for a longer time (89). In figure 15.6 these findings have been summarized in schematic form. It shows that among populations living in poor socio-economic conditions undernutrition in its acute or chronic manifestations (through its effect on maternal pregnancy status) leads to poor newborn nutritional status and reduced size at birth. Because of the cumulative effects of prenatal undernutrition, prenatal growth retardation, and chronic under-nutrition after birth linear growth during the postnatal period is slow and eventually leads to reduced adult body size.

## References


1. Bengoa, J. M. 1974. The problem of malnutrition. WHO Chron. 28:3–7.
2. Puffer, R. R., and C. V. Serrano. 1973. The role of nutritional deficiency in mortality. Boletin de la Oficina Sanitaria Panamericana 7:1–25.
3. Dobbing, J. 1968. Effects of experimental undernutrition in development of the nervous system. In N. S. Scrimshaw and J. F. Gordon, eds., Malnutrition, learning and behavior. Cambridge, MA: M.I.T. Press.
4. Dobbing, J., and J. Sands. 1972. Vulnerability of developing brain. Part 9. The effect of nutritional growth retardation on the timing of the brain growth-sprint. Biol. Neonate. 19:363–78.
5. Chase, H. P., J. Dorsey, and G. M. Mckhann. 1967. Malnutrition and the synthesis of myelin. Pediatrics 40:551–58.
6. Winick, M., and P. Rosso. 1969. Head circumference and cellular growth of the brain in normal and marasmic children. J. Pediatr. 74:774–78.
7. Winick, M., P. Rosso, and J. Waterlow. 1970. Cellular growth of cerebrum, cerebellum, and brain stem in normal and marasmic children. Exp. Neurol. 26: 393–98.
8. Winick, M., and A. Noble. 1966. Cellular response in rats during malnutrition at various ages. J. Nutr. 89:300–306.
9. Benton, J. W., H. W. Moser, P. R. Dodge, et al. 1966. Modification of the schedule of myelination in the rat by early nutritional deprivation. Pediatrics 38:801–7.
10. Winick, M. 1969. Malnutrition and brain development. Pediatrics 74:667–79.
11. Cravioto, J., and B. Robles. 1965. Evolution of adaptive and motor behavior during rehabilitation from kwashiorkor. Am. J. Orthopsychiatry 35:449–64.
12. Hertzig, M. E., H. G. Birch, S. A. Richardson, and J. Tizard. 1972. Intellectual levels of school children severely malnourished during the first two years of life. Pediatrics 49:814–24.
13. Hoorweg, J., and P. Stanfield. 1972. The influence of malnutrition on psychological and neurological development: Preliminary communication. In Nutrition, the nervous system and behavior. P.A.H.O. Scientific Publication 251. Washington, DC: Pan American Health Organization.
14. Nwuga, V. C. B. 1977. Effect of severe kwashiorkor on intellectual development among Nigerian children. Am. J. Clin. Nutr. 30:1423–30.
15. Winick, M. 1973. Relation of nutrition to physical and mental development. Bibl. Nutr. Dieta. 18:114–22.

16. Manocha, S. L. 1972. Malnutrition and retarded human development. Springfield, IL: Charles C. Thomas, Publisher.
17. Pollit, E., and C. Thomson. 1977. Protein-calorie malnutrition and behavior. A view from psychology. In R. J. Wurtman and J. J. Wurtman, eds., Nutrition and the brain. New York, NY: Raven Press.
18. Lloyd-Still, J. D. 1976. Clinical studies on the effects of malnutrition during infancy on subsequent physical and intellectual development. in J.D. Lloyd-Still, ed., Malnutrition and intellectual development. Littleton, MA: Publishing Sciences Group.
19. Puffer, R. P., and C. V. Serrano. 1973. Patterns of mortality in childhood. Scientific Publication 262. Washington, DC: Pan American Health Organization.
20. McGregor, I. A. 1964. Measles and child mortality in the Gambia. West Afr. Med. J. 13:251.
21. Morley, D. 1973. Paediatric priorities in the developing world. London: Butterworths.
22. Scrimshaw, N. S., J. B. Salomon, H. A. Bruch, and J. E. Gordon. 1966. Studies of diarrheal disease in Central America. Am. J. Trop. Med. Hyg. 15:625–31.
23. Mata, L. J. 1978. The children of Santa Maria Cauque: a prospective field study of health and growth. Cambridge, MA: MIT Press.
24. Sinha, D. P. 1977. Measles and malnutrition in a West Bengal village. Trop. Geogr. Med. 29:125–34.
25. Koster, F. T., G. C. Curlin, K. M. A. Aziz, and A. Haque. 1981. Synergistic impact of measles and diarrhea on nutrition and mortality in Bangladesh. Bull. WHO 59:901–8.
26. Morley, D. 1969. Severe measles in the tropics (I and II). Br. Med. J. 1:297–300, 363–65.
27. Salomon, J. B., L. J. Mata, and J. E. Gordon. 1968. Malnutrition and the common communicable diseases of childhood in rural Guatemala. Am. J. Pub. Health 3:505–16.
28. Suskind, R. M., ed. 1977. Malnutrition and the immune response. New York, NY: Raven Press.
29. Delgado, H. L., V. Valverde, J. M. Belizan, and R. E. Klein. 1983. Diarrheal diseases, nutritional status and health care: analyses of their interrelationships. Ecol. Food Nutr. 12:229–34.
30. Tomkins, A. 1981. Nutritional status and severity of diarrhea among pre-school children in rural Nigeria. Lancet 1:860–62.
31. Trowbridge, F. L., L. H. Newton, and C. C. Campbell. 1981. Letter to the editor. Lancet 1:1375.
32. Chen, L. C., E. Huq, and S. L. Huffman. 1981. A prospective study of the risk of diarrheal diseases according to the nutritional status of children. Am. J. Epidemiol. 114:284–92.
33. Black, R. E., K. H. Brown, and S. Becker. 1984. Malnutrition is a determining factor in diarrheal duration, but not incidence, among young children in a longitudinal study in rural Bangladesh. Am. J. Clin. Nutr. 39:87–94.
34. Rivera, J., and M. Martorell. 1988. Nutrition, infection, and growth Part II: Effects of malnutrition on infection and general conclusions. Clin. Nutr. 7:163–67.
35. Wray, J. D. 1978. Direct nutrition intervention and the control of diarrheal diseases in preschool children. Am. J. Clin. Nutr. 31:2073–82.

36. Chavez, A., and C. Martinez. 1982. Growing up in a developing community. Mexico City: Nueva Editorial Interamericana.
37. Mata, L. J., J. J. Urrutia, and A. Lechtig. 1971. Infection and nutrition of children of a low socioeconomic rural community. Am. J. Clin. Nutr. 24:249–59.
38. Scrimshaw, N. S. 1975. Interactions of malnutrition and infection: advances in understanding. In R. E. Olson, ed., Protein-calorie malnutrition. New York, NY: Academic Press.
39. Habicht, J-P., J. DaVanzo, and W. P. Butz. 1986. Does breastfeeding really save lives, or are apparent benefits due to biases? Am. J. Epidemiol. 123:279–90.
40. Sanhi, S., and R. K. Chandra. 1983. Malnutrition and susceptibility to diarrhea. In L. C. Chen and N. S. Scrimshaw, eds., Diarrhea and malnutrition. New York, NY: Plenum Press.
41. Martorell, R., J-P. Habicht, C. Yarbrough, A. Lechtig, and R. E. Klein. 1975. Acute morbidity and physical growth in rural Guatemalan children. Am. J. Dis. Child. 129:1296–1301.
42. Rowland, M. G. M., R. A. E. Barrell, R. G. Whitehead. 1978. Bacterial contamination in traditional Gambian weaning foods. Lancet 1:136–38.
43. Black, R. E., K. H. Brown, S. Becker, A. R. M. Abdul Alim, and M. N. Merson. Contamination of weaning foods and transmission of enterotoxigenic *Escherichia coli* diarrhea in children in rural Bangladesh. Trans. R. Soc. Trop. Med. Hyg. 73:259–64.
44. Black, R. E., M. W. Merson, and K. H. Brown. 1983. Epidemiological aspects of diarrhea associated with known enteropathogen in rural Bangladesh. In L. C. Chen and N. S. Scrimshaw, eds., Diarrhea and malnutrition. New York, NY: Plenum Press.
45. Rivera, J., and M. Martorell. 1988. Nutrition, infection, and growth Part I: Effects of infection on growth. Clin. Nutr. 7:156–62.
46. Scrimshaw, N. S., C. E. Taylor, and J. E. Gordon. 1968. Interactions of nutrition and infection. Geneva: World Health Organization.
47. Lutter, C. K., J. O. Mora, J-P. Habicht, K. M. Rasmussen, and D. S. Robson. Nutritional supplementation eliminates child stunting due to diarrhea. Am. J. Clin. Nutr. (in press.)
48. Cole, T. J., and J. M. Parkin. Infection and its effect on the growth of young children: a comparison of the Gambia and Uganda. Trans. R. Soc. Trop. Med. Hyg. 71:196–98.
49. Condon-Paoloni, D., J. Craviota, F. E. Johnston, E. R. de Licardie, and T. O. Scholl. 1977. Morbidity and growth of infants and young children in a rural Mexican village. Am. J. Pub. Health 67:651–56.
50. Draper, K. C., and C. C. Draper. 1960. Observations on the growth of African infants: with special reference to the effects of malaria control. J. Trop. Med. Hyg. 63:165–71.
51. Mata, L. J., J. J. Urrutia, and A. Lechtig. 1971. Infection and nutrition of children of a low socioeconomic rural community. Am. J. Clin. Nutr. 24:249–59.
52. Morley, D. 1969. Severe measles in the tropics (I and II). Br. Med. J. 1:297–300, 363–65.
53. Morley, D. , M. Woodland, and W. J. Martin. 1966. Whooping cough in Nigerian children. Trop. Geogr. Med. 18:169–82.
54. Mata, L. S., J. J. Urrutia, C. Albertazzi, O. Pellecer, and E. Arellano. 1972. Influence of recurrent infections on nutrition and growth of children in Guatemala. Am. J. Clin. Nutr. 25:1367–75.

55. Rowland, M. G. M., T. J. Cole, and R. G. Whitehead. 1977. A quantitative study into the role of infection in determining nutritional status in Gambian village children. Br. J. Nutr. 37:441–50.
56. Martorell, R., C. Yarbrough, S. Yarbrough, and R. E. Klein. 1980. The impact of ordinary illness on the dietary intakes of malnourished children. Am. J. Clin. Nutr. 33:345–50.
57. Chen, L. C. 1983. Interactions of diarrhea and malnutrition. Mechanisms and interventions. In L. C. Chen and N. S. Scrimshaw, eds., Diarrhea and nutrition. New York, NY: Plenum Press.
58. Molla, A., A. M. Molla, S. A. Sarker, M. Khatoon, and M. Mujibur Rahaman. 1983. Effects of acute diarrhea on absorption of macronutrients during disease and after recovery. In L. Chen, and N. S. Scrimshaw, eds., Diarrhea and malnutrition. New York, NY: Plenum Press.
59. Chandra, R. K. 1981. Immunodeficiency in undernutrition and overnutrition. Nutr. Rev. 39:225–31.
60. Chandra, R. K., and P. M. Newberne. 1977. Nutrition, immunity, and infection: Mechanisms of interactions. New York, NY: Plenum Press.
61. Sirisinha, S., R. Suskind, R. Eldeman, C. Charupatana, and R. Olson. 1973. Complement and C3-proactivator levels in children with protein-calorie malnutrition and effect on dietary treatment. Lancet 1:1016–20.
62. Keusch, G. T., J. J. Urrutia, R. Fernandez, O. Guerrero, and G. Castaneda. 1977. Humoral and cellular aspects of intracellular bacterial killing in Guatemalan children with PCM. In R. Suskind, ed., Malnutrition and the immune response. New York, NY: Raven Press.
63. McMurray, D. N., H. Rey, L. J. Casazza, and R. R. Watson. 1977. Effects of moderate malnutrition on concentrations of immunoglobulins and enzymes in tears and saliva of young Colombian children. Am. J. Clin. Nutr. 30:1944–48.
64. Edelman, R., R. Suskind, S. Sirisinha, and R. E. Olson. 1973. Mechanisms of defective cutaneous hypersensitivity in children with protein-calorie malnutrition. Lancet 1:506–8.
65. Suskind, R. M. 1977. Characteristics and causation of protein-calorie malnutrition in the infant and preschool child. In L. S. Green, ed., Malnutrition, behavior and social organization. New York, NY: Academic Press.
66. Neumann, C. G., G. J. Lawlor, Jr., E. R. Stichm, M. E. Swendseid, O. Newton, J. Herbert, A. J. Ammann, and M. Jacob. 1975. Immunologic responses in malnourished children. Am. J. Clin. Nutr. 28:89–104.
67. Mugerwa, J. W. 1971. Lymphoreticular system in kwashiorkor. J. Pathol. 105:105–9.
68. Work, T. N., A. Ikelwunigwe, D. B. Jellife, P. Jelliffe, and C. G. Neumann. 1973. Tropical problems in nutrition. Ann. Intern. Med. 79:701–11.
69. Smythe, P. M., M. Schonland, K. K. Breton-Stiles, H. M. Coovadia, H. J. Grace, W. E. K. Leoning, A. Mafoyane, M. A. Parent, and G. H. Vos. 1971. Thymolymphatic deficiency and depression of cell-mediated immunity in protein calorie malnutrition. Lancet 2:939–43.
70. Becker, W., W. E. Naude, A. Kipp, and D. McKenzie. 1963. Virus studies in disseminated herpes simplex infections: Association with malnutrition in children. S. Afr. Med. J. 37:74–76.
71. Nutrition and the immune response. 1985. Dairy Council Digest 3:7–12.
72. McMurray, D. N., S. A. Loomis, L. J. Casazza, H. Rey, and R. Miranda. 1981.

Development of impaired cell-mediated immunity in mild and moderate malnutrition. Am. J. Clin. Nutr. 34:68–77.
73. Reddy, V., V. Jagadeesan, N. Raharamulu, C. Bhaskaram, and S. G. Srikantia. 1976. Functional significance of growth retardation in malnutrition. Am. J. Clin. Nutr. 29:3–7.
74. Rivera, J. A., J-P. Habicht, N. Torres, et al. 1986. Decreased cellular immune response in wasted but not in stunted children. Nutr. Res. 6:1161–70.
75. Harland, P. S. 1965. Tuberculin reactions in malnourished children. Lancet 2:719–21.
76. Sinha, D. P., and F. B. Bang. 1976. Protein and calorie malnutrition, cell mediated immunity, and BCG vaccination in children from rural West Bengal. Lancet 2:531–34.
77. Kielmann, A. A., I. S. Uberoi, R. K. Chandra, and V. L. Mehra. 1976. The effect of nutritional status on immune capacity and immune responses in preschool children in a rural community in India. Bull. WHO 54:477–83.
78. Chandra, R. K. 1972. Immunocompetence in undernutrition. J. Pediatr. 81:1194–1200.
79. Green, F., and E. Heyworth. 1980. Immunoglobulin-containing cells in jejunal mucosa of children with protein-energy malnutrition and gastroenteritis. Arch. Dis. Child. 55:380–83.
80. Sirisinha, S., R. Suskind, R. Edelman, C. Asvapaka, and R. E. Olson. 1975. Secretory and serum IgA in children with protein calorie malnutrition. Pediatrics 55:166–70.
81. McMurray, D. N., H. Rey, L. J. Casazza, and R. R. Watson. 1977. Effect of moderate malnutrition on concentrations of immunoglobulins and enzymes in tears and saliva of young Colombian children. Am. J. Clin. Nutr. 30:1944–48.
82. Cravioto, J., and E. Delicardie. 1972. Environmental correlates of severe clinical malnutrition and language development in survivors from kwashiorkor or marasmus. In Nutrition, the nervous system and behavior. P.A.H.O. Scientific Publication 251. Washington, DC: Pan American Health Organization.
83. Niswander, K. R., and M. Gordon. 1972. The women and their pregnancies. Philadelphia, PA: W. B. Saunders Co..
84. Chase, H. C. 1969. Infant mortality and weight at birth: 1960 United States birth cohort. Am. J. Public Health 59:1618–28.
85. Eveleth, P. B., and J. M. Tanner. 1990. Worldwide variation in human growth. New York, NY: Cambridge University Press.
86. Bogin, B. 1990. Evolution of Human Growth. New York, NY: Cambridge University Press.
87. Frisancho, A. R., S. M. Garn, and W. Ascoli. 1970. Childhood retardation resulting in reduction of adult body size due to lesser adolescent skeletal delay. Am. J. Phys. Anthropol. 33:325–36.
88. Malcolm, L. A. 1970. Growth and development of the bundi child of the New Guinea highlands. Hum. Biol. 42:293–328.
89. McCance, R. A., and E. M. Widdowson. 1974. The determinants of growth and form. Proc. R. Soc. Lond. (Biol.) 185:1–17.
90. Blanco, R. A., R. M. Acheson, C. Canosa, and J. B. Salomon. 1975. Height, weight, and lines of arrested growth in young Guatemalan children. Am. J. Phys. Anthropol. 40:39–48.
91. Garn, S. M., C. G. Rohmann, M. Behar, F. Viteri, and M. A. Guzman. 1964. Compact bone deficiency in protein-calorie malnutrition. Science 145:144–45.

92. Frisancho, A. R., S. M. Garn, and W. Ascoli. 1970. Subperiosteal and endosteal bone apposition during adolescence. Hum. Biol. 42:639–64.
93. Burrell, R. J. W., J. M. Tanner, and M. J. R. Healy. 1961. Age at menarche in South Africa Bantu girls living in the Transkei reserve. Hum. Biol. 33:250–61.
94. Dreizen, S., C. N. Spirakis, and R. E. Stone. 1967. A comparison of skeletal growth and maturation in undernourished and well nourished girls before and after menarche. J. Pediatr. 70:256–64.
95. Chang, K. S. F. 1969. Growth and development of Chinese children and youth in Hong Kong. Hong Kong: University of Hong Kong.
96. Malina, R. M., C. Chumlea, C. D. Stepick, and F. G. Lopez. 1977. Age of menarche in Oaxaca, Mexico schoolgirls with comparative data for other areas of Mexico. Ann. Hum. Biol. 4:551–58.
97. Roberts, D. F., S. Chinn, B. Girija, and H. D. Singh. 1977. A study of menarcheal age in India. Ann. Hum. Biol. 4:171–78.
98. Pawson, I. G. 1977. Growth characteristics of populations of Tibetan origin in Nepal. Am. J. Phys. Anthropol. 47:473–82.
99. Stein, Z., and M. Susser. 1975. The Dutch famine 1944–45, and the reproductive process. Part 1. Effects on six indices at birth. Pediatr. Res. 9:70–75.
100. Stein, Z., and M. Susser. 1975. The Dutch famine, 1944–45, and the reproductive process. Part 2. Interrelations of caloric relations and six indices at birth. Pediatr. Res. 9:76–82.
101. Ellison, P. T. 1988. Human salivary steroids: methodological considerations and applications in physical anthropology. Yearbook of Phys. Anthrop. 31:115–42.
102. Teufel, N. I., and D. L. Dufour. 1990. Patterns of food use and nutrient intake of obese and non-obese. J. of Am. Diet. Assoc. 90:1229–35.
103. Hodge, L. G., and D. L. Dufour. 1991. Cross-sectional growth of young Shipibo Indian children in Eastern Peru. Am. J. of Physical Anthro. 84:35–41.
104. Dufor, D. L. 1992. Nutritional ecology in the tropical rain forests of Amazonia. Am. J. of Hum. Biol. 4:197–208.
105. Galvin, K. A. 1992. Nutritional ecology of Pastoralists in Dry Tropical Africa. Am. J. of Hum. Biol. 4:209–22.

# CHAPTER 16 Accommodation to Experimental Starvation and Chronic Undernutrition

## RESPONSES TO EXPERIMENTAL STARVATION

Our understanding of the effects of semistarvation and the responses to semistarvation have been derived from experimental studies conducted during World War II by Keys and co-investigators (1). Similarly, information has been derived from evaluations of the effects of famine on the civilian population in the concentration camps of Germany. Additional information has also been derived from studies of weight-loss programs that utilize a drastic reduction in dietary intakes. Findings derived from such studies are important for understanding the biological and behavioral responses to a drastic reduction of dietary energy intake. In this section the physiological and behavioral responses to experimental starvation and weight-loss programs will be discussed.

### The Minnesota Semistarvation Experiment

During World War II Keys and co-investigators (1) conducted their classic experimental study of the effects of starvation on healthy young men from Minnesota. The study design included 32 volunteers who subsisted for 6 months with a daily dietary intake that provided only 45% (1570 kcal) of the usual intake (3468 kcal). The general responses to semistarvation are summarized in table 16.1.

#### Body Weight and Body Composition

As shown in table 16.1 there is a 24% reduction of body weight. 67% of the reduced body weight is accounted for by a reduction in fat, 17% in fat-free mass. Figure 16.1 illustrates that during the first 12 weeks of starvation body

**TABLE 16.1. Changes in body composition and total energy exchange after 24 weeks of semistarvation diet (experiment) in Minnesota**

| Variables | Control | Experiment | Difference |
|---|---|---|---|
| Body weight (kg) | 70.0 | 53.2 | 24.0 |
| Body fat (kg) | 9.9 | 3.3 | 67.0 |
| Fat-free mass (kg) | 60.1 | 49.9 | 17.0 |
| Daily energy intake (kcal/d) | 3466.4 | 1569.6 | 55.0 |
| Basal metabolic rate (kcal/d) | 1593.5 | 962.8 | 39.6 |
| Estimated specific dynamic action (kcal/d) | 351.2 | 157.7 | 55.0 |
| Physical activity (kcal/d) | 1524.2 | 547.1 | 64.0 |
| Physical activity as percent of intake (%) | 44.0 | 34.9 | 21.0 |

*Source:* Adapted from A. Keys, J. Brozek, A. Henschel, O. Mickelsen, and H. L. Taylor. 1950. The biology of human starvation. Minneapolis: University of Minnesota Press.

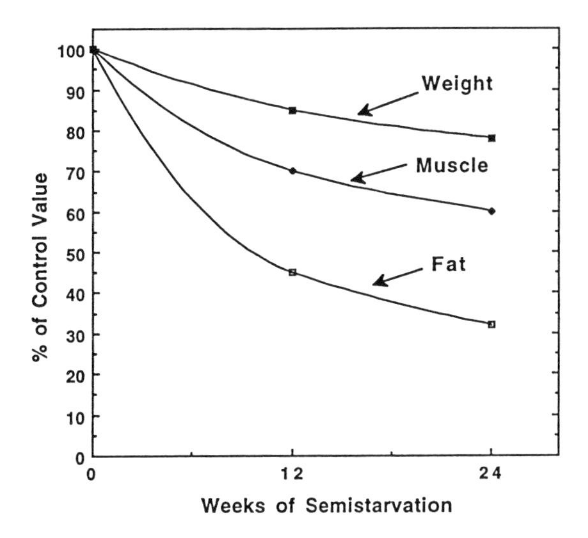


**Fig. 16.1.** Changes in body composition during 24 weeks of semistarvation. During semistarvation most of weight loss is accounted for the loss of fat, and only when the stored fat is depleted is muscle lost. (Adapted from A. Keys, J. Brozek, A. Henschel, O. Mickelsen, and H. L. Taylor. 1950. The biology of human starvation. Vol. I. Minneapolis, MN: University of Minnesota Press.)

fat decreased more markedly than muscle mass. The dramatic initial decrease in the fat content of the body indicates that during moderate starvation body fat is used preferentially as body fuel, but when starvation continues for as long as 24 weeks both fat and protein are used for metabolism.

## Metabolic Rate

Concomitant with the reduction in body mass and active tissue there is a BMR reduction of 40% (see fig. 16.2). The reduction in BMR per kg of active tissue is 17%. This implies that during starvation the metabolic efficiency has been increased by 17%. The observed increase in metabolic efficiency represents an adaptive response in that it conserves body energy in the face of reduced energy input.

**Diet-Induced Thermogenesis (DIT).** In the Minnesota experiment energy expenditure for diet-induced thermogenesis (DIT) (formerly referred to as specific dynamic action of food [SDA]) has been estimated at 7.5% of the daily energy intake. As shown in table 16.1 after starvation the wasteful loss of energy is reduced by as much as 55%.

## Behavior and Physical Activity

The most conspicuous manifestation of the calorie restrictions associated with starvation is the drastic reduction in energy expenditure from physical

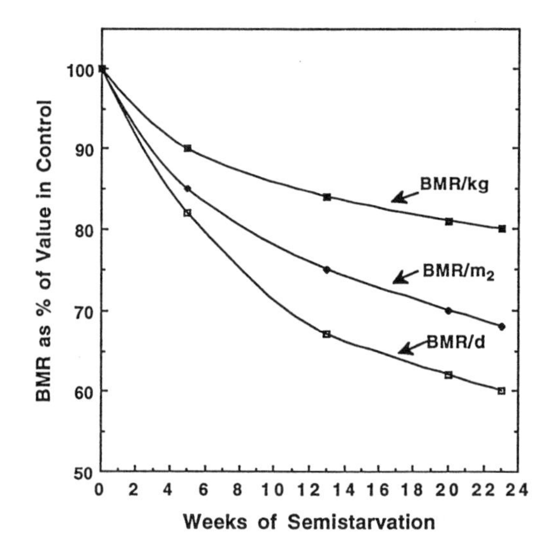


**Fig. 16.2.** Decrease in metabolic rate during semistarvation. The reduction in BMR in terms of active tissue is only 17%, while in terms of body weight it is only 40%, suggesting that during semistarvation the organism tries to diminish energy expenditure. (Adapted from A. Keys, J. Brozek, A. Henschel, O. Mickelsen, and H. L. Taylor. 1950. The biology of human starvation. Vol. I. Minneapolis, MN: University of Minnesota Press.)

activity. Table 16.1 shows an over-all reduction of 64% in the total calories available for work at the end of the 6 months of starvation. As a result, after starvation only 35% of the daily calorie intake is expended in physical activity, while before starvation more than 44% of it was expended in physical activity. In addition to a decrease in physical activity starvation is associated with: (1) apathy and reluctance to engage in any new activity, (2) unresponsiveness to external stimuli, (3) spontaneous reduction in all physical movements, (4) sensations of muscle weakness that are reflected in movements requiring muscular strength, (5) adoption of energy-sparing movements, (6) reduction in maximal oxygen intake or aerobic capacity (expressed as milliliters of oxygen intake per kilogram of body weight), and (7) reduction in grip strength. This decrease reflects the concomitant reduction in the efficiency of the cardiovascular, respiratory, and hematological systems, which results in a limited capacity to supply oxygen to the muscles. The result of these behavioral changes is a marked reduction in energy expenditure and thus a slowdown in the rate at which the starving organism deteriorates.

## Semistarvation in Weight-Loss Programs

### General Responses

Since the experimental studies of semistarvation were conducted on young and normal-weight men, an important question is whether the same changes occur among overweight individuals. In order to examine this question several studies have been conducted to evaluate the physiological responses to dietary restriction in overweight subjects (2–10). The research design of these studies includes either very low calorie diets (VLCD) that provide about 500 kcal/d for 10 weeks or moderate low calorie diets (MLCD) that provide about 1000 to 1200 kcal/d for 12 to 24 weeks. These different diets will be referred to as short-term and long-term semistarvation, respectively. These studies are by nature based on small sample sizes and as such the conclusions are not easily extrapolated to the population at large.

## Body Weight, Body Composition, and Resting Energy Expenditure

These studies indicate that short-term (less than 4 weeks) consumption of VLCD is associated with a decline in resting energy expenditure (17.4%) that is about three times as great as the corresponding decline in body weight (5.8%). In other words, the short-term decline in resting energy expenditure was related primarily to a reduction in calorie intake rather than a reduction in weight. On the other hand, during long-term semistarvation the resting energy expenditure was reduced by 11% which was paralleled by a 12% reduction in body weight (5). Studies conducted in Switzerland on overweight women (8) indicate that prolonged energy restriction (10 to 16 weeks on a hypocaloric diet) is not associated with a reduction of diet-induced thermogenesis. On the other hand, studies conducted on overweight men and women (10) indicate that 10–16 weeks of calorie restriction are associated with a 13% loss of body weight, 72% of which is accounted for by a reduction in fat. In addition to these changes the resting metabolic rate was reduced by an average of 9%. Similarly, as a consequence of the decrease in body weight, the energy cost of physical activity is decreased (11).

## Efficiency of Energy Use

An important source of individual variability in metabolic rate and energy requirements is the efficiency of energy utilization. The efficiency of energy utilization is measured as the ratio between the efficiency of use of dietary energy and efficiency of use of body energy (9).

After subsisting on a semistarvation diet (1000 kcal/d) for 8 weeks the relative efficiency of dietary energy utilization and the use of body energy is increased by 6% (see table 16.2). These findings suggest that when faced with energy restriction individuals use dietary energy more efficiently than they do body stores.

The general concensus from the above studies is that a decrease in the overall energy expenditure is accounted for by a loss of lean body mass, a

**TABLE 16.2. Changes in body composition and total energy exchange after 8 weeks of semistarvation diet in 12 overweight women**

| Variables | First Day | Starvation | % Control | % Diff. |
|---|---|---|---|---|
| Daily energy intake (kcal/d) | 2675.7 | 1003.0 | — | — |
| Body weight (kg) | 93.3 | 83.4 | 89.3 | 10.7 |
| Body fat (kg) | 40.86 | 32.03 | 78.4 | 21.6 |
| Fat free weight (kg) | 52.44 | 51.37 | 97.9 | 2.1 |
| 24h-EE (kcal/d)[a] | 2513.3 | 2130.9 | 84.8 | 15.2 |
| EE-sleep (kcal/min)[b] | 114.7 | 100.3 | 87.5 | 12.6 |
| EE-sedentary (kcal/min)[c] | 62.1 | 55.0 | 88.6 | 11.4 |
| EE-bicycling (kcal/min)[d] | 227.0 | 207.8 | 91.5 | 8.5 |
| Efficiency of dietary utilization[e] | | | | |
| per kg body weight | 0.88 | 0.93 | 105.7 | 5.7 |
| per kg fat-free weight | 0.87 | 0.95 | 109.2 | 9.2 |

*Source:* Modified from tables 2–4 De Boer et al. 1986. Adaptation of energy metabolism of overweight women to low-energy intake, studied with whole-body calorimeters. Am. J. Clin. Nutr. 44:585–95.

[a]24h-EE (kcal/d) = 24-hour energy expenditure.

[b]EE-sleep (kcal/min) = energy expenditure during sleep.

[c]EE-sedentary (kcal/min) = energy expenditure during sedentary activities.

[d]EE-bicycling (kcal/min) = energy expenditure while bicycling.

[e]Efficiency of dietary utilization = ratio of efficiency of dietary energy utilization and efficiency of body energy utilization.

decrease in diet induced thermogenesis, and a decrease in energy cost of physical activity. It should be noted, however, that a small increase in metabolic efficiency coupled with an increased efficiency during physical activity when maintained for long periods of time could potentially have important ramifications for energy balance. Therefore, in order to understand adaptation to caloric restriction one must also understand the physiological responses of populations who have subsisted under conditions of chronic undernutrition.

## ACCOMMODATION TO CHRONIC UNDERNUTRITION

Several studies indicate that the dietary intakes of some populations living in the Third World, particularly those of pregnant and lactating women (see table 16.3) are generally below the international recommended dietary allowances of the FAO/ WHO/UNU 1985 Committee (12). Furthermore, studies in Gambia reported drastic seasonal differences in energy intakes, and in some seasons the daily overall energy expenditure was higher than the energy intake (13). Therefore, there must be some accommodations that enable these populations to successfully carry on pregnancy and work under otherwise limited nutritional conditions. The postulated energy-sparing accommodations include: decreasing the metabolic rate, increasing energy utilization and increasing work efficiency, and decreasing activity levels, which can occur in women and men.

**TABLE 16.3. Estimates of energy intake during pregnancy and lactation in Third World and industrialized countries**

| Country | Pregnancy Dietary Intake kcal/day | Lactation Dietary Intake kcal/day |
|---|---|---|
| Third World | | |
| Gambia—wet season (Prentice et al., 1981) | 1299 | 1299 |
| Gambia—dry season (Prentice et al., 1981) | 1483 | 1681 |
| New Guinea (Oomen & Malcolm, 1958) | 1359 | — |
| New Guinea—lowland (Norgan et al., 1974) | 1414 | 1459 |
| New Guinea—highland (Norgan et al., 1974) | 1999 | 2166 |
| India (Gopalan, 1962) | 1400 | — |
| India (Venketachalam, 1962) | 1409 | — |
| India (Karmerkar et al., 1963) | — | 1299 |
| India (Devadas, 1978) | 1624 | — |
| India (Devadas & Murthy, 1977) | — | 1400 |
| India (Karmarkar et al., 1959) | — | 1440 |
| India (Rajalakshmi, 1971) | 1569 | 1619 |
| India (Bagchi & Bose, 1962) | 1918 | — |
| Guatemala (Lechtigit et al., 1972) | 1500 | — |
| Guatemala (Arroyave, 1975) | 1720 | 1600 |
| Guatemala (Mata et al., 1972) | 2059 | — |
| Guatemala (Schutz et al., 1980) | — | 1927 |
| Mexico (Martinez & Chavez, 1971) | — | 1949 |
| Colombia (Mora et al., 1978) | 1619 | — |
| Ethiopia (Gebre-Medhin & Gobezie, 1974) | 1538 | — |
| Tanzania (Maletnlema & Bavu, 1974) | 1849 | — |
| Iraq (Demarchi, 1966) | 1680 | — |
| Thailand (Thenangkul & Amatyakul, 1975) | 1980 | — |
| Mean ± SD | 1634 ± 232 | 1632 ± 288 |
| Industrialized | kcal/day | kcal/day |
| USA (Sims, 1978) | — | 2123 |
| England (Whicelow, 1976) | — | 2728 |
| England (Naismith & Ritchie, 1975) | — | 2928 |
| England (Smithells et al., 1977) | 1956 | — |
| England (Whitehead et al., 1981) | 1980 | 2293 |
| England (Darke et al., 1980) | 2152 | — |
| Sweden (Lunell et al., 1969) | 2152 | — |
| Sweden (Abrahmsson & Hofvander, 1977) | — | 2279 |
| Australia (English & Hitchcock, 1968) | 2090 | 2460 |
| Australia (Rattigan et al., 1981) | — | 2305 |
| Scotland (Thomson, 1958) | 2503 | — |
| Scotland (Thomson et al., 1970) | — | 2716 |
| Mean ± SD | 2138 ± 180 | 2479 ± 262 |

*Source:* A. Prentice, A. M. Prentice, and R. G. Whitehead. 1981. Breast-milk fat concentrations of rural African women. 1. Short-term variations within individuals. Br. J. Nutr. 45:483–94.

## Accommodation in Men

In general the three major components of energy expenditure are: (1) basal metabolic rate, (2) diet-induced thermogenesis, and (3) physical activity. Hence, studies of accommodation to variability in energy intake are usually evaluated with reference to these three parameters.

### Accommodation to Low Dietary Intakes Associated with Seasonal Differences

*Studies in Gambia, Africa*

In Keneba, a rural village of Gambia, investigators from the Dunn Nutrition Unit (Cambridge, UK) evaluated 24-hour energy expenditures (24-h EE) during the wet season, which is associated with a shortage of food, and the dry season, which is associated with increased food availability (14, 15). The study design included evaluations of Gambian men ($n = 18$ to 20) and European men ($n = 16$ to 20) under standardized conditions by using a respiration chamber. The method also permitted measurement of sleeping energy expenditure (SEE), basal metabolic rate (BMR), diet-induced thermogenesis (DIT), and the energy cost of a standardized exercise.

**Wet Season.** The metabolic characteristics during the wet season are summarized in table 16.4. From these characteristics it is evident that the 24-hour energy expenditure, basal metabolic rate, and energy expenditure during sleep were lower in Gambian than in European men. Furthermore, the diet-induced thermogenesis was blunted in Gambian men compared with European men. On the other hand, the net efficiency of walking was greater in Gambian men than in European men. Given the fact that the wet season is associated with a shortage of food, the low basal and sleeping EE, a reduced

**TABLE 16.4. Comparison of metabolic characteristics of 20 Gambian and 16 European men**

| Anthropometric Characteristics | | | | | | |
|---|---|---|---|---|---|---|
| Sample | Age yrs | Weight kg | Height cm | BMI kg/m2 | Body Fat % wt | FFM kg |
| Gambians | 23 ± 1 | 60.8 ± 1.4* | 175 ± 1* | 19.9 ± 0.4 | 9.8 ± 0.5 | 54.8 ± 1.2* |
| Europeans | 24 ± 1 | 66.9 ± 1.9 | 179 ± 2 | 20.8 ± 0.5 | 10.8 ± 0.6 | 59.7 ± 1.8 |

| Metabolic Characteristics | | | | | | | |
|---|---|---|---|---|---|---|---|
| | Energy Intake | | 24-hour Energy Expenditure | | | Basal Expenditure | |
| Sample | kcal/d | kcal/kg/d | kcal/d | kcal/kg/d | kcal/ gFFM/d | kcal/d | kcal/ kgFFM/d |
| Gambians | 2265 ± 59* | 37.5 ± 1* | 2047 ± 46* | 33.7 ± 0.6* | 37.5 ± 0.6* | 26.5 ± 0.4* | 27.8 ± 0.5* |
| Europeans | 2715 ± 74 | 40.6 ± 0.5 | 2635 ± 74 | 39.4 ± 0.5 | 44.2 ± 0.6 | 28.7 ± 0.4 | 30.3 ± 0.5 |

*Source:* Adapted from G. Minghelli, Y. Schutz, A. Charbonnier, R. Whitehead, and E. Jéquier. 1990. Twenty-four hour energy expenditure and basal metabolic rate measured in a whole-body indirect calorimeter in Gambian men. Am. J. Clin. Nutr. 51:536–70.

*Significantly different from Europeans, $p < 0.01$

DIT, and a high work efficiency are important energy-sparing mechanisms that enable the Gambian men to cope with a marginal level of dietary intake. Similarly, ergonometric studies conducted in India indicate that the energy cost of stepping in chronically energy deficient individuals is lower than in well subjects (39). As result of the lower energy expenditure the work efficiency in chronically energy deficient individuals is greater than in well subjects (39).

**Seasonal Differences.** Evaluations of energy expenditure were carried on: (1) at the end of the rainy season, characterized by low food availability; (2) during the dry season, associated with increased food availability; and (3) at the onset of the following rainy season. The study was designed to prospectively assess the seasonal influence of variations in food availability on energy metabolism, basal metabolic rate (BMR), sleeping metabolic rate, and the energy cost of exercise in young, rural Gambian men. As summarized in table 16.5, seasonal changes in food availability induced marked variations in body weight and metabolic activity in rural Gambian men. Thus, during the nutritionally favorable (dry) season body weight gain was accompanied by an increase in 24-h EE, which was due to increases in BMR and sleeping EE. Furthermore, diet-induced thermogenesis (DIT) and the energy cost when walking on a slope was increased during the dry season (15). Conversely, the DIT was reduced during the nutritionally unfavorable (wet) season.

## Seasonal Fluctuations in Body Weight, Physical Activity, and Energy Flow

### *Studies in India, Benin, and Ethiopia*

Using standardized protocols, the accommodations to seasonal fluctuations in energy balance of rural samples from India (16, 17), Benin (18), and Ethiopia (19, 20) were studied by Ferro-Lucci and associates. As inferred from observations in Ethiopia (21) the domestic flow of cash, the level of food stores, the agricultural potential of the households, and the work load as number of hours per day and per household were moved together in a smooth bimodal curve with two discreet periods: a low peak between May and August, and a high peak between September and April. During the low peak there was an average weight loss of 1.5 kg, suggesting that the subjects (adult men and women) were in negative energy balance during the hungry season and in positive balance during the postharvest season (21). Yearly observations of adult nonpregnant women in India, Benin, and Ethiopia (21) has quantified the energy stress, energy mobilization from the body, and the change in basal metabolic rate. In all three populations the energy debt was offset by a large (about 44% to 113% of energy) mobilization of energy from the body's adipose stores and by a modest (5–7%) decrease in BMR. These findings suggest that the energy stress caused by fluctuations in energy intake can be met in part by the mobilization of body energy and from the changes in BMR.

**TABLE 16.5. Anthropometric metabolic characteristics of 18 Gambian men studied in 3 seasons**

| | Anthropometry | | | | Energy Expenditure (EE) | | | |
|---|---|---|---|---|---|---|---|---|
| | Age | Height | Weight | Body Fat | 24hr | Daytime | Sleeping | BMR |
| Season[a] | yrs | cm | kg | % wt | kcal/d | kcal/min | kcal/min | kcal/min |
| 1 | 23 ± 1 | 175 ± 1 | 60.8 ± 1.5 | 9.8 ± 0.6 | 2044.0 ± 50.7 | 1.59 ± 0.04 | 1.00 ± 0.02 | 1.06 ± 0.02 |
| 2 | 23 ± 1 | 175 ± 1 | 63.6 ± 1.6* | 9.6 ± 0.5 | 2189.8 ± 53.5* | 1.70 ± 0.05* | 1.11 ± 0.02* | 1.12 ± 0.02* |
| 3 | 23 ± 1 | 175 ± 1 | 63.9 ± 1.5* | 10.6 ± 0.6 | 2088.0 ± 46.3* | 1.63 ± 0.04 | 1.02 ± 0.02 | 1.13 ± 0.02* |

*Source:* Adapted from G. Minghelli, Y. Schutz, R. Whitehead, and E. Jéquier. 1991. Seasonal changes 24-h and basal energy expenditures in rural Gambian men as measured in a respiration chamber. Am. J. Clin. Nutr. 53:14–20.

*Significantly different from season 1, $p < 0.01$.

[a] Season 1 = at the end of the rainy season, associated with shortage of food

Season 2 = during the dry season, associated with increased food availability

Season 3 = beginning of the rainy season, associated with beginning of food shortage

## Accommodation in Women

### Accommodation during Pregnancy and Lactation

As summarized in table 16.3 the dietary intakes during pregnancy and lactation in Third World countries are about 24% and 34% below those reported for industrialized countries. Yet these populations can successfully carry on pregnancy. This finding has led researchers to investigate the energy-sparing adaptive mechanisms that enable them to successfully reproduce. With this purpose in mind Prentice and associates (22–26) have conducted extensive studies in Keneba, a rural village of the Gambia in Africa. These investigations included evaluations of the 24-hour energy expenditure, basal metabolic rate (BMR), diet-induced thermogenesis (DIT), and the energy cost of a standardized exercise carried out by pregnant and lactating women under standardized conditions using a respiration chamber. The results of these studies will be reviewed along with those conducted by other researchers in India, China, Malay, Taiwan, and Guatemala.

### Energy Intake, Weight Gain, and Birth Weight

As summarized in table 16.6 dietary intake during the dry season (which is associated with increased food availability) is significantly greater than in the wet season (which is associated with reduced food availability). In addition to these differences, pregnancy weight gain and birth weight are significantly lower in the wet season.

### Basal Metabolic Rate

Figure 16.3 illustrates the resting metabolic rates (RMR) before and during pregnancy of well-nourished Swedish women, nonaffluent Scottish women,

**TABLE 16.6. Energy intake during pregnancy and lactation, and pregnancy weight gain of 143 Gambian women and their newborns' weight according to season**

| | Season | |
|---|---|---|
| | Dry[a] | Wet[b] |
| Characteristic | Mean ± SE | Mean ± SE |
| Pregnant (kcal/d) | 1483 ± 22 | 1417 ± 41 |
| Lactation1 (kcal/d) | 1773 ± 31 | 1474 ± 42 |
| Lactation2 (kcal/d) | 1662 ± 16 | 1413 ± 37 |
| Pregnancy weight gain (kg) | 1.4 | 0.4 |
| Birth weight (g) | 2940.0 ± 70.0 | 2780.0 ± 110.0 |

*Source:* Adapted from A. M. Prentice, R. Whitehead, S. B. Roberts, and A. A. Paul. 1981. Long-term energy balance in child-bearing Gambian women. Am. J. Clin. Nutr. 34: 2790–99.

[a] Dry season associated with increased food availability

[b] Wet season associated with reduced food availability

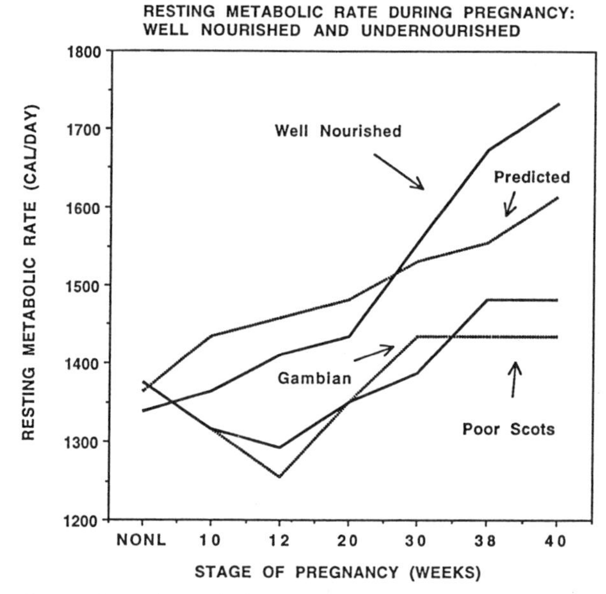


**Fig. 16.3.** Comparison of metabolic rate in well-nourished Western controls and undernourished women from Gambia. Among both the poor Scottish and undernourished Gambian women, the metabolic rate during first trimester declines and, during the later stages of pregnancy, increases to moderate levels. (Redrawn from data from A. M. Prentice, and R. G. Whitehead. 1987. The energetics of human reproduction. In A. S. I. Loudon and P. A. Racey, eds., Reproductive energetics in mammals. Oxford: Clarendon Press.)

and subsistence-farming Gambian women (25). From these data it is evident that the prepregnancy resting metabolic rate (RMR) in all three groups is similar. However, during pregnancy the well-nourished women increase their metabolic rate linearly with gestation. In contrast, the Gambian and nonaffluent Scottish women's RMR decreased markedly during the first 17 weeks of pregnancy. Subsequently, both groups showed a moderate increase in RMR, which was lower than that of the well-nourished women. As a result of the initial decrease in RMR, the net maintenance cost of pregnancy in the Scottish and Gambian women is only 3824 kcal, which is about 13% lower than predicted.

## Reduction of Energy during Physical Activity

**Pattern of Activity.** Another way of changing the energy requirements during pregnancy involves alterations in the patterns of physical activity. Studies of industrialized populations indicate that during pregnancy and the postpartum period there are no major alterations in activity patterns that may result in energy expenditure changes. On the other hand, populations subsisting on limited nutritional resources during pregnancy and lactation seem to decrease their energy expenditure, especially in tasks with high energy cost. Studies of 115 rural south Indian women (using 24 hr recall) indicate that pregnant and lactating women spend more time in personal activities (mainly resting) and less time in social activities, working in the fields, and travel than do nonpregnant, nonlactating women (26). This change can result in a difference of as much as 20% in energy availability. Thus, it would appear that, if sustained over several months, minor changes can play an important role in reducing requirements.

**Work Efficiency.** The increase in maternal body weight which occurs during pregnancy will proportionately increase the cost of weight bearing activities (27). Similarly, the oxygen consumption and the length of the recovery period (or oxygen debt) for women pedaling a bicycle ergometer during pregnancy compared to the postpartum period increases significantly (28). In contrast, studies of Chinese, Malay, and Indian women (29) showed no significant differences between 42 pregnant and 37 nonpregnant women in the energetic cost per unit of body weight of eight common daily activities. These data suggest that populations subsisting on limited nutritional resources are more efficient than those living under good nutritional conditions.

### Efficient Utilization of Energy Reserves

One of the purposes of accumulating fat during pregnancy is for use during lactation. Well-nourished women, because of their access to high dietary intakes, do not mobilize all their energy stored in the form of fat gain during lactation (30, 31). In contrast, among undernourished women from Gambia (22) fat utilization during lactation is directly related to the amount of pregnancy fat gain. Analysis of anthropometric data of Taiwanese women (32) indicates that women who had gained more fat during pregnancy seemed to lose more fat during the postpartum period compared to women who had gained less fat during pregnancy. Furthermore, the loss of fat during the postpartum period appears to be related to breast-feeding frequency (32), so that those women who lost fat in the postpartum period breast-fed their infants more frequently than women who gained fat during the postpartum period. Similarly, in a small sample ($n = 18$) of Guatemalan women it appears that the energetic cost of lactation was supplied almost entirely by fat tissue mobilization (rather than by increased dietary intakes) (33). These findings suggest that pregnancy fat gain is an emergency buffer which is efficiently used under conditions of increased energy demands as they occur during lactation.

### Inhibition of Diet-Induced Thermogenesis

Following food ingestion most individuals experience an increase in metabolic rate, which is referred to as diet-induced thermogenesis (DIT). DIT has two components: a relatively long-term adjustment in metabolic rate and heat dissipation following overfeeding (34), and a short-term postprandial increase in metabolic rate (specific dynamic action). Since postprandial increase in MR does not occur when energy intake is restricted, its effectiveness is evident among populations under conditions of nutritional restriction. A recent study of Gambian women (35) indicates a reduction of 289 kcal in DIT in lactating women when compared with bottle-feeding controls matched for the stage postpartum. This reduction has been attributed to an increased sensitivity to the action of insulin, which favours milk lipogenesis and reduced maternal fat oxidation (35).

### Increased Efficiency of Dietary Energy Absorption

In normal subjects the efficiency of dietary energy absorption varies between 80% and 100%. In adults under normal circumstances the daily fecal losses range from 8 to 146 kcal/day (36) and 74 to 199 kcal/day (37). Thus, while certain individuals probably absorb 100% of the nutrients others may absorb only 80% of the nutrients. Therefore, it is possible that a difference of this magnitude can contribute to existing energy-sparing mechanisms.

In summary, it is evident that women who subsist under conditions of chronic nutritional restriction are able to reproduce and carry on successfully throughout pregnancy. These various energy-sparing responses include a reduction of metabolic rate during the first 20 weeks of pregnancy, a slow increase in MR throughout pregnancy, a decrease in the time spent on tasks requiring expense of energy, and a decrease in energy expenditure per activity (increased work efficiency). All these responses along with the absence of diet-induced thermogenesis and perhaps increased nutrient absorption have enabled women to conceive, carry on pregnancy, and lactate with a dietary intake of about 24% and 34% less than the dietary intake of well-nourished women.

## OVERVIEW

Living organisms require a continuous supply of energy and nutrients to maintain their normal metabolic activities and continue tissue replacement. The responses to a drastic reduction of dietary intake are continuous processes. As such, the general findings in all these studies are similar; they differ in degree rather than in kind. These effects are reflected in the use of body tissue such as fat and muscle as fuel and a decreased energy expenditure and work capacity. Both acute and semiacute starvation severely affect the individual. At the behavioral level irritation and apathy are prominent features of severe starvation. In terms of body composition the available data indicate that during severe starvation fat is used as an energy source, whereas during prolonged semistarvation both fat and muscle are used as energy sources. The data from studies of starvation and semistarvation reveal the organism's hierarchical utilization of nutritional reserves stored in the forms of fat and muscle. The effects of short-term (fewer than 10 days) caloric restriction on physical performance are minimal, but prolonged semistarvation affects the individual's cardiovascular, respiratory, and circulatory systems resulting in a reduction of work capacity.

Populations living under conditions of chronic undernutrition have accommodated to this stress by: (1) reducing the cost of diet-induced thermogenesis (DIT), (2) increasing the cost efficiency of weight bearing activities, (3) mobilizing body energy, and (4) reducing the energy cost of basal metabolic rate. These responses become more evident during pregnancy and lactation so that even when the dietary intake is reduced by as much as 35% undernourished women are able reproduce as well as well-nourished women. The fact, that some populations, such as the African Pastoralists (40), have managed to survive in an otherwise inhospitable environment indicates great capacity of humans to adapt.

## References


1. Keys, A., J. Brozek, A. Henschel, O. Mickelsen, and H. L. Taylor. 1950. The biology of human starvation, Vol. I. Minneapolis, MN: University of Minnesota Press.
2. Rumpler, W. V., J. L. Seale, C. W. Miles, and C. E. Bodwell. 1991. Energy-intake restriction and diet-composition effects on energy expenditure in men. Am. J. Clin. Nutr. 53:430–36.
3. Geissler, C. A., D. S. Miller, and M. Shah. 1987. The daily metabolic rate of the post-obese and the lean. Am. J. Clin. Nutr. 45:914–20.
4. Hill, J. O., P. B. Sparling, T. W. Sheilds, and P. A. Heller. 1987. Effects of exercise and food restriction on body composition and metabolic rate in obese women. Am. J. Clin. Nutr. 46:622–30.
5. Foster, G. D., T. A. Wadden, I. D. Feurer, A. S. Jennings, A. J. Stunkard, L. O. Crosby, J. Ship, and J. L. Mullen. 1990. Controlled trial of the metabolic effects of a very-low-calorie diet: Short- and long-term effects. Am. J. Clin. Nutr. 51: 167–72.
6. Bray, G. A. 1969. Effect of caloric restriction on energy expenditure in obese patients. Lancet 1:397–98.
7. Welle, S. L., J. M. Amatruda, G. B. Forbes, and D. H. Lockwood. 1984. Resting metabolic rates of obese women after rapid weight loss. J. Clin. Endocrinol. Metab. 59:41–44.
8. Bessard, T., Y. Schutz, and E. Jéquier. 1983. Energy expenditure and postprandial thermogenesis in obese women before and after weight loss. Am. J. Clin. Nutr. 38:680–93.
9. De Boer, J. O., A. J. H. van Es, L. A. Roovers, J. M. A. van Raaij, and J. G. A. J. Hautvast. 1986. Adaptation of energy metabolism of overweight women

to low-energy intake, studied with whole-body calorimeters. Am. J. Clin. Nutr. 44:585–95.
10. Ravussin, E., B. Burnand, Y. Schutz, and E. Jéquier. 1985. Energy expenditure before and during energy restriction in obese patients. Am. J. Clin. Nutr. 41: 753–59.
11. Schutz, Y. E. Ravussin, R. Diethelm, and E. Jéquier. 1982. Spontaneous physical activity measured by radar in obese and control subjects studied in a respiration chamber. Int. J. Obesity 6:23–28.
12. FAO/WHO/UNU. 1985. Energy and protein requirements. Geneva: World Health Organization.
13. Paul, A. A., and E. M. Muller. 1980. Seasonal variations in dietary intake in pregnant and lactating women in a rural Gambian village. In H. Aebi and R. Whitehead, eds., Maternal nutrition during pregnancy and lactation. Bern: Hans Huber Verlag. 105–16.
14. Minghelli, G., Y. Schutz, A. Charbonnier, R. Whitehead, and E. Jéquier. 1990. Twenty-four hour energy expenditure and basal metabolic rate measured in a whole-body indirect calorimeter in Gambian men. Am. J. Clin. Nutr. 51:563–70.
15. Minghelli, G., Y. Schutz, R. Whitehead, and E. Jéquier. 1991. Seasonal changes in 24-h and basal energy expenditures in rural Gambian men as measured in a respiration chamber. Am. J. Clin. Nutr. 53:14–20.
16. Durnin, J. V. G. A., S. Drummond, and K. Satyanarayana. 1990. A collaborative EEC study on seasonality and marginal nutrition: the Glasgow-Hyderabad (India) Study. Eur. J. Clin. Nutr. 44, Suppl. 1:19–29.
17. Norgan, N. G., P. Shetty, T. Baskaran, T. Nandi, J. Rao, and A. Ferro-Luzzi. 1989. Seasonality in agriculture. Its nutritional and productive implications: seasonality in a Karnataka (South India) agricultural cycle: background and body weight changes. Washington, DC: International Food Policy Research Institute (IFPRI), Research Report. (In press.)
18. Schultink, W. J., W. Klaver, H. van Wijk, J. M. A. van Raaij, and J. G. A. J. Hautvast. 1990. Body weight changes and basal metabolic rates of rural Beninese women during seasons with different energy intakes. Eur. J. Clin. Nutr. 44, Suppl. 1:31–40.
19. Ferro-Luzzi, A., C. Scaccini, S. Taffese, B. Aberra, and T. Demeke. 1990. Seasonal energy deficiency in Ethiopian rural women. Eur. J. Clin. Nutr. 44, Suppl. 1: 7–18.
20. Ferro-Luzzi, A. 1990. Seasonal energy stress in marginally nourished rural women: Interpretation and integrated conclusions of a multicentre study in these developing countries. Eur. J. Clin. Nutr. 44, Suppl. 1:41–46.
21. Ferro-Luzzi, A. 1990. Social and public health issues in adaptation to low energy intakes. Am. J. Clin. Nutr. 51:309–15.
22. Prentice, A. M., R. Whitehead, S. B. Roberts, and A. A. Paul. 1981. Long-term energy balance in child-bearing Gambian women. Am. J. Clin. Nutr. 34:2790–99.
23. Prentice A., A. M. Prentice, and R. G. Whitehead. 1981. Breast-milk fat concentrations of rural African women. 1. Short-term variations within individuals. Br. J. Nutr. 45:483–94.
24. Prentice, A. M. 1984. Adaptations to long-term low energy intakes. In E. Pollitt and P. Amante, eds., Energy intake and activity. Current topics in nutrition and disease. New York, NY: Alan R. Liss. 2:3–31.
25. Prentice, A. M., and R. G. Whitehead. 1987. The energetics of human reproduction.

In A. S. I. Loudon and P. A. Racey, eds., Reproductive energetics in mammals. Oxford: Clarendon Press.
26. McNeill, G., and P. R. Payne. 1985. Energy expenditure of pregnant and lactating women. Lancet 2:1237–38.
27. Prentice, A. M., T. J. Cole, F. A. Foord, W. H. Lamb, and R. G. Whitehead. 1987. Increased birth weight after prenatal dietary supplementation of rural African women. Am. J. Clin. Nutr. 46:912–25.
28. Nagy, L. E., and J. C. King. 1983. Energy expenditure of pregnant women at rest or walking self-paced. Am. J. Clin. Nutr. 38:369–76.
29. Pernoll, M. L., J. Metcalfe, T. L. Schlenker, J. E. Welch, and J. A. Matsumoto. 1975. Oxygen consumption at rest and during exercise in pregnancy. Respir. Physio. 25:285–93.
30. Banerjee, B., K. S. Khew, and N. A. Saha. 1971. A comparative study of energy expenditure in some common daily activities of non-pregnant and pregnant Chinese, Malay and Indian women. Br. J. Obstet. Gynecol. 78:113–16.
31. Sadurskis, A., N. Kabir, J. Wager, and E. Forsum. 1988. Energy metabolism, body composition, and milk production in healthy Swedish women during lactation. Am. J. Clin. Nutr. 48:44–49.
32. Brewer, M. M., M. R. Bates, and L. P. Vannoy. 1989. Postpartum changes in maternal weight and body fat depots in lactating vs nonlactating women. Am. J. Clin. Nutr. 49:259–65.
33. Adair, L. S., E. Pollitt, and W. H. Mueller. 1983. Maternal anthropometric changes during pregnancy and lactation in a rural Taiwanese population. Hum. Biol. 55:771–87.
34. Schutz, Y., A. Lechtig, and R. B. Bradfield. 1980. Energy expenditures and food intakes of lactating women in Guatemala. Am. J. Clin. Nutr. 33:892–902.
35. Garrow, J. S., and P. M. Warwick. 1978. Diet and obesity. In J. Yudkin, ed., The diet of man: needs and wants. Barking, London: Applied Science Publishers. 127–144.
36. Illingworth, P. J., R. T. Jung, P. W. Howie, and T. E. Isles. 1987. Reduction in postprandial energy expenditure during pregnancy. Br. Med. J. 294:1573–76.
37. Heymsfield, S. B., J. Smith, S. Kasriel, J. Barlow, M. J. Lynn, D. Nixon, and D. H. Lawson. 1981. Energy malabsorption: Measurement and nutritional consequences. Am. J. Clin. Nutr. 34:1954–60.
38. Dallosso, H. M., P. R. Murgatroyd, and W. T. P. James. 1982. Feeding frequency and energy balance in adult males. Hum. Nutr.: Clin. Nutr. 36C:25–39.
39. Kulkarni, R. N., and P. S. Shetty. 1992. Net mechanical efficiency in chronically energy-deficient human subjects. Ann. Hum. Biol. 19:421–25.
40. Little, M. A. 1989. Human biology of African Pastoralists. Yearbook of Phys. Anthrop. 32:216–48.

# CHAPTER 17 Accommodation to Industrialization of Dietary Habits and Disease Expression

---

The advance of human civilization has created conditions which are in many ways unlike those which existed during the last 3 million years of human evolution. These new conditions have certainly had a beneficial impact on human survival, and, indeed, without these advances we would not have attained our present biological status. However, the advances of civilization have also created negative conditions that, if continued unchecked, may constitute a grave threat to the survival, let alone the well-being, of humanity. In terms of human adaptation and nutrition the most significant changes are quantitative and qualitative differences in food consumption and variations in food use. Knowing how the human organism is reacting biologically to these changes is important for understanding the expression of various diseases. Thus this chapter is limited to a discussion of the interaction of westernization of dietary habits and disease expression. Within this context the hunter-gatherer diet, diseases of breast cancer, colon cancer, and diverticulosis, beriberi, and sickle-cell anemia will be discussed.

## THE HUNTER-GATHERER DIET

Ever since humans became hunter-gatherers their diet has been based upon a mixture of animal foods and plants. To illustrate these characteristics the hunter-gatherers from Australia and Venezuela will be briefly discussed here.

### The Australian Aborigine

Before European settlement of Australia 200 years ago, aborigines lived as hunter-gatherers all over the continent, from the tropical coastal region of the north (latitude 11°S), through the vast arid regions of the center (latitude 20–30°S), to the cooler temperate regions of the south (30–43°S). Despite these major differences in climate and geographical location aborigines from all over Australia appear to have had a rich and varied diet in which animal foods were an important component (1).

As shown by measurements of all food hunted and gathered over a two-week period (89), the major characteristics of the hunter-gatherer Australian aborigine's diet are: (1) low in total fat; (2) particularly low in saturated fat; (3) a wide range of different polyunsaturated fatty acids (PUFA) including the physiologically important long chain PUFA of both the n-3 and n-6 series; (4) nutrient-rich (vitamins, minerals); (5) carbohydrates rich in fiber; and (6) low in sodium, high in potassium.

### The Hiwi from Venezuela

The Hiwi are Guahibo-speaking hunter-gatherers of southwestern Venezuela. They live in the extremely seasonal neotropical savannas of the Orinoco River

Basin. From 1985 to 1988 Hurtado and Hill (2) evaluated the dietary patterns of 87 individuals of the Hiwi band.

As shown in table 17.1 the dietary sources of the Hiwi include four major groups: (1) game and fish, (2) roots, (3) other vegetables, and (4) store-bought foods. The total dietary intake averaged about 2032 calories. About 69% of the total calorie intake is derived from game and feral animals and 27% is derived from vegetable food, and the remaining 4% of food energy is derived from store-bought foods and agricultural products.

In summary, it appears that among contemporary hunter-gatherers living in traditional conditions about 70% of the total daily calorie intake is derived from animal foods. The remainder of the diet is derived from gathered vegetable foods including tuberous roots, fruits, nuts, beans and seeds. Although the percentage of calories derived from animal foods is similar to the dietary pattern of industrialized populations, because wild game animals are leaner than domestic animals, the animal fat and especially the intake of saturated fat is low among most hunter-gatherers. The health implications of these changes, which will be discussed next, are a major concern among westernized populations.

## BREAST CANCER

### Dietary Factors

From a nutritional standpoint the most important changes associated with industrialization of dietary habits include increased consumption of animal protein and fat and a marked decrease in the intake of fiber, particularly cereal fiber. As shown in figure 17.1, in the United States between 1900 and 1970, the per capita consumption of meat, poultry, and fish increased by 50%, whereas the consumption of crude fiber derived from fruits and vegetables declined in a similar proportion (3, 4). Concomitant with these changes

**TABLE 17.1 The dietary sources for the Hiwi from southern Venezuela**

| Resources Name[a] | Amount (kg) | Energy[b] (cal/day/person) | Percent of Total Calories |
|---|---|---|---|
| Game and Fish | 4,537 | 1,392 | 69% |
| Vegetables | | | |
| (including roots, honey) | 2,859 | 392 | 26% |
| Cultivated foods | 48 | 11 | 1% |
| Store-bought food | 186 | 83 | 4% |
| Energy (cal/day/person) | | 2,032 | |

*Source:* Adapted from A. M. Hurtado, and K. R. Hill. 1991. Seasonal variation in the diet and sexual division of labor among Hiwi hunter-gatherers. J. Anthrop. Res. 46(3):293–346.

[a] Edible portion only.

[b] Kilograms × number of calories per edible kilogram.

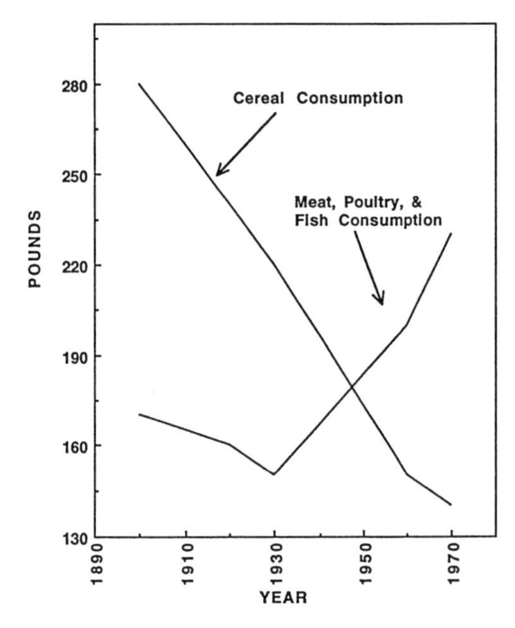


**Fig. 17.1.** Changes in per capita consumption of cereal, meat, poultry, and fish. As consumption of meat, poultry, and fish has increased, intake of cereals has decreased. (Adapted from U.S. Department of Agriculture, Economic Research Service. 1975.)

has been a drastic increase in the use of refined flour and the replacement of whole cereal grains with ready-made breakfast cereals (3,4). Dietary patterns in the Western world a century ago compared with today show major differences that are similar to those between foods consumed in the developing nations and in economically developed countries today (5–7). Recent epidemiological studies have called attention to these changes in diet being associated with the increased incidence of colon and breast cancer. The National Cancer Institute estimates that 60% of all cancers in women and 41% of cancers in men may be related to diet and nutrition (8–10). It has been postulated that this increase is related to overnutrition. The hypothesis that diet plays a role in breast cancer relates primarily to relative dietary excesses of fat and protein.

## Dietary Fat Intake

### *Experimental Evidence*

Animal experimental evidence implicating dietary fat intake with breast cancer has been derived from studies of both spontaneous mammary tumors and chemically-induced mammary tumors. Experimental studies indicate that dietary fat promotes the growth of populations of cancer-initiated cells in the multistage process of carcinogenesis (11). Furthermore, a high-fat diet accelerates the rates of growth and development of both spontaneous mammary tumors and those initiated by the chemical carcinogen 9,10-dimethyl-1,2-benzanthracene (DMBA) (12). On the other hand, reduction of dietary fat apparently reduces the tumor-promoting effect of a prior high-fat diet (13).

At present, whether the type of fat is important in the development of cancer is not well defined. While both saturated and unsaturated dietary fats have been found to have an effect on tumor development (14, 15), polyunsaturated fats (which are sources of the essential fatty acid linoleate) caused significantly more tumors than tallow or coconut oil (which are inadequate sources of linoleate) (16). On the other hand, it has been demonstrated that fatty acids derived from fats unique to marine animals may protect against the growth of mammary tumors in rats (17).

*Epidemiological Evidence*

Epidemiological studies show high correlations between national levels of breast cancer and per capita consumption of total fat, animal fat, total proteins animal protein, and eggs (18). For example, in Britain, breast cancer mortality is positively related to per capita intake of animal fat and protein during the decade prior to death (19). In the United States, age-adjusted cancer mortality is correlated to per capita intake of fat, protein, beef, milk, and total calories (20). Similarly, in Japan per capita fat intake of animal fat is positively correlated with breast cancer mortality by geographic prefecture (21). Furthermore, as inferred from a cohort study of 142,857 women followed for 10 years beginning at age 40, the risk of breast cancer among those with high dietary fat intake was 8.5 times greater than the risk among those with low fat intake (21).

### Dietary Protein Intake

*Experimental Evidence*

Experimental studies have shown an effect of increased dietary protein on tumor promotion in both spontaneous and chemically-induced tumors, including mammary tumors in rats (22–24). These studies have found an increasing incidence of spontaneous tumors, including mammary tumors, with increasing dietary protein intake in rats fed various isocaloric diets (24–26). Furthermore, a high protein intake early in life contributed to both a high tumor risk and a high mature weight in laboratory animals (22–24). This finding suggests that the effects of high protein intake are mediated by developmental factors. On the other hand, lower dietary protein apparently inhibits the growth of spontaneous and chemically-induced tumors (22–24).

*Epidemiological Evidence*

Epidemiological studies have shown a stronger association with animal protein than with total protein (possibly explained by the dietary association of animal protein consumption with fat consumption). Furthermore, studies in different populations and in the United States found that breast cancer mortality was correlated with per capita intake of animal protein (20, 27).

## Migration and Westernization Effects

### Japanese-Americans

Some of the strongest evidence for the role of environmental factors in breast cancer comes from epidemiological studies on Japanese immigrants to the United States and Japanese living in Japan. As shown in table 17.2 the incidence of breast cancer increases in successive generations of Japanese women in America compared to those in Japan (28, 29). Furthermore, epidemiological comparisons of breast cancer rates indicate breast cancer incidence is five times higher among women in the United States than in Japan. Furthermore, breast cancer rates have increased in succeeding generations of Japanese immigrants to the United States (28–30). As illustrated in figure 17.2, the breast cancer rates among the young Japanese-American women is as high as in the United States Caucasian women, but among the older Japanese-American women are lower (30). According to these studies persons who emigrate during childhood and adolescence experience the greatest changes in patterns of cancer (28, 29). Thus it would appear that the expression of breast cancer among immigrants depends upon the developmental stage at which the organism is exposed to the new environment.

### *Westernization in Japan*

As Japan's dietary habits are becoming westernized the incidence of breast cancer is also increasing (21). For example, the rate of death from breast cancer has steadily increased from 3.5 per 100,000 in 1955 to 5.8 per 100,000 in 1975. The increase has been mostly among women 45 to 59 years of age, and among higher socioeconomic groups in urban areas (21). The fact that the incidence of breast cancer among young Japanese-American women is almost as high as that among U.S. white women (31) suggests that the new generations are becoming as susceptible as the Americans.

**TABLE 17.2. Age-adjusted breast cancer incidence in first- and second-generation Japanese immigrants**

| Age | Japan | First Generation | Second Generation | U.S. Whites |
|---|---|---|---|---|
| 35–64 | 31.8 | 93.6 | 116.5 | 179.4 |
| 65–74 | 21.8 | 163.4 | — | 293.4 |

*Source:* Adapted from P. Buell. 1973. Changing incidence of breast cancer in Japanese-American women. J. Natl. Cancer Inst. 51:1457–79.

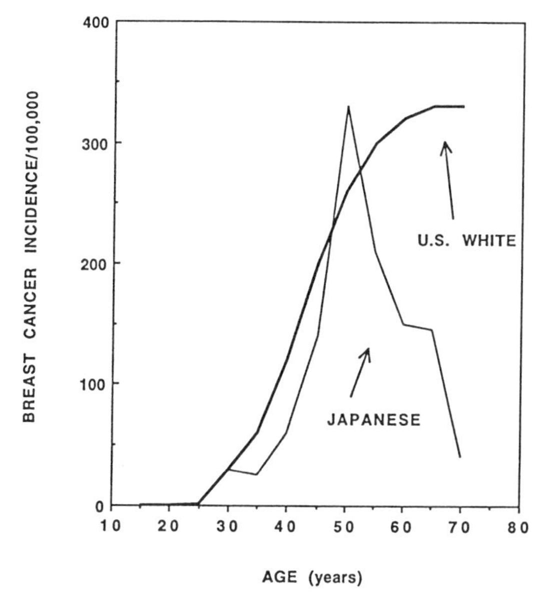


**Fig. 17.2.** Incidence of breast cancer among U.S. whites and Japanese women. Note that under the age of 50 years the incidence of breast cancer is similar in both groups, but thereafter it is much higher among U.S. whites. (Adapted from J. Waterhouse, P. Correa, C. Muir, and J. Powell. 1976. Cancer incidence in five continents. Vol. 3. Lyons, France: International Agency for Research on Cancer. New York: Springer-Verlag New York.)

### *Westernization in Hawaii*

Comparison of the diet of Japanese men whose wives had developed breast cancer with the diets of other Japanese men who had participated in the Japan-Hawaii Cancer Study (32) indicated that husbands of the breast cancer cases consumed a Western-style diet (more beef or other meat, butter, margarine, cheese, corn, and wieners), and less Japanese foods than the control group. Therefore, it has been inferred that the increasing incidence of breast cancer among Japanese-American women is associated with changes in lifestyle, including diet, in successive birth cohorts through time (33). Studies of Japanese-Hawaiians indicate that the longer the population lives in Hawaii the greater the cancer incidence. For example, the rates increased from 23 per 100,000 during the period 1960–64 (34) to 44.2 per 100,000 during 1967–71 (28), and 57.3 per 100,000 during 1973–77 (30).

## Genetic Factors

### Familial Factors

Although a large proportion of breast cancer may be attributed to environmental agents, not all persons similarly exposed develop cancer, indicating that genetic factors may account for this variability. There is abundant evidence that for women whose mothers or sisters had breast cancer the risk of developing breast cancer is about two or three times greater than for controls (35–36). Studies on twins indicate a higher rate of concordance of breast cancer in monozygotic than dizygotic twins (37). The influence of familial or hereditary factors may be expressed through differences in hormonal secretion. Studies of teenage daughters of breast cancer patients found higher prolactin levels during certain days of the menstrual cycle than in daughters of control women (38).

### Cerumen Type

Among the several genetic markers studied in relation to breast cancer, investigations of cerumen type have produced fruitful results. Human cerumen (earwax) has two phenotypic forms, wet and dry, which are inherited in a simple Mendelian fashion. It is well known that Asian countries have a lower incidence of breast cancer than Western countries (34). It has been postulated that the low risk of breast cancer in Asians is partially caused by their overall decreased breast secretion activity which is associated with dry cerumen (39–42). This hypothesis is based on the following associations. First, there is a direct relationship between the frequency of wet cerumen and world breast cancer mortality (39–42). Second, breast secretory activity of nonlactating, nonpregnant women is associated with differences in cerumen type; in women with dry cerumen there is a lower proportion of breast secretions (39–42). Third, the lowest proportion of breast secretory activity is found among Asians and American Indians.

Given the fact that earwax glands and breast glands are apocrine in structure and function, the association of breast cancer incidence with cerumen type is not implausible. Since breast gland secretions have been found to contain a variety of environmentally derived chemicals and nutrients (39–42) it is quite possible that a low or absent breast secretory activity may minimize contact of the breast with exogenous carcinogens (39–42). Viewed in this manner, the observed lower frequency of breast cancer in Asian women may be related to their lower breast secretory activity (39–42). However, this does not explain the increased frequency of breast cancer found among Japanese migrants to Hawaii and California and their American-born offspring (29). On the other hand, a relationship to changes in breast-feeding practices is possible; the decrease in breast-feeding may result in an increased retention of carcinogenic agents in the breast.

## Developmental Component

Studies of humans and experimental animals suggest that the relationship between types of dietary intakes and breast cancer may be mediated through developmental factors. The extent to which developmental factors affect the expression of breast cancer can be inferred from studies of factors determining variability in growth, maturation, and fat deposition. In general, overnutrition during childhood is associated with increased stature, accelerated rates of maturation, and increased deposition of adipose tissue. Epidemiological studies have shown that early maturation and increased fat deposition during growth are correlated with increased risk for breast cancer in later life. For example, women in Poland (43), Greece (44), Japan (45), and the United States (46) with early menarche have been found to have a higher risk of developing breast cancer than those with late menarche. Similarly, studies in the Netherlands and Japan found a direct association between weight and the incidence of breast cancer (47). It has been suggested that large body mass and breast cancer may be independent consequences of the same nutritional pattern early in life (48). Experimental studies on animals indicate that calorie restriction generally inhibits tumor formation and increases life expectancy (22–24, 49). A lower protein intake inhibits the development of spontaneous or chemically induced tumors (22–24, 49). Conversely, an increase of fat has been found to enhance the incidence of breast tumors, and the tumors also occurred earlier in the life of the animal (12). Furthermore, the incidence of tumors tends to be consistently greater in heavier rats than in lean rats (22–24).

The above studies suggest that the same nutritional pattern that is responsible for increased body size, increased deposition of body fat, and decreased age at menarche may be related to breast cancer risk in humans. These relationships are schematically depicted in figure 17.3. From these data the following points are important. First, as a result of the increased availability of fats and proteins, westernization leads to an increased storage of body fat and accelerated rate of growth and maturation, which in turn are associated with an increased incidence of breast cancer. Second, westernization concomitant with increased industrialization has resulted in an increased concentration of environmental carcinogens. These carcinogens reach the breast alveolar and ductal epithelium through circulation and breast fluid secretions, and as a result the risk of breast cancer increases. Third, because of low breast-feeding practices, westernization can also lead to increased contact of breast epithelium with carcinogenic agents and in turn affect the incidence of breast cancer.

## COLON CANCER, DIVERTICULOSIS, AND DIETARY FIBER

### Colon Cancer

Epidemiological data derived from world populations also indicate that the age-adjusted mortality rate from colonic and breast cancer is positively correlated with the per capita daily consumption of fats and oils and negatively

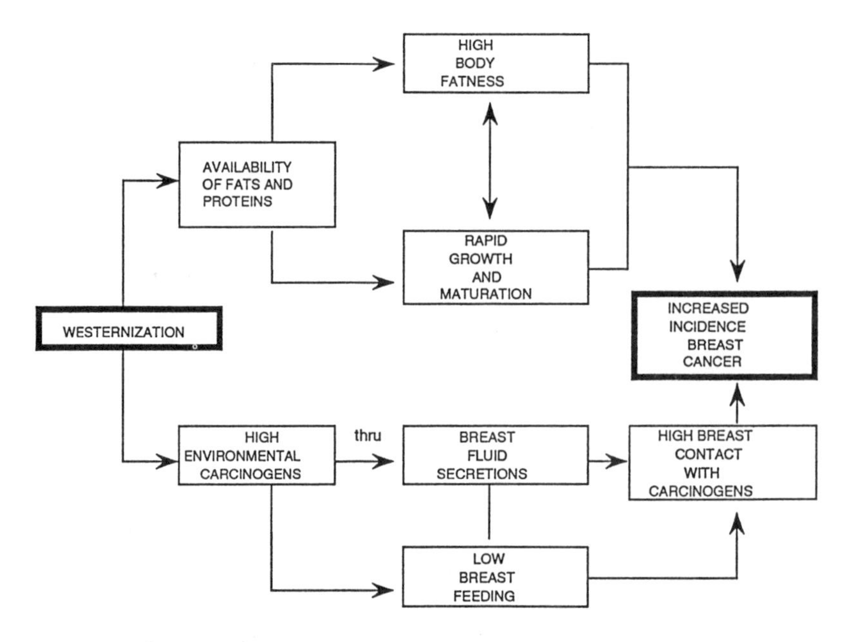


**Fig. 17.3.** Schematization of interaction of dietary, developmental, and environmental factors influencing increased incidence of breast cancer. Westernization as a result of industrialization has simultaneously increased availability of dietary fats and proteins and environmental carcinogens, and decreased the practice of breast-feeding. As a result of complex interaction of these factors, incidence of breast cancer increases.

correlated with *dietary fiber* intake (8, 9). That is, the higher the consumption of fats and oils and the lower the intake of dietary fiber, the greater the breast cancer mortality. In the United States the lowest rates of cancer incidence and mortality from all tumors are found among vegetarian populations such as the Seventh Day Adventists (50, 51), whose intake of protein is low and intake of dietary fiber and unrefined carbohydrates is high. Similarly, Mormons living in Utah have a lower incidence of cancer of all tumor types than non-Mormons also living in Utah (51–52), suggesting the possible role of dietary habits (52).

## Diverticulosis

Diverticular disease is characterized by small ''blow out'' protrusions (diverticula) in the large intestine, that become inflamed. The resulting condition is diverticulitis. Diverticulitis is generally associated with an increase in the segmenting pressures produced by the contracting colon. It has been postulated that the increased pressure results from a fiber-deficient diet, which causes colonic contents to be more viscous and harder to propel, as occurs in constipation (6, 7). Among those with a fiber-depleted diet, the colonic lumen is narrower because of the smaller bulk of the diet, which in turn causes a greater segmentation of the colon and the buildup of intraluminal pressure. This increased pressure eventually produces diverticula (6, 7). There appears to be a strong negative correlation between the epidemiology of diverticular disease and crude fiber intake (57, 58). The fiber deficiency theory of the etiology of diverticular disease is supported by clinical studies, which showed that symptoms of the disease are effectively alleviated following the addition of 20 to 30 g/day of wheat bran to the diet (59). Deficiency of dietary fiber has also been related to circulatory and metabolic diseases such as varicose veins, ischemic heart diseases, obesity, diabetes mellitus, and other diseases of the gastrointestinal tract (appendicitis, hemorrhoids, etc.).

### The Fecal Mutagen-Cancer Hypothesis

It has been hypothesized that diet, depending on its content of carcinogenic components, may modulate organ susceptibility and response to causative factors (53). According to this hypothesis the intestinal bacteria produce carcinogens (and mutagens) through metabolism of steroids associated with a high-fat diet (54). Constipation would result in greater contact time in the intestine and absorption into the enterohepatic circulation for these ''fecal mutagens.'' The fecal mutagen-cancer hypothesis is based on the interaction of fecal neutral steroids, bile salt and fat content, fecal ammonia, fecal pH, fecal enzyme activity, fecal transit time, stool weight and water content, and microbial flora. Stool analyses have found differences in fecal bacteria and dietary fiber intake between populations with high (16.2 per 100,000 in Denmark) and low (7.9 per 100,000 in Finland) rates of large bowel cancer (6–7). As illustrated in figure 17.4 according to this hypothesis an excessive fat intake modifies the metabolism of cholesterol, bile acids, and neutral steroids in the intestine, as well as the metabolism and secretion of steroid hormones in circulation (54). On one side the bile acids secreted in the intestines are degraded by bacteria growing in the intestine to form carcinogenic substances that may initiate colon cancer (55). This process of degradation and transformation of bile acids is modulated by the presence of fiber in the diet, which is known to decrease the food transit time, and as a result the contact between concentrated stool content and the mucosa is decreased. This in turn decreases the production of bacterial flora capable of forming carcinogens from the degradation of bile salts (56). Furthermore, the altered metabolism of steroid hormones could impose unnatural burdens on cellular receptors of specific target tissues, such as uterus or breast, again initiating cancer in those tissues (53).

## VITAMIN C AND CANCER

Along with the decline in the consumption of foods rich in fiber the dietary intake of vitamin C has also declined. For this reason, various studies are

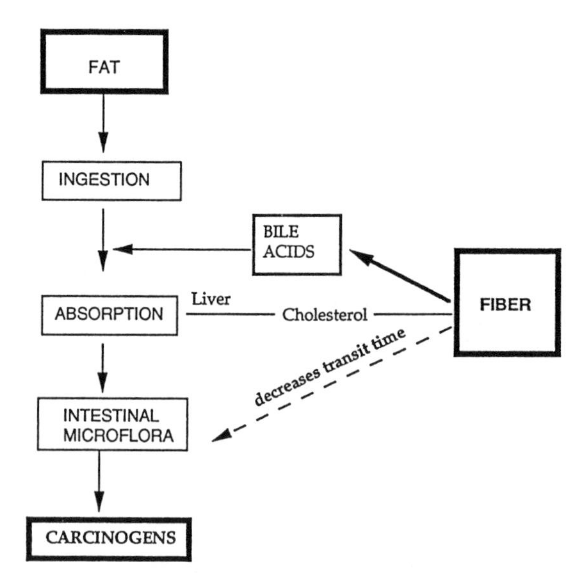


**Fig. 17.4.** Schematization of interaction of fat and fiber dietary intake and its influence on incidence of intestinal cancer. Dietary fiber, through its action on formation of bile acids and effect of food transit time, is said to influence rate of accumulation of carcinogens. (Modified from Diet, nutrition and cancer program status report. Sept. 30, 1977. Bethesda, MD: Diet, Nutrition and Cancer Program, National Cancer Institute, National Institute of Health.)

evaluating the role of vitamin C in the increased incidence of cancer observed in contemporary populations. Vitamin C or ascorbic acid is a dietary essential. Its deficiency leads to scurvy, which is characterized by damage of the skin, gums, and underlying tissues, impaired lung function and susceptibility to respiratory infections, and pain of the feet and legs. Scurvy was a major factor in shaping the course of history, destroying armies and navies and causing the death of many explorers (58).

## Synthesis

The importance of vitamin C lies in the fact that humans (and guinea pigs and monkeys), unlike other mammals, lack the enzyme necessary for the synthesis of vitamin C from glucose or galactose. Therefore, vitamin C has to be derived only from food. The most frequent sources of vitamin C are citrus fruits, potatoes, broccoli, spinach, cabbage, green pepper, tomato juice, strawberries, cantaloupe, etc.

Even though vitamin C has no clear-cut role as a coenzyme nor is it part of an enzyme or body component it plays an important role in almost all biological functions of the organism. The most important functions include the following.

## Biological Functions

**Collagen and Dentine Formation.** Vitamin C plays an important role in the formation of collagen in connective tissue, where it binds cells together in much the same way cement binds bricks (60). It is a major constituent of skin, teeth, and cartilage, and provides the structural framework of bone. The formation of the dentin layer of the teeth requires the presence of vitamin C and its lack leads to the formation of an uneven and weak surface that is unable to resist decay or mechanical injury.

**Synthesis of Neurotransmitters.** Vitamin C is also necessary for the conversion of tyrosine into the neurotransmitter norepinephrine, and the amino acid tryptophan into the neurotransmitter serotonin, both of which are indispensable for the organism's function (60).

**Utilization of Iron, Calcium, and Folacin.** Because it keeps iron in the reduced form Vitamin C facilitates iron absorption. It also assists in the transfer of iron from the blood to the liver where it is stored in the form of ferritin (60). Therefore, vitamin C plays an important role in hemoglobin's ability to retain iron and hence bind to oxygen. Thus, vitamin C is indirectly involved in the organism's oxygen transport system.

**Lipid Scavenging, Vitamin E, Regeneration, and Immunity.** Recent evidence indicates that vitamin C is an important antioxidant and free radical scavenger in plasma, and serves as protection against lipid peroxidation (61–62). It has also been shown to function in sparing or regenerating active vitamin E for protection of lipid membranes (62, 63). Furthermore, there is evidence that increasing the dosage of vitamin C increases the immunological activity (64).

## Cancer

In addition to such biological functions, epidemiologic studies indicate that increased incidence of different types of cancer is associated with low intake of vitamin C. As shown in table 17.3, 33 of 47 epidemiologic studies indicate that vitamin C plays a protective role against cancers of the esophagus, oral cavity, stomach, pancreas, cervix, rectum, breast, and even lung (65). Similarly, 21 of 29 epidemiologic studies found significant protection associated with frequent consumption of fruits.

In summary, given the fact that the consumption of foods rich in fiber has declined, the increased incidence of cancer observed in contemporary populations may also be related to changes in the consumption in foods rich in vitamin C. In the 1960s and 1970s the incidence of infant scurvy reappeared in the U.S. due to the fact that mothers had taken megadoses of vitamin C during pregnancy conditioning these infants to need more vitamin C than is supplied in milk or formula (60). Therefore, it is quite possible that the increased incidence of various types of cancer may also be related to an increased need for vitamin C resulting from the increased consumption of artificial sources of vitamin C that is typical in modern populations.

## BERIBERI

Beriberi is a disease caused by deficiencies of thiamine or vitamin $B_1$. There are two types of beriberi: dry beriberi and wet beriberi. Dry beriberi is characterized by muscular weakness, muscle pain, paralysis of the limbs, and sometimes mental confusion. Wet beriberi is typified by edema and cardiac failure. Infants born of thiamine-deficient mothers are liable to contract the disease because the mother's milk fails to supply sufficient thiamine to meet the infant's needs.

Beriberi in both its dry and wet forms is still common in Thailand and

**TABLE 17.3. Epidemiologic studies that reported on the protective role of vitamin C and cancer**

| | Proportion Significantly Protective | |
|---|---|---|
| Cancer Site | Vitamin C Index | Fruit |
| Nonhormone dependent | | |
| Oral cavity | 3/3 | 3/4 |
| Esophagus | 4/4 | 3/7 |
| Larynx | 1/1 | 1/1 |
| Lung | 5/10 | 2/2 |
| Pancreas | 1/2 | 4/4 |
| Stomach | 7/7 | 5/5 |
| Cervix | 3/4 | — |
| Rectum | 4/6 | — |
| Colon | 4/8 | 3/5 |
| Bladder | 0/1 | 0/1 |
| Brain | 1/1 | — |
| Total | 33/47 | 21/29 |
| Hormone-dependent | | |
| Ovary | 0/3 | 0/1 |
| Endometrium | — | 1/1 |
| Prostate | 0/6 | — |
| Total | 0/9 | 1/2 |
| Meta-analyses | | |
| Breast | + | + |

*Source:* Adapted from G. Block. 1991. Epidemiologic evidence regarding vitamin C and cancer. Am. J. Clin. Nutr. 54:1310–14S. (ref. 82).

other countries. Its origin is traced to consumption of polished or millet rice. Thiamine is located in the outer layer of cereal grain, which is removed during the polishing process. Thus the polished rice product is deficient in thiamine. With acculturation and westernization polished rice became preferred, and with it beriberi became a serious health problem (66). In other regions where Western patterns of food consumption have been acquired, the desire for white, highly refined cereal flour containing minimal amounts of thiamine is spreading, and concomitant with this practice symptoms of beriberi are manifested (67). Evidence that beriberi is caused by the loss of thiamine resulting from rice and not by actual vitamin deficiency in the cereal itself is derived from the fact that in other populations in which rice is consumed in parboiled form, beriberi is less frequent than in rice-eating areas where this is not a practice (68). In India, through the process of parboiling, in which unhusked rice is steamed and then dried before milling, the vitamin is not lost. This is because thiamine is water soluble, and through parboiling the vitamin from the outer layer of the grain is distributed more evenly in the inner parts of the rice. For this reason and to reduce the risk of vitamin deficiencies in the United States, all white flour used in the manufacture of bread is fortified with vitamins by law.

It must be noted, however, that beriberi might also occur when thiamine intake is adequate. Raw, fermented fish consumed daily (mean = 56 g) by the northeastern Thais contains about 225 U of antithiamine factor in the form of the thermolabile enzyme thiaminase I, which alters the structure and reduces the biological activity of the vitamin (69, 70). In addition, fermented tea leaves chewed as stimulants by the Thais also contain an antithiamine factor that has deleterious effects on the biological activity of thiamine (71). It is evident, then, that food consumption and cultural factors are symbiotically related to human health and well-being, now more than ever.

## SICKLE-CELL ANEMIA

The sickle-cell trait of red blood cells is one of the best examples of genetic adaptation to malaria (72, 73). In Africa and other malaria regions the homozygous normals develop malaria, many die early, and vitality of the survivors is impaired. On the other hand, the homozygotes for the sickle-cell trait develop sickle-cell anemia, whereas the heterozygous sicklers do not develop malaria. For this reason in Africa and other regions where malaria is endemic, the heterozygous condition occurs in up to 25% of the population (74–76). However, despite the high frequency of heterozygous sickle-cell trait in Africa, the West Indies, and the Mediterranean, the percentage of individuals who suffer from the clinical manifestations of sickle-cell anemia is much lower than in the United States (75, 77, 78). The usual explanation for the low frequency of sickle-cell anemia in Africa or other regions where malaria is endemic is that the majority of victims die in early infancy (72, 75, 79). Nevertheless, other studies suggest that the clinical manifestations of sickle-cell anemia may be diminished by foods containing cyanate, thiocyanate, folic acid, and ascorbic acid and iron-deficient foods (80, 81).

Various studies have shown that the clinical symptoms of sickle-cell anemia are decreased when treated with cyanate (82–85). Hematological improvement and a decrease in hemolytic anemia occur on oral administration of cyanate (83). However, since the cells inhibited from sickling are eventually replaced by new sicklers, cyanate oral therapy would have to be continued for life (83). It has been postulated that the low frequency of clinical manifestation of sickle-cell anemia in Africa and other regions is related to the high consumption of cyanate- and thiocyanate-containing foods such as cassava (manioc), yams, sorghum, millet grains, sugarcane, and dark varieties of lima beans (78, 80, 81). This is supported by the finding that Jamaican subjects with mild clinical manifestations of sickle-cell anemia, on migration to the United States or England, experience severe crises of sickle-cell anemia, which are eliminated on return to the original environment (78). Therefore there must be an environmental factor that is responsible for the mild symptoms of sickle-cell anemia in Jamaica (78). Specifically, the decrease in clinical manifestation of sickle-cell anemia in Jamaica is attributed to the high consumption of cassava, dark varieties of lima beans, and sugarcane (81) all of which contain large amounts of cyanate and thiocyanate. According to food analysis cassava flour contains about 70 to 80 mg/100 g of thiocyanate and yam flour between 50 to 60 mg/100 g of thiocyanate (86). Thus it would appear that the full expression of sickle-cell anemia in blacks in the United States represents an unrelieved nutritional dependency on cyanate precursors (80, 81). In other words, those with sickle-cell anemia constitute another such group in which the disease is rooted in a special nutritional need as the interface between heredity and pathology. Since present-day diets in the tropics provide foods considerably richer in cyanate and thiocyanate, or nitrilosides, than diets of the United States and other Western countries (87,88) the clinical manifestation of sickle-cell anemia may be viewed as the result of changed patterns of food consumption associated with westernization. It may also be noted that according to laboratory studies cyanate, to be effective in decreasing sickling, requires that about 50% of the hemoglobin tetrads be attached to cyanate. To achieve this goal the dosage reaches toxic levels. Oral administration of cyanate, to be effective, might have to be continued for the life span of the patient (83). Therefore natural sources, such as those found in the diet, might prove to be more practical. It would not be surprising to find that other natural sickling inhibitors exist in the diet of populations in which the genetic trait is prevalent.

Because foods rich in cyanate (which include sublethal cyanide, thiocyanate, and cyanate) are the staple dietary items of the West African populations exposure to cyanogens starts in utero and continues through childhood and adulthood (89). This is because thiocyanates are able to cross the placenta, and during breast-feeding infants ingest large amounts of cyanogens from breast milk during several years of nursing. These studies have shown that among samples uniformly exposed to holoendemic falciparum malaria there are significant differences in the frequency of hemoglobin S and these differences are associated with variability in the consumption of cyanate-rich foods (89). Thus the higher the consumption of cyanate-rich foods, the lower is the frequency of sickle-cell hemoglobin.

Laboratory and clinical studies indicate that when cyanogen derivatives enter into the blood they can combine with hemoglobin S and important proteins of plasmodium falciparum. When this happens the structures of both proteins are altered. One consequence of this alteration is the carbamylation of hemoglobin (whereby the carbamyl— $NH_2$-CO—of carbamylphosphateo— $H_2NCO$-$OPO_3H_2$—is transferred to an amino group), which in turn inhibits the sickling of hemoglobin S and thus functions like a normal hemoglobin A (89). An additional effect of this alteration is that the cyanogen derivatives hinder the intraerythrocytic growth and development of the plasmodium phalciparum (90) (see fig. 17.5). As a consequence, the viability of individuals with homozygous hemoglobin S increases and thus contributes to their increased frequency. These findings would suggest that the variability in the frequency of sickle-cell anemia is the result of not only selection against malaria but also the consequence of differences in cultural-nutritional practices that have characterized ongoing human evolution.

## SECULAR TREND IN SIZE AND MATURITY

One of the most clear changes that has characterized industrialized countries is the increased tendency to become progressively taller and mature earlier than nonindustrialized populations. As show in table 17.4, despite the earlier termination of growth, because of the faster growth rate, white populations in 1980 on the average are taller than the average in the 1940s. The increase in adult height has occurred at about 1 cm per decade (91). The acceleration of maturation is shown in the secular trend of age at menarche. As illustrated in figure 17.6, the age of menarche in the United States has declined from an average of 14.5 years in 1890 to about 12.5 years in 1980.

It is usually assumed that the secular trend is related to the increased consumption of animal foods rich in fat and protein and to improved public health conditions. The fact that the secular trend is confined to industrialized

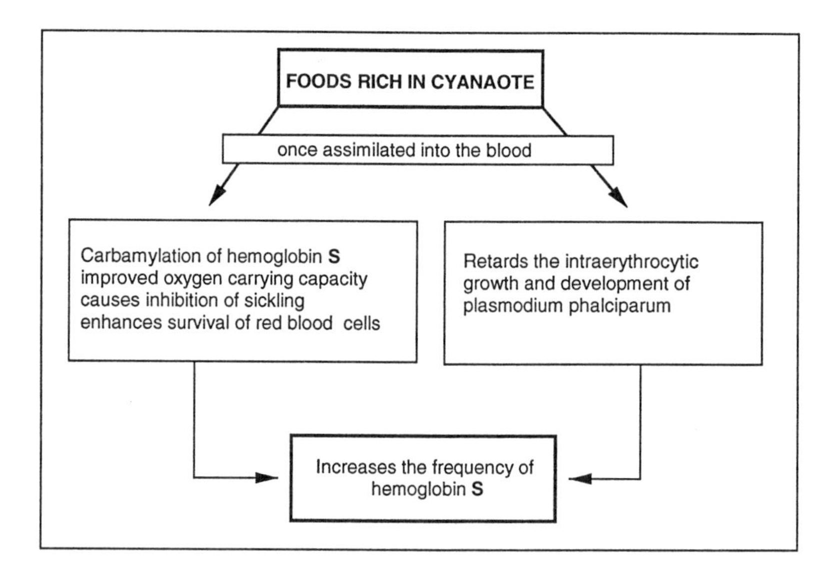


**Fig. 17.5.** Schematization of the role of consumption of foods rich in cyanaote on the expression of sickle-cell anemia. Foods rich in cyanote, such as cassava, improve the blood's oxygen-carrying capacity, resulting in an increased frequency of hemoglobin S. (Adapted from F. L. C. Jackson. 1990. Two evolutionary models for the interaction of dietary organic cyanogens, hemoglobins, and phalciparum malaria. A. J. Human Biology. 2:521–32.)

**TABLE 17.4. Secular trend in body size and body proportions among industrialized populations**

| Year Studied | Males | Females |
|---|---|---|
| | Age at Adolescence Spurt in Height (years) | |
| 1940 | 14–15 | 15–16 |
| 1980 | 13–14 | 14–16 |
| | Age of Attainment of Adult Height (years) | |
| 1940 | 18 | 22 |
| 1980 | 17 | 18 |
| | Adult Height (cm) | |
| 1940 | 155 | 165 |
| 1980 | 165 | 175 |
| | Sitting Height Index (%)[a] | |
| 1940 | 54 | 54 |
| 1980 | 51 | 52 |

[a] Sitting height index = (sitting height/standing height × 100)

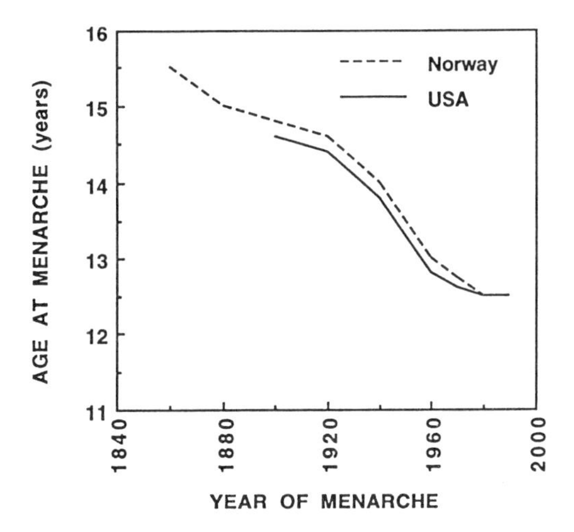


**Figure 17.6.** Secular trend in age at menarche from 1850 to 1990, United States and Norway. Values are plotted at year in which menarche took place. (Modified from J. M. Tanner. 1978. Foetus into man: physical growth from conception to maturity. Cambridge, MA: Harvard University Press.)

populations supports the nutritional health hypothesis. However, even in industrialized countries the secular trend in height appears to be more evident among the offspring of short than of tall individuals (92). This finding has been interpreted as an indication of the phenomenon of ''regression to the mean.'' It may also suggest that the short stature of the previous generation was related to the influence of negative environmental conditions, and given adequate conditions their offspring have been able to express their genetic growth potential, while the tall individuals in the previous and present generation already fulfilled their growth potential. The fact that populations undergoing deterioration of their socioeconomic and nutritional conditions are exhibiting a negative secular trend (93) supports the hypothesis that variability in height is to a great extent a function of the environmental conditions in which the individual develops.

## CHOLESTEROL AND CORONARY ARTERY DISEASE

### Cholesterol

Cholesterol is a fat that is derived from the combination of simple and compound fats and is found only in **animal tissue** that contains no fatty acids but exhibits some of the physical and chemical characteristics of fat. There are two sources of cholesterol: **endogenous cholesterol,** which is synthesized by the body and **exogenous cholesterol,** which is consumed in foods. About 70% of endogenous cholesterol is synthesized by the liver. The rest is produced by other tissues including the walls of the arteries and intestines. The rate of endogenous cholesterol synthesis may vary from 0.5 to 2.0 g per day, and, depending on the ingestion of exogenous cholesterol, even more can be synthesized. Exogenous cholesterol as the name implies is derived from the diet. The highest sources of exogenous or dietary cholesterol are egg yolks, brains, kidneys, liver of beef, certain dairy products (such as ice cream, cream cheese, butter, and whole milk), shellfish, and especially shrimp.

## Functions of Cholesterol

Cholesterol plays an important role in many complex bodily functions, including the building of cell membranes, the synthesis of vitamin D, and the production of adrenal gland hormones. Cholesterol is essential for the manufacture of sex hormones such as estrogen, androgen, and progesterone. Cholesterol also plays a role in the formation of the bile secretions that emulsify fat during digestion. When cholesterol is either ingested in excess or oversynthesized by the liver (endogenous cholesterol) it can lead to the formation of cholesterol-rich deposits called **plaque** on the inner lining of the medium and larger arteries. This process, which is called **atherosclerosis,** results in narrowing of these vessels and eventually to coronary artery disease (CAD). In recent years attention has been given to the lipoprotein fractions of cholesterol. The lipoproteins are formed primarily in the liver from the union of either triglycerides, phospholipids, or cholesterol with protein. Lipoproteins are usually classified into **high density lipoproteins (HDL), low density lipoproteins (LDL),** and **very low density lipoproteins (VLDL). HDL** is popularly known as the "good cholesterol." It assists the organism in the removal of cholesterol from cells and the blood. It then carries it to the bile in the liver and is subsequently excreted by the intestines. **VLDL** and **LDL,** popularly known as the "bad cholesterol," have a tendency to attach themselves to the arterial tissue, where they proliferate and, in the process, damage and narrow the artery.

## Industrialization and Coronary Artery Disease

Compared to developing nations the industrialized countries have a high incidence of coronary artery disease (CAD). The epidemiological and case-control studies in the United States and abroad have shown a strong independent inverse relation between HDL and CAD (94–97). These studies have shown that a 1% increase in LDL value is associated with slightly more than a 2% increase in CAD over 6 years and a 1% decrease in HDL value is associated with a 3 to 4% increase in CAD (94–97). Table 17.5 gives the total serum cholesterol concentration and cholesterol lipoprotein fractions of U.S. white

**TABLE 17.5. Mean values of serum total cholesterols, high-density lipoproteins (HDL), and low-density lipoproteins (LDL) cholesterols of U.S. whites and blacks derived from the second National Health and Nutrition Examination Survey of 1976–80**

| | Whites | | | | | Blacks | | | | |
|---|---|---|---|---|---|---|---|---|---|---|
| Age Group (years) | *N* | Cholesterol Mean ± SD (mg/dl) | HDL Mean ± SD (mg/dl) | *N* | LDL[1] Mean ± SD (mg/dl) | *N* | Cholesterol Mean ± SD (mg/dl) | HDL Mean ± SD (mg/dl) | *N* | LDL[1] Mean ± SD (mg/dl) |
| Males | | | | | | | | | | |
| 20–29 | 902 | 187 ± 38 | 45 ± 11 | 438 | 120 ± 35 | 130 | 182 ± 42 | 52 ± 14 | 62 | 115 ± 36 |
| 30–39 | 624 | 209 ± 44 | 43 ± 12 | 321 | 140 ± 41 | 77 | 205 ± 50 | 52 ± 16 | 38 | 130 ± 46 |
| 40–49 | 510 | 221 ± 42 | 43 ± 12 | 244 | 152 ± 41 | 48 | 221 ± 59 | 51 ± 14 | 28 | 129 ± 42 |
| 50–59 | 495 | 229 ± 44 | 44 ± 13 | 243 | 160 ± 39 | 51 | 225 ± 48 | 55 ± 19 | 27 | 139 ± 41 |
| 60–69 | 1,121 | 225 ± 44 | 45 ± 13 | 583 | 153 ± 41 | 113 | 225 ± 44 | 53 ± 16 | 56 | 151 ± 47 |
| 70–74 | 367 | 216 ± 39 | 45 ± 13 | 180 | 147 ± 34 | 43 | 220 ± 53 | 53 ± 17 | 21 | 148 ± 40 |
| Females | | | | | | | | | | |
| 20–29 | 1,005 | 187 ± 38 | 52 ± 13 | 458 | 110 ± 35 | 152 | 192 ± 42 | 56 ± 15 | 65 | 117 ± 38 |
| 30–39 | 739 | 196 ± 39 | 52 ± 13 | 384 | 126 ± 36 | 93 | 197 ± 42 | 55 ± 20 | 46 | 131 ± 43 |
| 40–49 | 568 | 218 ± 44 | 53 ± 14 | 277 | 142 ± 41 | 78 | 216 ± 51 | 57 ± 16 | 36 | 132 ± 38 |
| 50–59 | 552 | 241 ± 46 | 56 ± 16 | 278 | 158 ± 46 | 81 | 238 ± 52 | 58 ± 19 | 37 | 147 ± 41 |
| 60–69 | 1,251 | 248 ± 49 | 54 ± 15 | 588 | 163 ± 43 | 126 | 238 ± 46 | 58 ± 15 | 58 | 159 ± 43 |
| 70–74 | 448 | 246 ± 47 | 52 ± 14 | 226 | 163 ± 43 | 51 | 246 ± 54 | 60 ± 18 | 25 | 171 ± 51 |

Source: From Frisancho. ms. It includes women using estrogen pills.

[1] LDL = total cholesterol – HDL – (triglyceride × 0.16)

and black representative samples. From these data it is evident that the total cholesterol in males increases from about 187 mg/dl at around 25 years to about 229 mg/dl in the fifth decade, and after this age it declines somewhat. In females, on the other hand, it increases systematically with age from about 187 mg/dl attained in the mid-twenties to 246 mg/dl reached in the seventies. These data also show that in both males and females the concentration of HDL (the so-called good cholesterol) remains unchanged with age. This means that the increase in total cholesterol is due mostly to the increase of LDL (the so-called bad cholesterol). Based on the studies referred to above (94–97), it can be assumed that the risk of coronary artery disease is higher when the proportion of HDL is less than 25%. Applying this criteria, it would appear that only 23% of white males, 40% of white females, 45% of black males and 50% of black females would have *desirable* levels of serum cholesterols. Hence, once again industrialized economies while they have brought increased human productivity have also created a potentially new selective force to which humans must eventually adapt or be defeated by its own success.

## OVERVIEW

The advance of human civilization has created conditions that are in many ways unlike those which existed during the past 3 million years of human evolution. In terms of human adaptation and nutrition the most significant changes are quantitative and qualitative differences in food consumption and variations in food use. Ever since humans became hunter-gatherers their diet has been based upon a mixture of animal foods and plants. Although contemporary hunter-gatherers derive as much energy from animal food as westernized populations, because game animals are leaner than domestic ones westernized populations consume a proportionately greater amount of fat. With the emergence of industrialization, fat and cereal grains became major components of the diet. Thus the Western diet is characterized by a high level of fat (about 40% energy), of which 40-50% is saturated fat. The saturated fats are derived from animal fats and dairy products, or, indirectly, from manufactured foods containing these fats and industrially hydrogenated vegetable oils.

The dramatically changing rates of cancer among populations living in industralized countries suggest that the westernization of dietary habits is a contributory factor to the observed increased frequency of breast and colon cancer. Although it is often not possible to pinpoint the environmental factor(s) responsible for observed differences in cancer rates, both experimental and epidemiologic studies indicate that high intake of dietary fat and protein is associated with increased incidence of breast and colon cancer. The fact that breast and colon cancers are not more prevalent in occupational groups in which exposure to environmental carcinogens is high reinforces the view that environmental pollutants per se are not influential factors in the observed relationship between nutrition and cancer incidence and mortality.

Studies of humans and experimental animals suggest that the relationship between types of dietary intakes and breast cancer may be mediated through developmental factors. It is postulated that as a result of the increased availability of fats and proteins industrialization results in an increased storage of body fat and accelerated rate of growth and maturation, which in turn are associated with an increased incidence of breast cancer.

Epidemiological studies indicate that the increased incidence of the different types of cancer observed among industrialized populations may also be related to the decreased consumption of vitamin C resulting from the decreased consumption of fruits and vegetables. With respect to diverticulosis, the protective role of dietary fiber has been attributed to the fact that crude fiber provides bulk to the food and thereby affects intestinal functions such as transit time, fecal weight, and bowel habits. It has been observed that ingestion of large quantities of dietary fiber decreases the food transit time. Among Rhodesian subjects consuming a Western diet of highly refined carbohydrates, the food transit time was much longer than for those eating a traditional Bantu diet of mainly unrefined carbohydrates. However, the mechanisms by which the fiber exerts these effects presently are not known. The fact that the incidence of these diseases continues to increase among Western populations calls for intensive research to determine the specific mechanisms whereby a dietary factor can become a carcinogen.

Evidence suggests that among populations in which westernization of dietary habits is associated with refined flour replacing parboiled rice as the food of choice, the disease of beriberi, which is caused by a deficiency of vitamin $B_1$, is manifest.

Studies in West Africa and other tropical populations indicate that consumption of foods rich in cyanates (such as cassava and manioc) inhibits the sickling of hemoglobin and hence increases the oxygen carrying capacity of hemoglobin and also impairs the intraerythrocytic growth and development of plasmodium falciparum. As such, the viability of individuals who are homozygotes for hemoglobin S would increase their phenotypic and genetic frequency. These findings suggest that sickle-cell hemoglobin might have evolved in response to both selection against malaria and differences in cultural-nutritional practices. Viewed in this context, the clinical manifestations of homozygous sickle-cell anemia in blacks in the United States represent an unrelieved nutritional dependency on cyanate precursors. In other words, those with sickle-cell anemia constitute another such group in which the disease is rooted in a special nutritional need as the interface between heredity and pathology.

The increased productivity brought about by industrial economies has increased the availability of foods rich in protein and fats from animal sources and decreased the incidency of infections and chronic diseases. These changes translate themselves into an improvement in the health and nutritional status of children and adults. Reflecting these changes the age at menarche has declined throughout the industrial world at a rate of nearly 3 months per decade. Simultaneously, the growth rate in length and especially that of the legs has increased dramatically, so that the adult height has been augmented about 1 cm per decade. The earlier maturation experienced in the industrialized economies along with the break in traditional values is an important component for the increased incidence of adolescent pregnancy. Similarly, the increased consumption of food rich in fat along with the increased sedentarism of modern populations is an important contributing factor for the increased frequency of coronary artery disease experienced by industrialized populations. In addition, industrialization has increased the concentration of air pollutants, which are having negative effects at the prenatal and postnatal levels (98). As such industrialization is becoming a potentially strong selective force to which we must adapt.

## References

1. Sinclair, A. J., and K. O'Dea. 1990. Fats in human diets through history: Is the Western diet out of step? In: J. D. Wood and A. V. Fisher, eds., Reducing fat in meat animals. London: Elsevier Science Publishers. 1–47.
2. Hurtado, A. M., and K. R. Hill. 1991. Seasonal variation in the diet and sexual division of labor among Hiwi hunter-gatherers. J. Anthrop. Res. 46 (3):293–346.
3. Scala, J. 1974. Fiber, the forgotten nutrient. Food technology 28:34–36.
4. Department of Agriculture, Economic Research Service. 1975. (USDA/ERS).
5. Trowell, H. 1976. Definition of dietary fiber and hypotheses that it is a protective factor in certain diseases. Am. J. Clin. Nutr. 29:417–27.
6. Burkitt, D. P. 1976. Some mechanical effects of fibre-depleted diets. In W. W. Hawkins, ed., Dietary fibre. Ontario, Canada: The Nutrition Society of Canada and Mile Laboratories.
7. Burkitt, D. P. 1978. Symposium on the role of dietary fiber in health. Workshop V fiber and cancer: summary and recommendations. Am. J. Clin. Nutr. 31: 5213–15.
8. Wynder, E. L. 1975. The epidemiology of large bowel cancer. Cancer Res. 35: 3388–94.
9. Wynder, E. L., and G. B. Gori. 1977. Contribution of the environment to cancer incidence: an epidemiological exercise. J. Natl. Cancer Inst. 58:825–32.
10. U.S. Department of Health, Education and Welfare. 1975. Third national cancer survey: incidence data. DHEW Pub. (NIH)775–87.
11. Dao, T. L., and P. C. Chan. 1983. Hormones and dietary fat as promoters in mammary carcinogenesis. Environ. Health Perspect. 50:219–25.
12. Tannenbaum, A. 1942. The genesis and growth of tumors. Part 3. Effects of a high-fat diet. Cancer Res. 2:468–75.

13. Kalamegham, R., and K. K. Carroll. 1984. Reversal of promotional effect of high-fat diet on mammary tumorigenesis by subsequent lowering of dietary fat. Nutr. Cancer 6:22–30.
14. Chan, P. C., and L. A. Cohen. 1974. Effect of dietary fat, antistrogen, and anti-prolactin on the development of mammary tumors in rats. J. Natl. Cancer Inst. 52:25–30.
15. Carroll, K. K., and H. T. Khor. 1975. Dietary fat in relation to tumorigenesis. Prog. Biochem. Pharmacol. 10:308–53.
16. Carroll, K. K., and G. J. Hopkins. 1979. Dietary polyunsaturated fat versus saturated fat in relation to mammary carcinogenesis. Lipids 14:155–58.
17. Karmali, R. A., J. Marsh, and C. Fuchs. 1984. Effects of omega-3 fatty acids on growth of a rat mammary tumor, J. Natl. Cancer Inst. 73:457–61.
18. Drasar, B. S., and D. Irving. 1973. Environmental factors and cancer of the colon and breast. Br. J. Cancer 27:167–72.
19. Hems, G. 1980. Associations between breast cancer mortality rates, childbearing and diet in the United Kingdom. Br. J. Cancer 41:429–37.
20. Gaskill, S. P., W. I. L. McGuire, C. K. Osborne, and M. P. Stern. 1979. Breast cancer mortality and diet in the United States. Cancer Res. 39:3628–37.
21. Hirayama, T. 1978. Epidemiology of breast cancer with special reference to the role of diet. Prev. Med. 7:173–95.
22. Ross, M. H., and G. Bras. 1971. Lasting influence of early caloric restriction on prevalence of neoplasms in the rat. J. Natl. Cancer Inst. 47:1095–113.
23. Ross, M. H., G. Bras, and M. S. Ragbeer. 1970. Influence of protein and caloric intake upon spontaneous tumor incidence of the anterior pituitary gland of the rat. J. Nutr. 100:177–89.
24. Ross, M. H. 1977. Dietary behavior and longevity. Nutr. Rev. 35:257–65.
25. Wells, P., L. Aflergood, and R. B. Alfin-Slater. 1976. Effect of varying levels of dietary protein on tumor development and lipid metabolism in rats exposed to ilatoxin. J. Am. Oil Chem. Sec. 53:559–62.
26. Temcharoen, P., K. Anukarahanonta, and N. Bhamapravati. 1978. Influence of dietary protein and vitamin $B_{12}$ on the toxicity and carcinogenicity of aflatoxins in rat liver. Cancer Res. 38:2185–90.
27. Gray, G. E., M. C. Pike, and B. E. Henderson. 1979. Breast cancer incidence and mortality rates in different countries in relation to known risk factors and dietary practices. Br. J. Cancer 39:1–7.
28. Buell, P. 1973. Changing incidence of breast cancer in Japanese-American women. J. Natl. Cancer Inst. 51:1457–79.
29. Waterhouse, J., P. Correa, C. Muir, and J. Powell. 1976. Cancer incidence in five continents. Vol. 3. Lyons, France: International Agency for Research on Cancer. New York, NY: Springer Verlag New York.
30. Dunn, J. E. 1977. Breast cancer among American Japanese in the San Francisco Bay area. Nat.Cancer Inst. Monograph 47:157–160.
31. Nomura, A., B. E. Henderson, and J. I. Lee. 1978. Breast cancer and diet among the Japanese in Hawaii. Am. J. Clin. Nutr. 31:2020–25.
32. Moolgavkar, S. H., N. E. Day, and R. G. Stevens. 1980. Two-stage model for carcinogenesis: epidemiology of breast cancer in females. J. Natl. Cancer Inst. 65:559–69.
33. Doll, R., C. Muir, and J. Waterhouse, eds. 1970. Cancer incidence in three continents. Vol. 2. Lyons, France: International Agency for Research on Cancer. New York, NY: Springer Verlag New York.

34. Post, R. H. 1966. Breast cancer, lactation and genetics. Eugen. Quart. 13:1–28.
35. Tokuhata, G. K. 1969. Morbidity and mortality among offspring of breast cancer mothers. Am. J. Epidemiol. 89:139–53.
36. Kundson, A. G., I. C. Strong, Jr., and D. E. Anderson. 1973. Heredity and cancer in man. Prog. Med. Genet. 9:113–58.
37. Henderson, B. E., V. Gerkins, and I. Rosario. 1975. Elevated serum levels of estrogen and prolactin in daughters of patients with breast cancer. N. Engl. J. Med. 293:790–92.
38. Petrakis, N. L. 1971. Cerumen genetics and human breast cancer. Science 173:347–349.
39. Petrakis, N. L. 1977. Genetic factors in the etiology of breast cancer. Cancer 39:2709–15.
40. Petrakis, N. L., L. Mason, R. Lee, B. Sugimoto, S. Pawson, and F. Catchpool. 1975. Association of race, age, menopausal status, and cerumen type with breast fluid secretion in nonlactating women, as determined by nipple aspiration. J. Natl. Cancer Inst. 54:829–33.
41. Petrakis, N. L., L. D. Gruenke, T. C. Beeler, N. Castagnoli, Jr., and L. C. Craig. 1978. Nicotine in breast fluid of nonlactating women. Science 199:303–4.
42. Staszewski, J. 1971. Age at menarche and breast cancer. J. Natl. Cancer Inst. 47:935–40.
43. Valaoras, V. G., B. MacMahon, and D. Trichopoulos. 1969. Lactation and reproductive histories of breast cancer patients in greater Athens, 1965–1967. Int. J. Cancer 4:350–63.
44. Wells, P., L. Aflergood, and R. B. Alfin-Slater. 1976. Effect of varying levels of dietary protein on tumor development and lipid metabolism in rats exposed to ilatoxin. J. Am. Oil Chem. Soc. 53:559–62.
45. Yuasa, S., and B. MacMahon. 1970. Lactation and reproductive histories of breast cancer patients in Tokyo, Japan. Bull. W.H.O. 42:195–204.
46. Salber, E. J., D. Trichopoulos, and B. MacMahon. 1969. Lactation and reproductive histories of breast cancer patients in Boston, 1965–1966. J. Natl. Cancer Inst. 43:1013–24.
47. DeWaard, F., J. P. Cornelis, and K. Aichi. 1977. Breast cancer incidence according to weight and height in two cities of the Netherlands and Japan. Cancer 40: 1269–77.
48. MacMahon, B. 1975. Formal discussion of "Breast cancer incidence and nutritional status with particular reference to body weight and height." Cancer Res. 35: 3357–58.
49. Carroll, K. K. 1975. Experimental evidence of dietary factors and hormone-dependent cancers. Cancer Res. 35:3374–83.
50. Wynder, E. L., F. R. Lemon, and I. J. Bross. 1959. Cancer and coronary artery disease among Seventh Day Adventists. Cancer 12:1016–28.
51. Enstrom, J. E. 1974. Cancer mortality among Mormons. Cancer 36:825–41.
52. Phillips, R. L. 1975. Role of lifestyle and dietary habits in risk of cancer among Seventh Day Adventists. Cancer Res. 35:1162–65.
53. Gori, G. B. 1977. Diet, nutrition and cancer program. Status report of diet, nutrition and cancer program. Bethesda, MD: National Institutes of Health.
54. Hill, M. J. 1975. Metabolic epidemiology of dietary factors in large bowel cancer. Cancer Res. 35:3398–3402.
55. Reddy, B. S., A. Mastromarino, and E. L. Wynder. 1975. Further leads on metabolic epidemiology of large bowel cancer. Cancer Res. 35:3404–6.

56. Moore, W. E. C., and L. V. Holdeman. 1975. Discussion of current bacteriological investigations of the relationship between intestinal flora, diet, and colon cancer. Cancer Res. 35:3326–31.
57. Painter, N. S. 1975. Diverticular disease of the colon: a deficiency disease of western countries. London: Heinemann Educational Books.
58. Painter, N. S., and D. P. Burkitt. 1975. Diverticular disease of the colon, a 20th-century problem. Clin. Gastroenterol. 4:3–21.
59. Findlay, J. M., A. N. Smith, W. D. Mitchell, A. J. B. Anderson, and M. A. Eastwood. 1974. Effects of unprocessed bran on colon function in normal subjects and in diverticular disease. Lancet 1:146–49.
60. Guthrie, H. A. 1989. Introductory nutrition. Boston: Times Mirror.
61. Frei, B. 1991. Ascorbic acid protects lipids in human plasma and low density lipoprotein against oxidative damage. Am. J. Clin. Nutr. 54:1113–18S.
62. Niki, E. 1991. Action of ascorbic acid as a scavenger of active and stable oxygen radicals. Am. J. Clin. Nutr. 54:1119–24S.
63. Packer, J. E., T. F. Slater, and R. L. Wilson. 1979. Direct observation of a free radical interaction between vitamin E and vitamin C. Nature 278:737–38.
64. Anderson, R., R. Oosthuizen, R. Maritz, A. Theron, and A. J. Van Rensburg. 1980. The effects of increasing weekly doses of ascorbate on certain cellular and humoral immune functions in normal volunteers. Am. J. Clin. Nutr. 33:71–76.
65. Block, G. 1991. Epidemiologic evidence regarding vitamin C and cancer. Am. J. Clin. Nutr. 54:1310–14S.
66. Robson, J. R. K. 1972. Malnutrition, its causation and control. Vols. 1 and 2. New York, NY: Gordon and Breach, Science Publishers.
67. Latham, M. C. 1967. Present knowledge of thiamin. In M. Hegsted, C. O. Chichester, W. J. Barby, K. W. McNutt, R. M. Stalvey, and W. H. Stotz, eds., Present knowledge in nutrition. New York, NY: Nutrition Foundation.
68. Katsura, E., and T. Oiso. 1976. Beriberi. In G. H. Beaton and J. M. Bengoa, eds., Nutrition in preventive medicine. Geneva: World Health Organization.
69. Vimokesant, S., N. Nimitmongkol, P. Phuwastein, S. Nakornchai, S. Sripojanart, S. Dhanamitta, and D. M. Hilker. 1973. In Proceedings of the IX International Congress of Nutrition. Basel: S. Karger.
70. Vimokesant, L., S. Nakornchai, S. Dhanamita, and D. M. Hilker. 1974. Effect of tea consumption on thiamin status in man. Nutr. Rep. Int. 9:371–76.
71. Tamphaichitr, V. 1976. Thiamin. In M. Hegsted, C. O. Chichester, W. J. Barby, K. W. McNutt, R. M. Stalvey, and W. H. Stotz, eds., Present knowledge in nutrition. Washington, DC: Nutrition Foundation.
72. Neel, J. V. 1956. Genetics of human hemoglobin differences. Ann. Hum. Genet. 21:1–30.
73. Livingstone, F. B. 1967. Abnormal hemoglobins in human populations. Chicago, IL: Aldine Publishing Co.
74. Raper, A. B. 1950. Sickle-cell disease in Africa and America: a comparison. J. Trop. Med. 53:49–53.
75. Motulsky, A. G. 1973. Frequency of sickling disorders in U.S. blacks. N. Engl. J. Med. 288:31–33.
76. Bernstein, R. E. 1973. Mass screens for sickle cell disease. J.A.M.A. 288:31–33.
77. Song, J. 1971. Pathology of sickle cell disease. Springfield, IL: Charles C. Thomas, Publisher.
78. Serjeant, G. R. 1973. Sickle cell anemia: Clinical features in adulthood and old

age. In H. Abramson, J. F. Bertles, and D. L. Wethers, eds., Sickle cell disease. St. Louis, MO: C. V. Mosby Co.

79. Lambotte, C. 1970. Disorders of the blood and reticuloendothelial system—sickle cell anemia. In D. B. Jelliffe, ed., Diseases of children in the subtropics and tropics. London: Arnold.
80. Houston, R. G. 1973. Sickle cell anemia and dietary precursors of cyanate. Am. J. Clin. Nutr. 26:1261–64.
81. Houston, R. G. 1975. Sickle cell anemia and Vitamin $B_{17}$ a preventive model. Am. Laboratory. 7:51–63.
82. Cerami, A. 1972. Cyanate as an inhibitor of red cell sickling. N. Engl. J. Med. 287:807–12.
83. Gillette, P. N., C. M. Peterson, J. M. Manning, and A. Cerami. 1972. Decrease in the hemolytic anemia of sickle cell disease after administration of sodium cyanate. J. Clin. Invest. 51:36.
84. May, A., A. J. Bellingham, and E. R. Huehns. 1972. Effect of cyanate on sickling. Lancet 1:658–60.
85. Manning, J. M., A. Cerami, P. N. Gillette, F. G. De Furia, and D. R. Miller. 1972. Chemical and biological aspects of the inhibition of red blood cell sickling by cyanate. Adv. Exp. Med. Biol. 28:253–60.
86. Oke, O. L. 1969. The role of hydrocyanic acid in nutrition. World Rev. Nutr. Diet. 11:170–98.
87. Krebs, E. T., Jr. 1970. The laetriles-nitrilosides—in the prevention and control of cancer. Sausalito, CA: McNaughton Foundation.
88. Thomas, H. M. 1972. Some aspects of food and nutrition in Sierra Leone. World Rev. Nutr. Diet. 14:48–58.
89. Jackson, F. L. C. 1990. Two evolutionary models for the interaction of dietary organic cyanogens, hemoglobins, and falciparum malaria. A. J. Human Biology. 2:521–32.
90. Nagel, R. L., C. Raventos, H. B. Tanowitz, and M. Wittner. 1980. Effect of sodium cyanate of plasmodium falciparum in vitro. J. Parasit. 66:483–87.
91. Tanner, J. M. 1978. Foetus into man: physical growth from conception to maturity. Cambridge, MA: Harvard University Press.
92. Frisancho, A. R., P. E. Cole, and J. E. Klayman, 1977. Greater contribution to secular trend among offspring of short parents. Hum. Biol. 49:51–60.
93. Frisancho, A. R. 1978. Nutritional influences on human growth and maturation. Yearbook Phys. Anthrop. 21: 174–91.
94. Wilson, P. W. 1990. High-density lipoprotein, low-density lipoprotein and coronary artery disease. Am. J. Cardiol. 66(6): 7A–10A.
95. Rifkind, B. M. 1990. High-density lipoprotein cholesterol and coronary artery disease: survey of the evidence. Am. J. Cardiol. 66(6): 3A–6A.
96. LaRosa, J. C. 1992. Cholesterol and cardiovascular disease: how strong is the evidence? Clin. Cardiol. 11: III2–7; discussion III8–9.
97. Grundy, S. M. 1990. Cholesterol and coronary heart disease: future directions. JAMA 264(23): 3053–159.
98. Schell, L. M. 1991. Effects of pollutants on human prenatal and postnatal growth: noise, lead, polychlorobiphenyl compounds and toxic wastes. Yearbook of Phys. Anthrop. 34:157–210.

CHAPTER 18

# Obesity: Accommodation and Adaptation to Variability in Dietary Supply

---

**Definition and Measurements of Obesity**
- Weight-for-Height and Frame Size
- Body Mass Index
- Skinfold Thickness
- Types of Obesity
- Health Risk of Obesity

**Genetic Origins of Obesity**
- Evidence of Genetic Influence on Fatness

**Sociocultural Factors and Obesity**
- Preference for Fatness
- Acculturation and Decreased Activity
- Acculturation and Dietary Intake
- Socioeconomic Status and Fatness

**History of Dietary Restriction and Obesity**
- Experimental Studies in Animals
- Survivors of Famine
- Supplementation Studies
- Energy Expenditure during Exercise
- Uncertainty of Food Supply and Binge Eating

**Metabolic and Spontaneous Activities**
- Spontaneous Activities
- Metabolic Rate
- Diet-Induced Thermogenesis

**Overview**

Obesity, however defined, represents an excess accumulation of calories, and it occurs when the caloric intake exceeds energy expenditure. Although obesity has emerged as a major health concern in industrialized countries it is simultaneously one of the most important and one of the least understood health problems. It is a truism that obesity like all phenotypic traits is the result of the interaction of biological, behavioral, and cultural factors. The expression of this interaction becomes evident during development or during adulthood.

Hence, the purpose of this chapter is to discuss the evolutionary origins of variability in human fatness, the sociocultural factors that contribute to the expression of obesity, and the role of dietary restriction in the development of obesity. For clarity, however, first the various indices and measurements for defining obesity and the usual classifications of obesity will be briefly described.

## DEFINITION OF AND THE CLINICAL RISK OF OBESITY

Obesity is commonly defined with reference to: (i) weight-for-height tables, (ii) body mass index, and (iii) skinfold thickness.

### Weight-for-Height and Frame Size

Many investigators define obesity as weight above 20% of the weight-for-height Metropolitan Life Insurance tables. Despite their widespread usage evaluations of obesity based on weight alone are very imprecise because they do not provide specific information about body composition. Furthermore, the frame size categories given by the Metropolitan Life Insurance tables were not based on measurements of skeletal dimensions. To overcome this difficulty, elbow breadth measurements were made in order to establish new age- and gender-specific weight by frame size indices (1).

### Body Mass Index (BMI)

The BMI is calculated by dividing the weight kilograms (kg) by stature (m) squared ($m^2$), thus, $BMI = kg/m^2$. A BMI of 19–25 is considered a good weight for most people, and a BMI greater than 27 is associated with increasing risk (high) risk of developing health problems (2). While this approach is widely used, a BMI cut off is inappropriate when applied to individuals older than 40 years. As shown in figure 18.1 weight increases until about the fifth decade in males and sixth decade in females, while stature in both males and females

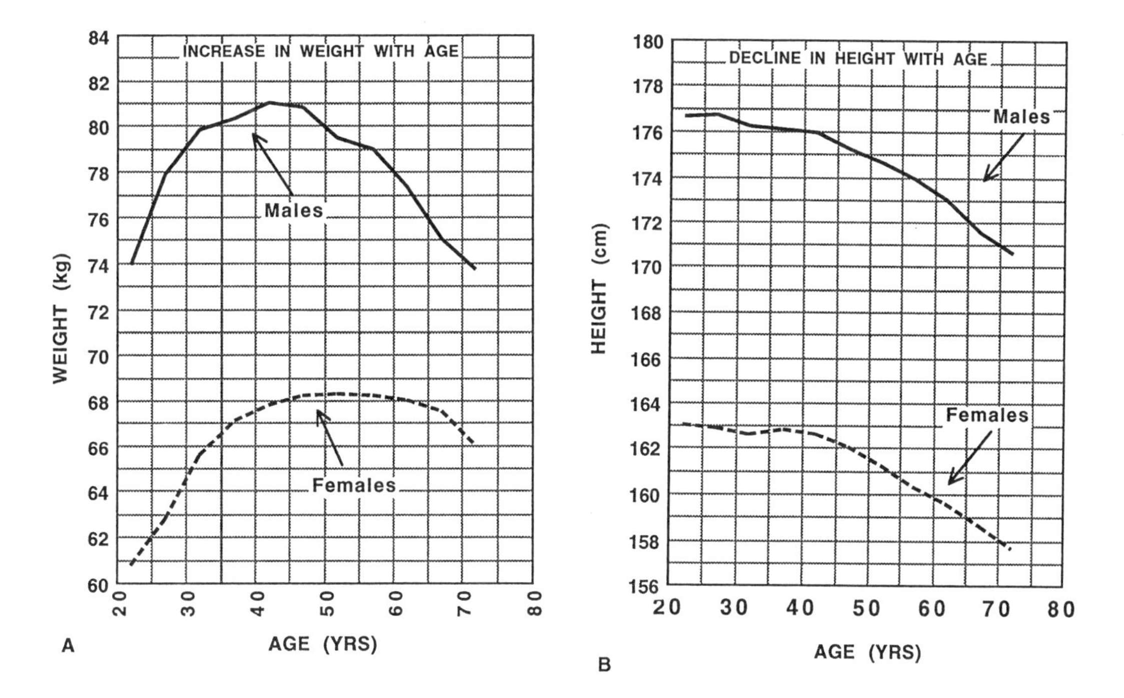


**Fig. 18.1.** Changes in mean weight and height with age among adults. Note that weight increases until about the fifth decade in males and sixth decade in females, while height begins to decline after the age of forty years in both males and females. For this reason, evaluations of obesity on body mass index need to be done with reference to age and gender. (Adapted from A. R. Frisancho. 1990. Anthropometric standards for the assessment of growth and nutritional status. Ann Arbor: University of Michigan Press.)

after the age of 40 years declines at a rate of about 1.75 cm/decade. For example, a 30-year-old male weighing 81 kg and 175 cm tall would have a BMI of 26.5 kg/m$^2$, which would place him in the normal range, but by the age of 60 years even if he maintained the same weight he would be classified as obese because his height would have declined to 171.5 cm, thereby giving him a BMI of 27.5 ($81/1.715^2$). For this reason, an appropriate definition of obesity must be done with reference to age- and gender-specific BMI values (1).

## Skinfold Thickness

Even when correcting for age and gender difference the BMI does not differentiate the components that contribute to the excess in weight. A productive way to obtain information about body composition is to measure skinfold thickness at different body sites and compare them to recent standards. Obesity, as given in recent anthropometric standards (1) is defined when the sum of skinfold thicknesses is above the age- and sex-specific 85th percentile of the sum of triceps and subscapular skinfold thicknesses (see fig. 18.2).

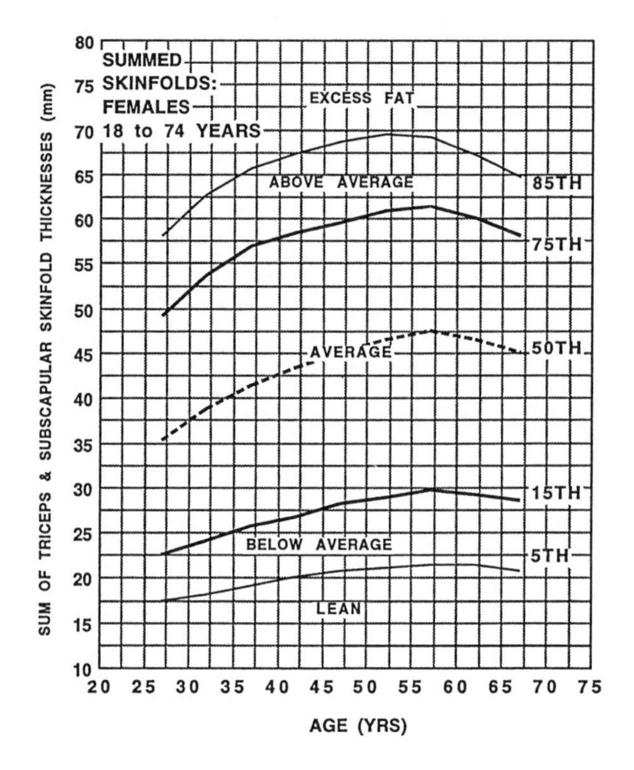


**Fig. 18.2.** Percentiles of sum of triceps and subscapular skinfold thicknesses for females ranging in age from 18 to 74 years. (Adapted from A. R. Frisancho. 1990. Anthropometric standards for the assessment of growth and nutritional status. Ann Arbor: University of Michigan Press.)

## Types of Obesity

Ever since Vague (3) noted that specific risk factors, especially those associated with diabetes differed depending on regional fat distribution great attention has been given to studying the relationship between regional fat distribution and clinical manifestations of cardiovascular function. Two types of fat distribution are recognized (see fig. 18.3).

### Central

This type of obesity is characterized by accumulation of body fat around the trunk. It is also referred to as upper body obesity, android type obesity, or apple shape obesity. It is more prevalent in men than in women.

### Peripheral

Peripheral obesity is characterized by greater amount of fat on the extremities and around the hip or femoral region, but it may also be distributed, to some

**Fig. 18.3.** Types of obesity according to distribution of body fat.

extent, in the arms. It is referred to as lower body obesity, gynecoid, pear shape obesity. Lower body obesity is more prevalent in women than in men.

Fat distribution can be established by using skinfold thickness measurements (4, 5):

(i) Trunk-Extremity Index = [Subscapular + Suprailiac Skinfold, mm) ÷ (Triceps + Calf Skinfold, mm)]
(ii) Extremity-Trunk Index = [(Triceps + Calf Skinfold, mm) ÷ (Subscapular + Suprailiac Skinfold, mm)]
(iii) Upper Lower Body Index = [(Subscapular + Triceps Skinfold, mm) ÷ (Calf Skinfold, mm)]

Fat distribution can also be determined by the waist-to-hip ratio (W:H). This technique is based on measurements of the circumferences of the waist and the hip and computing the waist-to-hip (W:H) ratio. Based on this information if the W:H ratio is less than 1 (i.e., the waist circumference is smaller than the circumference of the hip as it occurs in women), the distribution corresponds to peripheral or gynecoid fat distribution. In central or android (also called abdominal) fat distribution the W:H ratio is greater than 1 (i.e., the waist circumference is greater than the hip circumference). A recent analysis of anthropometric data indicate that the waist-to-thigh circumference ratio is a better indicator of fat distribution than the waist-to-hip circumference ratio (6).

### Health Risk of Obesity

No matter how obesity is defined the most common risk factors associated with obesity include: (i) elevation of very low-density lipoprotein (VLDL), triglycerides and low-density lipoprotein (LDL) cholesterol, and low concentrations of high-density lipoprotein (HDL) cholesterol; (ii) increase in the risk of developing cholesterol gallstones; (iii) increase in the risk of developing non-insulin-dependent diabetes mellitus (NIDDM); (iv) increase in hepatic secretion of glucose; and (v) increase in the risk for developing hypertension. Although obesity is associated with the above risk factors the mechanisms whereby it enhances disease risk are not fully understood. It should be noted that because many obese people do not have the above clinical risk factors, obesity per se does not produce a clinical disease. Therefore, obesity increases the risk only in individuals who already possess a metabolic weakness or ''defect'' in a given system (7) and in the absence of such underlying defects, obesity is well tolerated.

## GENETIC ORIGINS OF OBESITY

In an attempt to explain the increased incidence of diabetes mellitus, observed among traditional populations undergoing economic modernization such as the Pima Papago of the United States, Neel (8) postulated the ''thrifty gene'' hypothesis. As restated recently (9) it is hypothesized that during human evolution when periods of food abundance alternated with periods of near famine individuals who had the ability to remove glucose from the blood and convert it into fat had an advantage. This ability was made possible by a ''quick insulin trigger.'' The evolutionary advantage accruing to individuals with the ''quick insulin trigger'' was a more efficient utilization of food stuffs during periods of feast or famine. The ability to rapidly secrete insulin was an advantage because it allowed the rapid removal of energy from the blood and as such it helped ensure maximum food utilization. With the advent of refined carbohydrates and abundant and steady food supply, the high insulin level resulted in an overstimulation of the cellular insulin receptors, ultimately resulting in insulin resistance. In time the overworked $\beta$-cells lose their capacity to respond and the result is diabetes mellitus. When humans lived as hunter-gatherers food supplies fluctuated and activity levels were high. Thus, diabetes mellitus was not expressed. But as populations became economically modernized, the availability of energy increased and the energy expenditure in physical activity decreased, making diabetes mellitus much more common.

Although there is no evidence to prove or disprove the validity of the "thrifty genotype" of diabetes mellitus this hypothesis can be extended to explain the increased incidence of obesity observed in contemporary populations. Most contemporary hunter-gatherers and agriculturalists are exposed to seasonal availability of food (10–16). In fact, food shortages were always present in most economically nonmodernized societies (123). For example, of 118 nonindustrial societies (with hunting and gathering, pastoral, horticultural, and agricultural economies) nearly all of them experienced some form of food shortages. Roughly 50% of them experienced food shortages every year or every 2 to 3 years, and 29% experienced starvation and famine (17). Furthermore, the development of agriculture led to the emergence of social stratification and unequal access to food and nutritional resources (18). This inequality was accentuated during the famines that occurred during the frequent drought periods and that continue to occur in many societies today. Under conditions of food shortages, selection probably favored those individuals who could store calories in the form of fat during the times of surplus. Given the energetic cost of pregnancy and lactation, females with greater energy reserves in fat would have a selective advantage over their lean counterparts. On the other hand, under conditions where there is an abundance of food stuffs and low physical activity, it would result in excessive availability of energy and express itself as obesity. Thus, the emerging high frequency of obesity observed among traditional populations undergoing economic modernization could also be seen as a result of the "thrifty genotype" activated by the cultural and socioeconomic conditions brought about by industrial economies.

## Evidence of Genetic Influence on Fatness

Variability in body fat, like all anthropometric dimensions, is inherited in an additive and polygenic nature. Research addressed at quantifying the genetic components has been based on analysis of correlation coefficients and estimates of heritability in body weight, body mass index, and direct measures of subcutaneous fat and estimates of fat mass of twins, of parents and offspring, and siblings exposed to similar and different environmental conditions.

### Twins

**Body Mass Index.** A study of a sample of adult Danish adoptees found that the relation of the weight and the body mass index class of the adoptees and their biological parents was greater than those observed among foster parents and their adopted children (19). This finding is corroborated by a recent study of a sample of monozygotic and dizygotic twins from Sweden raised together and raised apart (20). This study showed that the parent-offspring correlation in body mass index was greater for monozygotic twins reared either apart or together ($r = 0.70$, $0.74$) than for dizygotic twins raised apart and raised together ($r = 0.15$, $0.33$). The difference between these two coefficients indicates a very strong genetic effect on body mass index. However, because the data are derived from twins they do not necessarily apply to the population at large.

**Skinfolds and Fat Mass.** As shown in table 18.1 correlations of sum of skinfolds and fat mass for monozygotic twins are nearly twice as much as that of dizygotic twins from Quebec, Canada (21).

**Overfeeding and Fat Gain.** Experimental studies have provided conclusive evidence indicating the role of genetic factors in fat gain and fat distribution. Experimental studies of 6 pairs of male monozygotic twins overfed by 1,000 kcal/day for 22 days (22) found a high within-pair resemblance for absolute changes in body weight and body fat. A similar finding has been reported in a long-term study (84 days) of 12 pairs of male identical twins overfed 1000 kcal/day (23). This study like the previous one found a high intrapair similarity in fat gain and fat distribution indicating the role of genetic factors (23).

## Parent-Offspring Similarities

**Body Mass Index.** Analysis of the data of the French Canadian samples (5) indicate that the parent-child correlations for biological offspring ($r = 0.23$) was similar to that of adopted children ($r = 0.22$), but adopted siblings showed less resemblance ($r = 0.08$) than biologically related siblings ($r = 0.26$). Analysis

**TABLE 18.1. Correlation coefficients for body composition for monozygotic (MZ) and dizygotic (DZ) twins living together**

| | Twins | |
|---|---|---|
| | MZ ($N = 87$ pairs) | DZ ($N = 69$ pairs) |
| Variable | r | r |
| BMI | 0.88 | 0.34 |
| Sum of skinfolds | 0.83 | 0.39 |
| % body fat | 0.73 | 0.21 |
| Fat mass | 0.76 | 0.31 |
| Fat-free mass | 0.93 | 0.53 |
| Distribution of fat | 0.80 | 0.40 |

*Source:* From C. Bouchard, R. Savard, J. P. Depres, A. Tremblay, and C. Leblanc. 1985. Body composition in adopted and biological siblings. Hum. Biol. 457:61–75.

of data from the Tecumseh population have found that the body mass index has a genetic heritability of about 30% (24). This finding has been confirmed by analysis of commingling of distributions of scores for body mass index (25). Similarly, analyses of anthropometric data derived from an Italian (26, 27) and a Norwegian sample (28) indicate that the body mass index has a heritability of about 30% to 40%. On the other hand, evaluations of data of more than 1,000 individuals participating in the Stanford Lipid Research Clinic Family Study indicate that the weight and the body mass index were mostly under environmental control (29).

**Skinfolds and Fat Mass.** Analysis of data sets derived from the Tecumseh population indicated that the parent-offspring correlations for the triceps skinfolds (30, 31) was 0.24 for the biological children, 0.24 for the singly adopted children (one of the parents is the biological parent), and 0.12 for the double adopted children (neither parent is the biological parent) (see table 18.2). The lower correlation in the double adopted children suggests the genetic influence on body fatness (32). Similarly, studies of a relatively large number of adoptive children from Quebec found that the parent-offspring correlations in fat mass were greater among biological offspring than among adoptees (21) (see table 18.3). On the other hand, analyses of a large sample

**TABLE 18.2. Parent-offspring correlations in skinfold thicknesses among biological and adoptive children**

| | Parent-Offspring | | |
|---|---|---|---|
| | | Adopted | |
| Skinfolds | Biological r (*N* = 6,234) | Singly[a] r (*N* = 155) | Doubly[b] r (*N* = 160) |
| Triceps | 0.21 | 0.23 | 0.11 |
| Subscapular | 0.19 | 0.21 | 0.09 |

*Source:* From S. M. Garn, P. E. Cole, and S. M. Bailey. 1979. Living together as a factor in family-line resemblances. H. Biol. 51:565–87.

[a] Singly adopted—one of the parents is the biological parent

[b] Neither parent is the biological parent

**TABLE 18.3. Parent-offspring correlation coefficients in body composition for biological and adopted children**

| | Parent-Offspring | |
|---|---|---|
| Variable | Biological r (*N* = 531) | Adopted r (*N* = 252) |
| % body fat | 0.23 | 0.13 |
| Fat mass | 0.22 | 0.16 |
| Fat-free mass | 0.24 | 0.06 |
| Distribution of fat | 0.31 | 0.20 |

*Source:* C. Bouchard, R. Savard, J. P. Depres, A. Tremblay, and C. Leblanc. 1985. Body composition in adopted and biological siblings. Hum. Biol. 457:61–75.

of nuclear families from Canada indicate that the sum of 5 skinfolds had a transmissibility ($t^2$) of only 0.37 (33). This means that environmental factors accounted for most of the variance in skinfolds.

In summary, although there is great variability between studies, the estimates of heritability of body mass and body fat are generally greater among identical twins than among dizygotic twins and also among biological offspring are greater than among adopted children. Furthermore, as inferred from studies of twins it appears that the sensitivity of individuals to gain fat when chronically exposed to positive energy balance (overfeeding) is subject to genotype-environment interaction effect.

## SOCIOCULTURAL FACTORS AND OBESITY

### Preference for Fatness

Since the paleolithic age, as shown by the so-called Venus figurines, humans have had a preference for fatness. This is evident among contemporary societies, in renaissance and classic art, and in the contemporary vocabulary.

**Ethnographic Evidence.** Ethnographic analyses indicate that out of 300 anthropological societies, 81% considered as desirable and attractive a woman whose characteristics included being ''plump'' or being ''filled out'' (34). Because fatness was the primary criterion of beauty traditional marriage preparation of Nigerian girls included up to 2 years of seclusion in ''fattening huts'' (35). Similarly, among the Havasupai of the American Southwest and the Tarahumara of northern Mexico fat thighs and legs were considered essential to beauty (36, 37). Even among contemporary populations such as the Kipsigis from Kenya, fatter brides (as well as early-maturing brides) demand signficantly higher bridewealth payments than do their leaner, late-blooming peers (38). In males from the Massas increased fatness is also associated with prestige. The Massas from Northern Cameroon frequently participate in the ''Guru Walla'' fattening tradition aimed at prestige acquisition, which includes a 2-month overfeeding representing twice or three fold the habitual daily intake (39, 40).

**Art.** In economically modernized societies, as shown by the paintings of the *Madona* of Goya (1746–1828), the *Joy of Life* of Matisse (1869–1954), *Mother and Child* of Picasso (1881–1974), etc., fatness in women and children was considered attractive.

**Contemporary Vocabulary.** Prior to the eradication of tuberculosis, fatness was considered an indicator of good health in the United States and fat babies were usually considered healthy. It is only in the last few decades that fat has not been viewed as desirable in the United States. However, in many other populations the term *fat* is still used in a positive form and may be interpreted as a sign of health and prosperity. In fact, the term *fat cat,* which is associated with wealth and power, is said to be a vestige of the idea that fatness is associated with success (41). Similarly, among South American populations the term *gordo* or *gorda* is used in a friendly fashion and studies of Puerto Ricans in the United States indicate that being fat does not have the social stigma of obesity given in other populations (42). For this reason, Mexican-Americans use the term of *gordura mala* (bad fatness) when describing fatness that is associated with a negative health risk (43). The San Antonio Heart study (42, 45) and analyses of the anthropometric data of the Hispanic Health and Nutritional Examination Survey (46) indicate that among Mexican-Americans the maintenance of traditional cultural values was associated with an increase in the incidence of obesity. This finding suggests that levels of fatness in Mexican-Americans, especially women, may reflect cultural preferences toward fatness.

### Acculturation and Decreased Activity

In general the prevalance of sedentary activity is high among traditional populations undergoing economic modernization. Comparative studies indicate that hunter-gatherers and agriculturalists have a greater work capacity than urbanized populations. For example, studies of work capacity indicate a 19% reduction in maximal aerobic capacity between the serious hunters ($VO_2$ max = 60.7 ml/kg/min) and the occasional Igloolik hunters ($VO_2$ max = 38.9 ml/kg/min) from Canada (47). Similarly, the urbanized Tarahuma Indians from Mexico had a 38.3% reduction in aerobic capacity ($VO_2$ max = 38.9 ml/kg/min) when compared to their traditional counterparts ($VO_2$ max = 63.0 ml/kg/min) (48). The decrease in work capacity is generally attributed to a decrease in overall activity pattern, as a consequence of reduction in the aerobic requirements or work and/or leisure activities (49, 50). The modification of activity patterns also has resulted in an excess energy storage in the form of fat. For example, the reduction (38.3%) in aerobic capacity of the urbanized Tarahuma Indians was associated with a 61.3% increase in body fat.

These studies indicate that intervention to reduce obesity among minorities and traditional populations undergoing economic modernization must take into account the role of physical activity.

### Acculturation and Dietary Intake

**Trends in Dietary Patterns.** From a nutritional standpoint the most important changes associated with westernization of dietary habits include increased consumption of animal protein and fat and a marked decrease in the intake of fiber, particularly cereal fiber. Ever since humans became hunter-gatherers their diet has been based on a mixture of animal foods and plants.

Although contemporary hunter-gatherers derive as much energy from animal food as westernized populations, they consume a proportionately lesser amount of fat because game animals are leaner than domestic ones consumed by traditional populations undergoing economic modernization. The average American diet derives about 16% of the total energy intake from protein, 44% from carbohydrate, and 40% from fat of which 40–50% is saturated fat. The saturated fats are derived directly from animal fats and dairy products, and indirectly from manufactured foods containing these fats, and from industrially hydrogenated vegetable oils. At present because obese individuals do not have a higher energy intake than nonobese (51) research has centered on determining the role of specific dietary constituents such as fat on obesity.

**Fat Intake and Obesity.** Experimental animal studies indicate that diet composition and particularly dietary fat intake is related with increased adiposity even when energy intake is controlled (52–55). Similarly, human correlational studies report that dietery fat intake is associated with increased fatness (59–59). The reasons for these associations have not been defined. It has been postulated that these associations are related to food selection. Short-term experiments indicate that obese subjects compared to normal individuals tend to show a preference for high-fat and high-lipid mixtures (60) and high-lipid content (61, 62). The association between fat intake and adiposity is probably related to the fact that dietary fat is stored in the body more efficiently than carboydrate (63). This study has demonstrated that dietery fat can join the triglyceride fat stores at a metabolic cost of only 3% of energy intake while the cost of storing carbohydrate as body fat requires the expenditure of 23% of ingested energy. Similarly, experimental studies in humans indicate that high fat diets are more efficiently used than low-fat diets relative to body fat (120). In other words it appears that dietary fat is more likley to be stored in the adipose tissue while carbohydrate is stored as fat in the adipose tissue only when there is an excess of energy and to convert the excess carbohydrate into fat requires more than that of the dietary fat.

## Socioeconomic Status and Fatness

Comparative studies indicate that variability in skinfold thicknesses in males is curvilinearly associated with socioeconomic status but is inversely associated with socioeconomic status in females (30, 31). Thus, upper-class males (depending on educational status) are fatter than lower-class males, but high socioeconomic status females are leaner than their low socioeconomic status counterparts. Investigators (64–70) have also reported that the frequency of obesity is higher in the lower social classes. On the other hand, in the developing nations where the availability of energy is lower, the frequency of

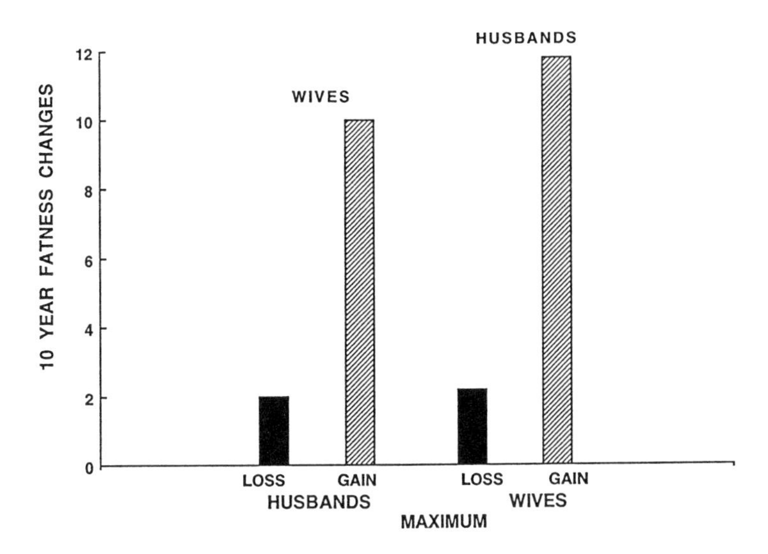


**Fig. 18.4.** Husband and wife similarities in gain and loss of fat during 10 years of cohabitation. (Adapted from S. M. Garn, P. E. Cole, and S. M. Bailey. 1979. Living together as a factor in family-line resemblances. H. Biol. 51:565–87.)

obesity is inversely related to socioeconomic status, i.e., the frequency of obesity if greater in the upper social classes (71–72).

## Husband-Wife Similarities in Fatness

In several publications based on analyses of patterns of likeness between married couples it has been demonstrated that with advancing number of years of marriage the couples' similarities in body fat increase (30, 31, 73–74). Furthermore, variability in body fat is profoundly influenced by the educational level of the spouse (74, 75) so that women with a high level of education married to men with a low level of education are fatter, while women with a high level of education married to men of the same level of education are leaner (see fig. 18.4). The correlation between spouses in fatness change is about 0.26, which means that about 7% of the variance in fatness change in one spouse is explained by the change in the other spouse (31, 30). It would thus appear that environmental factors as represented by socioeconomic status (SES) play a moderate role in spousal-likeness, and as such, the environment as represented by SES plays a moderate role in the expression of obesity.

## Developmental Experience with U.S. Economy

Several investigations have pointed out that minority populations such as the Samoans (76, 77), Pima Indians (78), North American Indians (79–81), and

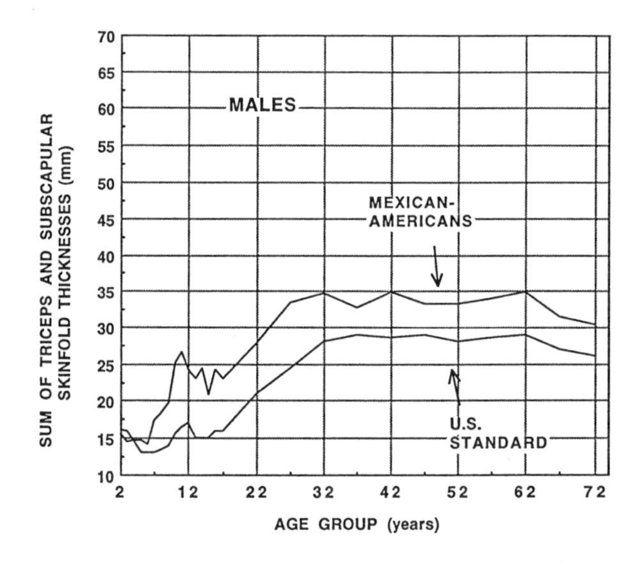


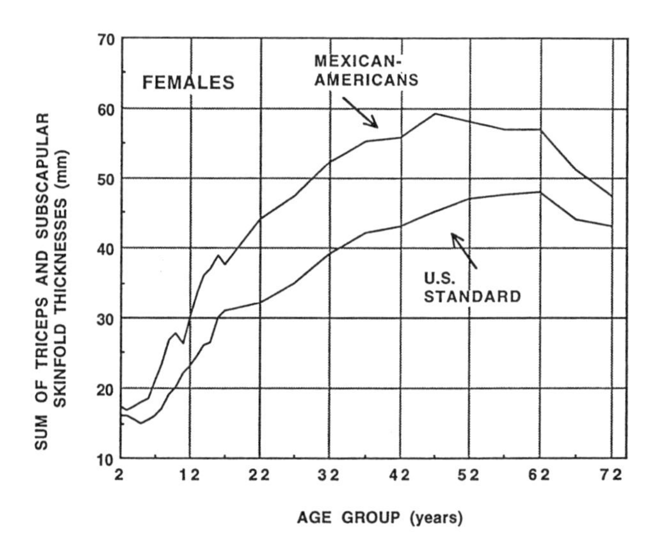


**Fig. 18.5.** Sum of triceps and subscapular skinfold thicknesses of Mexican-Americans. At every age and throughout adulthood the Mexican-Americans are significantly fatter than a U.S. reference.

Mexican-Americans (46, 82–86) have a high incidence of obesity (see fig. 18.5). The increased tendency toward obesity is occurring even in populations that in previous years were not considered to be at risk of obesity such as the Hmong children from Southeast Asia living in Minneapolis (87). A common denominator of these populations is that they are undergoing the process of economic modernization and a large proportion of them live under conditions of poverty.

In summary, there is abundant evidence indicating that environment plays an important role in the development of obesity. The persistence of the desirability of fatness among past and contemporary populations indicate that cultural factors did, and still do, contribute to the expression of the genetic susceptibility for obesity. The decreased energy expenditure in physical activity brought about by economic modernization along with an increased appetite and availability of foods rich in fats probably contributes toward the increased fatness seen among populations from developed countries and traditional populations undergoing economic modernization.

## HISTORY OF DIETARY RESTRICTION AND OBESITY

Another factor that is rarely considered is the role of dietary restriction in the etiology of obesity. In this section evidence from studies in animals and humans that had a history of dietary restriction will be discussed.

### Experimental Studies in Animals

Experimental studies with rats indicate that male offspring of rats that were food restricted during the first 2 weeks of pregnancy and then allowed to eat ad libitum (88–90), became obese by 5 weeks of age (see fig. 18.6). Studies of pigs who were growth retarded and protein-deficient during weaning found they became quite fat later in life (91). Similar investigations in rats who

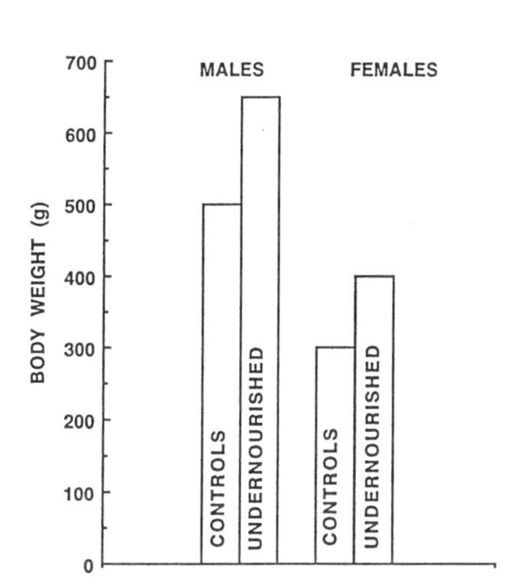


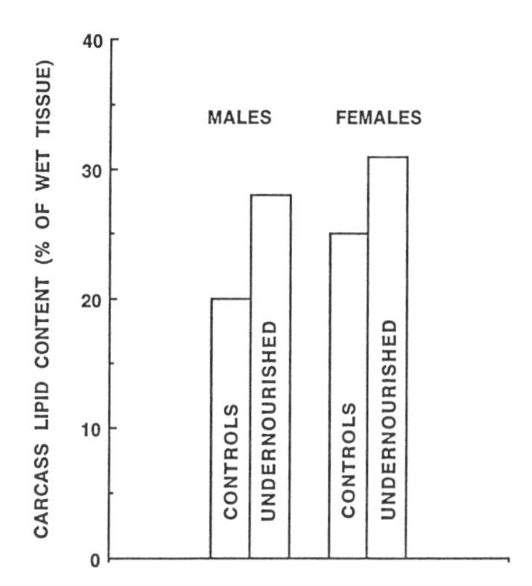


**Fig. 18.6.** Comparison of obesity in rats undernourished during pregnancy. Rats undernourished during pregnancy and allowed to eat ad libitum are fatter than controlled rats. (Adapted from A. P. Jones and M. I. Friedman. 1982. Obesity and adiposyte abnormalities in offspring of rats undernourished during pregnancy. Science 215:1518–19, and from A. P. Jones, E. L. Simson, and M. I. Friedman. 1984. Gestational undernutrition and the development of obesity in rats. J. Nutr. 114: 1484–92.)

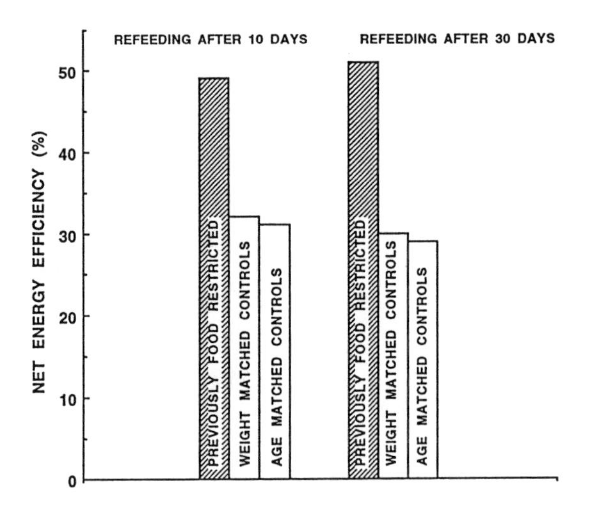


**Fig. 18.7.** Changes in net energy efficiency following caloric deprivation in rats. After thirty days of caloric deprivation the utilization of energy is significantly more efficient than in those not exposed to undernutrition. (Adapted from A. G. Dulloo, and L. Girardier. 1990. Adaptive changes in energy expenditure during refeeding following low-calorie intake: Evidence for a specific metabolic component favoring fat storage. Am. J. Clin. Nutr. 52:415–20.)

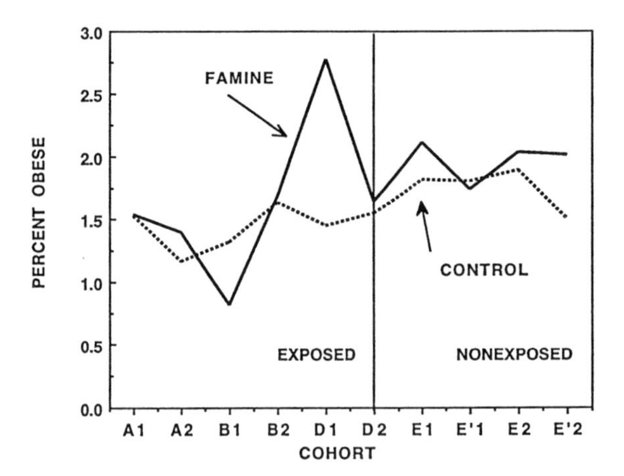


**Fig. 18.8.** Comparison of the prevalence of obesity among the survivors of the Dutch famine during World War II. The frequency of obesity during the first two trimesters of pregnancy is greater than those in the general population. (Adapted from G. Ravelli, Z. A. Stein, and M. W. Susser. 1976. Obesity in young men after famine exposure in utero and early infancy. N. Eng. J. Med. 295:350–53.)

were food restricted showed an elevated efficiency of energy utilization and a preferential accumulation of fat (92) (see fig. 18.7). The present findings suggest that decreased prenatal dietary availability is an important predisposing factor for the accumulation of excess body fat later in life. Whether this change is the consequence of the activation of energy-conserving mechanisms and improvement in the efficiency of energy utilization (92) remains to be determined.

### Survivors of Famine

Analyses of survivors of the Dutch famine of World War II found that the Dutch army draftees whose mothers had been deprived of food in the first 2 trimesters of pregnancy had a greater incidence of obesity than did the general population (93) (see fig. 18.8). These differences persisted when the sample was divided by occupation.

### Supplementation Studies

Supplementation studies conducted on growth retarded children (with an average age of 9.0 years) from Bundi, Papua-New Guinea (94) indicate that

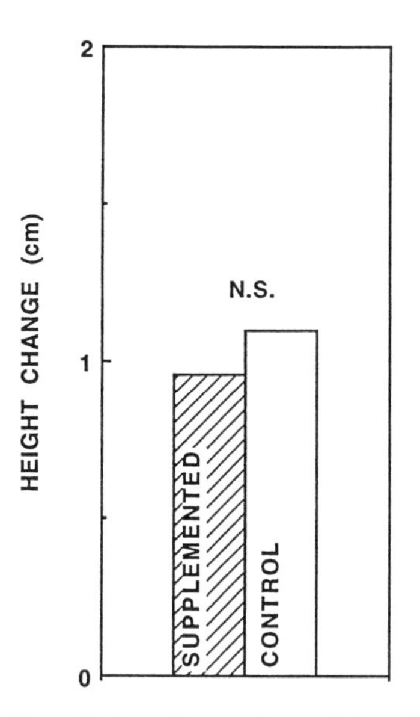


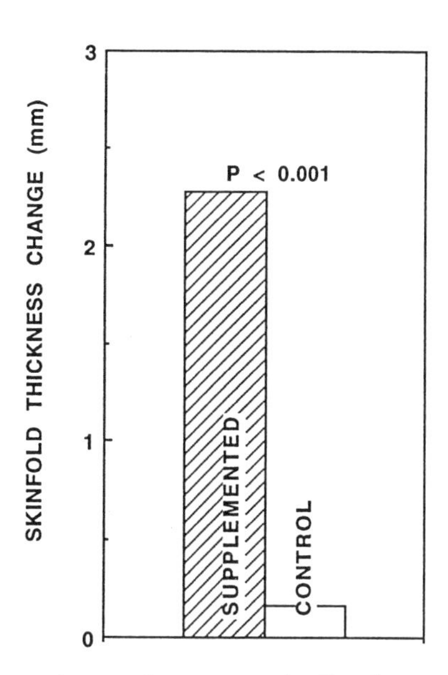


**Fig. 18.9.** Comparison of the effects of supplementation among the Bundi children from Papua New Guinea. Note that supplementation had no effect on height but had a significant effect on the increase in fatness. (Adapted from L. A. Malcolm. 1970. Growth and Development in New Guinea—a study of the Bundi people of the Madang district. Institute of Human Biology; Papua-New Guinea. Monograph Series No. 1. Australia: Surrey Beatty and Sons.)

protein and calorie supplementation for a 13-week period did not have a major effect on growth in body length. In contrast, supplementation did have a significant effect on increasing subscapular skinfold thickness (see fig. 18.9). These data would suggest that both protein and energy provided some extra energy which was used for fat deposition rather than growth. This conclusion is supported by recent studies conducted among Jamaican children. In these studies nutritional supplementation (calories and protein) of growth retarded children (with an average age of 18.5 months) for one year did not have a major effect on growth in length or mid-upper arm circumference (95), which is an indicator of body muscle mass. In contrast, supplementation did have a significant effect on increasing skinfold thickness.

These findings suggest that among previously malnourished children the extra energy derived from nutritional supplementation tends to be reflected in an increased propensity to accumulate fat; whether this is due to activation of energy-sparing mechanisms remains to be determined.

## Energy Expenditure during Exercise (EEE)

Studies of participants of weight loss programs indicates that the organism when challenged with dietary restriction responds with reduction of energy expenditure during exercise.

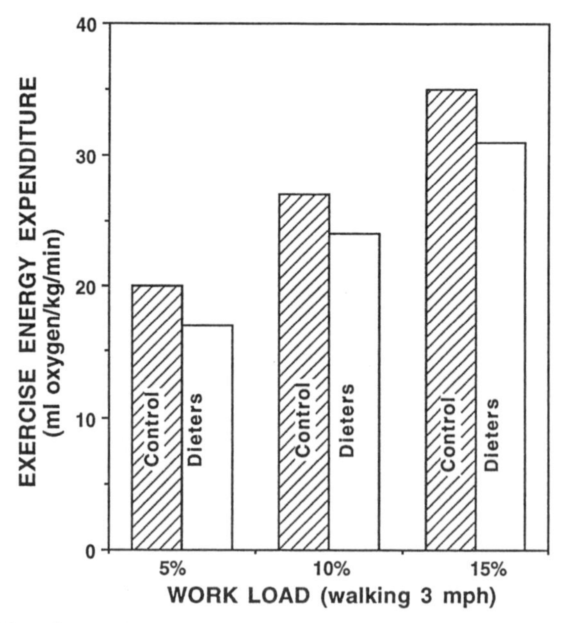


**Fig. 18.10.** Comparison of exercise energy expenditure between cyclic dieters and controls. Cyclic dieters expend less energy during exercise than controls. (Adapted from M. M. Manore, T. E. Berry, J. S. Skinner, and S. S. Carroll. 1991. Energy expenditure at rest and during exercise in nonobese female cyclical dieters and in nondieting control subjects. Am. J. Clin. Nutr. 54:41–46.)

**After Weight Loss.** After losing weight during a 10-week weight reduction program, the energy expenditure during exercise (EEE) of obese subjects is decreased by 15% (96). Furthermore, the reduction in EEE was greater than can be accounted for by the change in body weight and body composition. Likewise, the reduction in resting metabolic rate among obese women enrolled in a weight-loss program was greater than could be accounted for by the loss of fat-free mass (97). It has also been reported (98) that obese subjects who were obese since childhood (childhood-onset obesity), expend less energy per unit of fat-free mass than subjects who became obese during adulthood (adult-onset obesity). These findings together suggest that, in order to maintain their body weight, obese individuals utilize their body energy more efficiently than lean subjects (96) and if obesity begins early in life it may be associated with lesser energy expenditure than in normal-weight individuals.

**Cyclic Dieters.** The hypothesis that dietary restriction is associated with decreased EEE is supported by studies of cyclical dieters (99). Investigations of nonobese cyclical dieters (having dieted for ≥7–10 days four times/year) indicate that cyclical dieters when they exercised on a treadmill had a significantly lower energy expenditure (per unit of body weight) compared to controls with a similar calorie intake (99) (see fig. 18.10). Accordingly, the authors suggest that nonobese chronic dieters with similar lean body weights (but more body fat) use less energy to perform their daily activities than do nondieters (99).

## Uncertainty of Food Supply and Binge Eating

It has been postulated that under-eating predisposes to binge eating (100) and as such it contributes to the increased tendency of cyclic dieters to gain weight. This hypothesis is based upon the observation that in the Minnesota starvation experiment (101) after cessation of the semistarvation regimen some of the subjects exhibited a disproportionate increase in appetite and gained approximately 5% more weight than before the experiment. Ethnographic observations of hunter-gatherers indicate uncertainty of food supply is associated with binge eating. For example, among the nomads of eastern Bolivia, whose supply of food was rarely abundant and always insecure, when food becomes available the hunters go on a binge and gorge themselves consuming during 24 hours as much as 30 pounds of peccary meat (102). Likewise, once large game is obtained, the Ache hunter-gatherers from Paraguay will go on a binge and gorge themselves to the extent that they will lie down and go to sleep (12). The same response occurs in wild animals. It is well known that wolves eat several times a day (and rarely gorge) during the summer when an ample supply of small animals is available. However, when less game is available in the winter, they will gorge themselves after a kill of a large animal (100, 102). Within this context the increased appetite or so-called *delayed hunger* that is exhibited by many South American impoverished individuals that are exposed to positive conditions probably represents a compensatory adaptation to the possibility of uncertainty of food supply that is likely to occur under conditions of poverty.

In summary, the available experimental and epidemiological evidence suggests that, after exposure to dietary restriction there is an increased tendency to accumulate fat. In other words, individuals who had a past experience with dietary restriction are more likely to become fatter than before dieting or participating in weight loss programs.

## METABOLIC AND SPONTANEOUS ACTIVITIES

The tendency to gain fat may also be the result of differences in energy expenditure in spontaneous activities, metabolic rate, and thermic effect of food.

### Spontaneous Activities

Various investigations have noted that obese individuals have a greater economy of movement than nonobese. Observations of teenage high school girls playing volleyball together indicated that the obese girls exhibited less arm and leg movement than the lean ones (103). Recent studies done in calorimetry chambers, which allow for 24-hour observations of all activities, have shown that obese subjects have significantly fewer spontaneous activities (moving the arms, feet, legs, etc.) than lean individuals. The difference in energy expenditure amounted to as much as 1,000 kcal/day (104). Given the large individual variability in spontaneous (''fidgeting'') activity it is quite possible that energy spent in nonobligatory activities represents an important portion of the total energy expenditure and as such a decrease in this aspect may result in considerable energy saving.

### Metabolic Rate

An obvious assumption is that obesity is associated with a decrease in metabolic rate. However, most studies report that the obese have higher absolute metabolic rates than the nonobese (104, 105, 107–10, 121). This increase is related to the fact that obesity is usually associated with increased amounts of both fat mass and fat free mass. Nevertheless, it should be noted, that not all obese have increased metabolic rates. For example, in most studies between 10 to 15% of the obese males and between 10 to 30% of obese females have a metabolic rate and total energy expenditure that are below the average values seen in their lean counterparts. This implies that a proportion of obese individuals have a metabolic rate that is lower than is expected for their body size. Therefore, it is quite possible that the effect of dietary intake on metabolic rate and hence on body composition is not the same for the obese and the nonobese.

### Diet-Induced Thermogenesis (DIT)

Earlier investigations postulated that variability in diet-induced thermogenesis (DIT) is an important factor that contributes to variability in body fat. Some investigators have shown that the thermic response to food is identical for the obese and the nonobese (105–7). On the other hand, several investigators have demonstrated that obesity is associated with a reduction of diet-induced thermogenesis (108–15, 117, 118). These findings have been corroborated by studies conducted with large numbers of subjects that included adults, children, and adolescents (116, 122, 124). These studies have demonstrated that the ingestion of a standard meal in the obese is associated with blunted thermogenesis that amounts to as much as 25% when compared with nonobese (122). These findings together suggest that reduction in DIT is an important contributing factor of obesity.

In summary, differences in metabolic rate and DIT may not account for all the individual differences in the susceptibility to obesity. However, when considered with reference to other factors such as differences in spontaneous activity small differences in metabolic activities may become important in explaining variability in the propensity to accumulate fat.

## OVERVIEW

As schematized in figure 18.11 the expression of obesity is related to the interaction of both biological and sociocultural responses. The available evidence suggests that variability in body composition and especially in body fat is related to a strong genotype-environment interaction effect (21). Humans in the past and as it occurs today in some areas of the world are exposed to frequent conditions of food shortages. It is quite possible that the "genotype for obesity" or the ability to rapidly increase body fat stores has evolved as an adaptive response to the inevitability of food shortages. If such is the case one would expect that the expression of the genetic potential for obesity may be stimulated also by conditions of dietary restriction. The evidence

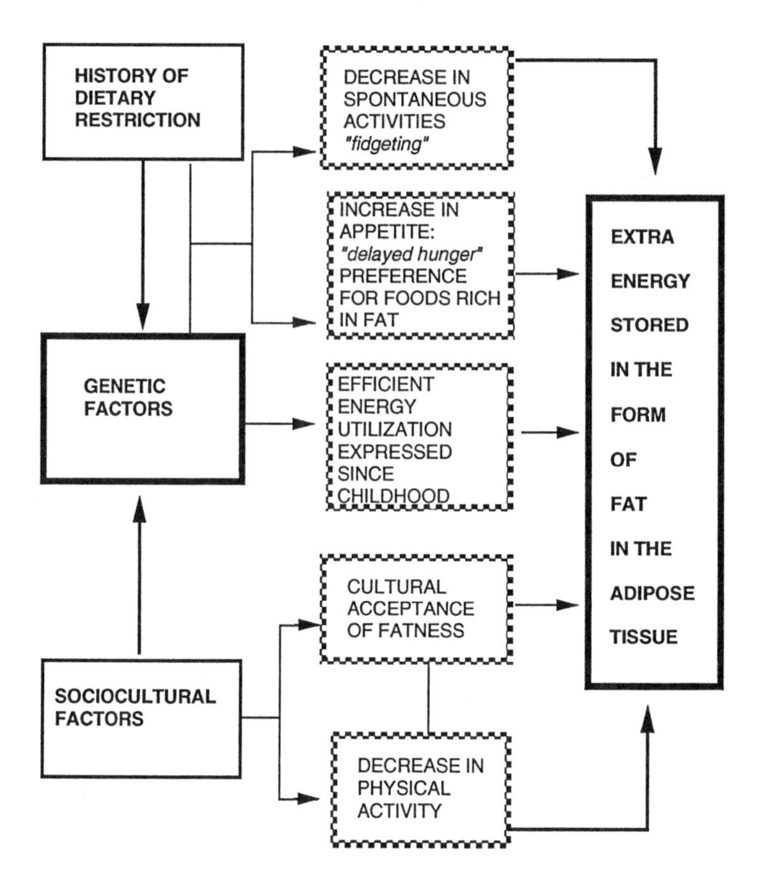


**Fig. 18.11.** Schematization of the components for the variability in body fatness. As individuals or populations are exposed to dietary restriction the genetic potential for fat storage is expressed in the form of obesity. This expression is facilitated by increased availability of food rich in fat and decreases in physical activity and cultural acceptance of fatness.

reviewed here indicates that exposure to dietary restriction during the period of development or adulthood stimulate the genetic potential to be expressed in the form of obesity. A contributing factor for the high storage of fat includes the increased consumption of foods rich in fats brought about by industrial economies and the culturally mediated acceptance of fatness, decrease in nonobligatory activities, as well as general decrease of energy expenditure in physical activity.

## References

1. Frisancho, A. R. 1990. Anthropometric standards for the assessment of growth and nutritional status. Ann Arbor: University of Michigan Press.

2. Bray, G. A. 1992. Pathophysiology of obesity. Am. J. Clin. Nutr. 55:488S–94S.
3. Vague, J. 1956. The degree of masculine differentiation of obesities: a factor determining predisposition to diabetes, atherosclerosis, gout, and uric calculous disease. Am. J. Clin. Nutr. 4:20–34.
4. Frisancho, A. R. and P. N. Fiegel. 1982. Advanced maturation associated with centripetal fat pattern. H. Biol. 54:717–27.
5. Bouchard, C., L. Perusse, C. Leblance, A. Tremblay, and G. Theriault. 1988. Inheritance of the amount and distribution of human body fat. International J. Obesity 12:205–15.
6. Mueller, W. H., A. Marbella, R. B. Harrist, H. J. Kaplowitz, J. A. Grunbaum, and D. B. Labarthe. 1990. Body circumferences as measures of body fat distribution in 10–14 year-old schoolchildren. Am. J. H. Biol. 2:117–24.
7. Grundy, S. M., and J. P. Barnett. 1990. Metabolic and health complications of obesity. Disease-a-Month 36 (12): 645–731.
8. Neel, J. V. 1962. Diabetes mellitus: a "thrifty" genotype rendered detrimental by "progress?" Am. J. Hum. Genet. 14:353–62.
9. Neel, J. V. 1982. The "thrifty" genotype revisited. In J. Kobberling and J. Tattersall, ed. The genetics of diabetes mellitus. New York: Academic Press. 49–60.
10. Wilmsen, E. 1982. Studies in diet, nutrition, and fertility among a group of Kung bushmen in Botswana. Social Science Information 21:95–126.
11. Meehan B. 1982. Shell bed to shell midden. Camberra: Australian Institute of Aboriginal Studies.
12. Hurtado, A. M., and K. R. Hill. 1991. Seasonal variation in the diet and sexual division of labor among Hiwi hunter-gatherers. J. Anthrop. Res. 46 (3): 293–346.
13. Sinclair, A. J. and K. O'Dea. 1990. Fats in human diets through history: is the western diet out of step? In J. D. Wood, A. V. Fisher, eds. Reducing Fat in Meat Animals, London: Elsevier Science Publishers. 1–47.
14. Huss-Ashmore R., and J. Good. 1988. Seasonality of work, weight, and body composition for women in highland Lessotho. In J. J. Curry and R. K. Hitchcock, eds. Coping with Seasonal Constraints. Philadelphia: University Museum, University of Pennsylvania.
15. Milton, K. 1991. Comparative aspects of diet in Amazonia forest-dweller. Phi. Trans. R. Soc. London B. 334:253–63.
16. Bailey R. C., G. Head, M. Jenike, B. Owen, R. Rechtman, and E. Zechenter. 1989. Hunting and gathering in tropical rain forest: it is possible? Am. Anthropol. 91:59–82.
17. Whiting, M. G. 1958. A cross-cultural nutrition survey. Doctoral dissertation, Harvard School of Public health, Cambridge.
18. Cohen, M. N. 1977. The food crisis in prehistory. New Haven: Yale University Press.
19. Stunkard, A. J., I. A. Sorensen, C. Hanis, T. W. Teasdale, R. Chakraborty, W. J. Schull, and F. Schulsinger. 1986. An adoption study of human obesity. N. E. J. Med. 314:193–98.
20. Stunkard, A. J., J. R. Harris, N. L. Pedersen, and G. E. McClearn. 1990. The body-mass index of twins who have been reared apart. N. Eng. J. Med. 322:1483–87.
21. Bouchard, C., R. Savard, J. P. Depres, A. Tremblay, and C. Leblanc. 1985. Body composition in adopted and biological siblings. Hum. Biol. 457:61–75.
22. Poehlman, E., A. Tremblay, J. P. Depres, E. Fontaine, L. Perusse, G. Theriault,

and C. Bouchard. 1986. Genotype-controlled changes in body composition and fat morphology following overfeeding in twins. Am. J. Clin. Nutr. 43: 723–31.
23. Bouchard, C., A. Tremblay, J. Depres, A. Nadeau, P. J. Lupien, G. Theriault, J. Dussault, S. Moorjani, S. Pinault, and G. Fournier. 1990. The response to long-term overfeeding in identical twins. N. Engl. J. Med. 322:1477–82.
24. Longini, I. M., Jr., M. W. Higgins, P. C. Hinton, P. P. Moll, and J. B. Keller. 1984. Genetic and environmental sources of familial aggregation of body mass in Tecumseh, Michigan. H. Biol. 56:733–57.
25. Province, M. A., J. Keller, M. Higgins, and D. C. Rao. 1991. A commingling analysis of obesity in the Tecumseh community health study. Am. J. H. Biol. 3:435–45.
26. Jayakar, S. D., L. Zonta-Sgaramella, P. Astolfi, A. Galanet, G. F. DeStaefano, and V. Pennetti. 1980. Selection of anthropometric variables and serum components in the definition of obesity. Coll. Anthropol 4:155–61.
27. Zonta, L. A., S. D. Jayakar, M. Boisisio, A. Galante, and V. Pennetti. 1987. Genetic analysis of human obesity in an Italian sample. Hum. Hered. 37:129–39.
28. Tambs, K., M. Torbjorn, L. Eaves, M. Neale, K. Midthjell, P. G. Lund-Larsen, S. Naess, and J. Holmen. 1991. Genetic and environmental contributions to the variance of the body mass index in a Norwegian sample of first- and second-degree relatives. Am. J. H. Biol. 3:257–67.
29. Karlin, S., P. T. Williams, S. Jensen, and J. W. Farquhar. 1981. Genetic analysis of the Stanford LRC family study data. Am. J. Epidemiol. 113:307–24.
30. Garn, S. M., P. E. Cole, and S. M. Bailey. 1979. Living together as a factor in family-line resemblances. H. Biol. 51:565–87.
31. Garn, S. M., S. M. Bailey, and I. T. T. Higgins. 1980. Effects of socioeconomic status, family line, and living together on fatness and obesity. Childhood Prevention of Atherosclerosis and Hypertension. 187–204.
32. Mueller, W. H. 1983. The genetics of human fatness. Yearbook Phys. Anthrop. 43:723–31.
33. Pérusse, L., C. LeBlanc, and C. Bouchard. 1988. Inter-generation transmission of physical fitness in the Canadian population. Can. J. Spt. Sci. 13:8–14.
34. Brown, P. J., and M. Konner. 1987. An anthropological perspective on obesity. In R. J. Wurtman and J. J. Wurtman, eds., Human obesity. Annals of the New York Academy of Sciences 499:29–46.
35. Malcom, L. W. G. 1925. Note on the seclusion of girls among the Efik at Old Calabar. Man 25:113–14.
36. Smithson, C. L. 1959. The Havasupai woman. Salt Lake City: University of Utah Press.
37. Bennett, W. C., and R. M. Zingg. 1935. The Tarahumara: an indian tribe of northern Mexico. Chicago: University of Chicago Press.
38. Borgerhoff, M. M. 1988. Kipsigis bridewealth payments. In L. Betzig, M. Bogerhoff Mulder, and P. Turke, eds., Human Reproductive Behavior. Cambridge: Cambridge University Press. 65–82.
39. De Garine, I; G. J. A. Koppert. 1991. Guru, fattening session among the Massa. Ecol. Food Nutr. 25:1–28.
40. Pasquet P., L. Brigant, A. Froment, G. A. Koppert, D. Bard, I. de Garine, and M. Apfelbaum. 1992. Massive overfeeding and energy balance in men: the Guru Walla model. Am. J. Clin. Nutr. 56:483–90.
41. Schwartz, H. 1986. Never satisfied: a cultural history of diets, fantasies, and fat. New York: Free Press.

42. Massara, E. B. 1989. Que gordita!: a study of weight among women in a Puerto Rican community. New York: AMS Press.
43. Ritenbaugh, C. 1982. Obesity as a culture-bound syndrome. Culture, Medicine and Psychiatry 6:347–61.
44. Hazuda, H. P., S. M. Haffner, M. P. Stern, and C. W. Eifler. 1988. Effects of acculturation and socioeconomic status on obesity and diabetes in Mexican Americans. Am. J. of Epidem. 128:1289–1301.
45. Haffner, S. M., M. P. Stern, B. D. Mitchell, and H. P. Hazuda. 1991. Predictors of obesity in Mexicans. Am. J. Clin. Nutr. 53:1571–76S.
46. Pawson, I. G., R. Martorell, and F. E. Mendoza. 1991. Prevalence of overweight and obesity in U.S. Hispanic populations. Am. J. Clin. Nutr. 53:1522S–8S.
47. Rode, A., and R. J. Shephard. 1971. Cardiorespiratory fitness of an Arctic community. J. App. Physiol. 31:519–26.
48. Aghemo, P., F. P. Limas, G. Pinera-Limas, and G. Sassi. 1971. Maximal aerobic power in primitive indians. Arbeitsphysiologie 29:337–42.
49. Shephard, R. J. 1978. Human physiological work capacity. Cambridge University Press: Cambridge.
50. Weitz, C. A., L. P. Greksa, R. B. Thomas, and C. M. Beall. 1989. An anthropological perspective on the study of work capacity. In M. A. Little and J. D. Haas, eds. Human Population Biology. New York: Oxford University Press. 113–27.
51. Lissner, L., J. Habicht, B. J. Strupp, D. A. Levitsky, J. D. Haas, and D. A. Roe. 1989. Body composition and energy intake: do overweight women overeat and underreport. Am J. Clin. Nutr. 49:320–25.
52. Herberg, L., W. Doppen, E. Major, F. A. Gries. 1974. Dietary-induced hypertrophic-hyperplastic obesity in mice. J. Lipid Res. 15:580–85.
53. Jen, K. L. C. 1988. Effects of diet composition on food intake and carcass composition in rats. Physiol. Behav. 42:551–56.
54. Levin, B. E., J. Triscari, A. C. Sullivan. 1984. Metabolic features of diet-induced obesity without hyperphagia in young rats. Am. J. Physiol. 251:R433–40.
55. Oscai, L. B., M. M. Brown, W. C. Miller. Effect of dietary fat on food intake, growth, and body composition in rats. Growth 48:415–24.
56. Romieu, L., W. C. Willett, M. J. Stampfer, et al. 1988. Energy intake and other determinants of relative weight. Am. J. Clin. Nutr. 47:406–12.
57. Dreon, D. M., B. Frey-Hewitt, N. Ellsworth, P. T. Williams, R. B. Terry, and P. D. Wood. 1988. Dietary fat: carbohydrate ratio and obesity in middle-aged men. Am. J. Clin. Nutr. 47:995–1000.
58. Miller, W. C., A. K. Linderman, J. Wallace, M. Niederpruem. 1990. Diet composition, energy intake, and exercise in relation to body fat in men and women. Am. J. Clin. Nutr. 52:426–30.
59. Tucker, L. A. and M. J. Kano. 1992. Dietary fat and body: a multivariate study. Am. J. Clin. Nutr. 56:616–22.
60. Hashim, S., and T. Van Itallie. 1965. Studies in normal and obese subjects with a monitored food-dispensing device. Ann. N.Y. Acad. Sci. 131:654–61.
61. Drewnowski, A., J. D. Brunzell, K. Sande, P. H. Iverius, M. R. C. Greenwood. 1985. Sweet tooth reconsidered: taste responsiveness in human obesity. Physiol. Behav. 35:617–22.
62. Ries, W. 1973. Feeding behavior in obesity. Proc. Nutr. Soc. 32:187–93.
63. Flatt, J. P. 1985. Energetics of intermediary metabolism. In J. S. Garrow and D. Halliday, eds. Substrate and energy metabolism in man, 58–69. London: John Libbey and Co Ltd.

64. Stunkard, A., E. D'Aquili, S. Fox, and R. D. L. Filin. 1972. Influence of social class on obesity and thinness in children. J.A.M.A. 221:579–84.
65. Goldblatt, P. E., N. E. Moore, and A. J. Stunkard. 1965. Social factor in obesity. J.A.M.A. 192:1039–44.
66. Silverstone, J. T. Obesity and social class. 1970. Psychother. Psychosom. 18:226–30.
67. Rona, R. J. and R. N. Morris. 1982. National study of health and growth: social and family factors and overweight in English-Scottish parents. Ann. Hum. Biol. 9:147–56.
68. Baeke, J. A. H., J. Burema, and J. E. R. Frijter. 1983. Obesity in young Dutch adults: I. Sociodemographic variables and body-mass index. Int. J. Obes. 7:1–12.
69. Kohrs, M. B., L. L. Wang, D. Eklund, B. Paulsen, and R. O'Neal. 1979. The association of obesity with socio-economic factors in Missouri. Am. J. Clin. Nutr. 32:2120–28.
70. Rose, G., and M. G. Marmot. 1981. Social class and coronary heart disease. Br. Heart J. 45:13-9.
71. Arteaga, H., J. E. Dos Santos, and J. E. Dutra de Oliveira. 1982. Obesity among schoolchildren of different socioeconomic levels in a developing country. Int. J. Obes. 6:291–97.
72. Sobal, J., and A. J. Stunkard. 1989. Socioeconomic status and obesity: a review of the literature. Psychol. Bull. 105:260–75.
73. Garn, S. M., T. V. Sullivan, and V. M. Hawthorne. 1988. Fatness and obesity among the parents of lean probands. Ecology of food and nutrition 22:277–83.
74. Garn, S. M., T. V. Sullivan, and V. M. Hawthorne. 1989. The education of one spouse and the fatness of the other spouse. Am. J. Hum. Biol. 1:233–38.
75. Garn, S. M., T. V. Sullivan, and V. M. Hawthorne. 1989. Educational level, fatness, and fatness differences between husbands and wives. Am. J. Clin. Nutr. 50:740–45.
76. Baker, P. T., and J. M. Hanna. 1986. The changing Samoans: behavior and health in transition. New York: Oxford University Press.
77. McGarvey, S. T. 1991. Obesity in Samoans and a perspective on its etiology in Polynesians. Am. J. Clin. Nutr. 53:1586S–94S.
78. Knowler: W. C., D. J. Pettit, M. F. Saad, M. A. Charles, R. G. Nelson, B. V. Howard, C. Bogardus and P. H. Bennett. 1991. Obesity in the Pima indians: Its magnitude and relationship with diabetes. Am. J. Clin. Nutr. 53:1543S–51S.
79. Broussard, B. A., A. Johnson, J. H. Himes, M. Stroy, R. Fichtner, F. Hauck, K. Bachman-Carter, J. Hayes, K. Frolich, N. Gray, S. Valway, and D. Gohdes. 1991. Prevalence of obesity in American Indians and Alaska Natives. Am. J. Clin. Nutr. 53:1535S–42S.
80. Szathmary, E. J. E. and N. Holt. 1983. Hyperglycemia in Dogrib indians of the Northwest Territories, Canada: association with age and a centripetal distribution of body fat. Hum. Biol. 55:493–515.
81. Young, T. K., and G. Sevenhuysen. 1989. Obesity in northern Canadian indians: patterns, determinants, and consequences. Am. J. Clin. Nutr. 49:786–93.
82. Kaplowitz, H., R. Martorell, and F. S. Mendoza. 1990. Fatness and fat distribution in Mexican-American children and youths for the Hispanic Health and Nutrition Examination Survey. Am. J. H. Biol. 1:631–48.
83. Villarreal, S. F., R. Martorell, and F. Mendoza. 1989. Sexual Maturation of Mexican-American adolescents. Am. J. H. Biol. 1:87–95.

84. Ryan, A. S., G. A. Martinez, R. N. Baumgartner, A. F. Roche, S. Guo, W. C. Chumlea, and R. J. Kuczmarski. 1990. Median skinfold thickness distributions and fat-wave patterns in Mexican-American children from the Hispanic Health and Nutritional Examination Survey (HHANES 1982–1984). Am. J. Clin. Nutr. 51:925S–35S.
85. Hazuda, H. P., B. D. Mitchell, S. M. Haffner, and M. P. Stern. 1991. Obesity in Mexican American subgroups: findings from the San Antonio Heart Study. Am. J. Clin. Nutr. 53:1529S–34S.
86. Georges, E., W. H. Mueller, and M. L. Wear. 1991. Body fat distribution: association with socioeconomic status in Hispanic Health and Nutritional Examination Survey. Am. J. Hum. Biol. 3:489–501.
87. Himes, J. H., M. Story, K. Czaplinski, and E. Dahlberg-Luby. 1992. Indications of early obesity in low-income Hmong children. A.J.D.C. 146:67–69.
88. Jones, A. P., and M. I. Friedman. 1982. Obesity and adiposyte abnormalities in offspring of rats undernourished during pregnancy. Science 215:1518–19.
89. Jones, A. P., E. L. Simson, and M. I. Friedman. 1984. Gestational undernutrition and the development of obesity in rats. J. Nutr. 114:1484–92.
90. Jones, A. P., S. A. Assimon, and M. I. Friedman. 1986. The effect of diet on food intake and adiposity in rats made obese by gestational undernutrition. Physiol. Behav. 37:381–86.
91. McCance, R. A. and E. M. Widdowson. 1974. The determinants of growth and form. Proc. R. Soc. Lond. 185:1–17.
92. Dulloo, A. G., and L. Girardier. 1990. Adaptive changes in energy expenditure during refeeding following low-calorie intake: evidence for a specific metabolic component favoring fat storage. Am. J. Clin. Nutr. 52:415–20.
93. Ravelli, G., Stein, Z. A., and Susser, M. W. 1976. Obesity in young men after famine exposure in utero and early infancy. N. Eng. J. Med. 295:350–53.
94. Malcolm, L. A. 1970. Growth and development in New Guinea—a study of the Bundi people of the Madang district. Institute of Human Biology, Papua-New Guinea. Monograph Series No. 1. Australia: Surrey Beatty and Sons.
95. Walker, S. P., C. A. Powell, S. M. Grantham-McGregor, J. H. Himes, and S. M. Chang. 1991. Nutritional supplementation, psychosocial stimulation, and growth of stunted children: the Jamaican study. Am. J. Clin. Nutr. 54:642–48.
96. De Boer, J. O., A. J. H. Van Es, L. A. Roovers, J. M. A. Van Raaj, and J. G. Hautvast. 1986. Adaptation of energy metabolism of overweight women to low-energy intake, studied with whole-body calorimeters. Am. J. Clin. Nutr. 44:585–95.
97. Heshka, S., M. Yang, J. Wang, P. Burt, and F. X. Pi-Sunyer. 1990. Weight loss and change in resting metabolic rate. Am. J. Clin. Nutr. 52:981–86.
98. Blair, D., and E. R. Buskirk. 1987. Habitual daily energy expenditure and activity levels of lean and adult-onset and child-onset obese women. Am. J. Clin. Nutr. 45:540–50.
99. Manore, M. M., T. E. Berry, J. S. Skinner, and S. S. Carroll. 1991. Energy expenditure at rest and during exercise in nonobese female, cyclical dieters and in nondieting control subjects. Am. J. Clin. Nutr. 54:41–46.
100. Callaway, C. W. 1988. Biological adaptations to starvation and semistarvation. In R. T. Frankle and M. U. Yang, eds. Obesity and weight control: the health professional's guide to understanding and treatment. Rockville, Md: Aspen Publishers. 97–108.
101. Keys, A., J. Brozek, A. Henschel, O. Mickelsen, and H. L. Taylor. 1950. The biology of human starvation. Minneapolis: University of Minnesota Press.

102. Holmberg, A. 1950. Nomads of the long bow: the Siriono of eastern Bolivia. Smithsonian Institution, Institute of Social Anthropology, Publication No. 10. Washington, D.C.: U.S. GPO.
103. Jonhson, M. I., B. S. Burke, and J. Meyer. 1956. Relative importance of inactivity and overeating in energy balance of obese high school girls. Am. J. Clin. Nutr. 4:37–44.
104. Ravussin, E., S. Lillioja, and T. E. Anderson. 1966. Determinants of 24 hour energy expenditure in man: methods and results using a respiratory chamber. J. Clin. Invest. 78:1568–78.
105. Felig, P., J. Cunningham, M. Levitt, R. Hendler, and E. Nadel. 1983. Energy expenditure in obesity in fasting and postprandial state. Am. J. Physiol. 244:E45–51.
106. D'Alessio, D. A., E. C. Kaavle, and M. A. Mozzoli. 1988. Thermic effect of food in lean and obese men. J. Clin. Invest. 81:1781–89.
107. Diaz, E. O., A. M. Prentice, G. R. Goldberg, P. R. Murgatroyd, and W. A. Coward. 1992. Metabolic response to experimental overfeeding and overweight healthy volunteers. Am. J. Clin. Nutr. 56:641–55.
108. Bessard, T., Y. Schutz, and E. Jequier. 1983. Energy expenditure and postprandial thermogenesis in obese women before and after weight loss. Am. J. Clin. Nutr. 38:680–93.
109. Ravussin, E., B. Burnard, Y. Schutz, and E. Jequier. 1982. Twenty-four-hour energy expenditure and resting metabolic rate in obese, moderate obese, and control subjects. Am. J. Clin. Nutr. 35:566–73.
110. Bandini, L. G., D. A. Schoeller, W. H. Dietz. 1990. Energy expenditure in obese and nonobese adolescents. Pediatr. Res. 27:982–83.
111. Schutz, Y., T. Bessard, and E. Jequier. 1984. Diet-induced thermogenesis measured over a whole day in obese and nonobese women. Am. J. Cl. Nutr. 40:542–52.
112. Blasa, S., and J. Garrow. 1983. Thermogenic response to temperature, exercise and food stimuli in lean and obese women studied by 24 hour direct calorimetry. Br. J. Nutr. 49:171–80.
113. D'Alessio, D. A., E. C. Kavle, M. Mozzoli, K. J. Smalley, M. Polansky, Z. V. Kendrik, L. R. Owen, M. C. Bushamn, G. Boden, and O. E. Owen. 1988. Thermic effect of food in lean and obese men. J. Clin. Invest. 81:1781–89.
114. Schutz, Y. I., A. Golay, J. P. Felber, and E. Jequier. 1984. Decreased glucose-induced thermogenesis after weight loss in obese subjects: a predisposing factor for relapse of obesity. Am. J. Clin. Nutr. 39:380–87.
115. Shettey, P. S., R. T. Jung, W. P. T. James, M. A. Barrand, and B. A. Callingham. 1981. Postprandial thermogenesis in obesity. Clin. Sci. (London) 60:519–25.
116. Segal, K. R., B. Gutin, J. Albu, and F. X. Pi-Sunyer. 1987. Thermic effects of food and exercise in lean and obese men of similar lean body mass. Am. J. Physiol. 252:E110–17.
117. Katseff, H. L. 1988. Energy metabolism and thermogenesis in obesity. In R. T. Frankle and M. U. Yang, eds. Obesity and weight control: the health professional's guide to understanding and treatment. Rockville, MD: Aspen Publishers. 55–70.
118. Maffeis, C., Y. Schutz, and L. Pinelli. 1992. Meal-induced thermogenesis in lean and obese prepubertal children. Am. J. Clin. Nutr. (in press).
119. Nelson, K. M., R. L. Weinsier, C. L. Long, and Y. Schutz. 1992. Prediction of resting energy expenditure from fat-free mass and fat mass. Am. J. Clin. Nutr. 56:848–56.
120. Rumpler, W. V., J. L. Seale, C. W. Miles, and C. E. Bodwell. 1991. Energy-intake

restriction and diet-composition effects on energy expenditure in men. Am. J. Clin. Nutr. 53:430–36.

121. Welle, S., G. B. Forbes, M. Statt, R. R. Barnard, and J. M. Amatruda. 1992. Energy expenditure under free-living conditions in normal-weight and overweight women. Am. J. Clin. Nutr. 55:14–21.

122. Katch, V. L., C. C. Marks, M. D. Becque, C. Moorehead, and A. Rocchini. 1990. Basal metabolism of obese adolescents: evidence for energy conservation compared to normal and lean adolescents. Am. J. H. Biol. 2:543–51.

123. Little, M. A. 1989. Human biology of African Pastoralists. Yearbook of Phys. Anthrop. 32:216–48.

124. Maffeis, C., Y. Schutz, L. Zoccate, R. Micciolo, and L. Pinelli. 1993. Meal-induced thermogenesis in lean and obese prepubertal children. Am. J. Clin. Nutr. 57:481–85.

# Appendices

APPENDIX A

# Coversion Factors for Measurements Used in the Biological Sciences

| Blood Chemistries | Present Symbol | Conversion Factor | SI Unit Symbol | Significant Digits | Suggested Minimum Increment |
|---|---|---|---|---|---|
| Alpha-fetoprotein (S) | ng/ml | 1.00 | μg/l | XX | 1 μg/l |
| Alpha-fetoprotein (Amf) | mg/dl | 10 | mg/l | XX | 1 mg/l |
| Chloride (S) | mEq/l | 1.00 | mmol/l | XXX | 1 mmol/l |
| Cholesterol (P) | mg/dl | 0.02586 | mmol/l | X.XX | 0.05 mmol/l |
| Ethanol (P) | mg/dl | 0.2171 | mmol/l | XX | 1 mmol/l |
| Ferritin (S) | ng/ml | 1.00 | μg/l | XXO | 10 μg/l |
| Folate (S) | ng/ml | 2.266 | nmol/l | XX | 2 nmol/l |
| Glucagon (S) | pg/ml | 1 | ng/l | XXO | 10 ng/l |
| Glucose (P,S) | mg/dl | 0.05551 | mmol/1 | XX.X | 0.1 mmol/l |
| Iron (S) | μg/l | 0.1791 | μmol/l | XX | 1 μmol/l |
| Lead (B) | mg/dl | 48.26 | mmol/l | X.XX | 0.05 mmol/l |
| Lead (U) | μg/24h | 0.004826 | μmol/d | X.XX | 0.05 μmol/d |
| Lipoproteins (P) | | (10) | (g/l) | | |
| [LDL][HDL] | mg/d1 | 0.02586 | mmol/1 | X.XX | 0.05 mmol/1 |
| Protoporphyrin (Erc) | mg/dl | 0.0177 | mmol/1 | X.XX | 0.02 mmol/l |
| Sodium (S) | mEq/l | 1.00 | mmol/l | XXX | 1 mmol/l |
| Sodium (U) | mEq/24h | 1.00 | mmol/d | XXX | 1 mmol/d |
| Thiocyanate (P) | mg/dl | 0.1722 | mmol/l | X.XX | 0.1 mmol/l |
| Triglycerides (P) (as triolein) | mg/dl | 0.1129 | mmol/l | X.XX | 0.02 mmol/l |
| Vitamin A (retinol) (P,S) | μg/dl | 0.03491 | μmol/l | X.XX | 0.05 μmol/l |
| Vitamin $B_1$ (thiamine hydrochloride) (U) | μg/24 h | 0.002965 | μmol/d | X.XX | 0.01 μmol/l |

| Blood Chemistries | Present Symbol | Conversion Factor | SI Unit Symbol | Significant Digits | Suggested Minimum Increment |
|---|---|---|---|---|---|
| Vitamin $B_2$ (S) (riboflavin) | μg/dl | 26.57 | nmol/l | XXX | 5 nmol/l |
| Vitamin $B_6$ (B) (pyridoxal) | ng/ml | 5.982 | nmol/l | XXX | 5 nmol/l |
| Vitamin $B_{12}$ (P,S) (cyanocobalamin) | ng/dl | 7.378 | pmol/l | XXO | 10 pmol/l |
| Vitamin C (B,P,S) (ascorbate) | mg/dl | 56.78 | μmol/l | XO | 10 μmol/l |
| Vitamin $D_3$ (P) (cholecalciferol) | μg/ml | 2.599 | nmol/l | XXX | 5 nmol/l |
| Vitamin E (P,S) (alpha-tocopherol) | mg/dl | 23.22 | μmol/l | XX | 1 μmol/l |
| Hematology | | | | | |
| Erythrocyte count (B) | $10^6$/mm$^3$ | 1 | $10^{12}$/l | X.X | 0.1 $10^{12}$/l |
| Hematocrit | 0/0 | 0.01 | (1) | O.XX | 0.01 |
| Hemoglobin (B) (mass concentration) | g/dl | 10 | g/l | XXX | 1 g/l |

**Symbols**

| | | | |
|---|---|---|---|
| (1) | Number one | m | Milli ($10^{-3}$) |
| Amf | Amniotic fluid | mol | Mole |
| B | Blood | n | Nano ($10^{-9}$) |
| d | Deci ($10^{-1}$) | p | Pico ($10^{-12}$) |
| Erc | Erythrocyte | P | Plasma |
| g | Gram | S | Serum |
| l | Liter | μ | Micro ($10^{-6}$) |

*Source:* Adapted from the American Journal of Public Health. 1987. 77:1398.

APPENDIX B **Conversion Factors**

| TO CONVERT | INTO | MULTIPLY BY |
|---|---|---|
| **A** | | |
| Abcoulomb | Statcoulombs | $2.998 \times 10^{10}$ |
| Acre | Sq chain (Gunters) | 10 |
| Acre | Rods | 160 |
| Acre | Square links (Gunters) | $1 \times 10^{5}$ |
| Acre | Hectare or sq hectometer | .4047 |
| Acres | Sq feet | 43,560.0 |
| Acres | Sq meters | 4,047.0 |
| Acres | Sq miles | $1.562 \times 10^{-3}$ |
| Acres | Sq yards | 4,840. |
| Acre-feet | Cu feet | 43,560.0 |
| Acre-feet | Gallons | $3.259 \times 10^{5}$ |
| Amperes/sq cm | Amps/sq in | 6.452 |
| Amperes/sq cm | Amps/sq meter | $10^{4}$ |
| Amperes/sq in | Amps/sq cm | 0.1550 |
| Amperes/sq in | Amps/sq meter | 1,550.0 |
| Amperes/sq meter | Amps/sq cm | $10^{-4}$ |
| Amperes/sq meter | Amps/sq in | $6.452 \times 10^{-4}$ |
| Ampere-hours | Coulombs | 3,600.0 |
| Ampere-hours | Faradays | 0.03731 |
| Ampere-turns | Gilberts | 1.257 |
| Ampere-turns/cm | Amp-turns/in | 2.540 |
| Ampere-turns/cm | Amp-turns/meter | 100.0 |
| Ampere-turns/cm | Gilberts/cm | 1.257 |
| Ampere-turns/in | Amp-turns/cm | 0.3937 |
| Ampere-turns/in | Amp-turns/meter | 39.37 |
| Ampere-turns/in | Gilberts/cm | 0.4950 |
| Ampere-turns/meter | Amp-turns/cm | 0.01 |
| Ampere-turns/meter | Amp-turns/in | 0.0254 |
| Ampere-turns/meter | Gilberts/cm | 0.01257 |

| TO CONVERT | INTO | MULTIPLY BY |
|---|---|---|
| Angstrom unit | Inch | $3937 \times 10^{-9}$ |
| Angstrom unit | Meter | $1 \times 10^{-10}$ |
| Angstrom unit | Micron or mu | $1 \times 10^{-4}$ |
| Are | Acre U.S. | .02471 |
| Ares | Sq yards | 119.60 |
| Ares | Acres | 0.02471 |
| Ares | Sq meters | 100.0 |
| Astronomical Unit | Kilometers | $1.495 \times 10^{8}$ |
| Atmospheres | Ton/sq inch | .007348 |
| Atmospheres | Cms of mercury | 76.0 |
| Atmospheres | Ft of water (at 4°C) | 33.90 |
| Atmospheres | In of mercury (at 0°C) | 29.92 |
| Atmospheres | Kgs/sq cm | 1.0333 |
| Atmospheres | Kgs/sq meter | 10,332. |
| Atmospheres | Pounds/sq in | 14.70 |
| Atmospheres | Tons/sq ft | 1.058 |
| **B** | | |
| Barrels (U.S., dry) | Cu inches | 7056. |
| Barrels (U.S., dry) | Quarts (dry) | 105.0 |
| Barrels (U.S., liquid) | Gallons | 31.5 |
| Barrels (oil) | Gallons (oil) | 42.0 |
| Bars | Atmospheres | 0.9869 |
| Bars | Dynes/sq cm | $10^{6}$ |
| Bars | Kgs/sq meter | $1.020 \times 10^{4}$ |
| Bars | Pounds/sq ft | 2,089. |
| Bars | Pounds/sq in | 14.50 |
| Baryl | Dyne/sq cm | 1.000 |
| Bolt (U.S. cloth) | Meters | 36.576 |
| Btu | Liter-atmosphere | 10.409 |
| Btu | Ergs | $1.0550 \times 10^{10}$ |
| Btu | Foot-lbs | 778.3 |
| Btu | Gram-calories | 252.0 |
| Btu | Horsepower-hrs | $3.931 \times 10^{-4}$ |
| Btu | Joules | 1,054.8 |
| Btu | Kilogram-calories | 0.2520 |
| Btu | Kilogram-meters | 107.5 |
| Btu | Kilowatt-hrs | $2.928 \times 10^{-4}$ |
| Btu/hr | Foot-pounds/sec | 0.2162 |
| Btu/hr | Gram-cal/sec | 0.0700 |
| Btu/hr | Horsepower-hrs | $3.929 \times 10^{-4}$ |

| TO CONVERT | INTO | MULTIPLY BY |
|---|---|---|
| Btu/hr | Watts | 0.2931 |
| Btu/min | Foot-lbs/sec | 12.96 |
| Btu/min | Horsepower | 0.02356 |
| Btu/min | Kilowatts | 0.01757 |
| Btu/min | Watts | 17.57 |
| Btu/sq ft/min | Watts/sq in | 0.1221 |
| Bucket (Br. dry) | Cubic cm | $1.818 \times 10^{4}$ |
| Bushels | Cu ft | 1.2445 |
| Bushels | Cu in | 2,150.4 |
| Bushels | Cu meters | 0.03524 |
| Bushels | Liters | 35.24 |
| Bushels | Pecks | 4.0 |
| Bushels | Pints (dry) | 64.0 |
| Bushels | Quarts (dry) | 32.0 |
| **C** | | |
| Calories, gram (mean) | Btu (mean) | $3.9685 \times 10^{-3}$ |
| Candle/sq cm | Lamberts | 3.142 |
| Candle/sq inch | Lamberts | .4870 |
| Centares (centiares) | Sq meters | 1.0 |
| Centigrade | Fahrenheit | $(C° \times 1.8) + 32$ |
| Centigrams | Grams | 0.01 |
| Centiliter | Ounce fluid (U.S.) | .3382 |
| Centiliter | Cubic inch | .6103 |
| Centiliter | Drams | 2.705 |
| Centiliters | Liters | 0.01 |
| Centimeters | Feet | $3.281 \times 10^{-2}$ |
| Centimeters | Inches | 0.3937 |
| Centimeters | Kilometers | $10^{-5}$ |
| Centimeters | Meters | 0.01 |
| Centimeters | Miles | $6.214 \times 10^{-6}$ |
| Centimeters | Millimeters | 10.0 |
| Centimeters | Mils | 393.7 |
| Centimeters | Yards | $1.094 \times 10^{-2}$ |
| Centimeters-dynes | Cm-grams | $1.020 \times 10^{-3}$ |
| Centimeter-dynes | Meter-kgs | $1.020 \times 10^{-8}$ |
| Centimeter-dynes | Pound-feet | $7.376 \times 10^{-8}$ |
| Centimeter-grams | Cm-dynes | 980.7 |
| Centimeter-grams | Meter-kgs | $10^{-5}$ |
| Centimeter-grams | Pound-feet | $7.233 \times 10^{-5}$ |
| Centimeters of mercury | Atmospheres | 0.01316 |

| TO CONVERT | INTO | MULTIPLY BY |
|---|---|---|
| Centimeters of mercury | Feet of water | 0.4461 |
| Centimeters of mercury | Kgs/sq meter | 136.0 |
| Centimeters of mercury | Pounds/sq ft | 27.85 |
| Centimeters of mercury | Pounds/sq in | 0.1934 |
| Centimeters/sec | Feet/min | 1.1969 |
| Centimeters/sec | Feet/sec | 0.03281 |
| Centimeters/sec | Kilometers/hr | 0.036 |
| Centimeters/sec | Knots | 0.1943 |
| Centimeters/sec | Meters/min | 0.6 |
| Centimeters/sec | Miles/hr | 0.02237 |
| Centimeters/sec | Miles/min | $3.728 \times 10^{-4}$ |
| Centimeters/sec/sec | Feet/sec/sec | 0.03281 |
| Centimeters/sec/sec | Kms/hr/sec | 0.036 |
| Centimeters/sec/sec | Meters/sec/sec | 0.01 |
| Centimeters/sec/sec | Miles/hr/sec | 0.02237 |
| Chain | Inches | 792.00 |
| Chain | Meters | 20.12 |
| Chains (surveyors' or Gunter's) | Yards | 22.00 |
| Circular mils | Sq cms | $5.067 \times 10^{-6}$ |
| Circular mils | Sq mils | 0.7854 |
| Circumference | Radians | 6.283 |
| Circular mils | Sq inches | $7.854 \times 10^{-7}$ |
| Cords | Cord feet | 8 |
| Cord feet | Cu feet | 16 |
| Coulomb | Statcoulombs | $2.998 \times 10^{9}$ |
| Coulombs | Faradays | $1.036 \times 10^{-5}$ |
| Coulombs/sq cm | Coulombs/sq in | 64.52 |
| Coulombs/sq cm | Coulombs/sq meter | $10^{4}$ |
| Coulombs/sq in | Coulombs/sq cm | 0.1550 |
| Coulombs/sq in | Coulombs/sq meter | 1,550. |
| Coulombs/sq meter | Coulombs/sq cm | $10^{-4}$ |
| Coulombs/sq meter | Coulombs/sq in | $6.452 \times 10^{-4}$ |
| Cubic centimeters | Cu feet | $3.531 \times 10^{-5}$ |
| Cubic centimeters | Cu inches | 0.06102 |
| Cubic centimeters | Cu meters | $10^{-6}$ |
| Cubic centimeters | Cu yards | $1.308 \times 10^{-6}$ |
| Cubic centimeters | Gallons (U.S. liq) | $2.642 \times 10^{-4}$ |
| Cubic centimeters | Liters | 0.001 |
| Cubic centimeters | Pints (U.S. liq) | $2.113 \times 10^{-3}$ |
| Cubic centimeters | Quarts (U.S. liq) | $1.057 \times 10^{-3}$ |

| TO CONVERT | INTO | MULTIPLY BY |
|---|---|---|
| Cubic feet | Bushels (dry) | 0.8036 |
| Cubic feet | Cu cms | 28,320.0 |
| Cubic feet | Cu inches | 1,728.0 |
| Cubic feet | Cu meters | 0.02832 |
| Cubic feet | Cu yards | 0.03704 |
| Cubic feet | Gallons (U.S. liq) | 7.48052 |
| Cubic feet | Liters | 28.32 |
| Cubic feet | Pints (U.S. liq) | 59.84 |
| Cubic feet | Quarts (U.S. liq) | 29.92 |
| Cubic feet/min | Cu cms/sec | 472.0 |
| Cubic feet/min | Gallons/sec | 0.1247 |
| Cubic feet/min | Liters/sec | 0.4720 |
| Cubic feet/min | Pounds of water/min | 62.43 |
| Cubic feet/sec | Million gals/day | 0.646317 |
| Cubic feet/sec | Gallons/min | 448.831 |
| Cubic inches | Cu cms | 16.39 |
| Cubic inches | Cu feet | $5.787 \times 10^{-4}$ |
| Cubic inches | Cu meters | $1.639 \times 10^{-5}$ |
| Cubic inches | Cu yards | $2.143 \times 10^{-5}$ |
| Cubic inches | Gallons | $4.329 \times 10^{-3}$ |
| Cubic inches | Liters | 0.01639 |
| Cubic inches | Mil-feet | $1.061 \times 10^{5}$ |
| Cubic inches | Pints (U.S. liq) | 0.03463 |
| Cubic inches | Quarts (U.S. liq) | 0.01732 |
| Cubic meters | Bushels (dry) | 28.38 |
| Cubic meters | Cu cms | $10^{6}$ |
| Cubic meters | Cu feet | 35.31 |
| Cubic meters | Cu inches | 61,023.0 |
| Cubic meters | Cu yards | 1.308 |
| Cubic meters | Gallons (U.S. liq) | 264.2 |
| Cubic meters | Liters | 1,000.0 |
| Cubic meters | Pints (U.S. liq) | 2,113.0 |
| Cubic meters | Quarts (U.S. liq) | 1,057. |
| Cubic yards | Cu cms | $7.646 \times 10^{5}$ |
| Cubic yards | Cu feet | 27.0 |
| Cubic yards | Cu inches | 46,656.0 |
| Cubic yards | Cu meters | 0.7646 |
| Cubic yards | Gallons (U.S. liq) | 202.0 |
| Cubic yards | Liters | 764.6 |
| Cubic yards | Pints (U.S. liq) | 1,615.9 |
| Cubic yards | Quarts (U.S. liq) | 807.9 |

| TO CONVERT | INTO | MULTIPLY BY |
|---|---|---|
| Cubic yards/min | Cubic ft/sec | 0.45 |
| Cubic yards/min | Gallons/sec | 3.367 |
| Cubic yards/min | Liters/sec | 12.74 |
| **D** | | |
| Dalton | Gram | $1.650 \times 10^{-24}$ |
| Days | Seconds | 86,400.0 |
| Decigrams | Grams | 0.1 |
| Deciliters | Liters | 0.1 |
| Decimeters | Meters | 0.1 |
| Degrees (angle) | Quadrants | 0.01111 |
| Degrees (angle) | Radians | 0.01745 |
| Degrees (angle) | Seconds | 3,600.0 |
| Degrees/sec | Radians/sec | 0.01745 |
| Degrees/sec | Revolutions/min | 0.1667 |
| Degrees/sec | Revolutions/sec | $2.778 \times 10^{-3}$ |
| Dekagrams | Grams | 10.0 |
| Dekaliters | Liters | 10.0 |
| Dekameters | Meters | 10.0 |
| Drams (apothecaries' or troy) | Ounces (avoirdupois) | 0.1371429 |
| Drams (apothecaries' or troy) | Ounces (troy) | 0.125 |
| Drams (U.S., fluid or apoth.) | Cubic cm | 3.6967 |
| Drams | Grams | 1.7718 |
| Drams | Grains | 27.3437 |
| Drams | Ounces | 0.0625 |
| Dyne/cm | Erg/sq millimeter | .01 |
| Dyne/sq cm | Atmospheres | $9.869 \times 10^{-7}$ |
| Dyne/sq cm | Inch of Mercury at 0°C | $2.953 \times 10^{-5}$ |
| Dyne/sq cm | Inch of Water at 4°C | $4.015 \times 10^{-4}$ |
| Dynes | Grams | $1.020 \times 10^{-3}$ |
| Dynes | Joules/cm | $10^{-7}$ |
| Dynes | Joules/meter (newtons) | $10^{-5}$ |
| Dynes | Kilograms | $1.020 \times 10^{-6}$ |
| Dynes | Poundals | $7.233 \times 10^{-5}$ |
| Dynes | Pounds | $2.248 \times 10^{-6}$ |
| Dynes/sq cm | Bars | $10^{-6}$ |
| **E** | | |
| Ell | Cm | 114.30 |

| TO CONVERT | INTO | MULTIPLY BY |
|---|---|---|
| Ell | Inches | 45 |
| Em, Pica | Inch | .167 |
| Em, Pica | Cm | .4233 |
| Erg/sec | Dyne-cm/sec | 1.000 |
| Ergs | Btu | $9.480 \times 10^{-11}$ |
| Ergs | Dyne-centimeters | 1.0 |
| Ergs | Foot-pounds | $7.367 \times 10^{-8}$ |
| Ergs | Gram-calories | $0.2389 \times 10^{-7}$ |
| Ergs | Gram-cms | $1.020 \times 10^{-3}$ |
| Ergs | Horsepower-hrs | $3.7250 \times 10^{-14}$ |
| Ergs | Joules | $10^{-7}$ |
| Ergs | Kg-calories | $2.389 \times 10^{-11}$ |
| Ergs | Kg-meters | $1.020 \times 10^{-8}$ |
| Ergs | Kilowatt-hrs | $0.2778 \times 10^{-13}$ |
| Ergs | Watt-hours | $0.2778 \times 10^{-10}$ |
| Ergs/sec | Btu/min | $5{,}688 \times 10^{-9}$ |
| Ergs/sec | Ft-lbs/min | $4.427 \times 10^{-6}$ |
| Ergs/sec | Ft-lbs/sec | $7.3756 \times 10^{-8}$ |
| Ergs/sec | Horsepower | $1.341 \times 10^{-10}$ |
| Ergs/sec | Kg-calories/min | $1.433 \times 10^{-9}$ |
| Ergs/sec | Kilowatts | $10^{-10}$ |
| **F** | | |
| Fahrenheit | Centigrade | (F° – 32)/1.8 |
| Farads | Microfarads | $10^{6}$ |
| Faraday/sec | Ampere (absolute) | $9.6500 \times 10^{4}$ |
| Faradays | Ampere-hours | 26.80 |
| Faradays | Coulombs | $9.649 \times 10^{4}$ |
| Fathom | Meter | 1.828804 |
| Fathoms | Feet | 6.0 |
| Feet | Centimeters | 30.48 |
| Feet | Kilometers | $3.048 \times 10^{-4}$ |
| Feet | Meters | 0.3048 |
| Feet | Miles (naut) | $1.645 \times 10^{-4}$ |
| Feet | Miles (stat) | $1.894 \times 10^{-4}$ |
| Feet | Millimeters | 304.8 |
| Feet | Mils | $1.2 \times 10^{4}$ |
| Feet of water | Atmospheres | 0.02950 |
| Feet of water | In of mercury | 0.8826 |
| Feet of water | Kgs/sq cm | 0.03048 |
| Feet of water | Kgs/sq meter | 304.8 |

| TO CONVERT | INTO | MULTIPLY BY |
|---|---|---|
| Feet of water | Pounds/sq ft | 62.43 |
| Feet of water | Pounds/sq in | 0.4335 |
| Feet/min | Cms/sec | 0.5080 |
| Feet/min | Feet/sec | 0.01667 |
| Feet/min | Kms/hr | 0.01829 |
| Feet/min | Meters/min | 0.3048 |
| Feet/min | Miles/hr | 0.01136 |
| Feet/sec | Cms/sec | 30.48 |
| Feet/sec | Kms/hr | 1.097 |
| Feet/sec | Knots | 0.5921 |
| Feet/sec | Meters/min | 18.29 |
| Feet/sec | Miles/hr | 0.6818 |
| Feet/sec | Miles/min | 0.01136 |
| Feet/sec/sec | Cms/sec/sec | 30.48 |
| Feet/sec/sec | Kms/hr/sec | 1.097 |
| Feet/sec/sec | Meters/sec/sec | 0.3048 |
| Feet/sec/sec | Miles/hr/sec | 0.6818 |
| Feet/100 feet | Per cent grade | 1.0 |
| Foot-candle | Lumen/sq meter | 10.764 |
| Foot-pounds | Btu | $1.286 \times 10^{-3}$ |
| Foot-pounds | Ergs | $1.356 \times 10^{7}$ |
| Foot-pounds | Gram-calories | 0.3238 |
| Foot-pounds | Hp-hrs | $5.050 \times 10^{-7}$ |
| Foot-pounds | Joules | 1.356 |
| Foot-pounds | Kg-calories | $3.24 \times 10^{-4}$ |
| Foot-pounds | Kg-meters | 0.1383 |
| Foot-pounds | Kilowatt-hrs | $3.766 \times 10^{-7}$ |
| Foot-pounds/min | Btu/min | $1.286 \times 10^{-3}$ |
| Foot-pounds/min | Foot-pounds/sec | 0.01667 |
| Foot-pounds/min | Horsepower | $3.030 \times 10^{-5}$ |
| Foot-pounds/min | Kg-calories/min | $3.24 \times 10^{-4}$ |
| Foot-pounds/min | Kilowatts | $2.260 \times 10^{-5}$ |
| Foot-pounds/sec | Btu/hr | 4.6263 |
| Foot-pounds/sec | Btu/min | 0.07717 |
| Foot-pounds/sec | Horsepower | $1.818 \times 10^{-3}$ |
| Foot-pounds/sec | Kg-calories/min | 0.01945 |
| Foot-pounds/sec | Kilowatts | $1.356 \times 10^{-3}$ |
| Furlongs | Miles (U.S.) | 0.125 |
| Furlongs | Rods | 40.0 |
| Furlongs | Feet | 660.0 |

| TO CONVERT | INTO | MULTIPLY BY |
|---|---|---|
| **G** | | |
| Gallons | Cu cms | 3,785.0 |
| Gallons | Cu feet | 0.1337 |
| Gallons | Cu inches | 231.0 |
| Gallons | Cu meters | $3.785 \times 10^{-3}$ |
| Gallons | Cu yards | $4.951 \times 10^{-3}$ |
| Gallons | Liters | 3.785 |
| Gallons (liq Br. Imp.) | Gallons (U.S. liq) | 1.20095 |
| Gallons (U.S.) | Gallons (Imp) | 0.83267 |
| Gallons of water | Pounds of water | 8.3453 |
| Gallons/min | Cu ft/sec | $2.228 \times 10^{-3}$ |
| Gallons/min | Liters/sec | 0.06308 |
| Gallons/min | Cu ft/hr | 8.0208 |
| Gausses | Lines/sq in | 6.452 |
| Gausses | Webers/sq cm | $10^{-8}$ |
| Gausses | Webers/sq in | $6.452 \times 10^{-8}$ |
| Gausses | Webers/sq meter | $10^{-4}$ |
| Gilberts | Ampere-turns | 0.7958 |
| Gilberts/cm | Amp-turns/cm | 0.7958 |
| Gilberts/cm | Amp-turns/in | 2.021 |
| Gilberts/cm | Amp-turns/meter | 79.58 |
| Gills (British) | Cubic cm | 142.07 |
| Gills | Liters | 0.1183 |
| Gills | Pints (liq) | 0.25 |
| Grade | Radian | .01571 |
| Grains | Drams (avoirdupois) | 0.03657143 |
| Grains (troy) | Grains (avdp) | 1.0 |
| Grains (troy) | Grams | 0.06480 |
| Grains (troy) | Ounces (avdp) | $2.0833 \times 10^{-3}$ |
| Grains (troy) | Pennyweight (troy) | 0.04167 |
| Grains/U.S. gal | Parts/million | 17.118 |
| Grains/U.S. gal | Pounds/million gal | 142.86 |
| Grains/Imp. gal | Parts/million | 14.286 |
| Grams | Dynes | 980.7 |
| Grams | Grains | 15.43 |
| Grams | Joules/cm | $9.807 \times 10^{-5}$ |
| Grams | Joules/meter (newtons) | $9.807 \times 10^{-3}$ |
| Grams | Kilograms | 0.001 |
| Grams | Milligrams | 1,000. |
| Grams | Ounces (avdp) | 0.03527 |

| TO CONVERT | INTO | MULTIPLY BY |
|---|---|---|
| Grams | Ounces (troy) | 0.03215 |
| Grams | Poundals | 0.07093 |
| Grams | Pounds | $2.205 \times 10^{-3}$ |
| Grams/cm | Pounds/inch | $5.600 \times 10^{-3}$ |
| Grams/cu cm | Pounds/cu ft | 62.43 |
| Grams/cu cm | Pounds/cu in | 0.03613 |
| Grams/cu cm | Pounds/mil-foot | $3.405 \times 10^{-7}$ |
| Grams/liter | Grains/gal | 58.417 |
| Grams/liter | Pounds/1,000 gal | 8.345 |
| Grams/liter | Pounds/cu ft | 0.062427 |
| Grams/liter | Parts/million | 1,000.0 |
| Grams/sq cm | Pounds/sq ft | 2.0481 |
| Gram-calories | Btu | $3.9683 \times 10^{-3}$ |
| Gram-calories | Ergs | $4.1868 \times 10^{7}$ |
| Gram-calories | Foot-pounds | 3.0880 |
| Gram-calories | Horsepower-hrs | $1.5596 \times 10^{-6}$ |
| Gram-calories | Kilowatt-hrs | $1.1630 \times 10^{-6}$ |
| Gram-calories | Watt-hrs | $1.1630 \times 10^{-3}$ |
| Gram-calories/sec | Btu/hr | 14.286 |
| Gram-centimeters | Btu | $9.297 \times 10^{-8}$ |
| Gram-centimeters | Ergs | 980.7 |
| Gram-centimeters | Joules | $9.807 \times 10^{-5}$ |
| Gram-centimeters | Kg-cal | $2.343 \times 10^{-8}$ |
| Gram-centimeters | Kg-meters | $10^{-5}$ |
| **H** | | |
| Hand | Cm | 10.16 |
| Hectares | Acres | 2.471 |
| Hectares | Sq feet | $1.076 \times 10^{5}$ |
| Hectares | Sq m | 10,000 |
| Hectograms | Grams | 100.0 |
| Hectoliters | Liters | 100.0 |
| Hectometers | Meters | 100.0 |
| Hectowatts | Watts | 100.0 |
| Henries | Millihenries | 1,000.0 |
| Hogsheads (British) | Cubic ft | 10.114 |
| Hogsheads (U.S.) | Cubic ft | 8.42184 |
| Hogsheads (U.S.) | Gallons (U.S.) | 63 |
| Horsepower | Btu/min | 42.44 |
| Horsepower | Foot-lbs/min | 33,000. |
| Horsepower | Foot-lbs/sec | 550.0 |

| TO CONVERT | INTO | MULTIPLY BY |
|---|---|---|
| Horsepower (metric) (542.5 ft lb/sec) | Horsepower (550 ft lb/sec) | 0.9863 |
| Horsepower (550 ft lb/sec) | Horsepower (metric) (542.5 ft lb/sec) | 1.014 |
| Horsepower | Kg-calories/min | 10.68 |
| Horsepower | Kilowatts | 0.7457 |
| Horsepower | Watts | 745.7 |
| Horsepower (boiler) | Btu/hr | 33,479 |
| Horsepower (boiler) | Kilowatts | 9.803 |
| Horsepower-hrs | Btu | 2,547. |
| Horsepower-hrs | Ergs | $2.6845 \times 10^{13}$ |
| Horsepower-hrs | Foot-lbs | $1.98 \times 10^{6}$ |
| Horsepower-hrs | Gram-calories | 641,190. |
| Horsepower-hrs | Joules | $2.684 \times 10^{6}$ |
| Horsepower-hrs | Kg-calories | 641.1 |
| Horsepower-hrs | Kg-meters | $2.737 \times 10^{5}$ |
| Horsepower-hrs | Kilowatt-hrs | 0.7457 |
| Hours | Days | $4.167 \times 10^{-2}$ |
| Hours | Weeks | $5.952 \times 10^{-3}$ |
| Hundredweights (long) | Pounds | 112 |
| Hundredweights (long) | Tons (long) | 0.05 |
| Hundredweights (short) | Ounces (avoirdupois) | 1600 |
| Hundredweights (short) | Pounds | 100 |
| Hundredweights (short) | Tons (metric) | 0.0453592 |
| Hundredweights (short) | Tons (long) | 0.0446429 |
| **I** | | |
| Inches | Centimeters | 2.540 |
| Inches | Meters | $2.540 \times 10^{-2}$ |
| Inches | Miles | $1.578 \times 10^{-5}$ |
| Inches | Millimeters | 25.40 |
| Inches | Mils | 1,000.0 |
| Inches | Yards | $2.778 \times 10^{-2}$ |
| Inches of mercury | Atmospheres | 0.03342 |
| Inches of mercury | Feet of water | 1.133 |
| Inches of mercury | Kgs/sq cm | 0.03453 |
| Inches of mercury | Kgs/sq meter | 345.3 |
| Inches of mercury | Pounds/sq ft | 70.73 |
| Inches of mercury | Pounds/sq in | 0.4912 |
| Inches of water (at 4°C) | Atmospheres | $2.458 \times 10^{-3}$ |
| Inches of water (at 4°C) | Inches of mercury | 0.07355 |

| TO CONVERT | INTO | MULTIPLY BY |
|---|---|---|
| Inches of water (at 4°C) | Kgs/sq cm | $2.540 \times 10^{-3}$ |
| Inches of water (at 4°C) | Ounces/sq in | 0.5781 |
| Inches of water (at 4°C) | Pounds/sq ft | 5.204 |
| Inches of water (at 4°C) | Pounds/sq in | 0.03613 |
| International Ampere | Ampere (absolute) | .9998 |
| International Volt | Volts (absolute) | 1.0003 |
| International volt | Joules (absolute) | $1.593 \times 10^{-19}$ |
| International volt | Joules | $9.654 \times 10^{4}$ |
| **J** | | |
| Joules | Btu | $9.480 \times 10^{-4}$ |
| Joules | Ergs | $10^{7}$ |
| Joules | Foot-pounds | 0.7376 |
| Joules | Kg-calories | $2.389 \times 10^{-4}$ |
| Joules | Kg-meters | 0.1020 |
| Joules | Watt-hrs | $2.778 \times 10^{-4}$ |
| Joules/cm | Grams | $1.020 \times 10^{4}$ |
| Joules/cm | Dynes | $10^{7}$ |
| Joules/cm | Joules/meter (newtons) | 100.0 |
| Joules/cm | Poundals | 723.3 |
| Joules/cm | Pounds | 22.48 |
| **K** | | |
| Kilograms | Dynes | 980,665. |
| Kilograms | Grams | 1,000.0 |
| Kilograms | Joules/cm | 0.09807 |
| Kilograms | Joules/meter (newtons) | 9.807 |
| Kilograms | Poundals | 70.93 |
| Kilograms | Pounds | 2.205 |
| Kilograms | Tons (long) | $9.842 \times 10^{-4}$ |
| Kilograms | Tons (short) | $1.102 \times 10^{-3}$ |
| Kilograms/cu meter | Grams/cu cm | 0.001 |
| Kilograms/cu meter | Pounds/cu ft | 0.06243 |
| Kilograms/cu meter | Pounds/cu in | $3.613 \times 10^{-5}$ |
| Kilograms/cu meter | Pounds/mil-foot | $3.405 \times 10^{-10}$ |
| Kilograms/meter | Pounds/ft | 0.6720 |
| Kilogram/sq cm | Dynes | 980,665 |
| Kilograms/sq cm | Atmospheres | 0.9678 |
| Kilograms/sq cm | Feet of water | 32.81 |
| Kilograms/sq cm | Inches of mercury | 28.96 |
| Kilograms/sq cm | Pounds/sq ft | 2,048. |

| TO CONVERT | INTO | MULTIPLY BY |
|---|---|---|
| Kilograms/sq cm | Pounds/sq in | 14.22 |
| Kilograms/sq meter | Atmospheres | $9.678 \times 10^{-5}$ |
| Kilograms/sq meter | Bars | $98.07 \times 10^{-6}$ |
| Kilograms/sq meter | Feet of water | $3.281 \times 10^{-3}$ |
| Kilograms/sq meter | Inches of mercury | $2.896 \times 10^{-3}$ |
| Kilograms/sq meter | Pounds/sq ft | 0.2048 |
| Kilograms/sq meter | Pounds/sq in | $1.422 \times 10^{-3}$ |
| Kilograms/sq mm | Kgs/sq meter | $10^{6}$ |
| Kilogram-calories | Btu | 3.968 |
| Kilogram-calories | Foot-pounds | 3,088. |
| Kilogram-calories | Hp-hrs | $1.560 \times 10^{-3}$ |
| Kilogram-calories | Joules | 4,186. |
| Kilogram-calories | Kg-meters | 426.9 |
| Kilogram-calories | Kilojoules | 4.186 |
| Kilogram-calories | Kilowatt-hrs | $1.163 \times 10^{-3}$ |
| Kilogram meters | Btu | $9.294 \times 10^{-3}$ |
| Kilogram meters | Ergs | $9.804 \times 10^{7}$ |
| Kilogram meters | Foot-pounds | 7.233 |
| Kilogram meters | Joules | 9.804 |
| Kilogram meters | Kg-calories | $2.342 \times 10^{-3}$ |
| Kilogram meters | Kilowatt-hrs | $2.723 \times 10^{-6}$ |
| Kilolines | Maxwells | 1,000.0 |
| Kiloliters | Liters | 1,000.0 |
| Kilometers | Centimeters | $10^{5}$ |
| Kilometers | Feet | 3,281. |
| Kilometers | Inches | $3.937 \times 10^{4}$ |
| Kilometers | Meters | 1,000.0 |
| Kilometers | Miles | 0.6214 |
| Kilometers | Millimeters | $10^{6}$ |
| Kilometers | Yards | 1,094. |
| Kilometers/hr | Cms/sec | 27.78 |
| Kilometers/hr | Feet/min | 54.68 |
| Kilometers/hr | Feet/sec | 0.9113 |
| Kilometers/hr | Knots | 0.5396 |
| Kilometers/hr | Meters/min | 16.67 |
| Kilometers/hr | Miles/hr | 0.6214 |
| Kilometers/hr/sec | Cms/sec/sec | 27.78 |
| Kilometers/hr/sec | Ft/sec/sec | 0.9113 |
| Kilometers/hr/sec | Meters/sec/sec | 0.2778 |
| Kilometers/hr/sec | Miles/hr/sec | 0.6214 |
| Kilowatts | Btu/min | 56.92 |

| TO CONVERT | INTO | MULTIPLY BY |
|---|---|---|
| Kilowatts | Foot-lbs/min | $4.426 \times 10^{4}$ |
| Kilowatts | Foot-lbs/sec | 737.6 |
| Kilowatts | Horsepower | 1.341 |
| Kilowatts | Kg-calories/min | 14.34 |
| Kilowatts | Watts | 1,000.0 |
| Kilowatt-hrs | Btu | 3,413. |
| Kilowatt-hrs | Ergs | $3.600. \times 10^{13}$ |
| Kilowatt-hrs | Foot-lbs | $2.655 \times 10^{6}$ |
| Kilowatt-hrs | Gram-calories | 859,850. |
| Kilowatt-hrs | Horsepower-hrs | 1.341 |
| Kilowatt-hrs | Joules | $3.6 \times 10^{6}$ |
| Kilowatt-hrs | Kg-calories | 860.5 |
| Kilowatt-hrs | Kg-meters | $3.671 \times 10^{5}$ |
| Kilowatt-hrs | Pounds of water evaporated from and at 212° F. | 3.53 |
| Kilowatt-hrs | Pounds of water raised from 62° to 212° F. | 22.75 |
| Knots | Feet/hr | 6,080. |
| Knots | Kilometers/hr | 1.8532 |
| Knots | Nautical miles/hr | 1.0 |
| Knots | Statute miles/hr | 1.151 |
| Knots | Yards/hr | 2,027. |
| Knots | Feet/sec | 1.689 |
| **L** | | |
| League | Miles (approx.) | 3.0 |
| Light year | Miles | $5.9 \times 10^{12}$ |
| Light year | Kilometers | $9.46091 \times 10^{12}$ |
| Lines/sq cm | Gausses | 1.0 |
| Lines/sq in | Gausses | 0.1550 |
| Lines/sq in | Webers/sq cm | $1.550 \times 10^{-9}$ |
| Lines/sq in | Webers/sq in | $10^{-8}$ |
| Lines/sq in | Webers/sq meter | $1.550 \times 10^{-5}$ |
| Links (engineer's) | Inches | 12.0 |
| Links (surveyor's) | Inches | 7.92 |
| Liters | Bushels (U.S. dry) | 0.02838 |
| Liters | Cu cm | 1,000.0 |
| Liters | Cu feet | 0.03531 |
| Liters | Cu inches | 61.02 |
| Liters | Cu meters | 0.001 |

| TO CONVERT | INTO | MULTIPLY BY |
|---|---|---|
| Liters | Cu yards | $1.308 \times 10^{-3}$ |
| Liters | Gallons (U.S. liq) | 0.2642 |
| Liters | Pints (U.S. liq) | 2.113 |
| Liters | Quarts (U.S. liq) | 1.057 |
| Liters | Milliliters | 1000 |
| Liters/min | Cu ft/sec | $5.886 \times 10^{-4}$ |
| Liters/min | Gals/sec | $4.403 \times 10^{-3}$ |
| Lumens/sq ft | Foot-candles | 1.0 |
| Lumen | Spherical candle power | .07958 |
| Lumen | Watt | .001496 |
| Lumen/sq ft | Lumen/sq meter | 10.76 |
| Lux | Foot-candles | 0.0929 |
| **M** | | |
| Maxwells | Kilolines | 0.001 |
| Maxwells | Webers | $10^{-8}$ |
| Megalines | Maxwells | $10^{6}$ |
| Megohms | Microhms | $10^{12}$ |
| Megohms | Ohms | $10^{6}$ |
| Meters | Centimeters | 100.0 |
| Meters | Feet | 3.281 |
| Meters | Inches | 39.37 |
| Meters | Kilometers | 0.001 |
| Meters | Miles (naut) | $5.396 \times 10^{-4}$ |
| Meters | Miles (stat) | $6.214 \times 10^{-4}$ |
| Meters | Millimeters | 1,000.0 |
| Meters | Yards | 1.094 |
| Meters | Varas | 1.179 |
| Meters/min | Cms/sec | 1.667 |
| Meters/min | Feet/min | 3.281 |
| Meters/min | Feet/sec | 0.05468 |
| Meters/min | Kms/hr | 0.06 |
| Meters/min | Knots | 0.03238 |
| Meters/min | Miles/hr | 0.03728 |
| Meters/sec | Feet/min | 196.8 |
| Meters/sec | Feet/sec | 3.281 |
| Meters/sec | Kilometers/hr | 3.6 |
| Meters/sec | Kilometers/min | 0.06 |
| Meters/sec | Miles/hr | 2.237 |
| Meters/sec | Miles/min | 0.03728 |
| Meters/sec/sec | Cms/sec/sec | 100.0 |

| TO CONVERT | INTO | MULTIPLY BY |
|---|---|---|
| Meters/sec/sec | Ft/sec/sec | 3.281 |
| Meters/sec/sec | Kms/hr/sec | 3.6 |
| Meters/sec/sec | Miles/hr/sec | 2.237 |
| Meter-kilograms | Cm-dynes | $9.807 \times 10^7$ |
| Meter-kilograms | Cm-grams | $10^5$ |
| Meter-kilograms | Pound-feet | 7.233 |
| Microfarad | Farads | $10^{-6}$ |
| Micrograms | Grams | $10^{-6}$ |
| Microhms | Megohms | $10^{-12}$ |
| Microhms | Ohms | $10^{-6}$ |
| Microliters | Liters | $10^{-6}$ |
| Microns | Meters | $1 \times 10^{-6}$ |
| Miles (naut) | Feet | 6,080.27 |
| Miles (naut) | Kilometers | 1.853 |
| Miles (naut) | Meters | 1,853. |
| Miles (naut) | Miles (statute) | 1.1516 |
| Miles (naut) | Yards | 2,027. |
| Miles (statute) | Centimeters | $1.609 \times 10^5$ |
| Miles (statute) | Feet | 5,280. |
| Miles (statute) | Inches | $6.336 \times 10^4$ |
| Miles (statute) | Kilometers | 1.609 |
| Miles (statute) | Meters | 1,609. |
| Miles (statute) | Miles (naut) | 0.8684 |
| Miles (statute) | Yards | 1,760. |
| Miles/hr | Cms/sec | 44.70 |
| Miles/hr | Feet/min | 88. |
| Miles/hr | Feet/sec | 1.467 |
| Miles/hr | Kms/hr | 1.609 |
| Miles/hr | Kms/min | 0.02682 |
| Miles/hr | Knots | 0.8684 |
| Miles/hr | Meters/min | 26.82 |
| Miles/hr | Miles/min | 0.1667 |
| Miles/hr/sec | Cms/sec/sec | 44.70 |
| Miles/hr/sec | Feet/sec/sec | 1.467 |
| Miles/hr/sec | Kms/hr/sec | 1.609 |
| Miles/hr/sec | Meters/sec/sec | 0.4470 |
| Miles/min | Cms/sec | 2,682. |
| Miles/min | Feet/sec | 88. |
| Miles/min | Kms/min | 1.609 |
| Miles/min | Knots/min | 0.8684 |
| Miles/min | Miles/hr | 60.0 |

| TO CONVERT | INTO | MULTIPLY BY |
|---|---|---|
| Mil-feet | Cu inches | $9.425 \times 10^{-6}$ |
| Milliers | Kilograms | 1,000. |
| Millimicrons | Meters | $1 \times 10^{-9}$ |
| Milligrams | Grains | 0.01543236 |
| Milligrams | Grams | 0.001 |
| Milligrams/liter | Parts/million | 1.0 |
| Millihenries | Henries | 0.001 |
| Milliliters | Liters | 0.001 |
| Milliliters | Quarts (U.S. liq) | 0.001057 |
| Millimeters | Centimeters | 0.1 |
| Millimeters | Feet | $3.281 \times 10^{-3}$ |
| Millimeters | Inches | 0.03937 |
| Millimeters | Kilometers | $10^{-6}$ |
| Millimeters | Meters | 0.001 |
| Millimeters | Miles | $6.214 \times 10^{-7}$ |
| Millimeters | Mils | 39.37 |
| Millimeters | Yards | $1.094 \times 10^{-3}$ |
| Million gals/day | Cu ft/sec | 1.54723 |
| Mils | Centimeters | $2.540 \times 10^{-3}$ |
| Mils | Feet | $8.333 \times 10^{-5}$ |
| Mils | Inches | 0.001 |
| Mils | Kilometers | $2.540 \times 10^{-8}$ |
| Mils | Yards | $2.778 \times 10^{-5}$ |
| Miner's inches | Cu ft/min | 1.5 |
| Minims (British) | Cubic cm | 0.059192 |
| Minims (U.S., fluid) | Cubic cm | 0.061612 |
| Minutes (angles) | Degrees | 0.01667 |
| Minutes (angles) | Quadrants | $1.852 \times 10^{-4}$ |
| Minutes (angles) | Radians | $2.909 \times 10^{-4}$ |
| Minutes (angles) | Seconds | 60.0 |
| Myriagrams | Kilograms | 10.0 |
| Myriameters | Kilometers | 10.0 |
| Myriawatts | Kilowatts | 10.0 |
| **N** | | |
| Nepers | Decibels | 8.686 |
| Newton | Dynes | $1 \times 105$ |
| **O** | | |
| OHM (International) | OHM (absolute) | 1.0005 |
| Ohms | Megohms | $10^{-6}$ |

| TO CONVERT | INTO | MULTIPLY BY |
|---|---|---|
| Ohms | Microhms | $10^6$ |
| Ounces | Drams | 16.0 |
| Ounces | Grains | 437.5 |
| Ounces | Grams | 28.349527 |
| Ounces | Pounds | 0.0625 |
| Ounces | Ounces (troy) | 0.9115 |
| Ounces | Tons (long) | $2.790 \times 10^{-5}$ |
| Ounces | Tons (metric) | $2.835 \times 10^{-5}$ |
| Ounces (fluid) | Cu inches | 1.805 |
| Ounces (fluid) | Liters | 0.02957 |
| Ounces (troy) | Grains | 480.0 |
| Ounces (troy) | Grams | 31.103481 |
| Ounces (troy) | Ounces (avdp) | 1.09714 |
| Ounces (troy) | Pennyweights (troy) | 20.0 |
| Ounces (troy) | Pounds (troy) | 0.08333 |
| Ounce/sq inch | Dynes/sq cm | 4309 |
| Ounces/sq in | Pounds/sq in | 0.0625 |
| **P** | | |
| Parsec | Miles | $19 \times 10^{12}$ |
| Parsec | Kilometers | $3.084 \times 10^{13}$ |
| Parts/million | Grains/U.S. gal | 0.0584 |
| Parts/million | Grains/lmp. gal | 0.07016 |
| Parts/million | Pounds/million gal | 8.345 |
| Pecks (British) | Cubic inches | 554.6 |
| Pecks (British) | Liters | 9.091901 |
| Pecks (U.S.) | Bushels | 0.25 |
| Pecks (U.S.) | Cubic inches | 537.605 |
| Pecks (U.S) | Liters | 8.809582 |
| Pecks (U.S.) | Quarts (dry) | 8 |
| Pennyweights (troy) | Grains | 24.0 |
| Pennyweights (troy) | Ounces (troy) | 0.05 |
| Pennyweights (troy) | Grams | 1.55517 |
| Pennyweights (troy) | Pounds (troy) | $4.1667 \times 10^{-3}$ |
| Pints (dry) | Cu inches | 33.60 |
| Pints (liq) | Cu cms | 473.2 |
| Pints (liq) | Cu feet | 0.01671 |
| Pints (liq) | Cu inches | 28.87 |
| Pints (liq) | Cu meters | $4.732 \times 10^{-4}$ |
| Pints (liq) | Cu yards | $6.189 \times 10^{-4}$ |
| Pints (liq) | Gallons | 0.125 |

| TO CONVERT | INTO | MULTIPLY BY |
|---|---|---|
| Pints (liq) | Liters | 0.4732 |
| Pints (liq) | Quarts (liq) | 0.5 |
| Planck's quantum | Erg-second | $6.624 \times 10^{-27}$ |
| Poise | Gram/cm sec | 1.00 |
| Pounds (avoirdupois) | Ounces (troy) | 14.5833 |
| Poundals | Dynes | 13,826. |
| Poundals | Grams | 14.10 |
| Poundals | Joules/cm | $1.383 \times 10^{-3}$ |
| Poundals | Joules/meter (newtons) | 0.1383 |
| Poundals | Kilograms | 0.01410 |
| Poundals | Pounds | 0.03108 |
| Pounds | Drams | 256. |
| Pounds | Dynes | $44.4823 \times 10^{4}$ |
| Pounds | Grains | 7,000. |
| Pounds | Grams | 453.5924 |
| Pounds | Joules/cm | 0.04448 |
| Pounds | Joules/meter (newtons) | 4.448 |
| Pounds | Kilograms | 0.4536 |
| Pounds | Ounces | 16.0 |
| Pounds | Ounces (troy) | 14.5833 |
| Pounds | Poundals | 32.17 |
| Pounds | Pounds (troy) | 1.21528 |
| Pounds | Tons (short) | 0.0005 |
| Pounds (troy) | Grains | 5,760. |
| Pounds (troy) | Grams | 373.24177 |
| Pounds (troy) | Ounces (avdp) | 13.1657 |
| Pounds (troy) | Ounces (troy) | 12.0 |
| Pounds (troy) | Pennyweights (troy) | 240.0 |
| Pounds (troy) | Pounds (avdp) | 0.822857 |
| Pounds (troy) | Tons (long) | $3.6735 \times 10^{-4}$ |
| Pounds (troy) | Tons (metric) | $3.7324 \times 10^{-4}$ |
| Pounds (troy) | Tons (short) | $4.1143 \times 10^{-4}$ |
| Pounds of water | Cu feet | 0.01602 |
| Pounds of water | Cu inches | 27.68 |
| Pounds of water | Gallons | 0.1198 |
| Pounds of water/min | Cu ft/sec | $2.670 \times 10^{-4}$ |
| Pound-feet | Cm-dynes | $1.356 \times 10^{7}$ |
| Pound-feet | Cm-grams | 13,825. |
| Pound-feet | Meter-kgs | 0.1383 |
| Pounds/cu ft | Grams/cu cm | 0.01602 |
| Pounds/cu ft | Kgs/cu meter | 16.02 |

| TO CONVERT | INTO | MULTIPLY BY |
|---|---|---|
| Pounds/cu ft | Pounds/cu in | $5.787 \times 10^{-4}$ |
| Pounds/cu ft | Pounds/mil-foot | $5.456 \times 10^{-9}$ |
| Pounds/cu in | Gms/cu cm | 27.68 |
| Pounds/cu in | Kgs/cu meter | $2.768 \times 10^{4}$ |
| Pounds/cu in | Pounds/cu ft | 1,728. |
| Pounds/cu in | Pounds/mil-foot | $9.425 \times 10^{-6}$ |
| Pounds/ft | Kgs/meter | 1.488 |
| Pounds/in | Gms/cm | 178.6 |
| Pounds/mil-foot | Gms/cu cm | $2.306 \times 10^{6}$ |
| Pounds/sq ft | Atmospheres | $4.725 \times 10^{-4}$ |
| Pounds/sq ft | Feet of water | 0.01602 |
| Pounds/sq ft | Inches of mercury | 0.01414 |
| Pounds/sq ft | Kgs/sq meter | 4.882 |
| Pounds/sq ft | Pounds/sq in | $6.944 \times 10^{-3}$ |
| Pounds/sq in | Atmospheres | 0.06804 |
| Pounds/sq in | Feet of water | 2.307 |
| Pounds/sq in | Inches of mercury | 2.036 |
| Pounds/sq in | Kgs/sq meter | 703.1 |
| Pounds/sq in | Pounds/sq ft | 144.0 |
| **Q** | | |
| Quadrants (angle) | Degrees | 90.0 |
| Quadrants (angle) | Minutes | 5,400.0 |
| Quadrants (angle) | Radians | 1.571 |
| Quadrants (angle) | Seconds | $3.24 \times 10^{5}$ |
| Quarts (dry) | Cu inches | 67.20 |
| Quarts (liq) | Cu cms | 946.4 |
| Quarts (liq) | Cu feet | 0.03342 |
| Quarts (liq) | Cu inches | 57.75 |
| Quarts (liq) | Cu meters | $9.464 \times 10^{-4}$ |
| Quarts (liq) | Cu yards | $1.238 \times 10^{-3}$ |
| Quarts (liq) | Gallons | 0.25 |
| Quarts (liq) | Liters | 0.9463 |
| **R** | | |
| Radians | Degrees | 57.30 |
| Radians | Minutes | 3,438. |
| Radians | Quadrants | 0.6366 |
| Radians | Seconds | $2.063 \times 10^{5}$ |
| Radians/sec | Degrees/sec | 57.30 |
| Radians/sec | Revolutions/min | 9.549 |

| TO CONVERT | INTO | MULTIPLY BY |
|---|---|---|
| Radians/sec | Revolutions/sec | 0.1592 |
| Radians/sec/sec | Revs/min/min | 573.0 |
| Radians/sec/sec | Revs/min/sec | 9.549 |
| Radians/sec/sec | Revs/sec/sec | 0.1592 |
| Revolutions | Degrees | 360.0 |
| Revolutions | Quadrants | 4.0 |
| Revolutions | Radians | 6.283 |
| Revolutions/min | Degrees/sec | 6.0 |
| Revolutions/min | Radians/sec | 0.1047 |
| Revolutions/min | Revs/sec | 0.01667 |
| Revolutions/min/min | Radians/sec/sec | $1.745 \times 10^{-3}$ |
| Revolutions/min/min | Revs/min/sec | 0.01667 |
| Revolutions/min/min | Revs/sec/sec | $2.778 \times 10^{-4}$ |
| Revolutions/sec | Degrees/sec | 360.0 |
| Revolutions/sec | Radians/sec | 6.283 |
| Revolutions/sec | Revs/min | 60.0 |
| Revolutions/sec/sec | Radians/sec/sec | 6.283 |
| Revolutions/sec/sec | Revs/min/min | 3,600.0 |
| Revolutions/sec/sec | Revs/min/sec | 60.0 |
| Rod | Chain (Gunters) | .25 |
| Rod | Meters | 5.029 |
| Rods (Surveyors' meas) | Yards | 5.5 |
| Rods | Feet | 16.5 |
| **S** | | |
| Seconds (angle) | Degrees | $2.778 \times 10^{-4}$ |
| Seconds (angle) | Minutes | 0.01667 |
| Seconds (angle) | Quadrants | $3.087 \times 10^{-6}$ |
| Seconds (angle) | Radians | $4.848 \times 10^{-6}$ |
| Slug | Kilogram | 14.59 |
| Slug | Pounds | 32.17 |
| Sphere | Steradians | 12.57 |
| Square centimeters | Circular mils | $1.973 \times 10^{5}$ |
| Square centimeters | Sq feet | $1.076 \times 10^{-3}$ |
| Square centimeters | Sq inches | 0.1550 |
| Square centimeters | Sq meters | 0.0001 |
| Square centimeters | Sq miles | $3.861 \times 10^{-11}$ |
| Square centimeters | Sq millimeters | 100.0 |
| Square centimeters | Sq yards | $1.196 \times 10^{-4}$ |
| Square feet | Acres | $2.296 \times 10^{-5}$ |
| Square feet | Circular mils | $1.833 \times 10^{8}$ |

| TO CONVERT | INTO | MULTIPLY BY |
|---|---|---|
| Square feet | Sq cms | 929.0 |
| Square feet | Sq inches | 144.0 |
| Square feet | Sq meters | 0.09290 |
| Square feet | Sq miles | $3.587 \times 10^{-8}$ |
| Square feet | Sq millimeters | $9.290 \times 10^{4}$ |
| Square feet | Sq yards | 0.1111 |
| Square inches | Circular mils | $1.273 \times 10^{6}$ |
| Square inches | Sq cms | 6.452 |
| Square inches | Sq feet | $6.944 \times 10^{-3}$ |
| Square inches | Sq millimeters | 645.2 |
| Square inches | Sq mils | $10^{6}$ |
| Square inches | Sq yards | $7.716 \times 10^{-4}$ |
| Square kilometers | Acres | 247.1 |
| Square kilometers | Sq cms | $10^{10}$ |
| Square kilometers | Sq ft | $10.76 \times 10^{6}$ |
| Square kilometers | Sq inches | $1.550 \times 10^{9}$ |
| Square kilometers | Sq meters | $10^{6}$ |
| Square kilometers | Sq miles | 0.3861 |
| Square kilometers | Sq yards | $1.196 \times 10^{6}$ |
| Square meters | Acres | $2.471 \times 10^{-4}$ |
| Square meters | Sq cms | $10^{4}$ |
| Square meters | Sq feet | 10.76 |
| Square meters | Sq inches | 1,550. |
| Square meters | Sq miles | $3.861 \times 10^{-7}$ |
| Square meters | Sq millimeters | $10^{6}$ |
| Square meters | Sq yards | 1.196 |
| Square miles | Acres | 640.0 |
| Square miles | Sq feet | $27.88 \times 10^{6}$ |
| Square miles | Sq kms | 2.590 |
| Square miles | Sq meters | $2.590 \times 10^{6}$ |
| Square miles | Sq yards | $3.098 \times 10^{6}$ |
| Square millimeters | Circular mils | 1,973. |
| Square millimeters | Sq cms | 0.01 |
| Square millimeters | Sq feet | $1.076 \times 10^{-5}$ |
| Square millimeters | Sq inches | $1.550 \times 10^{-3}$ |
| Square mils | Circular mils | 1.273 |
| Square mils | Sq cms | $6.452 \times 10^{-6}$ |
| Square mils | Sq inches | $10^{-6}$ |
| Square yards | Acres | $2.066 \times 10^{-4}$ |
| Square yards | Sq cms | 8,361. |
| Square yards | Sq feet | 9.0 |

| TO CONVERT | INTO | MULTIPLY BY |
|---|---|---|
| Square yards | Sq inches | 1,296. |
| Square yards | Sq meters | 0.8361 |
| Square yards | Sq miles | $3.228 \times 10^{-7}$ |
| Square yards | Sq millimeters | $8.361 \times 10^{5}$ |
| **T** | | |
| Temperature (°C) + 273 | Absolute temperature (∞C) | 1.0 |
| Temperature (°F) + 460 | Absolute temperature (°F) | 1.0 |
| Tons (long) | Kilograms | 1,016. |
| Tons (long) | Pounds | 2,240. |
| Tons (long) | Tons (short) | 1.120 |
| Tons (metric) | Kilograms | 1,000. |
| Tons (metric) | Pounds | 2,205. |
| Tons (short) | Kilograms | 907.1848 |
| Tons (short) | Ounces | 32,000. |
| Tons (short) | Ounces (troy) | 29,166.66 |
| Tons (short) | Pounds | 2,000. |
| Tons (short) | Pounds (troy) | 2,430.56 |
| Tons (short) | Tons (long) | 0.89287 |
| Tons (short) | Tons (metric) | 0.9078 |
| Tons (short)/sq ft | Kgs/sq meter | 9,765. |
| Tons (short)/sq ft | Pounds/sq in | 2,000. |
| Tons of water/24 hrs | Pounds of water/hr | 83.333 |
| Tons of water/24 hrs | Gallons/min | 0.16643 |
| Tons of water/24 hrs | Cu ft/hr | 1.3349 |
| **V** | | |
| Volt/inch | Volt/cm | .39370 |
| Volt (absolute) | Statvolts | .003336 |
| **W** | | |
| Watts | Btu/hr | 3.4129 |
| Watts | Btu/min | 0.05688 |
| Watts | Ergs/sec | 107. |
| Watts | Foot-lbs/min | 44.27 |
| Watts | Foot-lbs/sec | 0.7378 |
| Watts | Horsepower | $1.341 \times 10^{-3}$ |
| Watts | Horsepower (metric) | $1.360 \times 10^{-3}$ |
| Watts | Kg-calories/min | 0.01433 |
| Watts | Kilowatts | 0.001 |
| Watts (Abs) | Btu (mean)/min | 0.056884 |

| TO CONVERT | INTO | MULTIPLY BY |
|---|---|---|
| Watts (Abs) | Joules/sec | 1 |
| Watt-hours | Btu | 3.413 |
| Watt-hours | Ergs | $3.60 \times 10^{10}$ |
| Watt-hours | Foot-pounds | 2,656. |
| Watt-hours | Gram-calories | 859.85 |
| Watt-hours | Horsepower-hrs | $1.341 \times 10^{-3}$ |
| Watt-hours | Kilogram-calories | 0.8605 |
| Watt-hours | Kilogram-meters | 367.2 |
| Watt-hours | Kilowatt-hrs | 0.001 |
| Watt (International) | Watt (absolute) | 1.0002 |
| Webers | Maxwells | $10^{8}$ |
| Webers | Kilolines | $10^{5}$ |
| Webers/sq in | Gausses | $1.550 \times 10^{7}$ |
| Webers/sq in | Lines/sq in | $10^{8}$ |
| Webers/sq in | Webers/sq cm | 0.1550 |
| Webers/sq in | Webers/sq meter | 1,550. |
| Webers/sq meter | Gausses | $10^{4}$ |
| Webers/sq meter | Lines/sq in | $6.452 \times 10^{4}$ |
| Webers/sq meter | Webers/sq cm | $10^{-4}$ |
| Webers/sq meter | Webers/sq in | $6.452 \times 10^{-4}$ |
| **Y** | | |
| Yards | Centimeters | 91.44 |
| Yards | Kilometers | $9.144 \times 10^{-4}$ |
| Yards | Meters | 0.9144 |
| Yards | Miles (naut) | $4.934 \times 10^{-4}$ |
| Yards | Miles (stat) | $5.682 \times 10^{-4}$ |
| Yards | Millimeters | 914.4 |

APPENDIX C **Glossary**

---

The terminology and concepts used in the study of human adaptation continue to evolve. The following definitions incorporate the experience gained from the past as well as those agreed upon by the American Physiological Society.

**acclimation** physiological change in response to a single experimentally induced stress; if the adaptive traits are acquired during the growth period of the organism, the process is referred to as *developmental acclimation*

**acclimatization** biological changes occurring within the lifetime of an organism that reduce the strain caused by stressful changes in the natural climate or by complex environmental stresses; if the adaptive traits are acquired during the growth period of the organism, the process is referred to as either *developmental adaptation* or *developmental acclimatization*; differences in acclimatization may have a phenotypic or a genetic basis

**acclimatization to pressure** decreasing susceptibility to decompression sickness induced by repeated exposures to decompression

**accommodation** responses to environmental stresses that are not wholly successful because, even though they favor survival of the individual, they also result in significant losses in some important functions

**acculturation** process of change from a traditional culture to an urbanized, westernized lifestyle; control of the microenvironment by behavioral means (clothing, heating, cooling, etc.)

**acidity** concentration of free, unbound hydrogen ions in a solution; the higher the $H^+$ concentration, the greater the acidity

**acidosis** any situation in which arterial $H^+$ concentration is elevated

**acoustic** pertaining to sound, its generation, propagation, or effects (primarily audible sound); acoustics is the science of sound

**acoustic trauma** damage to the auditory receptor resulting from sound exposure

**acquired immune deficiency syndrome (AIDS)** disease caused by *human immunodeficiency virus* (HIV) and characterized by profound inability to resist many infections; major deficit is lack of helper T cells

**active immunity** resistance to reinfection acquired by contact with microorganisms, their toxins, or other antigenic material

**active transport** energy-requiring, carrier-mediated transport system that can move molecules across a membrane against an electrochemical gradient

**acute** brief or transient exposure to a high concentration or intense level of some stress

**adaptation** any change in an organism resulting from exposure to an altered environment that enables the organism to function more efficiently in the new environment; differences in adaptation may have a phenotypic or a genetic basis; among geneticists differences in adaptation reflect genetic differences

**adenosine diphosphate (ADP)** two-phosphate product of ATP breakdown

**adenosine monophosphate (AMP)** monophosphate derivative of ATP; nucleotide in RNA

**adenosine triphosphate (ATP)** major molecule that transfers energy from metabolism to cell functions during its breakdown to ADP and $P_i$

**adipose tissue** tissue composed largely of fat-storing cells

**adjustment** any change in an organism produced by alteration of the environment, without regard to whether the change is beneficial to the organism

**adrenal cortex** endocrine gland that forms outer shell of each adrenal gland; secretes mainly cortisol, aldosterone, and androgens

**adrenal gland** one of a pair of endocrine glands above each kidney; each gland consists of outer *adrenal cortex* and inner *adrenal medulla*

**adrenaline** British name for epinephrine

**adrenal medulla** endocrine gland that forms inner core of each adrenal gland; main secretion is epinephrine

**adrenergic** pertaining to norepinephrine or epinephrine; compound that acts like norepinephrine or epinephrine

**adrenocorticotropic hormone (ACTH)** polypeptide hormone secreted by anterior pituitary; stimulates adrenal cortex to secrete cortisol; also called *corticotropin*

**aerobic** in presence of oxygen

**aerosol** suspension of solid particles or liquid droplets in air; man-made aerosols are of many types, may be harmless, noxious, obnoxious, or even therapeutic, and may be a component of polluted air

**afterbirth** placenta and associated membranes expelled from uterus after delivery of infant

**aldosterone** mineralocorticoid steroid hormone secreted by adrenal cortex; regulates electrolyte balance

**alkaline** having $H^+$ concentration lower than that of pure water; that is, having a pH greater than 7

**alkalosis** any situation in which arterial blood $H^+$ concentration is reduced

**Allen's rule** within a polytypic warm-blooded species, or closely related species, there tends to be an increase of body surface area by an enlargement of the relative size of protruding body structures, such as ears, legs and tail, with increasing temperature of the habitat, insofar as they play a role in temperature regulation

**allergen** a substance or material (of protein or nonprotein nature) capable of inducing an allergic reaction; allergic reactions are characteristically very specific and may involve only minute amounts of the allergen

**allergy** acquired, specific immune reactivity to environmental antigens involving IgE antibodies; hypersensitivity

**altitude** vertical distance above sea level; usually measured in feet or meters; the term is often used to indicate high altitude

**alveolar dead space** volume of inspired air that reaches alveoli but cannot undergo gas exchange with blood

**alveolar gas** the gas mixture present within gas-exchanging regions of the lungs reflecting the effects of respiratory gas exchange; alveolar gas is assumed to be saturated with water vapor at 37° C

**alveolar gas volume** the volume of the regions of the lung within which respiratory gas exchange occurs

**alveolar ventilation** volume of atmospheric air entering alveoli each minute

**alveolus** thin-walled, air-filled ''outpocketing'' from terminal air passageways in lungs; cell cluster at end of duct in secretory gland

**Alzheimer's disease** degenerative brain disease that is most common cause of declining intellectual function in late life

**ambient** prevailing environmental condition

**ambient temperature** average temperature of a gaseous or liquid environment (usually air or water) surrounding a body

**amine hormone** hormone derived from amino acid tyrosine; includes thyroid hormone, epinephrine, and norepinephrine

**amino acid** molecule containing amino group, carboxyl group, and side chain attached to a carbon atom; molecular subunit of protein

**amino group** $NH_2$

**ammonia** $NH_3$; produced during amino acid breakdown; converted in liver to urea

**amnesia** memory loss

**amniotic membrane** membrane surrounding fetus in utero

**amphetamine** drug that increases transmission at catecholamine-mediated synapses in brain

**anabolic steroid** testosterone-like agent that increases protein synthesis

**anabolism** cellular synthesis of organic molecules

**anaerobic** in absence of oxygen

**anaerobic metabolism** transformation of matter and energy without uptake of oxygen

**analgesia** removal of pain

**anatomic dead space** space in respiratory tract airways whose walls do not permit gas exchange with blood

**androgen** any chemical with testosterone-like actions

**anemia** reduction in total blood hemoglobin

**anemic hypoxia** hypoxia with normal arterial $O_2$ pressure but reduced total blood oxygen content

**angina pectoris** chest pain associated with inadequate blood flow to heart muscle

**angiotensin I** peptide generated in plasma by renin's action on angiotensinogen

**angiotensin II** hormone formed by enzymatic action on angiotensin I; stimulates aldosterone secretion from adrenal cortex, vascular smooth-muscle contraction, and thirst

**angiotensinogen** plasma protein precursor for angiotensin I

**angstrom (Å)** unit of length equal to $10^{-4}\mu m$

**anion** negatively charged ion

**anoxia** total lack of oxygen

**antagonist** something whose action opposes intended function or movement

**anterior pituitary** anterior portion of pituitary gland; synthesizes, stores, and releases various hormones including ACTH, GH, TSH, prolactin, FSH, and LH

**antibody** immunoglobulin that both functions as antigen receptor on B cell and is secreted by plasma cell; combines with type of antigen that stimulated its production; directs attack against antigen or cell bearing it

**antigen** any foreign molecule that stimulates a specific immune response

**antihistamine** chemical that blocks histamine action

**antithrombin III** plasma anticlotting protein that inactivates thrombin

**antrum** (gastric) lower portion of stomach, that is, region closest to pyloric sphincter; (ovarian) fluid-filled cavity in maturing ovarian follicle

**aorta** largest artery in body; carries blood from left ventricle of heart to thorax and abdomen

**aortic body chemoreceptor** chemoreceptor located near aortic arch; sensitive to arterial blood $0_2$ pressure and $H^+$ concentration

**aortic valve** valve between left ventricle of heart and aorta

**aphasia** specific language deficit not due to mental retardation or muscular weakness

**apnea** cessation of respiration

**appendix** small fingerlike projection from cecum of large intestine

**arachidonic acid** polyunsaturated fatty acid precursor of eicosanoids

**area, total body** the area of the outer surface of a body, assumed smooth; surface area is usually estimated from the formula as that of DuBois, which relates to the total surface area in $m^2$ based on height and weight; Surface Area $(m^2) = 0.202 + \text{Weight, kg}^{0.425} \times \text{Height, m}^{0.725}$

**arrhythmia** any variation from normal heartbeat rhythm

**arterial baroreceptor** nerve endings sensitive to stretch or distortion produced by arterial blood pressure changes; located in carotid sinus or aortic arch arteries; also called the *carotid sinus* and *aortic arch baroreceptors*

**arteriole** blood vessel between artery and capillary, surrounded by smooth muscle; primary site of vascular resistance

**artery** thick-walled, elastic vessel that carries blood away from heart to arterioles

**arthralgia** joint pain; a symptom of dysbarism, or decompression sickness, and rapid compression; in the latter case it is thought to be caused by osmotically induced fluid shifts during blood and tissue uptake of inert gas

**asthma** disease characterized by severe airway constriction and plugging of the airways with mucus

**atelectasis** collapse of a lung or lung subunit following obstruction of an airway. Obliteration of pulmonary air spaces and apparent consolidation occur as gas is absorbed

**atherosclerosis** disease characterized by thickening of arterial walls with abnormal smooth-muscle cells, cholesterol deposits, and connective tissue; results in narrowing of vessel lumen

**atmosphere** mantle of gases surrounding the earth (and other planets)

**atmospheric pressure** air pressure surrounding the body (760 mm Hg at sea level and 33 ft of sea water)

**atom** smallest unit of matter that has unique chemical characteristics; has no net charge; combines with other atoms to form molecules

**atomic mass** relative value that indicates an atom's mass relative to the mass of other types of atoms, based on the assignment of a value of 12 to carbon

**atrioventricular (AV) node** region at base of right atrium near interventricular septum, containing specialized cardiac-muscle cells through which electrical activity must pass to go from atria to ventricles

**atrioventricular (AV) valve** valve between atrium and ventricle of heart; AV valve on right side of heart is the tricuspid valve, and that on left is the mitral valve

**atrium** chamber of heart that receives blood from veins and passes it on to ventricle on same side of heart

**atrophy** wasting way; decrease in size

**audiofrequency range** interval of sound or vibration frequency spectrum from 20 to 20,000 Hz, which is the extent of the audible range for human beings

**audiogram** chart, table, or graph showing hearing level for pure tones as a function of frequency

**audiometer** instrument for measuring hearing acuity

**auditory** pertaining to sense of hearing

**auditory cortex** region of cerebral cortex that receives nerve fibers from auditory pathways that is secreted into extracellular fluid and acts upon cell that secreted it

**autoimmune disease** disease produced by antibody-mediated or cell-mediated attack against body's own cells that results in damage or alteration of cell function

**autonomic balance** relative predominance in function of the adrenergic and cholinergic branches of the autonomic nervous system

**b-lipotropin** endorphin peptide; synthesized as part of pro-opiomelanocortin by anterior pituitary and cosecreted with ACTH; upon cleavage gives rise to beta-endorphin

**barometeorism** expansion of intestinal gas on ascent to altitude

**barometric pressure** the pressure exerted by the column of air extending vertically from the reference point to the upper limit of the atmosphere; measured in millimeters of mercury or in torr (1 torr 1/760 atmosphere 1 mm Hg)

**basal metabolic rate** the rate of energy exchange of human beings per unit of body surface area recorded under conditions of rest and thermal neutrality

**basophil** polymorphonuclear granulocytic leukocyte whose granules stain with basic dyes; enters tissues and becomes mast cell

**bends** painful sensation at or near the joints caused by rapid ascent from area of high pressure as it occurs among divers and is relieved by recompression; synonym is *caisson disease* disorder experienced by divers and caisson workers on transition from high to normal pressure

**Bergmann's rule** within a polytypic warm-blooded species, or closely related species, the body size of the subspecies usually increases as the temperature of its habitat decreases

**beta-adrenergic receptor** plasma-membrane receptor for epinephrine and norepinephrine that utilizes cAMP second-messenger system

**beta oxidation** series of reactions that transfers hydrogens from fatty acid breakdown to oxidative phosphorylation for ATP synthesis

**beta rhythm** low, fast EEG oscillations in alert adults who are paying attention to (or thinking hard about) something

**bicarbonate** $HCO_{3-}$

**bile** fluid secreted by liver; contains bicarbonate, bile salts, cholesterol, lecithin, bile pigments, metabolic end products, and certain trace metals

**bile canaliculi** small ducts adjacent to liver cells into which bile is secreted

**bile duct** carries bile from liver and gallbladder to small intestine

**bile pigment** colored substance, derived from breakdown of heme group or hemoglobin, secreted in bile

**bile salt** one of a family of steroid molecules secreted in bile by the liver; promotes solubilization and digestion of fat in small intestine

**bilirubin** yellow substance resulting from heme breakdown; excreted in bile as a bile pigment

**binding site** region of protein to which a specific ligand binds

**biological clock** means by which organisms provide the observable highly dependable timing for their overt physiologic and behavioral rhythmic patterns that, in nature, correlate with geophysical periods

**biologic rhythm** periodic fluctuation or variation of a biologic process as a function of time

**biosphere** region of the earth's surface, including air and water, in which life can exist

**blackbody** substance that completely absorbs the radiant energy falling on its surface; at thermal equilibrium all the energy absorbed is emitted

**blastocyst** early embryonic stage consisting of ball of developing cells around a central cavity

**blood-brain barrier** group of anatomical barriers and transport systems that controls kinds of substances entering brain extracellular space from blood and their rates of entry

**blood coagulation** blood clotting

**blood sugar** glucose

**blood-testis barrier** barrier that limits chemical movements between blood and lumen of seminiferous tubules

**blood type** blood classification according to presence of A, B, or O antigens

**body mass index (BMI)** calculated as weight in kilograms divided by square of height in meters

**bone marrow** highly vascular, cellular substance in central cavity of some bones; site of erythrocyte, leukocyte, and platelet synthesis

**bony labyrinth** the bony portion of the inner ear apparatus

**Boyle's law** pressure of a fixed amount of gas in a container is inversely proportional to container's volume

**bradycardia** slow heart rate; by definition a heart rate less than 60 beats/min in a resting adult human subject

**brainstem** brain subdivision consisting of medulla oblongata, pons, and midbrain and located between spinal cord and forebrain

**Broca's area** region of left frontal lobe associated with speech production

**calorie** the quantity of heat required to increase the temperature of 1 gm of water at 15° C by 1° C

**capsule environment** the microenvironment within a diver's suit or a submarine chamber

**carbamino compound** compound resulting from combination of carbon dioxide and protein amino groups, particularly in hemoglobin

**carbohydrate** substance composed of carbon, hydrogen, and oxygen according to general formula $Cn(H_20)n$, where *n* is any whole number

**carbon dioxide narcosis** the impaired state of awareness, ranging to unconsciousness, produced by hypercapnia

**carbon dioxide output** the rate at which carbon dioxide is exhaled from the lungs as determined by measurement and analysis of expired gas, with correction for any inspired carbon dioxide; expressed as a rate in volume units per unit time

**carbon dioxide pressure ($PCO_2$)** the tension, or partial pressure, of carbon dioxide in a defined region, system, or space; a measure of carbon dioxide fugacity, or escaping tendency

**carbon dioxide production** the rate at which carbon dioxide is produced by the metabolizing tissues

**carbonic acid ($H_2CO_3$)** acid formed from $H_20$ and $CO_2$

**carbonic anhydrase** enzyme that catalyzes the reaction $CO_2 + H_20 - > H_2CO_3$

**carbon monoxide (CO)** gas that reacts with hemoglobin and decreases blood oxygen-carrying capacity

**carboxypeptidase** enzyme secreted into small intestine by exocrine pancreas as precursor, procarboxypeptidase; breaks peptide bond at carboxyl end of protein

**carcinogen** chemical substance or physical agent capable of inducing cancer

**cardiac cycle** one contraction-relaxation sequence of heart

**cardiac muscle** heart muscle

**cardiac output** blood volume pumped by each ventricle per minute (total output pumped by both ventricles)

**cardiovascular center** neuron cluster in brainstem medulla that serves as a major integrating center for reflexes affecting heart and blood vessels

**cardiovascular system** heart and blood vessels

**carotid** pertaining to two major arteries (carotid arteries) in neck that convey blood to head

**carotid body chemoreceptor** chemoreceptor near main branching of carotid artery; sensitive to blood $O_2$ pressure and $H^+$ concentration

**carotid sinus** dilatation of internal carotid artery just above main carotid branching; location of carotid baroreceptors

**Cartesian coordinate system** mathematical coordinate system of mutually perpendicular axes, either in two or three dimensions

**catabolism** cellular breakdown of organic molecules

**catalyst** substance that accelerates chemical reactions but does not itself undergo any net chemical change during the reaction

**catecholamine** dopamine, epinephrine, or norepinephrine, all of which have similar chemical structures

**cation** ion having net positive charge

**cell-mediated immune response** type of specific immune response mediated by cytotoxic T and NK cells; major defense against intracellular viruses and cancer cells

**centripetal acceleration** acceleration directed toward the center of curvature of the path of a body moving along a curvilinear path

**cerebrospinal fluid (CSF)** fluid that fills cerebral ventricles and the subarachnoid space surrounding brain and spinal cord

**cholesterol** particular steroid molecule; precursor of steroid hormones and bile salts and a component of plasma membranes

**cholinergic** pertaining to acetylcholine; a compound that acts like acetylcholine

**chorionic gonadotropin (CG)** protein hormone secreted by trophoblastic cells of blastocyst and placenta; maintains secretory activity of corpus luteum during first 3 months of pregnancy

**chromatid** one of two identical strands of chromatin resulting from DNA duplication during mitosis or meiosis

**chromatin** combination of DNA and nuclear proteins that is the principle component of chromosomes

**chronic** continuous or frequently recurring long-term exposure; often a long-term exposure to low concentrations or levels of some agent

**chronobiology** the science of biologic rhythms

**circadian rhythm** regular fluctuation of body chemistry, physiologic activity, or behavior, related to the 24-hour or day-night cycle

**clitoris** small body of erectile tissue in female external genitalia; homologous to penis

**$CO_2$** carbon dioxide

**coenzyme** organic cofactor; generally serves as a carrier that transfers atoms or small molecular fragments from one reaction to another; is not consumed in the reaction and can be reused

**collagen** strong, fibrous protein that functions as extracellular structural element in connective tissue

**compression** transition to an environment of higher pressure, particularly in diving or in a hyperbaric chamber

**concentration** amount of material per unit volume of solution

**concentration gradient** gradation in concentration that occurs between two regions having different concentrations

**conditioning** the transfer of an existing response to a new stimulus

**conducting system** network of cardiac-muscle fibers specialized to conduct electrical activity to different areas of heart

**conduction** heat exchange by transfer of thermal energy during collisions of adjacent molecules

**congestive heart failure** set of signs and symptoms associated with decreased contractility of heart and engorgement of heart, veins, and capillaries

**constant environmental conditions** approximate experimental constancy of known environmental variables as specified

**contact force** force directly applied to the surface of a body

**contaminant** foreign or extraneous substance or material; often one that impairs the quality or reduces the life-supporting capacity of an environment

**creatinine** waste product derived from muscle creatine

**creatinine clearance** plasma volume from which creatinine is removed by renal filtration per unit time; approximates glomerular filtration rate

**critical period** time during development when a system is most readily influenced by environmental factors

**curie** a unit of radioactivity; one curie is that quantity of radioactive nuclide, or isotope, that disintegrates at the rate of $3.700 \times 10^{10}$ atoms per second; thus a microcurie (one millionth of a curie) is that quantity that disintegrates at $3.7 \times 10^4$ atoms per second

**2,3-diphosphoglyceraldehyde (DPG)** substance produced by erythrocytes during glycolysis; binds reversibly to hemoglobin, causing it to release oxygen

**deceleration** negative acceleration

**decibel (dB)** one tenth of a bel; scale for measurement of level (intensity) of pressure or power; decibel is the ratio between two quantities, one of which is an arbitrary reference level; for sound pressure, 0 dB = 0.0002 microbar

**decompression** transition to an environment of lower pressure, either from depth to sea level (hyperbaric decompression) or from sea level to altitude (hypobaric decompression)

**decompression chamber** a chamber in which the barometric pressure may be reduced by means of a vacuum pump

**decompression sickness** disorder associated with transition from higher to lower pressure in the environment; synonyms are *dysbarism, aeroembolism, caisson disease*

**density** the ratio of mass to volume (weight/volume)

**deoxyhemoglobin (Hb)** hemoglobin not combined with oxygen; reduced hemoglobin

**deoxyribonucleic acid (DNA)** nucleic acid that stores and transmits genetic information; consists of double strand of nucleotide subunits that contain deoxyribose

**desert equivalent temperature (DET)** the same as effective temperature except that values are for a water vapor density of 10 gm/m$^3$ (typical for a dry desert in summer) instead of water vapor saturation

**diabetes insipidus** disease due to defective control of urine concentration by ADH; marked by great thirst and excretion of a large volume of dilute urine

**diabetes mellitus** disease in which plasma glucose control is defective because of insulin deficiency or decreased target-cell response to insulin

**diabetic ketoacidosis** acute life-threatening emergency in type I diabetes; characterized by increased plasma glucose and ketones, marked urinary losses, and metabolic acidosis

**dialysis** process of altering the concentration of substances in the blood by using concentration differences between plasma and a bathing solution separated by a semipermeable membrane

**diaphragm** dome-shaped skeletal-muscle sheet that separates the abdominal and thoracic cavities; principal muscle of respiration

**diastole** period of cardiac cycle when ventricles are not contracting

**diastolic pressure** minimum blood pressure during cardiac cycle

**diencephalon** core of anterior part of brain; lies beneath cerebral hemispheres and contains thalamus and hypothalamus

**differentiation** process by which cells acquire specialized structural and functional properties

**diffusion** random movement of molecules from one location to another; net diffusion always occurs from a region of higher concentration to a region of lower concentration

**dissipation** process by which energy is lost in the form of heat

**dissociation** separation from

**dissolved gas** gas that is in simple physical solution, as distinguished from that which is chemically combined or which has reacted chemically with solutes or solvent

**diuresis** increased urine excretion

**diuretic** substance that inhibits fluid reabsorption in renal tubule, thereby increasing urine excretion

**dust** fine dry particles of a substance; often small enough to be easily suspended in the atmosphere

**dysbarism** any morbid condition or disease resulting from exposure to a change of ambient pressure and, therefore, usually sudden exposure or great change

**dyspnea** sensation of shortness of breath

**ecology** branch of the science of biology; the study of the complex web of interactions, interrelationships, interdependencies, and processes that link living organisms to each other and to their natural environment

**edema** effusion of serous fluid between cells in tissue spaces or into body cavities

**ejaculation** discharge of semen from penis

**ejaculatory duct** continuation of vas deferens after it is joined by seminal vesicle duct; joins urethra in prostate gland

**elasticity** property that enables a body, when deformed by external forces, to recover without assistance its normal configuration as the forces of deformation are removed

**electroencephalogram (EEG)** brain electrical activity as recorded from scalp

**electrolyte** substance that dissociates into ions when in aqueous solution

**electron** subatomic particle that carries one unit of negative charge

**electromagnetic radiation** radiation composed of waves with electric and magnetic components

**electron-volt (ev)** energy acquired by an electron moving through a potential difference of 1 volt

**electrostatic** pertaining to electric charges at rest

**embolus** (pl. *emboli*) undissolved material impacted in some vascular tissue space such as tissue fragments, gas bubbles, and the like

**embryo** organism during early stages of development; in human beings, the first 2 months of intrauterine life

**emphysema** lung disease characterized by alveolar wall destruction and consequent impairment of gas exchange

**emulsification** fat-solubilizing process in which large lipid droplets are broken into smaller ones

**endocrine gland** group of cells that secretes into the extracellular space hormones that then diffuse into blood stream; also called a *ductless gland*

**endocrine system** all the body's hormone-secreting glands

**endogenous carbon dioxide** carbon dioxide produced within the body by metabolic processes

**endogenous pyrogen (EP)** one of a family of monokines (tumor necrosis factor, interleukin 1, and interleukin 6) that acts in the brain to cause fever

**endometrium** glandular epithelium lining uterine cavity

**environment** the totality of all elements that interact directly or indirectly with an organism at any level of biological and sociological organization

**environmental physiology** branch of the science of physiology concerned with the physiologic responses to environmental change and environmental stressors

**enzyme** highly specific biologic catalyst of protein nature produced by living cells and necessary for in vivo catalysis of biochemical reactions under ordinary conditions of life

**epithelium** tissue covering outer surfaces of the body and inner surfaces of hollow organs and tissues

**erythrocyte** red blood cell

**erythropoiesis** erythrocyte formation

**erythropoietin** hormone secreted mainly by kidney cells; stimulates red-blood-cell production

**esophagus** portion of digestive tract that connects throat (pharynx) and stomach

**essential amino acid** amino acid that cannot be formed by the body at all (or at rate adequate to meet metabolic requirements) and must be obtained from diet

**essential nutrient** substance required for normal or optimal body function but synthesized by the body either not at all or in amounts inadequate to prevent disease

**estradiol** steroid hormone of estrogen family; major ovarian estrogen

**estriol** steroid hormone of estrogen family; major estrogen secreted by placenta during pregnancy

**estrogen** group of steroid hormones that have effects similar to estradiol on female reproductive tract

**exosphere** outermost layer of the atmosphere

**expiration** movement of air out of lungs

**expiratory reserve volume** volume of air that can be exhaled by maximal contraction of expiratory muscles after a normal expiration

**exposure, acute** brief or transient exposure; in practice, often a brief exposure to a high concentration or intense level of some agent

**exposure, chronic** continuous or frequently recurring long-term exposure; in practice, often a long-term exposure to low concentrations or levels of some agent

**facilitated diffusion** carrier-mediated transport system that moves molecules from high to low concentration across a membrane; energy not required

**facultative** organism capable of living under conditions other than the usual

**fat mobilization** increased breakdown of triacylglycerols and release of glycerol and fatty acids into blood

**fat-soluble vitamin** vitamin that is soluble in nonpolar solvents and insoluble in water; vitamin A, D, E, or K

**fatty acid** carbon chain with carboxyl group at one end through which chain can be linked to glycerol to form triacylglycerol

**feedback** characteristic of control systems in which output response influences input to system

**ferritin** iron-binding protein that stores iron in body

**fertilization** union of sperm and ovum

**fetus** period of human development from second month of intrauterine life until birth

**fever** increased body temperature due to setting of ''thermostat'' of temperature-regulating mechanisms at higher-than-normal level

**fibrillation** rapid, unsynchronized cardiac-muscle contractions that prevent effective pumping of blood

**fibrin** protein polymer resulting from enzymatic cleavage of fibrinogen; can turn blood into gel (clot)

**fight-or-flight response** activation of sympathetic nervous system during stress

**folic acid** vitamin of B-complex group; essential for nucleotide formation

**follicle-stimulating hormone (FSH)** protein hormone secreted by anterior pituitary in males and females; a gonadotropin

**follicular phase** that portion of menstrual cycle during which a follicle and ovum develop to maturity prior to ovulation

**forced vital capacity (FVC)** maximal expiration as fast as possible following maximal inspiration

**forebrain** large, anterior brain subdivision consisting of right and left cerebral hemispheres (the cerebrum) and diencephalon

**frequency** number of complete cycles in a unit of time (units: the hertz [Hz] = 1 cps; also for convenience in dealing with high-frequency vibrations in acoustics, ultrasonics, and electronics, the kilo-hertz, 1 kHz = 1,000 Hz; and the megahertz = 1,000,000 Hz)

**functional residual capacity** lung volume after relaxed expiration

**G** the acceleration of gravity; 978 $cm/sec^2$

**galactose** six-carbon monosaccharide; present in lactose (milk sugar)

**gallbladder** small sac under the liver; concentrates bile and stores it between meals; contraction of gallbladder ejects bile into small intestine

**gallstone** precipitate of cholesterol (and occasionally other substances) in gallbladder or common bile duct

**gamete** germ cell or reproductive cell; sperm in male and ovum in female

**gametogenesis** gamete production

**gamma globulin** immunoglobulin G (IgG), most abundant class of plasma antibodies

**gamma (g) rays and X rays** short wavelength electromagnetic radiation emitted from the nucleus; gamma rays are produced by nuclear processes, whereas X rays result from the interaction of high-speed electrons with atoms

**ganglion cell** retinal neuron that is postsynaptic to bipolar cells; axons of ganglion cells form optic nerve

**ganglion** generally reserved for cluster of neuron cell bodies outside CNS

**gas** state of matter in which the atoms or molecules are almost unrestricted by cohesive forces and as such its molecules tend to disperse uniformly throughout the entire available space

**gas density** mass of gas per unit volume

**gas embolism** obstruction of blood vessels by bubbles of gas that form in or gain access to the bloodstream (usually systemic veins) through wounds or during decompression

**gastrointestinal system** gastrointestinal tract plus salivary glands, liver, gallbladder, and pancreas

**gastrointestinal tract** mouth, pharynx, esophagus, stomach, and small and large intestines

**gene** unit of hereditary information; portion of DNA containing information required to determine a protein's amino acid sequence

**genetic code** sequence of nucleotides in gene, three nucleotides indicating the location of one amino acid in protein specified by that gene

**germ cell** cell that gives rise to male or female gametes (the sperm and ova)
**globin** polypeptide chains of hemoglobin molecule
**globulin** one of a family of proteins found in blood plasma
**glomerular filtration** movement of an essentially protein-free plasma from renal glomerular capillaries into Bowman's capsule
**glomerular filtration rate (GFR)** volume of fluid filtered from renal glomerular capillaries into Bowman's capsule per unit time
**glomerulus** structure at beginning of kidney nephron consisting of vascular component and Bowman's capsule
**glucagon** peptide hormone secreted by A cells of pancreatic islets of Langerhans; leads to rise in plasma glucose
**glucocorticoid** steroid hormone produced by adrenal cortex and having major effects on glucose metabolism
**gluconeogenesis** formation of glucose by the liver from pyruvate, lactate, glycerol, or amino acids
**glucose** major monosaccharide (carbohydrate) in the body; six-carbon sugar, $C_6H_{12}O_6$; also called blood sugar
**glucose sparing** switch from glucose to fat utilization by most cells during postabsorptive state
**glucose-6-phosphate** first intermediate in glycolytic pathway
**glutamate** formed from the amino acid glutamic acid; a major excitatory CNS neurotransmitter
**glycerol** three-carbon carbohydrate; forms backbone of triacylglycerol
**glycerophosphate** three-carbon molecule; combines with free fatty acids in triacylglycerol synthesis
**glycine** an amino acid; a neurotransmitter at some inhibitory synapses in CNS
**glycogen** highly branched polysaccharide composed of glucose subunits; major carbohydrate storage form in body
**glycogenolysis** glycogen breakdown to glucose
**glycolysis** metabolic pathway that breaks down glucose to two molecules of pyruvic acid (aerobically) or two molecules of lactic acid (anaerobically)
**gonad** gamete-producing reproductive organ, that is, testes in male and ovaries in female
**gonadotropic hormone** hormone secreted by anterior pituitary; controls gonadal function; FSH or LH
**gonadotropin-releasing hormone (GnRH)** hypothalamic hormone that controls LH and FSH secretion by anterior pituitary in males and females
**gray matter** area of brain and spinal cord and consists mainly of cell bodies and unmyelinated portions of nerve fibers
**growth factor** one of a group of peptides that is highly effective in stimulating mitosis and/or differentiation of certain cell types

**growth hormone (GH)** peptide hormone secreted by anterior pituitary; stimulates somatomedin release; enhances body growth by acting on carbohydrate and protein metabolism

**growth hormone releasing hormone (GRH)** hypothalamic hormone that stimulates growth hormone secretion by anterior pituitary

**growth-inhibiting factor** one of a group of peptides that modulates growth by inhibiting mitosis in specific tissues

**habituation** gradual reduction of responses to or perception of repeated stimulation; diminution of normal neural responses, for example, the decrease of sensations such as pain; such changes can be generalized for the whole organism (general habituation) or can be specific for a given part of the organism (specific habituation); habituation necessarily depends on learning and conditioning, which enable the organism to transfer an existing response to a new stimulus

**heart murmur** heart sound caused by turbulent blood flow through narrowed or leaky valves or through hole interventricular or interatrial septum

**heart rate** number of heart contractions per minute

**heart sound** noise that results from vibrations due to closure of atrioventricular valves (first heart sound) or pulmonary and aortic valves (second heart sound)

**heartburn** pain that seems to occur in region of heart but is due to pain receptors in esophageal wall stimulated by acid refluxed from stomach

**heat** form of energy associated with the random and chaotic motion of the individual atoms or molecules of which matter is composed; it is measured in mechanical energy units, such as ergs or joules

**heat exhaustion** state of collapse due to hypotension because of plasma volume depletion, secondary to sweating, and extreme skin blood-vessel dilation; thermoregulatory centers still function

**heat loss** loss of heat considered primarily as a function of the thermodynamic properties of gas mixtures subjected to hyperbaric pressures

**heat stroke** condition caused by an excessive rise in body temperature as the result of overloading or failure of the thermoregulatory system during exposure to heat stress; it is characterized by a sudden and sustained loss of consciousness and may be preceded by vertigo, nausea, headache, muscular cramps, and cessation of sweating

**Heimlich maneuver** forceful elevation of diaphragm produced by rescuer's fist against choking victim's abdomen; causes sudden sharp increase in alveolar pressure to expel obstructing material that is causing choking

**helper T cell** T cell that enhances antibody production and cytotoxic T and NK cell function

**hematocrit** percentage of blood volume occupied by erythrocytes

**heme** iron-containing organic molecule bound to each of the four polypeptide chains of hemoglobin or to cytochromes

**hemoglobin** protein composed of four polypeptide chains, heme, and iron; located in erythrocytes and transports most blood oxygen

**hemoglobin saturation** percent of oxygen binding sites in hemoglobin combined with oxygen

**hemophilia** disorder in which a clotting factor is absent, resulting in excessive bleeding

**hemorrhage** bleeding

**hemostasis** stopping blood loss from a damaged vessel

**Henry's Law** at constant temperature the mass of a gas that dissolves in unit volume of a solvent is proportional to the partial pressure of the gas in the gas phase that is in equilibrium with the solvent

**heparin** anticlotting agent found on endothelial-cell surfaces; binds antithrombin III; an anticoagulant drug

**hepatic portal vein** vein that conveys blood from capillaries in the intestines, portions of the stomach, and pancreas to capillaries in the liver

**high-density lipoprotein (HDL)** plasma lipid-protein aggregate having low proportion of lipid; promotes removal of cholesterol from cells

**hippocampus** portion of limbic system associated with learning and emotions

**histamine** inflammatory chemical messenger secreted mainly by mast cells; monoamine neurotransmitter

**histotoxic hypoxia** hypoxia in which cell cannot utilize oxygen because a toxic agent has interfered with its metabolic machinery

**homeostasis** relatively stable condition of extracellular fluid that results from regulatory system actions

**homeostatic system** physiologic regulatory system or hierarchy of systems and subsystems; generally, a feedback control mechanism

**homeotherm** an animal that has relatively constant normal body temperature that is maintained despite environmental temperature change within limits

**homeothermic** capable of maintaining body temperature within very narrow limits

**homologous** corresponding in origin, structure, and position

**hormone** chemical messenger synthesized by specific endocrine gland in response to certain stimuli and secreted into the blood, which carries it to target cells

**humoral immune response** immune response mediated by antibodies; major protection against bacteria and viruses in extracellular fluid

**hydrochloric acid (HCl)** strong acid secreted into stomach lumen by parietal cells

**hydrogen bond** weak chemical bond between two molecules or parts of the same molecule, in which negative region of one polarized substance is electrostatically attracted to a positively polarized hydrogen atom in the other

**hydrogen ion ($H^+$)** single proton; $H^+$ concentration of a solution determines its acidity

**hydrolysis** breaking of chemical bond with addition of elements of water (– H and – OH) to the products formed; also called a hydrolytic reaction

**hydrostatic pressure** pressure exerted by fluid

**hyperbaria** atmospheric pressure higher than that at sea level

**hypercalcemia** increased plasma calcium

**hypercapnia** abnormally high carbon dioxide pressure, usually within an organism

**hyperglycemia** plasma glucose concentration increased above normal levels

**hyperosmotic** having total solute concentration greater than plasma

**hyperoxia** a condition with higher partial pressure of oxygen than in air at sea level; an antonym is hypoxia

**hypertension** chronically increased arterial blood pressure

**hyperthermia** increased body temperature regardless of cause

**hypertrophy** enlargement of a tissue or organ due to increased cell size rather than increased cell number

**hyperventilation** increased ventilation without similar increase in $O_2$ consumption or $CO_2$ production

**hypobaria** atmospheric pressure less than that at sea level

**hypocalcemic tetany** skeletal-muscle spasms due to a low extracellular calcium concentration

**hypocapnia** abnormally low carbon dioxide pressure within an organism

**hypoglycemia** low blood sugar (glucose) concentration

**hypoosmotic** having total solute concentration less than that of plasma

**hypotension** low blood pressure

**hypothalamic releasing hormone** hormone released from hypothalamic neurons into hypothalamopituitary portal vessels to control release of anterior pituitary hormone

**hypothalamus** brain region below thalamus; responsible for integration of many basic behavioral patterns involving correlation of neural and endocrine functions, especially those concerned with regulation of internal environment

**hypotonia** abnormally low muscle tone

**hypotonic** containing a lower concentration of effectively membrane-impermeable solute particles than cells contain

**hypoventilation** decrease in ventilation without similar decrease in $O_2$ consumption or $CO_2$ production

**hypoxemia** low blood oxygen pressure and low oxyhemoglobin saturation

**hypoxia** reduced partial pressure of oxygen

**immunity** physiological mechanisms that allow body to recognize materials as foreign or abnormal and to neutralize or eliminate them

**immunoglobulin** (Ig) synonym for antibody; five classes are: (i) *IgG* gamma globulin, most abundant class of antibodies; (ii) *IgA* and (iii) *IgD,* class of antibodies secreted by, and acting locally in, lining of gastrointestinal, respiratory, and urinary tracts; (iv) *IgE,* class of antibodies that mediates immediate hypersensitivity (allergy) and resistance to parasites; (v) *IgM,* class of antibodies that, along with IgG, provides major specific humoral immunity against bacteria and viruses

**impedance (Z)** ratio of generalized effort to generalized flow; has been defined for electric, mechanical, and acoustic cases; impedance indicates the degree to which a stimulus (input) produces a response (output)

**inertia** property manifested by all matter characterized by resistance to any change of motion as to either speed or direction; inertial force acts through the center of mass of the body

**infrared** band of electromagnetic radiation with wavelengths longer than those of visible light (about 0.78 $\mu$m); for practical purposes infrared is divided into a solar band (0.78 to 3.0 $\mu$m) and a low temperature terrestrial band (4 to 50 $\mu$m); other infrared bands of natural origin are weak at ground level

**inner ear (cochlea)** portion of the ear that contains the sensorineural elements that transduce displacement into electrochemical energy

**insensible water loss** water loss of which a person is unaware; that is, loss by evaporation from skin and respiratory passages

**inspiration** air movement from atmosphere into lungs

**inspiratory reserve volume** maximal air volume that can be inspired above resting tidal volume

**insulin** peptide hormone secreted by B cells of pancreatic islets of Langerhans; has metabolic and growth-promoting effects; stimulates glucose and amino-acid uptake by most cells and stimulates protein, fat, and glycogen synthesis

**insulin-like growth factor** one of a group of peptides that have growth-promoting effects

**insulin-like growth factor 1 (IGF-1)** insulin-like growth factor that mediates mitosis-stimulating effect of growth hormone on bone and possibly other tissues; also known as *somatomedin C*

**insulin resistance** hyporesponsiveness of insulin's target cells to circulating insulin due to altered insulin receptors or intracellular processes

**interstitial fluid** extracellular fluid surrounding tissue cells; excludes plasma, which is extracellular fluid surrounding blood cells

**intracellular fluid** fluid in cells; cytosol plus fluid in cell organelles, including nucleus

**inversely proportional** relationship in which, as one factor increases by a given amount, the other decreases by a proportional amount

**ion** atom or small molecule containing unequal number of electrons and protons and, therefore, carrying a net positive or negative electric charge

**ionic bond** strong electrical attraction between two oppositely charged ions

**ionization** process of removing electrons from or adding them to an atom or small molecule to form an ion

**ionizing radiation** any electromagnetic or particulate radiation capable of producing ions, directly or indirectly, in passing through matter

**ionosphere** layer of earth's atmosphere above the stratosphere where particles are in ionized form

**islet of Langerhans** cluster of pancreatic endocrine cells; different islet cells secrete insulin, glucagon, somatostatin, and pancreatic polypeptide

**isocaloric** of or having equal calories

**isometric contraction** contraction of muscle under conditions in which it develops tension but does not change length

**isosmotic** having the same osmotic pressure, thus the same solute-particle concentration

**ketone** product of fatty acid metabolism that accumulates in blood during starvation and in untreated diabetes mellitus; acetoacetic acid, acetone, or b-hydroxybutyric acid; also called ketone body

**kilocalorie (kcal)** amount of heat required to heat 1 l water 1° C; calorie used in nutrition; also called Calorie and large calorie

**kiloelectron-volt (kev)** one thousand electron-volts, or $10^3$ ev

**kinetics** the branch of dynamics that pertains to the turnover, or rate of change of a specific factor, commonly expressed as units of amount per unit time

**Krebs cycle** mitochondrial metabolic pathway that utilizes fragments derived from carbohydrate, protein, and fat breakdown and produces carbon dioxide, hydrogen, and small amounts of ATP; also called *tricarboxylic acid cycle* or *citric acid cycle*

**labyrinth** the system of intercommunicating cavities and canals that constitute the inner ear

**lactase** small-intestine enzyme that breaks down lactose (milk sugar) into glucose and galactose

**lactation** production and secretion of milk by mammary glands

**lacteal** blind-ended lymph vessel in center of each intestinal villus

**lactose** disaccharide composed of glucose and galactose; also called milk sugar

**lactose intolerance** inability to digest lactose because of lack of intestinal lactase; leads to accumulation of large amounts of gas and fluid in large intestine, which causes pain and diarrhea

**latent heat of vaporization** the quantity of heat necessary to change 1 g of a liquid to its vapor without change of temperature; cal/g or j/g

**leukocyte** white blood cell

**Leydig cell** testosterone-secreting endocrine cell that lies between seminiferous tubules of testes; also called *interstitial cell*

**lipid** molecule composed primarily of carbon and hydrogen and characterized by insolubility in water

**lipolysis** triacylglycerol breakdown

**lipoprotein** lipid aggregate that is partially coated by protein; involved in lipid transport in blood

**lipoprotein lipase** capillary endothelial enzyme that hydrolyzes triacylglycerol to glycerol and fatty acids

**lipoxygenase** enzyme that acts on arachidonic acid and leads to leukotriene formation

**low-density lipoprotein (LDL)** protein-lipid aggregate that is major carrier of plasma cholesterol to cells

**lung compliance** change in lung volume caused by given change in transpulmonary pressure; the greater the lung compliance, the more stretchable the lung wall

**luteal phase** last half of menstrual cycle; corpus luteum is active ovarian structure

**luteinizing hormone (LH)** peptide gonadotropin hormone secreted by anterior pituitary; increase in female at mid menstrual cycle initiates ovulation; also called *interstitial cell-stimulating hormone* (ICSH) in male

**lymph** fluid in lymphatic vessels

**lymphatic system** network of vessels that conveys lymph from tissues to blood, and lymph nodes along these vessels

**lymph node** small organ, containing lymphocytes, located along lymph vessel; site of lymphocyte formation and storage and immune reactions

**lymphocyte** type of leukocyte that is responsible for specific immune defenses; mainly B cells and T cells

**lymphocyte activation** mitosis and differentiation of lymphocytes

**lymphoid organ** lymph node, spleen, thymus, tonsil, or aggregate of lymphoid follicles

**macula densa** portion of renal tubule where loop of Henle joins distal tubule; component of juxtaglomerular apparatus

**major histocompatibility complex (MHC)** group of genes that code for *major histocompatibility complex proteins*, which are important for immune function

**malignant** tending to become worse and result in death; opposite of benign

**mammary gland** milk-secreting gland in breast; also used synonymously with breast

**mean arterial pressure (MAP)** average blood pressure during cardiac cycle; approximately diastolic pressure plus one-third pulse pressure

**meiosis** process of cell division leading to gamete (sperm and ova) formation; daughter cells receive only half the chromosomes present in original cell

**melatonin** candidate hormone secreted by pineal gland; suspected role in puberty onset and control of body rhythms
**menarche** onset, at puberty, of menstrual cycling
**menopause** cessation of menstrual cycling
**menstrual cycle** cyclic rise and fall in female reproductive hormones and processes
**menstruation** flow of menstrual fluid from uterus; also called menstrual period
**messenger RNA (mRNA)** ribonucleic acid that transfers genetic information from DNA to ribosome
**metabolic acidosis** acidosis due to any cause other than accumulation of carbon dioxide
**metabolic alkalosis** alkalosis resulting from any cause other than excessive respiratory removal of carbon dioxide
**metabolic end product** waste product from a metabolic reaction or series of reactions
**metabolic pathway** sequence of enzyme-mediated chemical reactions by which molecules are synthesized and broken down in cells
**metabolic rate, basal (BMR)** the rate of metabolic free energy production calculated from measurements of heat production or oxygen consumption in an organism in a rested, awake, fasting, and thermoneutral state; expressed as $kcal \times m^{-2} \times hr^{-1}$ or $W/m^2$
**metabolism** aggregate of chemical reactions that occur in a living organism
**metabolite** substance produced by metabolism
**metabolize** change by chemical reactions
**microclimate** immediate environment in the habitat of plants and animals, or the environment under the clothing worn by humans
**micron** unit of length equal to one thousandth of a millimeter
**micturition** urination
**milliliter (ml)** volume equal to 0.001 l
**millimole** concentration equal to 0.001 mole
**millivolt (mV)** electric potential equal to 0.001 V
**mineral** inorganic substance, that is, without carbon; major minerals in body are calcium, phosphorous, potassium, sulfur, sodium, chloride, and magnesium
**mineralocorticoid** steroid hormone produced by adrenal cortex that has major effect on sodium and potassium balance; major mineralocorticoid is aldosterone
**minute ventilation** total ventilation per minute; equals tidal volume times respiratory rate
**mitochondrion** rod-shaped or oval cytoplasmic organelle that produces most of cell's ATP; site of Krebs cycle and oxidative-phosphorylation enzymes

**mitosis** process in cell division in which DNA is duplicated and an identical set of chromosomes is passed to each daughter cell
**mitral valve** valve between left atrium and left ventricle of heart
**mole** number of molecules of substance; number of moles = weight in grams/ molecular mass
**monoamine** class of neurotransmitters having the structure of R-$NH_2$, where R is molecule remainder; by convention, excludes peptides and amino acids
**monoamine oxidase** enzyme that inactivates catecholamine neurotransmitters and serotonin
**monocyte** type of leukocyte; leaves bloodstream and is transformed into a macrophage
**monosaccharide** carbohydrate consisting of one sugar molecule, which generally contains five or six carbon atoms
**mutation** any change in base sequence of DNA that changes genetic information
**myelin** insulating material covering axons of many neurons; consists of layers of myelin-forming cell plasma membrane wrapped around axon
**myoglobin** muscle-fiber protein that binds oxygen
**nanometer** unit of length equal to $10^{-9}$ m; formerly called millimicron
**nausea** sickness referred to "stomach" or epigastric area
**neurotransmitter** chemical messenger used by neurons to communicate with each other or with effectors
**$NH_3$** ammonia
**$NH_4^+$** ammonium ion
**nitrogen narcosis** psychologic and physiologic changes, including euphoria, intoxication, loss of ability to concentrate, stupefaction, and unconsciousness, associated with exposure to high-pressure air atmospheres
**nystagmus** involuntary rapid motion of the eyeball, either from side to side, up and down, circular, or in a combined motion
**Ohm's Law** current $I$ is directly proportional to voltage $E$ and inversely proportional to resistance $R$ such that $I = E/R$
**olfactory mucosa** mucous membrane in upper part of nasal cavity containing receptors for sense of smell
**ontogenesis** ontogeny; a history of the process of growth and development in the early life of an organism
**oogenesis** gamete production in female
**oogonium** primitive ovum that, upon mitotic division, gives rise to additional oogonia or to primary oocyte
**optimal** most favorable, referring to real environment
**oscillation** repetitive backward and forward or up and down motion
**osmolarity** measure of water concentration in that the higher the solution osmolarity, the lower the water concentration

**osmosis** net diffusion of water across a selective barrier from region of low solute concentration (high water concentration) to region of high solute concentration (low water concentration)

**osmotic pressure** pressure that must be applied to a solution on one side of a membrane to prevent osmotic flow of water across the membrane from a compartment of pure water

**osteoblast** cell type responsible for laying down protein matrix of bone

**ovarian follicle** ovum and its encasing granulosa and theca cells prior to ovulation

**ovary** gonad in female

**ovulation** release of ovum, surrounded by its zona pellucida and cumulus, from ovary

**ovum** gamete of female

**oxidation** combining, or causing a substance to combine, with oxygen

**oxidative deamination** reaction in which an amino group from an amino acid is replaced by oxygen to form a keto acid

**oxidative fiber** muscle fiber that has numerous mitochondria and, therefore, a high capacity for oxidative phosphorylation

**oxidative phosphorylation** process by which energy derived from reaction between hydrogen and oxygen to form water is transferred to ATP during its formation

**oxygen consumption, maximum** maximum rate at which the lungs can take up oxygen during maximal or strenuous excercise

**oxygen uptake** the rate at which oxygen is removed by the blood from alveolar gas; also termed *oxygen intake*

**oxyhemoglobin ($HbO_2$)** hemoglobin combined with oxygen

**oxytocin** peptide hormone synthesized in hypothalamus and released from posterior pituitary; stimulates mammary glands to release milk and uterus to contract

**ozone layer** layer of the stratosphere with increased ozone concentration

**pacemaker** neurons that set rhythm of biological clocks independent of external cues; any nerve or muscle cell that has an inherent autorhythmicity and determines activity pattern of other cells

**pacemaker potential** spontaneous depolarization to threshold of some nerve and muscle cells' plasma membranes

**pancreas** gland in abdomen near stomach; connected by duct to small intestine; contains endocrine gland cells, which secrete insulin, glucagon, and somatostatin into the bloodstream, and exocrine gland cells, which secrete digestive enzymes and bicarbonate into intestine

**pancreatic lipase** enzyme secreted by exocrine pancreas; acts on triacylglycerols to form 2-monoglycerides and free fatty acids

**partial pressure** that part of total gas pressure due to molecules of one gas species; measure of concentration of a gas in a gas mixture

**passive immunity** resistance to infection resulting from direct transfer of antibodies or sensitized T cells from one person (or animal) to another

**pepsin** family of several protein-digesting enzymes formed in the stomach; breaks protein down to peptide fragments

**pepsinogen** inactive precursor of pepsin; secreted by chief cells of gastric mucosa

**peptide** short polypeptide chain; by convention, having less than 50 amino acids

**perception** understanding of objects and events of external worlds that we acquire from neural processing of sensory information

**perfusion, pulmonary** pulmonary capillary blood flow, supplying terminal airways and alveoli

**pericardium** connective-tissue sac surrounding heart

**peripheral chemoreceptor** carotid or aortic body; responds to changes in blood $O_2$ and $CO_2$ pressures and $H^+$ concentration

**peripheral lymphoid organ** lymph node, spleen, tonsil, or lymphocyte accumulation in gastrointestinal, respiratory, urinary, or reproductive tracts

**peripheral nervous system** nerve fibers extending from CNS

**peripheral thermoreceptor** cold or warm receptor in skin or certain mucous membranes

**pH** measure of the acidity of an aqueous solution; the negative common logarithm of the hydrogen ion concentration (activity); pH decreases as acidity increases

**phagocyte** any cell capable of phagocytosis

**phagocytosis** engulfment of particles by a cell followed by the particles' digestion

**phospholipid** lipid subclass similar to triacylglycerol except that a phosphate group($-PO_4^{2-}$) and small nitrogen-containing molecule are attached to third hydroxyl group of glycerol; major component of cell membranes

**phosphoric acid** acid generated during catabolism of phosphorus-containing compounds; dissociates to form inorganic phosphate and hydrogen ions

**phosphorylation** addition of phosphate group to an organic molecule

**photon** smallest and indivisible unit of electromagnetic radiation

**photoperiodism** regulation of the biologic activities of animals and plants by the cyclic annual change of the relative durations of daylight and darkness

**physiology** branch of biology dealing with the mechanisms by which the body functions

**pigmentation, solar, delayed** persistent skin pigmentation occurring several days after exposure to solar UV-B

**pigmentation, solar, immediate** brownish red pigment that appears promptly in the skin of some subjects during exposure to the sun; it is a response to visible light and ultraviolet radiation

**pituitary** endocrine gland that lies in bony pocket below hypothalamus; includes anterior pituitary and posterior pituitary

**placenta** interlocking fetal and maternal tissues that serve as organ of molecular exchange between fetal and maternal circulations

**placental lactogen** hormone that is produced by placenta and has effects, particularly in the mother, similar to growth hormone

**plasma** liquid portion of blood; component of extracellular fluid

**plasma membrane** membrane that forms outer surface of cell and separates cell's contents from extracellular fluid

**plasma protein** albumins, globulins, and fibrinogen

**plasticity** ability of neural tissue to change its responsiveness to stimulation because of its past history of activation

**pollutant** foreign or extraneous substance, material, or agent that impairs the quality or diminishes the life-supporting capacity of an environment

**polypeptide** polymer consisting of amino acid subunits joined by peptide bonds; also called peptide and protein

**polysaccharide** large carbohydrate formed by linking monosaccharide subunits together

**polyunsaturated fatty acid** fatty acid that contains more than one double bond

**portal vessel** blood vessel that links two capillary networks

**postabsorptive state** period during which nutrients are not present in gastrointestinal tract and energy must be supplied by body's endogenous stores

**posterior pituitary** portion of pituitary from which oxytocin and vasopressin are released

**postganglionic** autonomic-nervous-system neuron or nerve fiber whose cell body lies in ganglion and whose axon terminals form neuroeffector junctions; conducts impulses away from ganglion toward periphery

**postsynaptic neuron** neuron that conducts information away from a synapse

**postsynaptic potential** local potential that arises in postsynaptic neuron in response to activation of synapses upon it

**postural reflex** reflex that maintains or restores upright, stable posture

**potential** (or potential difference) voltage difference between two points

**potentiation** presence of one agent enhances response to a second agent such that final response is greater than the sum of the two individual responses

**ppm** parts per million

**predisposition** the condition of being unusually susceptible to the action of a stressor, substance, or agent

**pressure cabin** aircraft compartment pressurized by compressors from ambient air or by ram pressure

**primary oocyte** female germ cell that undergoes first meiotic division to form secondary oocyte and polar body

**primary spermatocyte** male germ cell derived from spermatogonia; undergoes meiotic division to form two secondary spermatocytes

**primary visual cortex** first part of visual cortex to be activated by visual pathways

**progestagen** progesterone-like substance

**progesterone** steroid hormone secreted by corpus luteum and placenta; stimulates uterine-gland secretion, inhibits uterine smooth-muscle contraction, and stimulates breast growth

**prolactin** peptide hormone secreted by anterior pituitary; stimulates milk secretion by mammary glands

**prolactin release-inhibiting hormone (PIH)** dopamine, which serves as hypothalamic hormone to inhibit prolactin secretion by anterior pituitary

**prolactin releasing hormone (PRH)** one or more hypothalamic hormones that stimulate prolactin release from anterior pituitary

**proliferative phase** stage of menstrual cycle between menstruation and ovulation during which endometrium repairs itself and grows

**prostaglandin** one of a group of eicosanoids that function mainly as paracrine or autocrine

**prostate gland** large gland encircling urethra in the male; secretes fluid into urethra

**protein** large polymer consisting of one or more sequences of amino acid subunits joined by peptide bonds

**proton** an elementary particle having a single positive charge equivalent to the negative charge of the electron but a mass approximately 1,837 times as great

**psi** pounds per square inch, a unit of pressure equal to 51.7 torr

**psychrometry** a method using wet- and dry-bulb thermometers to determine water vapor tension, or humidity

**puberty** attainment of sexual maturity when conception becomes possible; as commonly used, refers to 3 to 5 years of sexual development that culminate in sexual maturity

**pulmonary circulation** circulation through lungs; portion of cardiovascular system between pulmonary trunk, as it leaves the right ventricle, and pulmonary veins, as they enter the left atrium

**pulmonary edema** fluid accumulation in lung interstitium and air sacs

**pulmonary perfusion** pulmonary capillary blood flow, supplying terminal airways and alveoli

**pulmonary ventilation** volume of air moved into or out of the lungs per unit time; it is calculated as the product of breathing frequency and tidal volume

**pulse pressure** difference between systolic and diastolic arterial blood pressures

**Purkinje fiber** specialized myocardial cell that constitutes part of conducting system of heart; conveys excitation from bundle branches to ventricular muscle

**pyrogen** any substance that causes fever

**pyruvic acid** three-carbon intermediate in glycolytic pathway that, in absence of oxygen, forms lactic acid or, in presence of oxygen, enters Krebs cycle

**QRS complex** component of electrocardiogram corresponding to ventricular depolarization

**rad** acronym for radiation absorbed dose; a unit of measurement of the absorbed dose of ionizing radiation; an energy transfer of 100 ergs/gm of mass of the irradiated absorbing material

**radiation** emission of heat from object's surface in form of electromagnetic waves

**radon** heavy radioactive gaseous element formed by disintegration of radium; radon belongs to a group of chemically inert gases

**receptor** cell, tissue, or organ that is the site of reaction with a toxicant, substance, or agent

**recombinant DNA** DNA formed by joining portions of two DNA molecules previously fragmented by a restriction enzyme

**recruitment** activation of additional cells in response to increased stimulus strength

**rectum** short segment of large intestine between sigmoid colon and anus

**red muscle fiber** muscle fiber having high oxidative capacity and large amount of myoglobin

**reflectance** ratio of reflected to incident radiation above a reflecting surface

**reflex** biological control system linking stimulus with response and mediated by a reflex arc

**refractory period** time during which an excitable membrane does not respond to a stimulus that normally causes response

**relaxin** polypeptide hormone secreted mainly by corpus luteum; softens cervix prior to parturition

**renal** pertaining to kidneys

**renin** enzyme secreted by kidneys; catalyzes splitting off of angiotensinogen I from angiotensinogen in plasma

**replicate** duplicate

**repolarize** return transmembrane potential to its resting level

**residual lung volume** air volume remaining in the lungs after maximal expiration; measured either by helium dilution or carbon dioxide washout

**respiration** (cellular) oxygen utilization in metabolism of organic molecules; (respiratory system) oxygen and carbon dioxide exchange between organism and external environment

**respiratory acidosis** increased arterial $H^+$ concentration due to carbon dioxide retention

**respiratory alkalosis** decreased arterial $H^+$ concentration when carbon dioxide elimination from lungs exceeds its production

**respiratory distress syndrome** disease of premature infants in whom surfactant-producing cells do not function adequately

**respiratory heat loss** the loss of body heat associated with pulmonary ventilation

**respiratory "pump"** effect on venous return of changing intrathoracic and intraabdominal pressures associated with respiration

**respiratory quotient (RQ)** ratio of carbon dioxide produced to oxygen consumed during metabolism

**respiratory rate** number of breaths per minute

**respiratory system** structures involved in gas exchange between blood and external environment, that is, lungs, tubes leading to lungs, and chest structures responsible for breathing

**respiratory zone** portion of airways from beginning of respiratory bronchi to alveoli; contains alveoli across which gas exchange occurs

**resting membrane potential** voltage difference between inside and outside of cell in absence of excitatory or inhibitory stimulation; also called resting potential

**restriction nuclease** bacterial enzyme that splits DNA into fragments, acting at different loci in the two DNA strands

**restrictive lung disease** disease characterized by normal airway resistance but impaired respiratory movement

**reticular formation** extensive neuron network extending through brainstem core; receives and integrates information from many afferent pathways and from other CNS regions

**retina** thin layer of neural tissue lining back of eyeball; contains receptors for vision

**retinal** form of vitamin A that forms chromophore component of photopigment

**retrograde amnesia** loss of memory for events immediately preceding a memory-disturbing trauma such as a blow to the head

**retrovirus** virus with RNA core and enzyme capable of transcribing that RNA to DNA, which is then incorporated into host's DNA

**reversible reaction** chemical reaction in which reactants are converted to products and, simultaneously, products are converted to reactants

**Rh factor** group of erythrocyte plasmamembrane antigens that may ($Rh^+$) or may not ($Rh^-$) be present

**rhythm method** contraceptive technique in which couples refrain from sexual intercourse near time of ovulation

**ribonucleic acid (RNA)** single-stranded nucleic acid involved in transcription of genetic information and translation of that information into protein structure; contains the sugar ribose

**ribosomal RNA (rRNA)** type of RNA used in ribosome assembly; becomes part of ribosome

**ribosome** cytoplasmic particle that mediates linking together of amino acids to form proteins; attached to endoplasmic reticulum as *bound ribosome*, or suspended in cytoplasm as *free ribosome*

**rickets** disease in which new bone matrix is inadequately calcified due to 1,25-dihydroxyvitamin $D_3$ deficiency

**rigidity** hypertonia with normal muscle reflexes

**rigor mortis** stiffness of skeletal muscles after death resulting from ATP loss

**RNA polymerase** enzyme that forms RNA by joining together appropriate nucleotides after they have base-paired to DNA

**RNA processing** enzymatic removal of intron sequences from newly formed RNA

**rod** one of two receptor types for photic energy; contains the photopigment rhodopsin

**roentgen** a unit of measurement of ionizing radiation; that amount of ionizing radiation that produces 1 electrostatic unit of ions per cubic centimeter of volume

**saliva** watery solution of salts and proteins, including mucins and amylase, secreted by salivary glands

**sarcomere** repeating structural unit of myofibril; composed of thick and thin filaments; extends between two adjacent Z lines

**sarcoplasmic reticulum** endoplasmic reticulum in muscle fiber; site of storage and release of calcium ions

**satiety signal** input to food-control centers that causes hunger to cease and sets time period before hunger returns

**saturated fatty acid** fatty acid whose carbon atoms are linked by single covalent bonds

**saturation** degree to which protein-binding sites are occupied by ligands

**scala tympani** fluid-filled inner-ear compartment that receives sound waves from basilar membrane and transmits them to round window

**scala vestibuli** fluid-filled inner-ear compartment that receives sound waves from oval window and transmits them to basilar membrane and cochlear duct

**schizophrenia** disease, or family of diseases, characterized by altered motor behavior, distorted perceptions, disturbed thinking, altered mood, and abnormal interpersonal behavior

**Schwann cell** nonneural cell that forms myelin sheath in peripheral nervous system

**scrotum** sac that contains testes and epididymides

**scuba** self-contained underwater breathing apparatus; an open-circuit breathing system supplying compressed air to a diver through a mouthpiece

**secondary active transport** active transport in which energy released during transmembrane movement of one substance from higher to lower concentration

**secondary peristalsis** esophageal peristaltic waves not immediately preceded by pharyngeal phase of swallow

**secondary sexual characteristics** external differences between male and female not directly involved in reproduction

**secondary spermatocyte** male germ cell derived from primary spermatocyte as a result of the first meiotic division

**secretin** peptide hormone secreted by upper small intestine; stimulates pancreas to secrete bicarbonate into small intestine

**secretion** elaboration and release of organic molecules, ions, and water by cells in response to specific stimuli

**secretory phase** stage of menstrual cycle following ovulation during which secretory type of endometrium develops

**secretory vesicle** membrane-bound vesicle produced by Golgi apparatus; contains protein to be secreted by cell

**section** cut surface; slice

**segmentation** series of stationary rhythmic contractions and relaxations of rings of intestinal smooth muscle; mixes intestinal contents

**semen** sperm-containing fluid of male ejaculate

**semiconductors** crystals, including the metallic oxides and sulfides, cupric oxide, zinc oxide, and lead sulfide and the elements germanium, silicon, and selenium that have feeble electronic conductivity which increases extremely rapidly with increasing temperature

**seminal vesicle** one of pair of glands in males that secrete fluid into vas deferens

**seminiferous tubule** tubule in testis in which sperm production occurs; contains Sertoli cells

**sensation** awareness based on a specific sensory stimulus without the aid of higher centers

**sensitizer** a substance or material that renders an organism sensitive, hypersensitive, or unusually susceptible to its action or to the action of another substance upon subsequent exposure

**sensorimotor cortex** areas of cerebral cortex that play a role in skeletal-muscle control, including primary motor, somatosensory, and parts of parietal-lobe association cortex and the premotor area

**sensory deprivation** experimental situation in which subject is isolated as completely as possible from sensory stimulation; includes restriction of movements

**sensory information** information that originates in stimulated sensory receptors

**sensory system** parts of nervous system that receive, conduct, or process information that leads to perception of a stimulus

**sensory unit** afferent neuron plus receptors it innervates

**septum** forebrain area anterior to hypothalamus; involved in emotional behavior

**serosa** connective-tissue layer surrounding outer surface of stomach and intestines

**serotonin** monoamine neurotransmitter; paracrine in blood platelets and digestive tract; also called 5-hydroxytryptamine, or 5-HT

**Sertoli cell** cell intimately associated with developing germ cells in seminiferous tubule; creates "blood-testis barrier," secretes fluid into seminiferous tubule, and mediates hormonal effects on tubule

**serum** blood plasma from which fibrinogen and other clotting proteins have been removed as result of clotting

**sex chromatin** nuclear mass not usually found in cells of males; consists of condensed X chromosome

**sex chromosome** X or Y chromosome

**sex determination** genetic basis of individual's sex, XY determining male and XX, female

**sex differentiation** development of male or female reproductive organs

**sex hormone** estrogen, progesterone, testosterone, or related hormones

**sickle-cell anemia** disease in which an amino acid in hemoglobin is abnormal, and at low oxygen concentrations erythrocytes assume sickle shapes or other bizarre forms that block capillaries

**sigmoid colon** S-shaped terminal portion of colon

**simulated altitude** an artificial environment in which the oxygen pressure has been reduced by lowering the barometric pressure in a chamber; less often, oxygen pressure is reduced by decreasing oxygen concentration of a gas mixture at atmospheric pressure

**sinoatrial (SA) node** region in right atrium of heart containing specialized cardiac-muscle cells that depolarize spontaneously faster than other heart cells; determines heart rate

**sister chromatid** one of two identical DNA threads joined together during meiosis

**skeletal muscle** striated muscle attached to bones or skin and responsible for skeletal movements and facial expression; controlled by somatic nervous system

**skeletal-muscle "pump"** pumping effect of contracting skeletal muscles on blood flow through underlying vessels

**skeletomotor fiber** primary skeletal-muscle fiber, as opposed to modified fibers in muscle spindle

**sleep center** neuron cluster in brainstem whose activity periodically opposes the awake state and induces cycling of slow-wave and paradoxical sleep

**slow fiber** muscle fiber whose myosin has low ATPase activity
**slow virus** virus that replicates very slowly; may become associated with cell's DNA and be passed to daughter cells upon cell division
**slow-wave sleep** sleep state associated with large, slow EEG waves and considerable postural-muscle tone but not dreaming
**smooth muscle** nonstriated muscle that surrounds hollow organs and tubes; controlled by autonomic nervous system, hormones, and paracrines
**soft palate** nonbony region at back of roof of mouth
**solute** substances dissolved in a liquid
**solution** liquid (solvent) containing dissolved substances (solutes)
**solvent** liquid in which substances are dissolved
**somatic** pertaining to the body; related to body's framework or outer walls, including skin, skeletal muscle, tendons, and joints
**somatic nervous system** component of efferent division of peripheral nervous system; innervates skeletal muscle
**somatic receptor** neural receptor that responds to mechanical stimulation of skin or hairs and underlying tissues, rotation or bending of joints, temperature changes, or painful stimuli
**somatosensory cortex** strip of cerebral cortex in parietal lobe in which nerve fibers transmitting somatic sensory information synapse
**somatostatin (SS)** hypothalamic hormone that inhibits growth hormone and TSH secretion by anterior pituitary; possible neurotransmitter; also in stomach and pancreatic islets
**sound wave** air disturbance due to variations between regions of high air-molecule density (*compression*) and low density (*rarefaction*)
**spasm** sustained involuntary muscle contraction
**spasticity** hypertonia with increased responses to motor reflexes
**specific heat** ratio of thermal capacity of a substance to that of water at 15° C
**specific immune response** response that depends upon recognition of specific foreign material for reaction to it
**specificity** selectivity; ability of binding site to react with only one, or a limited number of, types of molecules
**sperm** male gamete; also called spermatozoa
**spermatid** immature sperm
**spermatogenesis** sperm formation
**spermatogonium** undifferentiated germ cell that gives rise to primary spermatocyte
**sperm capacitation** process by which sperm in female reproductive tract gains ability to fertilize ovum
**sphincter** smooth-muscle ring that surrounds a tube, closing tube as muscle contracts

**sphygmomanometer** device consisting of inflatable cuff and pressure gauge for measuring arterial blood pressure

**spinal nerve** one of 86 (43 pairs) peripheral nerves that join spinal cord

**spinal reflex** reflex whose afferent and efferent components are in spinal nerves; can occur in absence of brain control

**spindle fiber** (muscle) modified skeletal muscle fiber in muscle spindle; also called intrafusal fibers; (mitosis) microtubule that connects chromosome to centriole during mitosis

**spleen** largest lymphoid organ; located between stomach and diaphragm

**stable balance** net loss of substance from body equals net gain, and body's amount of substance neither increases nor decreases

**stapes** third middle-ear bone; transmits sound waves to scala vestibuli of inner ear

**starch** moderately branched plant polysaccharide composed of glucose subunits

**state of consciousness** degree of mental alertness; that is, whether awake, drowsy, asleep, and so on

**steroid** lipid subclass; molecule consists of four interconnected carbon rings to which polar groups may be attached

**stimulus** detectable change in environment

**STPD** standard temperature, pressure, and dry; an abbreviation of the physical conditions used to define the mass of a gas contained in a stated volume; standard physical conditions are 0° C, 760 torr, and dry

**strain** condition of the animal body resulting from exposure to stress, with reversible or irreversible changes in performance, physiological or biochemical characteristics

**stratosphere** the layer of the atmosphere between the limits 10 and 20 km above the surface of the earth

**stress** loosely used to refer to any stressful environmental condition capable of producing a physiologic response or deterioration of performance

**striated muscle** muscle having transverse banding pattern due to repeating sarcomere structure

**stroke** brain damage due to blood stoppage because of occlusion or rupture of cerebral vessel

**stroke volume** blood volume ejected by a ventricle during one heartbeat

**strong acid** acid that ionizes completely to form hydrogen ions and corresponding anions when dissolved in water

**subcortical nucleus** neuron cluster deep in brain; includes basal ganglia

**sublingual gland** salivary gland under tongue

**submandibular gland** salivary gland in lower jaw; formerly called submaxillary gland

**submucosa** connective tissue layer under mucosa in gastrointestinal tract

**submucous plexus** nerve-cell network in submucosa of esophageal, stomach, and intestinal walls

**substrate** reactant in enzyme-mediated reaction

**substrate phosphorylation** direct transfer of phosphate group from metabolic intermediate to ADP to form ATP

**subsynaptic membrane** that part of postsynaptic neuron's plasma membrane under synaptic knob

**sucrose** disaccharide composed of glucose and fructose; also called table sugar

**sulfuric acid** acid generated during catabolism of sulfur-containing compounds; dissociates to form sulfate and hydrogen ions

**superior vena cava** large vein that carries blood from upper half of body to right atrium of heart

**supersensitivity** increased responsiveness of target cell to given messenger due to upregulation

**suppressor T cell** T cell that inhibits antibody production and cytotoxic T-cell function

**suprathreshold stimulus** any agent capable of depolarizing membrane more than its threshold potential, that is, closer to zero

**surface tension** attractive forces between water molecules at surface resulting in net force that acts to reduce surface area

**surfactant** detergent-like phospholipid produced by pulmonary alveolar cells; reduces surface tension of fluid film lining alveoli

**sympathetic nervous system** portion of autonomic nervous system whose preganglionic fibers leave CNS at thoracic and lumbar portions of spinal cord

**synapse** anatomically specialized junction between two neurons where electrical activity in one neuron influences excitability of second

**systemic circulation** circulation from left ventricle through all organs except lungs and back to heart

**systole** period of ventricular contraction

**systolic pressure** maximum arterial blood pressure during cardiac cycle

**T cell** lymphocyte derived from precursor that differentiated in thymus

**T wave** component of electrocardiogram corresponding to ventricular repolarization

**tachycardia** rapid heart rate; by definition a heart rate faster than 100 beats/min in a resting adult human subject

**target cell** cell influenced by a certain hormone

**taste bud** sense organ that contains chemoreceptors for taste

**teleology** explanation of events in terms of ultimate purpose served by them

**temperature, dew-point** temperature at which condensation first occurs when an air-water vapor mixture is cooled at constant pressure

**temperature, dry bulb** temperature of a gas or mixture of gases indicated by a thermometer shielded from radiation

**temperature, globe** temperature of a blackened hollow sphere of thin copper (usually 0.15-m diameter) as measured by a thermometer at its center

**temperature, mean body** sum of the products of the heat capacity and temperature of all the tissues of the body divided by the total heat capacity of the organism

**temperature, mean radiant** temperature of an imaginary isothermal ''black'' enclosure in which a solid body or occupant would exchange the same amount of heat by radiation as in the actual nonuniform enclosure

**temperature, mean skin** sum of the products of the area of each regional surface element and its mean temperature divided by the total area of body surface

**temperature regulation** maintenance of the temperature or temperatures of a body within a restricted range under conditions involving variable internal and/or external heat loads

**temperature sensor** neuronal structure that is differentially sensitive to temperature and that responds to a maintained temperature with a characteristic sustained impulse frequency

**temperature survival limit, lower** environmental temperature below which thermal balance cannot be maintained for a long period and animals become progressively hypothermic

**temperature survival limit, upper** environmental temperature above which thermal balance cannot be maintained for a long period, and animals become progressively hyperthermic

**temperature, wet bulb** lowest temperature to which a sample of air can be cooled by evaporating water adiabatically

**temporal lobe** region of cerebral cortex where primary auditory cortex and Wernicke's speech center is located

**temporal summation** membrane potential produced as two or more inputs, occurring at different times, are added together; potential change is greater than that caused by single input

**tendon** collagen fiber bundle that connects muscle to bone and transmits muscle contractile force to the bone

**termination code word** three-nucleotide sequence in DNA that signifies gene end

**testis** gonad in male

**testosterone** steroid hormone produced in interstitial cells of testes; major male sex hormone; essential for spermatogenesis and maintains growth and development of reproductive organs and secondary sexual characteristics of male

**tetanus** maintained mechanical response of muscle to high-frequency stimulation; the disease lockjaw

**tetrad** grouping of two homologous chromosomes, each with its sister chromatid, during meiosis

**thalamus** subdivision of diencephalon; integrating center for sensory input on its way to cerebral cortex; also contains motor nuclei
**theca** cell layer that surrounds ovarian-follicle granulosa cells; formed from follicle and connective-tissue cells
**thermal conductivity** time rate of transfer of heat by conduction, through unit thickness
**thermal equilibrium** condition under which the rate of heat production of an animal is equal to the rate of heat loss, so that body temperature does not change
**thermogenesis** heat generation
**thermogenesis, nonshivering** increase in the rate of heat production during cold exposure due to processes which do not involve contractions of voluntary muscles, i.e. increased heat production by processes other than tone, microvibrations, or clonic contractions of skeletal muscles
**thermogenesis, shivering** increase in the rate of heat production during cold exposure due to increased contractile activity of skeletal muscles not involving voluntary movements and external work
**thermoneutral zone** range of ambient temperature within which metabolic heat production equals heat loss, and within which temperature regulation is achieved by nonevaporative physical processes alone
**thermoreceptor** sensory receptor for temperature and temperature changes, particularly in low (*cold receptor*) or high (*warm receptor*) range
**thoracic cavity** chest cavity
**thoracic wall** chest wall
**thorax** closed body cavity between neck and diaphragm; contains lungs, heart, thymus, large vessels, and esophagus; also called the chest
**thrombin** enzyme that catalyzes conversion of fibrinogen to fibrin
**thrombosis** clot formation in body
**thromboxane** eicosanoid closely related to prostaglandins; synthesized from arachidonic acid
**thromboxane $A_2$** thromboxane that, among other effects, stimulates platelet aggregation in blood clotting
**thrombus** blood clot
**thymine (T)** pyrimidine base in DNA but not RNA
**thymosin** group of hormones secreted by thymus; also called *thymopoietin*
**thymus** lymphoid organ in upper part of chest; site of T lymphocyte differentiation and thymosin secretion
**thyroglobulin** large protein to which thyroid hormones bind in thyroid gland; storage form of thyroid hormones
**thyroid gland** paired endocrine gland in neck; secretes thyroid hormones and calcitonin
**thyroid hormones (TH)** collective term for amine hormones released from thyroid gland, that is thyroxine ($T_4$) and triiodothyronine ($T_3$)

**thyroid-stimulating hormone (TSH)** glycoprotein hormone secreted by anterior pituitary; induces secretion of thyroid hormone; also called thyrotropin

**thyrotropin-releasing hormone (TRH)** hypothalamic hormone that stimulates thyrotropin and prolactin secretion by anterior pituitary

**thyroxine ($T_4$)** tetraiodothyronine; iodine-containing amine hormone secreted by thyroid gland

**tidal volume** air volume entering or leaving lungs with single breath during any state of respiratory activity

**tissue** aggregate of differentiated cells of similar type united in performance of particular function; also denotes general cellular fabric of a given organ

**tolerance** condition in which increasing drug doses are required to achieve effects that initially occurred in response to a smaller dose

**tonsil** one of several small lymphoid organs in pharynx

**torr** unit of pressure equal to 1/760 of a standard atmosphere; named for the Italian mathematician E. Torricelli (1608–47)

**total carbon dioxide** sum total of dissolved carbon dioxide, bicarbonate, and carbamino-$C0_2$

**total dead space** alveolar and anatomic dead space

**total energy expenditure** sum of external work done plus heat produced plus any energy stored by body

**total lung capacity** the amount of air in the lung usually measured as the sum of forced vital capacity and residual lung volume

**total peripheral resistance (TPR)** total resistance to flow in systemic blood vessels from beginning of aorta to end of venae cavae

**toxemia of pregnancy** disease occurring in pregnant women and associated with fluid retention, urinary protein, hypertension, and possibly convulsions; also called eclampsia

**toxin** poison

**trace element** mineral present in body in extremely small quantities

**trachea** single airway connecting larynx with bronchi

**transamination** reaction in which an amino-acid amino group ($-NH_2$) is transferred to a keto acid, the keto acid thus becoming an amino acid

**transferrin** iron-binding protein carrier for iron in plasma

**transpulmonary pressure** difference between alveolar and intrapleural pressures; force that holds lungs open

**triacylglycerol** subclass of lipids composed of glycerol and three fatty acids; also called fat, neutral fat, or triglyceride

**tricuspid valve** valve between right atrium and right ventricle of heart

**tricyclic antidepressant** drug that interferes with reuptake of norepinephrine and serotonin by presynaptic endings, thereby increasing amount of neurotransmitter in synaptic cleft

**triiodothyronine ($T_3$)** iodine-containing amine hormone secreted by thyroid gland

**triplet code** three-base sequence in DNA and RNA that specifies particular amino acid

**trophoblast** outer layer of blastocyst; gives rise to placental tissues

**tropic** growth-promoting

**tropic hormone** hormone that stimulates the secretion of another hormone

**troposphere** layer of the atmosphere extending from the surface of the earth upward to a distance of 10 km

**trypsin** enzyme secreted into small intestine by exocrine pancreas as precursor trypsinogen; breaks certain peptide bonds in proteins and polypeptides

**trypsinogen** inactive precursor of trypsin; secreted by exocrine pancreas

**tryptophan** essential amino acid; serotonin precursor

**tubular reabsorption** transfer of materials from kidney-tubule lumen to peritubular capillaries

**tubular secretion** transfer of materials from peritubular capillaries to kidney-tubule lumen

**tumorigenic** capacity of a substance to cause or induce tumors or neoplasia

**tumor necrosis factor (TNF)** monokine that kills cells, stimulates inflammation, and mediates many systemic acute-phase responses

**twitch** mechanical response of muscle to single action potential

**tympanic membrane** membrane stretched across end of ear canal; also called ear drum

**type I diabetes** diabetes in which insulin is completely or almost completely absent; also called *insulin-dependent diabetes* or *childhood onset diabetes*

**type II diabetes** diabetes in which insulin is present at near normal or even above normal levels, but the tissues do not respond optimally to it; also called *insulin-independent diabetes* or *adult onset diabetes*

**ulcer** erosion or sore, as in stomach or intestinal wall

**ultrafiltrate** essentially protein-free fluid formed from plasma as it is forced through capillary walls by pressure gradient

**ultrasonic** having a frequency above the audible frequency range

**ultraviolet** band of electromagnetic radiation with wavelengths shorter than those of visible light; ultraviolet is divided into three regions: (i) UV-A, 320 to 400 nm; this region has only weak photochemical effects on man. It is visible and causes some sunburn and immediate pigmentation. Ultraviolet of this wavelength is naturally abundant at ground level; (ii) UV-B, 290 to 320 nm; this invisible region produces the typical natural ultraviolet effects on skin and eye. In nature at ground level its intensity is relatively low and highly variable; (iii) UV-C, 240 to 290 nm; this region produces most of the effects of UV-B but penetrates only superficially; at ground level it is available only from artificial sources

**umbilical vessel** artery or vein transporting blood between fetus and placenta
**unsaturated fatty acid** fatty acid containing one or more double bonds
**upper motor neuron** neuron of the motor cortex or descending pathway
**uracil** pyrimidine base; present in RNA but not DNA
**urea** major nitrogenous waste product of protein breakdown and amino acid catabolism
**uremia** general term for symptoms of profound kidney malfunction
**ureter** tube that connects renal pelvis to urinary bladder
**urethra** tube that connects urinary bladder to outside of body
**uric acid** waste product derived from nucleic acid catabolism
**urinary bladder** thick-walled sac composed of smooth muscle; stores urine prior to urination
**uterine tube** one of two tubes that carries ovum from ovary to uterus; also called *fallopian tube, oviduct*
**uterus** hollow organ in pelvic region of females; houses fetus during pregnancy; also called *womb*
**vagina** canal leading from uterus to outside of body; also called *birth canal*
**vagus nerve** cranial nerve X; major parasympathetic nerve
**valence** number of hydrogen atoms (or its equivalent) that an ion can hold in combination or displace in a reaction
**Valsalva maneuver** forced exhalation against a closed glottis; also a procedure used to vent the middle ears during compression or descent from altitude by forced exhalation against the closed nose and mouth
**vapor** the gaseous state of an element or compound that is liquid or solid under ordinary conditions, for example, water vapor
**varicosity** swollen region of axon; contains neurotransmitter-filled vesicles; analogous to presynaptic ending
**vas deferens** one of paired male reproductive ducts that connect epididymis of testis to urethra; also called *ductus deferens*
**vasectomy** cutting and tying off of both ductus (vas) deferens, which results in sterilization of male without loss of testosterone
**vasoconstriction** decrease in blood-vessel diameter due to vascular smooth-muscle contraction
**vasodilation** increase in blood-vessel diameter due to vascular smooth-muscle relaxation
**vasopressin** peptide hormone synthesized in hypothalamus and released from posterior pituitary; increases water permeability of kidneys, collecting ducts; also called *antidiuretic hormone (ADH)*
**vein** any vessel that returns blood to heart
**vena cava** (plural **venae cavae**) one of two large veins that returns systemic blood to heart

**venous return (VR)** blood volume flowing to heart per unit time
**ventilation** air exchange between atmosphere and alveoli; alveolar air flow
**ventilation, pulmonary** volume of air moved into or out of the lungs per unit time; it is calculated as the product of breathing frequency and tidal volume
**ventilation rate** volume of air breathed in and out of the lungs per minute; expressed as l/min
**ventral** toward or at the front of body
**venule** small vessel that carries blood from capillary network to vein
**vertigo** sensation or false perception that the external environment is rotating around the subject; vertigo should not be confused with dizziness
**very low density lipoprotein (VLDL)** lipid-protein aggregate having high proportion of fat
**vesicle** small, membrane-bound organelle
**vestibular system** sense organ in temporal bone of skull; consists of three semicircular canals, a utricle, and a saccule; also called vestibular apparatus, sense organ of balance
**villi** finger-like projections from highly folded surface of small intestine; covered with single layered epithelium
**virilism** development of masculine physical characteristics in a woman
**virus** causitive agent of an infectious disease; it lacks enzyme machinery for energy production and ribosomes for protein synthesis, thus cannot survive or reproduce except inside other cells whose biochemical apparatus it uses
**viscera** organs in thoracic and abdominal cavities
**viscosity** property of fluid that makes it resist flow; fluid thickness
**visual field** that part of world being viewed at a given time
**vital capacity** maximal amount of air that can be expired, regardless of time required, following maximal inspiration
**vital capacity, forced** maximal amount of air that can be expired following a maximal inspiration, regardless of time
**vitamin** organic molecule that is required in trace amounts for normal health and growth and must be supplied by diet; classified as *water-soluble* (vitamins C and the B complex) and *fat-soluble* (vitamins A, D, E, and H)
**vocal cord** one of two elastic-tissue bands stretched across laryngeal opening and caused to vibrate by air movement past them, producing sounds
**volt (V)** unit of measurement of electric potential between two points
**voltage** measure of potential of separated electric charges to do work; measure of electric force between two points
**vomiting center** neurons in brainstem medulla that coordinate vomiting reflex
**vulva** female external genitalia; the mons pubis, labia majora and minora, clitoris, vestibule of the vagina, and vestibular glands

**wavelength** distance between two wave peaks in oscillating medium

**weak acid** acid whose molecules do not completely ionize to form hydrogen ions when dissolved in water

**weightlessness** any object undergoing free fall in a vacuum is weightless, as is an unaccelerated satellite in orbit about the earth

**white matter** portion of CNS that appears white in unstained specimens and contains primarily myelinated nerve fibers

**white muscle fiber** muscle fiber lacking appreciable amounts of myoglobin

**withdrawal** physical symptoms (usually unpleasant) associated with cessation of drug use

**Wolffian duct** part of embryonic duct system that, in male, remains and develops into reproductive-system ducts, but in female, degenerates

**zona pellucida** thick, clear layer separating ovum from surrounding granulosa cells

# Index

**A. ROBERTO FRISANCHO, Ph.D.,** is a biological anthropologist with a distinguished career. He was born in Cuzco, Perú. He is a research scientist at the Center for Human Growth and Development and professor of anthropology at the University of Michigan. He graduated from the Ciensas High School, and in 1962 he obtained a Bachelor of Humanities degree from the University of Cuzco, Perú. In 1966 he obtained a Master in Anthropology from Pennsylvania State University, and in 1969 he obtained a Ph.D. in biological anthropology from the same university.

Dr. Frisancho's research includes more than 120 publications in the fields of human growth, nutrition, ecology, and environmental physiology. He has: (1) demonstrated that the low birth weight associated with adolescent pregnancy is the result of the maternal-fetal growth competition for nutrients; (2) documented the effects of the high-altitude environment during the life cycle; (3) expanded the utility of anthropometry for the evaluation of growth and nutritional status; (4) developed the conceptual framework of developmental adaptation, which maintains that interpopulation differences in phenotypic traits are the result of environmental influences and adaptive responses that the organism makes during the period of growth and development. As president of the American Human Biology Council, he founded the *American Journal of Human Biology.*